ESO ASTROPHYSICS SYMPOSIA
European Southern Observatory

Series Editor: Philippe Crane

Springer-Verlag Berlin Heidelberg GmbH

Luiz Nicolaci da Costa   Alvio Renzini (Eds.)

Galaxy Scaling Relations:

Origins, Evolution and Applications

Proceedings of the ESO Workshop
Held at Garching, Germany,
18–20 November 1996

Springer

Volume Editors

Luiz Nicolaci da Costa
Alvio Renzini
European Southern Observatory
Karl-Schwarzschild-Strasse 2
D-85748 Garching, Germany
email: ldacosta@eso.org
      arenzini@eso.org

Series Editor

Philippe Crane
European Southern Observatory
Karl-Schwarzschild-Strasse 2
D-85748 Garching, Germany

Cataloging-in-Publication data applied for
Die Deutsche Bibliothek - CIP-Einheitsaufnahme

Galaxy scaling relations : origins, evolution and applications ;
proceedings of the ESO workshop, held at Garching, Germany, 18 -
20 November 1996 / L. N. da Costa ; A. Renzini (ed.). - Berlin ;
Heidelberg ; New York ; Barcelona ; Budapest ; Hong Kong ;
London ; Milan ; Paris ; Santa Clara ; Singapore ; Tokyo : Springer,
1997
    (ESO astrophysics symposia)

ISBN 978-3-662-21934-8   ISBN 978-3-540-69654-4 (eBook)
DOI 10.1007/978-3-540-69654-4

© Springer-Verlag Berlin Heidelberg 1977

Softcover reprint of the hardcover 1st edition 1977

Originally published by Springer-Verlag Berlin Heidelberg New York in 1977.

Typesetting: Camera ready by authors/editors
Cover design: *design & production* GmbH, Heidelberg
SPIN: 10552376   55/3142-543210 - Printed on acid-free paper

Preface

At close inspection every galaxy appears to have its own individuality. A galaxy can be warped, lop-sided, doubly-nucleated, boxy or disky, ... in its own specific, *peculiar* way. Hence, for a complete description, galaxy taxonomy may ask for finer and finer classification schemes. However, for some applications it may be more fruitful to let details aside and focus on some global properties of galaxies. One is then seeking to measure just a few quantities for each galaxy, a minimum set of global observables that yet captures some essential aspect of these objects.

One very successful example of this approach is offered by the scaling relations of galaxies, the subject of the international workshop held at ESO headquarters in Garching on November 19–21, 1996. Discovered in the late 1970's, the Tully–Fisher relation for the spirals and the Faber–Jackson relation, or its more recent version the Fundamental Plane, for ellipticals have now become flourishing fields of astronomical research in their own right, as well as being widely used tools for a broad range of astronomical investigations. The workshop was designed to address three key issues on galaxy scaling relations, i.e., their **Origins**, **Evolution**, and **Applications** in astronomy.

The **Origins** of galaxy scaling relations still escape our full understanding. For such relations to exist, galaxies must be in a state of dynamical equilibrium, which is not surprising. Yet, this is not the whole story, as for both ellipticals and spirals the scaling relations show that structural/dynamical observables (such as effective radius, velocity dispersion, or width) are tightly related to the galaxy luminosity, which is a quantity of an entirely different nature. Luminosity is produced by stars, and therefore it reflects the stellar population content of each galaxy, which brings into the scene an *age coordinate*. Structural/dynamical quantities are instead *timeless* entities. Now, the success of the scaling relations is due to the fact that they are indeed very *good* relations, i.e., they are very tight. What strikes most is how tiny is the fraction of the allowed parameter space actually occupied by galaxies. Ellipticals cluster around their Fundamental Plane with astonishingly small scatter, which can be interpreted as implying a dispersion in luminosity of less than 10% for a given galaxy mass. Similarly, the luminosity dispersion of spirals is less than only $\sim 20\%$ for a given rotational velocity.

This means that for a given mass, galaxies appear to be very similar to each other. In turn, similar luminosities imply similar star formation histories (SFH), separately for present-day passively evolving galaxies (ellipticals) and for still actively star-forming galaxies (spirals). Such required uniformity, and tight mass-SFH connection, pose together a major challenge to current theories of galaxy formation and evolution. Scenarios of galaxy formation based on the stochastic agglomeration of smaller galaxies may lead to dispersions exceeding those observed. Hence, understanding the origin of these relations will bring us directly into the core of the galaxy formation problem.

Cold dark matter theories of galaxy formation are very successful in predicting the properties of dark matter halos, but are forced to introduce parameterized algorithms when dealing with the dissipative, star-forming, *feedbacking*, i.e., highly nonlinear, baryonic component. Attempts are currently under way to tune these algorithms so as to generate synthetic galaxies that match the scaling relations of nearby galaxies. While partial success has been achieved, efforts to improve this approach are likely to continue in the near future. To some extent it is not surprising if – given the many degrees of freedom of the problem – different scenarios provide equivalent fits to the data. Moreover, fits are obtained by adjusting algorithms and parameters precisely to match the scaling relations of $z = 0$ galaxies. Hence, in such an approach, theory is at best able to *reproduce* such relations, not to *predict* them.

The **Evolution** of the scaling relations, i.e., their change with redshift (lookback time), offers a powerful way of discriminating among competing galaxy formation scenarios that may equally well *fit* local galaxies. This offers an attractive chance of breaking a degeneracy that it is difficult to circumvent when restricting to nearby galaxies. A theory adjusted to reproduce $z = 0$ galaxies automatically implies, in fact *predicts*, a specific evolution with redshift, and these predictions can be subjected to observational test. Over the last two years – thanks especially to HST spatial resolution and Keck faint object spectroscopy – scaling relations have been established for galaxies at progressively larger redshifts, and we should expect this field of research to expand during the first several years of the ESO Very Large Telescope (VLT). The primary goal is to establish how the slope, zero point, and width of the scaling relations evolve with redshift. One distinct advantage comes from moving to high redshift, as the discriminating power offered by the scaling relations dramatically increases. For example, the tightness of the distribution of elliptical galaxies about the Fundamental Plane at $z = 0$ implies that over 90% of their stars formed at $z > 2$. If the tightness of the distribution remains the same at $z \simeq 0.4$ (which seems to be the case), then the formation epoch is pushed back to $z \gtrsim 5$ (!). Moreover, at $z \gtrsim 0.1$ redshift is a very good distance indicator, which allows one, for example, to compare the scaling relations of cluster and field galaxies, a much more difficult task in the local universe. Again, some theories predict that cluster and field galaxies should differ in some fundamental aspects, and – although based on meager direct evidence – there has been a widespread perception that this is actually the case, at least for ellipticals. Observations will soon show whether such differences do really exist, at low as well as at high redshift.

The **Applications** of the scaling relations as distance indicators are possible thanks to their intrinsic tightness. As is well known, the Fundamental Plane of ellipticals was used to map local deviations from the Hubble flow immediately after its discovery. Yet, scaling relations can only be used for this purpose if cluster and field ellipticals are equally old. Otherwise, they would not be good standard candles, and the inferred peculiar velocities would be a mere artifact of different SFHs in different patches of the sky. Rather than real peculiar motions, the use of scaling relations would trace something else, e.g., local overdensities, or whatever.

As long as there are no significant environmental effects, and no compelling evidence currently exists for that, the scaling relations are an extremely powerful cosmological tool with many important applications. For instance, they can be used to map the peculiar velocity field from which, in turn, one can reconstruct the mass density field in the local universe under the assumption that gravity is responsible for the growth of structures. One can then, at least in principle, measure the *mass* power-spectrum on intermediate scales ($\lesssim 60h^{-1}$ Mpc) complementing the information from measurements of the cosmic microwave background on $\sim 1°$ scales and by COBE on very large scales. Moreover, the *mass* density field and peculiar velocity field of galaxies can be used in conjunction with complete local redshift surveys to investigate to which extent galaxies map the underlying (dark matter) mass distribution. Finally, comparison of the measured peculiar velocity field with that predicted from redshift surveys can be used to constrain the cosmological density parameter Ω_0, or more precisely the parameter $\beta = \Omega_0^{0.6}/b$, where b is the linear biasing factor. Until recently, large scale flow studies pointed to large values of β, while cluster abundances, small-scale velocity field studies, and the tightness of scaling relations themselves favored smaller values. Finding consistency amongst these various estimates would reinforce the underlying assumption that the scaling relations are universal, at least at $z = 0$. In conclusion, we need to look at the redshift evolution of scaling relations in order to understand their origin. We can then return to the local universe, and with accrued confidence, use these relations to map distances and peculiar velocities of galaxies, and the *mass* distribution in the local universe.

All these topics are likely to continue to attract a great deal of attention in the foreseeable future, as demonstrated by the wide and highly qualified attendance at the workshop. This workshop was part of the ESO conference series meant to survey the current status of the most active areas of astronomical research in view of the rapidly approaching beginning of the VLT scientific operations. The success of the meeting clearly demonstrated how eagerly this event is being awaited by the ESO community, and we would like to thank all the attendees for their active participation in the Workshop. Special thanks go to those who contributed to the lively debate at the end of the meeting, to the Local Organizing Committee for their invaluable help, to the Scientific Organizing Committee for having directed us in putting together an attractive scientific programme, and last but not least to Pamela Bristow and Christina Stoffer for their efficient handling of the local organization and preparation of these proceedings.

Garching, September 1997   *Luiz Nicolaci da Costa and Alvio Renzini*

Contents

x

PART 2. EVOLUTION

PART 3. APPLICATIONS

PART 4. POSTERS

List of Participants

Name	Institution
ADAMO, Angelo	Università di Bologna, Dip. Astronomia lau19@astbo3.bo.astro.it
ALONSO, Maria	Observatório Nacional, Rio de Janeiro vicky@oac.uncor.edu
ANTONUCCIO–DELOGU, Vincenzo	Osservatorio Astrofisico di Catania van@sunct.ct.astro.it, antonucc@tac.dk
ARIMOTO, Nobuo	University of Tokyo, Institute of Astronomy arimoto@mtk.ioa.s.u-tokyo.ac.jp
BAUGH, Carlton	University of Durham, Dept. of Physics c.m.baugh@durham.ac.uk
BELLONI, Paola	Universitätssternwarte München belloni@usm.uni-muenchen.de
BENDER, Ralf	Universitätssternwarte München bender@usm.uni-muenchen.de
BERGERON, Jacqueline	ESO, Sci. Div., Garching jbergero@eso.org
BEST, Philip	Sterrewacht Leiden pbest@strw.LeidenUniv.nl
BEUING, Jan	Universitätssternwarte München beuing@usm.uni-muenchen.de
BINGGELI, Bruno	University of Basel, Astronomical Institute binggeli2@ubaclu.unibas.ch
BLANCHARD, Alain	Observatoire de Strasbourg blanchar@wirtz.u-strasbg.fr
BOSELLI, Alessandro	Laboratoire d'Astronomie Spatiale, Marseille boselli@astrsp-mrs.fr
BRANCHINI, Enzo	University of Durham, Dept. of Physics e.f.branchini@durham.ac.uk

BURKERT, Andreas

MPI für Astronomie, Heidelberg
burkert@mpia-hd.mpg.de

BUSARELLO, Giovanni

Osservatorio Astronomico di Capodimonte
busarello@astrna.na.astro.it

BUSON, Lucio

Osservatorio Astronomico di Capodimonte
buson@astrna.na.astro.it

CANNON, Russell

ESO, Garching / AAO
rdc@aaoepp.aao.gov.au

CAPPI, Alberto

Osservatorio Astronomico di Bologna
cappi@astbo3.bo.astro.it

CHIOSI, Cesare

Università di Padova, Dip. di Astronomia
chiosi@astrpd.pd.astro.it

CIOTTI, Luca

Osservatorio Astronomico di Bologna
ciotti@astbo3.bo.astro.it

CÔTÉ, Stéphanie

ESO, Sci. Div., Garching
scote@eso.org

D'ODORICO, Sandro

ESO, INS Div., Garching
sdodoric@eso.org

DA COSTA, Luiz

ESO, Sci. Div., Garching
ldacosta@eso.org

DANZIGER, John

ESO, Garching /
Osservatorio Astronomico di Trieste
jdanzige@eso.org / danziger@oat.ts.astro.it

DE CARVALHO, Reinaldo

Caltech
reinaldo@astro.caltech.edu

DE MARCHI, Guido

ESO, Sci. Div., Garching
demarchi@eso.org

DEKEL, Avishai

Hebrew University /
University of California, Berkeley
dekel@ucolick.org / dekel@astro.huji.ac.il

DI NELLA, Helene

MPI für Astronomie, Heidelberg
dinella@mpia-hd.mpg.de

DICKINSON, Mark

Space Telescope Science Institute, Baltimore
med@stsci.edu

DOUBLIER, Vanessa

ESO, Sci. Div., Garching
vdoublie@eso.org

DUC, Pierre–Alain

ESO, Sci. Div., Garching
pduc@eso.org

EGAMI, Eiichi

MPI für extraterrestrische Physik, Garching
egami@mpe-garching.mpg.de

EISENSTEIN, Daniel   Institute for Advanced Study, Princeton
eisenste@sns.ias.edu

EMSELLEM, Eric   ESO, Sci. Div., Garching
eemselle@eso.org

EVRARD, August   Institut d'Astrophysique, Paris /
University of Michigan
evrard@iap.fr

FASANO, Giovanni   Osservatorio Astronomico di Padova
fasano@astrpd.pd.astro.it

FRANX, Marijn   Kapteyn Astronomical Institute, Groningen
franx@astro.rug.nl

FREUDLING, Wolfram   ST–ECF, Garching
wfreudli@eso.org

GAVAZZI, Giuseppe   Università di Milano, Dip. di Astrofisica
gavazzi@brera.mi.astro.it

GEIGER, Bernhard   MPI für Astrophysik, Garching
bernhard@mpa-garching.mpg.de

GIOVANELLI, Riccardo   Cornell University, Department of Astronomy
riccardo@astrosun.tn.cornell.edu

GUZMÁN, Rafael   UCSC/Lick Observatory
rguzman@ucolick.org

HABE, Asao   Hokkaido University
habe@phys.hokudai.ac.jp

HELD, Enrico V.   Osservatorio Astronomico di Padova
held@astrpd.pd.astro.it

HENSLER, Gerhard   Universität Kiel
hensler@astrophysik.uni-kiel.de

HJORTH, Jens   NORDITA, Copenhagen
jens@nordita.dk

HOPP, Ulrich   Universitätssternwarte München
hopp@usm.uni-muenchen.de

HUDSON, Mike   University of Victoria / CITA
hudson@uvastro.phys.uvic.ca

JABLONKA, Pascale   Observatoire de Paris, Meudon – DAEC
jablonka@gin.obspm.fr

JØRGENSEN, Inger   McDonald Observatory
inger@roeskva.as.utexas.edu

KRITSUK, Alexei   St. Petersburg University,
Astronomical Observatory
agk@aispbu.spb.su, agk@agk.usr.pu.ru

KUNTH, Daniel · Institut d'Astrophysique, Paris
kunth@iap.fr

KUNTSCHNER, Harald · University of Durham, Dept. of Physics
Harald.Kuntschner@durham.ac.uk

LANZONI, Barbara · Osservatorio Astronomico di Capodimonte
lanzoni@astrna.na.astro.it,
lanzoni@cerere.na.astro.it

LEIBUNDGUT, Bruno · ESO, Sci. Div., Garching
bleibund@eso.org

LIMA NETO, Gastão B. · Observatoire de Lyon
gastao@obs.univ-lyon1.fr

MARASTON, Claudia · Università di Bologna, Dip. Astronomia
claudiam@astbo3.bo.astro.it

MEHLERT, Doerte · Universitätssternwarte München
mehlert@usm.uni-muenchen.de

MÉNDEZ, René · ESO, Sci. Div., Garching
rmendez@eso.org

MEYLAN, Georges · ESO, Sci. Div., Garching
gmeylan@eso.org

MILVANG–JENSEN, Bo · Niels Bohr Institute for Astronomy,
Copenhagen
milvang@astro.ku.dk

MO, Houjun · MPI für Astrophysik, Garching
hom@mpa-garching.mpg.de

MOESSNER, Richhild · MPI für Astrophysik, Garching
bernhard@mpa-garching.mpg.de

MOULD, Jeremy · Mt. Stromlo and Siding Spring Observatories
jrm@mso.anu.edu.au,
director@mso.anu.edu.au

MULLER, Karen · ESO, Sci. Div., Garching
kmueller@eso.org

NUSSER, Adi · MPI für Astrophysik, Garching
adi@mpa-garching.mpg.de

PAHRE, Michael · Caltech
map@astro.caltech.edu

PELETIER, Reynier · Kapteyn Astronomical Institute, Groningen
peletier@astro.rug.nl, rpelet@ll.iac.es

PELLEGRINI, Silvia · Università di Bologna, Dip. Astronomia
pellegrini@astbo3.bo.astro.it

PIERINI, Daniele	Univ. of Milan, Astronomical Observatory Brera pierini@brera.mi.astro.it
PIERRE, Marguerite	C.E. Saclay – Service d'Astrophysique mpierre@cea.fr
PRUNET, Simon	Institut d'Astrophysique Spatiale, Université de Paris XI prunet@ias.fr
PUDDU, Emanuella	Osservatorio Astronomico di Capodimonte puddu@astrna.na.astro.it
QUINN, Peter	ESO, DMD, Garching pjq@eso.org
RASMUSSEN, Per Kjaergaard	Copenhagen University Observatory per@astro.ku.dk
RENZINI, Alvio	ESO, Garching arenzini@eso.org
RÖTTGERING, Huub	Sterrewacht Leiden rottgeri@reusel.strw.LeidenUniv.nl
SADLER, Elaine	University of Sydney, School of Physics ems@physics.usyd.edu.au
SAGLIA, Roberto	Universitätssternwarte München saglia@usm.uni-muenchen.de
SAVAGLIO, Sandra	ESO, Sci. Div., Garching ssavagli@eso.org
SCHADE, David	University of Toronto schade@astro.utoronto.ca
SCHNEIDER, Peter	MPI für Astrophysik, Garching peter@mpa-garching.mpg.de
SCODEGGIO, Marco	ESO, Sci. Div., Garching mscodegg@eso.org, scodeggi@astrosun.tn.cornell.edu
SECCO, Luigi	Osservatorio Astronomico di Padova secco@astrpd.pd.astro.it
SMITH, Russell J.	University of Durham, Dept. of Physics R.J.Smith@durham.ac.uk
TULLY, Brent	Institute for Astronomy, Univ. of Hawaii tully@uhifa.ifa.hawaii.edu
VAN ALBADA, Tjeerd	Kapteyn Astronomical Institute, Groningen albada@astro.rug.nl

VAN DOKKUM, Pieter — Kapteyn Astronomical Institute, Groningen
dokkum@astro.rug.nl

VAN LOON, Jacco — ESO, Sci. Div., Garching
jvloon@eso.org

VAZDEKIS, Alexandre — Instituto de Astrofisica de Canarias (IAC)
asv@ll.iac.es

VERHEIJEN, Marc — Kapteyn Astronomical Institute, Groningen
verheyen@astro.rug.nl

VILLUMSEN, Jens — MPI für Astrophysik, Garching
jens@mpa-garching.mpg.de

WARMELS, Rein — ESO, DMD, Garching
rwarmels@eso.org

WHITE, Simon — MPI für Astrophysik, Garching
swhite@mpa-garching.mpg.de

YAN, Lin — ESO, Sci. Div., Garching
lyan@eso.org

ZIEGLER, Bodo — Universitätssternwarte München
ziegler@usm.uni-muenchen.de

Part 1

ORIGINS

The Physical Origin of Galaxy Scaling Relations

Simon D.M. White

Max-Planck-Institut für Astrophysik, Karl-Schwarzschild-Straße 1,
D-85740 Garching bei München, Germany

Abstract. A standard paradigm is now available for the recent evolution ($z < 10$) of structure on galactic and larger scales. Most of the matter is assumed to be dark and dissipationless and to cluster hierarchically from gaussian initial conditions. Gas moves under the gravitational influence of this dark matter, settling dissipatively at the centres of dark halos to form galaxies. The evolution of the dark matter component has been studied extensively by N-body simulations. The abundances, density profiles, shape and angular momentum distributions and the formation histories of the dark halo population can all be predicted reliably for any hierarchical cosmogony. The systematic variation of these properties with halo mass can produce scaling relations in the galaxy population. Simple hypotheses for how galaxies condense within dark halos lead to characteristic luminosities, sizes, and spins which are close to those of real spiral and elliptical galaxies. Furthermore, correlations similar to the Tully-Fisher, Faber-Jackson and luminosity-metallicity relations arise quite naturally. A quantitative explanation of the fundamental plane of elliptical galaxies appears within reach.

1 Introduction

The current popularity of the Cold Dark Matter (CDM) model and its variants stems from a variety of sources. Most of the mass in the Universe appears to be in a dark collisionless form which is concentrated towards galaxies but extends far beyond their visible boundaries. If this matter has interacted only gravitationally since early times it is possible to reconcile the very small observed amplitude of fluctuations in the Microwave Background with the existence of massive nonlinear structures in the present Universe. Furthermore, the large-scale distribution of galaxies looks quite similar to the patterns which result from the gravitational amplification of gaussian density fluctuations. This is a simple and natural initial condition in CDM-like models, and in such models the galaxies are indeed expected to trace the dark matter distribution on large scales.

The idea that hierarchical clustering under gravity has given rise to galactic and larger structures predates the CDM model (e.g. Peebles 1980), and almost two decades of simulation using N-body methods have provided a reasonably complete understanding of the structure produced by this process. If galaxy formation results from the dissipative collapse of gas within the potential wells provided by dark halos, then it is the internal structure of such halos and their formation history which must regulate the global properties of galaxies. In recent years there has been substantial progress in understanding the predictions of hierarchical clustering models for these and other aspects of halo structure. In the next section I summarise the aspects of this work most relevant for a discussion of galaxy scaling relations.

The structure and evolution of dark halos may determine the mass and angular momentum of the material available for galaxy formation, as well as the rate of interactions between galaxies. The global properties of galaxies must in addition depend on how gas cools to form dense clouds, on how star-formation proceeds in such clouds, and on how this star formation affects the surrounding material through the injection of heavy elements and energy. These processes interact in a highly nonlinear way and involve a very wide range of scales; there is little hope of simulating them realistically. If, however, they are parametrised by an appropriate set of efficiencies, to be assigned physically reasonable values by comparison with available observational or simulation data, then it is possible to make simple "semi-analytic" models which can predict a very wide range of properties of the galaxy population for any specific cosmogony, for example any of the popular CDM variants (e.g. luminosity functions, colours, sizes, morphologies, gas contents, and the dependence of all of these on environment and redshift). In sections 3 and 4 I discuss how such models can be used to predict galaxy scaling relations, and I emphasise, in particular, the inferences which can be drawn from the observed tightness of some of these relations.

2   Structure and Evolution of Dark Halos

N-body simulations have provided a good understanding of the structure and evolution of the dark halos which form through hierarchical clustering of dissipationless matter. For example, the angular momentum of dark halos, best characterised by the dimensionless spin parameter,

$$\lambda \equiv J|E|^{1/2}/GM^{5/2}, \tag{1}$$

where J, E and M are the total angular momentum, energy and mass of the halo, is found to have a broad distribution with a median near $\lambda \sim 0.04$. This distribution appears "universal" in the sense that it has no strong dependence on the mass of the halo or on the parameters of the particular cosmogony in which clustering occurs (e.g. initial power spectrum, $P(k)$, cosmic density, Ω, cosmological constant, Λ,...; see Barnes & Efstathiou 1987, Frenk et al 1988, Warren et al 1992, Cole & Lacey 1996). For a cold rotationally-supported self-gravitating disk one finds $\lambda \sim 0.4$. Hence $\lambda \sim 0.04$ implies a system in which rotation velocities are an order of magnitude smaller than needed for centrifugal support. Any gas component in such a system must shrink in radius by a similar factor if it is to make a centrifugally supported disk (see below). Within individual dark halos rotational streaming usually varies quite weakly with radius, but there are large variations from halo to halo.

The axial ratios of dark halos have also been extensively studied and also appear to show a broad distribution which depends at most weakly on mass or cosmogony (Frenk et al 1988; Warren et al 1992; Cole & Lacey 1996). Nearly spherical halos are quite uncommon. There is a slight preference for near-prolate over near-oblate shapes, and major-to-minor axis ratios in excess of two are common. It is interesting to ask whether such a distribution is consistent with the fact that deviations from axisymmetry in observed disks are typically quite small (e.g. Rix & Zaritsky 1995). I will not pursue this question further here.

A third regularity in the structure of dark halos has emerged only recently. Navarro et al (1996, 1997; hereafter NFW) studied halo density profiles in a wide variety of hierarchical cosmogonies. Their work is distinguished from earlier studies in that they simulated the evolution of each halo separately. This allowed them to set the resolution limits in mass and in linear scale to be constant fractions of the characteristic mass and radius of each halo, even though these characteristic values ranged over several orders of magnitude. NFW found the remarkable result that the spherically averaged density profiles of halos of *all* masses in *all* the cosmogonies they considered could be adequately represented by a suitable scaling of the same analytic form:

$$\rho(r)/\rho_{crit} = \delta_c r_s^3 / r(r + r_s)^2. \tag{2}$$

In this formula r_s sets the "core" radius of the halo and δ_c is its characteristic density in units of the critical density, ρ_{crit}. Thus the inner regions have a density cusp with $\rho \propto 1/r$ while at larger radii the profile steepens towards $\rho \propto 1/r^3$. The bounding radius of a virialised halo is conventionally defined as the radius r_{200} within which the mean density is 200 times the critical value; the "total" halo mass is then the mass within this radius M_t. Defining the concentration of a halo to be $c \equiv r_{200}/r_s$, one immediately finds c to be determined implicitly from

$$\delta_c = \frac{200}{3} \frac{c^3}{[\ln(1 + c) - c/(1 + c)]}, \tag{3}$$

and, of course, we have

$$M_t = \frac{4\pi}{3} \, 200 \rho_{crit} \, (cr_s)^3. \tag{4}$$

NFW found that for any particular hierarchical cosmogony the two parameters δ_c and r_s are strongly correlated with each other and so with M_t. This correlation is always in the sense that lower mass halos have higher characteristic densities and so greater concentrations. It turns out that this correlation can be understood as a reflection of the fact that smaller mass halos form earlier, and indeed, for a suitable definition of the formation epoch of a halo z_f, NFW showed that all the halos in all their cosmogonies obey the simple relation

$$\delta_c \approx 5 \times 10^3 \Omega_0 (1 + z_f)^3. \tag{5}$$

To a good approximation it seems that equilibrium dark halos in all hierarchical cosmogonies have similar density profiles and furthermore that the characteristic density of a halo is just proportional to the density of the universe at the time it formed. It is hard to imagine a simpler situation.

Of course, the properties of the galaxies within a dark halo depend not only on its current structure but also on the details of its formation history. There has been substantial progress over the last five years in understanding how individual nonlinear objects are built up by hierarchical clustering. This is primarily a result of the discovery that extensions of the original argument of Press & Schechter (1974) can provide a remarkably detailed and accurate description of the statistics of merging and accretion in N-body simulations of hierarchical clustering (Bond et al 1991; Bower 1991; Lacey & Cole 1993,1994). Indeed, these formulae provide a basis for Monte Carlo realisations of the full merging tree which describes how any particular object, for example a rich cluster, is built up by successive merging of smaller systems (Kauffmann & White 1993).

Armed with such a tree one can attempt to model all the additional processes which determine the galaxy population within a dark halo (gas cooling, star formation, energy injection from young stars, chemical enrichment, stellar population evolution, galaxy (as opposed to halo) merging, etc.). A major success of recent galaxy formation studies has been the demonstration that even very simple physical models for these processes lead to explanations not only for the luminosities, colours, morphologies, metallicities and abundances of galaxies, and for scaling relations between these properties, but also for the fact that $10^{15}M_\odot$ halos typically contain many bright early-type galaxies while $10^{12}M_\odot$ halos typically contain a single central spiral and a few satellites (Kauffmann et al 1993; Baugh et al 1996a). Furthermore, since such models automatically specify the full history of the galaxy population, they can be compared directly with observational indicators of galaxy evolution, for example with counts and redshift distributions of faint galaxies (Cole et al 1994; Kauffmann et al 1994; Heyl et al 1995; Baugh et al 1996b) or with the properties of damped Lyα systems in QSO spectra (Kauffmann 1996a). The results so far are encouraging, and it seems that a reasonably complete, if schematic, picture of galaxy formation is now in place. In the next two sections I discuss the scaling relations this picture predicts for disk and elliptical galaxies.

3   Scaling Relations for Disks

The defining characteristic of galaxy disks is that they are made of stars which are almost all on near-circular orbits confined close to the disk plane. Since it appears impossible to create a thin centrifugally supported disk without very substantial dissipation, one draws two immediate conclusions:
(i) Galaxy disks were assembled while still gaseous – their stars were all formed *in situ*. Of course, this does not preclude disk growth through gas infall after the formation of many of the stars.
(ii) galaxy disks cannot have been violently disturbed since formation of the bulk of their stars, otherwise they would no longer be thin.
Another critical observation is that the outer rotation curves of most spirals are approximately flat and appear to supported primarily by the gravity of their dark halos. This suggests that the properties of disks may be determined by

those of their dark halos.

The standard model for disk formation was set out by Fall & Efstathiou (1980) in an extension of the ideas of White & Rees (1978). After a dark halo comes to equilibrium, much of its baryonic material is supposed to remain as diffuse gas with a distribution similar to that of the dark matter. Subsequently this gas radiates its binding energy but retains its angular momentum and so flows inwards until it settles into a rotationally supported disk. Fall & Efstathiou showed that an extended dark halo is required, and that little angular momentum can be lost if disks similar to observed spirals are to form. Their scheme is the basis of most recent modelling of spiral galaxy formation (e.g. Kauffmann 1996a; Dalcanton et al 1997) but has not yet been shown to work in any numerical simulation of hierarchical galaxy formation. The difficulty is that inclusion of feedback from young stars is critical. Without it gas cools into small dense clumps at early times, and these lose most of their angular momentum as they merge at the centres of massive dark halos; the resulting disks are then too small to represent real galaxies (Navarro & Benz 1991; Navarro et al 1995; Navarro & Steinmetz 1997).

Let me work through a simple example to show how this scheme can be used to derive scaling relations for galaxy disks. If we model a halo as a singular isothermal sphere of circular velocity V_c, then its mass, kinetic energy and angular momentum within $r_{200} = V_c/10H(z)$ can be written as

$$M_t = \frac{V_c^3}{10GH(z)}, \quad E = M_t V_c^2/2, \quad \text{and} \quad J_h = \sqrt{2}\lambda M_t r_{200} V_c, \tag{6}$$

where $H(z)$ is the Hubble constant at the redshift when the halo is identified and its central disk is made. Assume a fraction F of the mass of the halo is in the form of gas with the *same* specific angular momentum as the dark matter. Assume further that this gas sinks to the centre conserving its angular momentum and forms an exponential disk of mass M_d, central surface density S_0 and scale radius r_d. If we neglect the contribution of the self-gravity of the disk to its rotation curve then we find

$$M_d = \frac{FV_c^3}{10GH(z)}, \quad r_d = \frac{\lambda V_c}{10\sqrt{2}H(z)} \quad \text{and} \quad S_0 = \frac{10FV_cH(z)}{\pi\lambda^2 G}. \tag{7}$$

These relations have some immediate consequences. If the stellar mass-to-light ratio of disks is assumed to be a constant value $\Upsilon(z)$ at each redshift, then the first relation gives a Tully-Fisher-like relation, $L \propto V_c^3$, which is *independent* of λ. On the other hand r_d and S_0 depend strongly on λ; slowly rotating halos produce compact and high surface brightness disks. This is encouraging because the exponent of the observed Tully-Fisher relation is not far from 3, and furthermore this relation appears to hold independent of galaxy surface brightness (de Blok & McGaugh 1996; Tully, this meeting). In addition, the proportionality constant seems reasonable. If, following McGaugh & de Blok (1997), we adopt $\Upsilon_B = 2.5h$, then the zero-point of the observed T-F relation, $L_B = 1.5 \times 10^{10}h^{-2}L_\odot$ at $V_c = 200$km/s (e.g. Strauss & Willick 1995), agrees with the prediction provided

$F = 0.02H_0/H(z)$, i.e. $F \approx 0.02$ if disks are assembled near $z = 0$ and $F \approx 0.05$ if disks are assembled near $z = 1$.

The predicted characteristic sizes of disks also seem reasonable. For a "typical" halo with $\lambda = 0.05$ and $V_c = 200$km/s the predicted scale radius is $r_d = 7H_0/H(z)\ h^{-1}$kpc, or $R_d \approx 7h^{-1}$kpc for assembly near $z = 0$ and $R_d \approx 3h^{-1}$kpc for assembly near $z = 1$. Notice that the redshift dependence in these equations is quite strong. It does not appear possible to make substantial numbers of big disks at high redshifts. Thus if damped Lyα absorbers in QSO spectra at $z \sim 3$ are indeed equilibrium disk systems with circular velocities of order 200km/s, then they must be quite small, $r_d \sim 1$ to 2 kpc, if they are to be explained in a hierarchical clustering model. Notice also, as mentioned above, that there cannot be much transfer of angular momentum from gas to dark matter during disk formation, otherwise the resulting disks will be too small for *any* assumed redshift of assembly.

The strong λ-dependence of r_d and S_0 together with the broad λ-distribution resulting from hierarchical clustering implies that galaxy disks are predicted to have a wide range of sizes and surface brightnesses at any given luminosity or circular velocity. A recent discussion of the observational data by Dalcanton et al (1997) suggests that this may indeed be the case. "Disks" formed from the low λ tail of the distribution are predicted to be so compact, however, that they should perhaps be identified with observed spheroids. In these objects the baryonic component should dominate strongly over the dark matter, and this may, perhaps, lead to violent instabilities which prevent thin disk formation.

A final important issue concerns the tightness of the observed T-F relation. This obviously implies some considerable uniformity in the formation of disk galaxies. As we have seen, the broad spin distribution does not, of itself, induce scatter. Variations in assembly time can do so through the $H(z)$ dependence of M_t. In combination with the size constraints already discussed, this suggests that most disks were assembled well after $z = 1$. Variations in the actual structure of halos of given mass and assembly epoch must also be sufficiently small to avoid excessive scatter in the $M_t - V_c$ relation. For the halos simulated by Navarro et al (1996, 1997) this relation is indeed tight enough. Finally, small scatter is required in the fraction F of the halo mass which condenses into a disk, in the disk mass-to-light ratio Υ, and in the disk contribution to the observed V_c values (which will vary with λ). The observed colours of disk galaxies are quite uniform, suggesting that Υ may not vary too much, and recent observations favour small Υ values, thus helping to satisfy the last condition (e.g. McGaugh & de Blok 1997). Since the required F-values are smaller than observed in galaxy clusters (e.g. White and Fabian 1995), the uniformity of F suggests that some feedback process lowers the condensation efficiency in a way which depends only on V_c. Substantial feed-back appears necessary to account for the apparent global inefficiency of galaxy formation (e.g. White & Rees 1978; White & Frenk 1991) and a variation with V_c can induce a metallicity-luminosity relation (Larson 1972; Dekel & Silk 1986). In particular, feedback from star formation in CDM-like cosmologies can plausibly explain the observed metallicities both of present-day disks and of high redshift damped Lyα systems (Kauffmann 1996a).

A more careful analysis of many of the ideas in this section, together with applications to specific hierarchical cosmologies can be found in Mo et al (1997) and Dalcanton et al (1997). The latter paper compares its predictions in some detail with the observed sizes and surface brightnesses of disk galaxies.

4   Scaling Relations for Ellipticals

The properties of elliptical galaxies, particularly those of ellipticals in rich clusters, show some remarkable regularities. Most have very nearly elliptical isophotes and a luminosity profile which is well described by de Vaucouleurs' empirical fitting function. There is a tight relation, known as the fundamental plane, between the characteristic size of a galaxy, its total luminosity, and its central velocity dispersion. In addition there are tight relations between the luminosities of ellipticals and their colours and metallicities. The simplest interpretation requires

(i) that all ellipticals are made of old stars,

(ii) that they all formed in a similar way,

(iii) that the initial mass functions of their stellar populations (and so their M/L ratios at given age) are similar or at least vary only slowly with mass, and

(iv) that their metallicity increases (and so their colour reddens) systematically with mass.

The fundamental plane then reflects the virial relation $M \sim R\sigma^2$ with a slight tilt arising from the systematic variation of M/L with mass. Recent data on the evolution of ellipticals support this interpretation in that they are consistent with the fading in luminosity expected for a passively evolving population of equilibrium galaxies (see other contributions to this volume). An indication that the real picture may be more complex comes, however, from dynamical analyses which suggest that much of the mass within the luminous regions of ellipticals may in fact be pregalactic dark matter (e.g. Rix et al 1997).

More than twenty years ago Toomre (1977) remarked that star formation is observed only in galaxy disks, and further that the final state of pairs of interacting spirals must be something resembling an elliptical galaxy. In view of this he suggested that *all* star formation might occur in disk systems, and that ellipticals might *all* be formed by the merger of stellar disks. Although remaining controversial, these suggestions have gained much theoretical and observational support since they were made. Direct simulations of mergers between systems resembling disk galaxies have shown that they do indeed evolve into objects with a structure very like that of ellipticals (e.g. Barnes 1988). Furthermore, a number of transition cases have been found which seem to demonstrate empirically that merging spirals end up as ellipticals (e.g. Schweizer 1990). Finally it is still true that substantial star formation has been seen only in galaxy disks, or in starbursts either in the nuclear regions of gas-rich galaxies or in interacting disk systems.

The strongest objections to Toomre's proposal have come:

(i) from the tight systematic relations between E-galaxy properties – tight cor-relations seem intuitively surprising if ellipticals are produced by the stochastic accumulation of smaller units,

(ii) from the fact that ellipticals are denser and more strongly bound than spirals – their progenitors must then have been more compact and more tightly bound than present-day disks, and

(iii) from the fact that most disk galaxies have central bulges which resemble ellipticals in many of their properties – how could mergers produce a central "elliptical" without disturbing the surrounding disk.

Semi-analytic models of hierarchical galaxy formation generally adopt the hy-pothesis that all star formation occurs in quiescent or interacting disks, and can address the above objections directly because they keep track of how and where disks grow and of how they merge together. It is therefore possible to trace the formation history of each elliptical galaxy, and to ask how it depends on lu-minosity and environment. The first detailed models of this kind were able to reproduce the characteristic luminosities and colours of ellipticals, the distribu-tion of bulge-to-disk ratios of spirals, and the environmental segregation between ellipticals and spirals (Kauffmann et al 1993; Baugh et al 1996a). Objects with little or no disk are predicted to occur primarily in clusters and to have old stellar populations. They form by the merger of disks which were assembled well before $z = 1$ and so were compact (equ. 7). Present-day disks form late by accretion of new gas onto small "ellipticals" produced by the merging of earlier generations of disks.

In Figure 1 (adapted from Kauffmann 1996b) I illustrate when star-formation and merging are predicted to occur for cluster ellipticals in an $\Omega = 1$ CDM cos-mogony normalised to give the correct abundance of rich clusters. The modelling scheme assumes that all objects with disk-to-bulge ratios less than 0.67 are clas-sified as ellipticals, and for this plot the elliptical population in clusters of mass $10^{15} M_\odot$ is analysed. The solid histogram shows the formation times of the stars which end up in these ellipticals. More than 40% form before $z = 3$, about 60% before $z = 2$, and more than 80% before $z = 1$. Very few stars have formed in these objects over the last few billion years. Thus cluster ellipticals are pre-dicted to be red and to show little scatter in their colour-luminosity relation. More detailed study shows that ellipticals in high-z clusters are predicted to form their stars earlier on average than present-day ellipticals, and as a result the scatter in the luminosity-colour relation remains small out to redshifts of or-der unity (Kauffmann 1996b). The dashed histogram in Fig. 1 shows when these ellipticals underwent their last major merger. This is predicted to be quite late – more than 70% were assembled after $z = 1$. Somewhat later star-formation and merger times are predicted for ellipticals in groups rather than clusters, and similar patterns are predicted in other cosmogonies – formation is somewhat earlier in low density universes and somewhat later in $\Omega = 1$ cosmogonies with less small-scale power than CDM (e.g. mixed dark matter).

A natural prediction of hierarchical cosmogonies is that small things form first. As Figure 2 demonstrates, however, this effect is barely detectable for el-

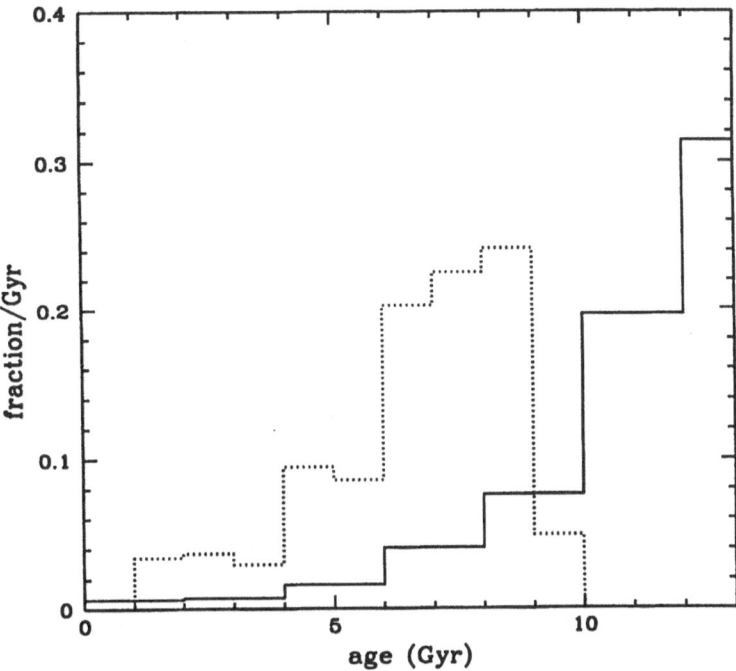

Fig. 1. The solid histogram gives the distribution of formation times for the stars in elliptical galaxies in a $10^{15} M_\odot$ cluster. The semi-analytic model assumes a standard CDM cosmogony normalised to $\sigma_8 = 0.67$. The dotted histogram gives the distribution of the times when these elliptical galaxies underwent their last major merger (data taken from Kauffmann 1996b).

lipticals in clusters. In this plot the mean stellar age of ellipticals is shown as a function of their total stellar mass for the same cosmogony analysed in Fig. 1. Ellipticals of all masses are made of old stars, and the decrease in age with increasing mass is less than the (small) age scatter between galaxies. If metallicity effects are ignored, the colours of ellipticals are predicted to be essentially independent of luminosity and to have small scatter (e.g. Kauffmann 1996b; Baugh et al 1996a). The inclusion of chemical enrichment effects can plausibly produce the observed colour-luminosity relation because: (a) more massive ellipticals are predicted to form from the merging of more massive disks, and (b) as a result of feed-back effects, the metallicity of disks is predicted to increase strongly with their mass (e.g. Kauffmann 1996a). The first effect is illustrated in Fig. 2 which gives the ratio of mean progenitor mass to final mass as a function of final mass. The mean progenitor mass is defined by tagging each star with the mass of the disk galaxy in which it formed, and then averaging this mass over all the stars in the final elliptical. The stars in a $2.5 \times 10^{12} M_\odot$ elliptical typically formed in disks which were more than ten times as massive as those which merged to make a $10^{10} M_\odot$ elliptical. It will be possible to check whether the resulting metallicity-luminosity relation reproduces the observed colour-luminosity plots as soon as reliable population synthesis models are available for a wide range of metallicities.

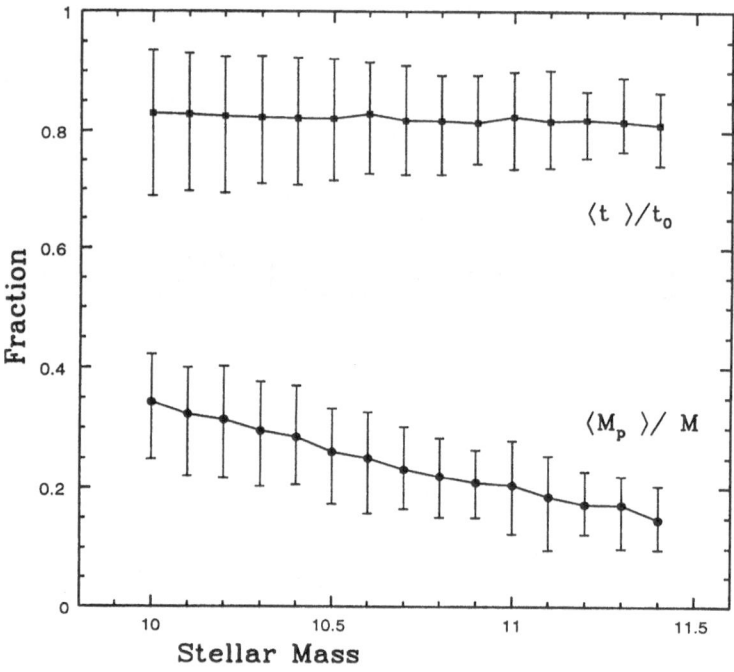

Fig. 2. The upper points give the mean ages of cluster ellipticals (in units of the age of the Universe) as a function of their stellar mass for the same model plotted in Fig. 1. The lower points give the mean progenitor masses of these same ellipticals in units of their total mass. In both cases the error bars join the upper and lower 5% points of the galaxy to galaxy scatter in these quantities.

According to these models the stellar population of present-day ellipticals formed in compact disks with scale radii $\sim$ 1kpc and circular velocities $\sim$ 200km/s. These disks were among the most rapidly star-forming objects at $z \sim 2.5$, and should presumably be identified with the galaxies recently discovered by Steidel and collaborators (Steidel et al 1996; Giavalisco et al 1996). The observed objects have roughly the correct size, abundance, star formation rate and internal characteristic velocity, but it is not yet clear whether they are indeed disk-like. Very recent studies of the abundance of such objects during $0 < z < 5$ suggest that the overall rate of unobscured star formation in the Universe actually peaked at $1 < z < 2$, that star formation in this mode could possibly account for *all* the observed stars in galaxies, and that most stars formed after $z \sim 1$ (Madau et al 1997). Such late star formation is one of the most robust and controversial predictions of hierarchical models (e.g. White 1989; Cole et al 1994) but observational verification is difficult since the conversion from observed UV flux to star-formation rate is uncertain by at least a factor of 2. Thus one cannot tell whether all stars formed in the observed unobscured mode or only 30% to 50% of them, for example the stars in present-day disks. A direct proof of recent elliptical formation could come from a survey of the Universe at, say, $z = 2$, which showed the current population to be absent at that epoch. Recent deep redshift surveys selected at I and K allow complete samples of early-type galaxies to be identified to $z \sim 1$. V/V_{max} tests applied to these samples show unambiguously that the early-type population does not follow standard passive evolution models (Kauffmann et al 1996). In fact, roughly two thirds of the present population appears to be missing at $z = 1$; either the galaxies were actively forming stars or they were in several pieces at that time. If further deep surveys confirm this result, we may conclude that the bulk of galaxy formation has already been observed, and that we now have a crude quantitative understanding of the origin and evolution of the basic properties of galaxies.

Acknowledgements I thank my collaborators S. Charlot G. Kauffmann, S. Mao and H.J. Mo for many helpful discussions of the material in this review. G. Kauffmann also provided the model data plotted in figures 1 and 2.

References

Barnes J., 1988, ApJ, 331, 699
Barnes J., Efstathiou G.P., 1987, ApJ, 319, 575
Baugh C.M., Cole S., Frenk C.S., 1996a, MNRAS, 283, 1361
Baugh C.M., Cole S., Frenk C.S., 1996b, MNRAS, 282, L27
de Blok W.J.G., McGaugh, S.S., 1996, ApJ, 469, L89
Bond J.R., Cole S., Efstathiou G.P., Kaiser N., 1991, ApJ, 379, 440
Bower R.J., 1991, MNRAS, 248, 332
Cole S. et al 1994, MNRAS, 271, 781
Cole S., Lacey C.G., 1996, MNRAS, 281, 716
Dalcanton, J.J., Spergel, D.N., Summers F.J., 1997, ApJ, in press
Dekel A., Silk J.I., 1986, ApJ, 303, 39
Fall S.M., Efstathiou G., 1980, MNRAS, 193, 189
Frenk C.S., White S.D.M., Davis M., Efstathiou G.P., 1988, ApJ, 327, 507
Giavalisco M., Steidel C.C., Macchetto F.D., 1996, ApJ, 470, 189
Heyl J.S., Cole S., Frenk C.S., Navarro J.F., 1995, MNRAS, 274, 755
Kauffmann G., 1996a, MNRAS, 281, 475
Kauffmann G., 1996b, MNRAS, 281, 487
Kauffmann G., Charlot S., White S.D.M., 1996, MNRAS, 283, L117
Kauffmann G., Guiderdoni B., White S.D.M., 1994, MNRAS, 267, 981

Kauffmann G., White S.D.M., 1993, MNRAS, 261, 921

Kauffmann G., White S.D.M., Guiderdoni B., 1993, MNRAS, 264, 201

Lacey, C.G., Cole, S., 1993, MNRAS, 262, 627

Lacey, C.G., Cole, S., 1994, MNRAS, 271, 676

Larson R.B., 1974, MNRAS, 169, 229

Madau P. et al, 1996, MNRAS, 283, 1388

McGaugh, S.S., de Blok W.J.G., 1997, ApJ, in press

Mo H.J., Mao S., White S.D.M., 1997, in preparation

Navarro J.F., Benz W., 1991, ApJ, 380, 320

Navarro J.F., Frenk C.S., White S.D.M., 1995, MNRAS, 275, 56

Navarro J.F., Frenk C.S., White S.D.M., 1996, ApJ, 462, 563

Navarro J.F., Frenk C.S., White S.D.M., 1997, ApJ, submitted

Navarro J.F., Steinmetz M., 1997, ApJ, in press

Peebles P.J.E., 1980, Physical Cosmology, Princeton Univ. Press

Press W.H., Schechter P.L., 1974, ApJ, 187, 425

Rix H.-W., et al, 1997, astroph/9702126

Rix H.-W., Zaritsky D.F., 1995, ApJ, 447, 82

Schweizer F., 1990, in Dynamics and Interactions of Galaxies (ed. R. Wielen), Springer, p60.

Steidel C.C. et al, 1996, ApJ, 462, L17

Strauss M.A, Willick J.A., 1995, Phys.Rep., 261, 271

Toomre A., 1977, in Tinsley B.M., Larson R.B., eds, Evolution of Galaxies and Stellar Populations, Yale Univ. Obs, p401

Warren M.S., Quinn P.J., Salmon J.K., Zurek W.H., 1992, ApJ, 399, 405

White D.A., Fabian A.C., 1995, MNRAS, 273, 72

White S.D.M., 1989, in The Epoch of Galaxy Formation (ed. C.S. Frenk et al), Kluwer, p15

White S.D.M., Frenk, C.S., 1991, ApJ, 379, 52

White S.D.M., Rees M.J., 1978, MNRAS, 183, 341

Can the Tully-Fisher Relation Be the Result of Initial Conditions?

Daniel J. Eisenstein[1,2,3] and Abraham Loeb[2,4]

[1] Institute for Advanced Study, Olden Lane, Princeton, NJ 08540, USA
[2] Astronomy Department, Harvard University, 60 Garden St., Cambridge, MA 02138, USA
[3] eisenste@sns.ias.edu
[4] aloeb@cfa.harvard.edu

Abstract. We use Monte Carlo realizations of halo formation histories and a spherical accretion model to calculate the expected scatter in the velocity dispersions of galactic halos of a given mass due to differences in their formation times. Assuming that the rotational velocity of a spiral galaxy is determined by the velocity dispersion of its halo and that its luminosity is related to its total baryonic mass, this scatter translates to a minimum intrinsic scatter in the Tully-Fisher relation. For popular cosmological models we find that the scatter due to variations in formation histories is by itself greater than allowed by observations. Unless halos of spiral galaxies formed at high redshift ($z \gtrsim 1$) and did not later accrete any significant amount of mass, the Tully-Fisher relation is not likely to be the direct result of cosmological initial conditions but rather a consequence of a subsequent feedback process.

1 Introduction

The tight power-law correlation between the luminosities and rotational velocities of spiral galaxies (Tully & Fisher 1977; Aaronson, Huchra, & Mould 1979) is one of the most striking and useful results of extragalactic astronomy. The small scatter around $L \propto v_c^\alpha$ allows one to accurately measure distances and peculiar velocities (Strauss & Willick 1995, and references therein). While differences in observational technique, such as bandpass, sample selection, and the method for observing the rotational velocity, affect the power-law slope and degree of scatter, most large samples find a scatter of about 0.4 mag in the relation over two decades in luminosity (Mathewson & Ford 1994; Willick et al. 1995, 1996a,b; Giovanelli, this volume). Other samples report scatter as low as 0.1 mag (Bernstein et al. 1994). The Tully-Fisher relation seems to be at most weakly dependent on morphological type (Willick et al. 1996b) and central surface brightness (Zwaan et al. 1995; Sprayberry et al. 1995).

Despite the popularity of the relation, the reason behind its existence is still unknown. It is particularly surprising that the complex processes of gravitational collapse and star formation do not act to blur this tight relation. Here we address the most basic question in this context, namely: *Could the low scatter observed in the Tully-Fisher relation be the direct result of cosmological initial conditions?* We therefore wish to relate the two quantities in the relation, the circular velocity and the luminosity, to properties of the dark matter halos in which the spiral galaxies form. The former depends on the shape and depth of the halo potential-well, while the latter depends on the total baryonic mass. If the cosmological scatter in these properties is greater than allowed by observations, then the Tully-Fisher relation would imply a subsequent feedback process that regularizes gas dynamics and star formation in galaxies according to the depth of their potential wells.

We have studied the spread in the potential depth of an ensemble of galaxy halos of a given mass due to the differences in their formation times. We construct Monte Carlo realizations of the formation histories of halos as provided by the excursion set formalism (Bond et al. 1991, hereafter BCEK; Kauffmann & White 1993; Lacey & Cole 1993, 1994, hereafter LCa,b). The history of a given halo is used to estimate its current binding energy and velocity dispersion under the most regularizing assumption that the halo relaxes to an isothermal configuration through spherical accretion. This method and its minimum scatter estimates are described in §2 for different halo masses and cosmological models. In §3 we relate these results to the observed scatter in the Tully-Fisher relation. We conclude that the implied minimum scatter due to the variations in halo formation histories is too high to be reconciled with observations. A full account of this work is given in Eisenstein & Loeb (1996).

2   Method and Results

The universality of flat rotation curves and the Tully-Fisher relation imply that spiral galaxies obey a regular behaviour. In order to show that this regularity cannot naturally emerge from a random set of initial conditions, it is sufficient to consider the most optimistic model in which all galaxy halos share the same density profile. Indeed, such a trend has been found in collisionless N-body simulations (Navarro et al. 1996a,b). Here, we approximate the halo density profiles as truncated isothermal spheres; the deviations of this profile from more realistic profiles are not important for this calculation.

The potential depth of an isothermal halo is simply characterized by its one-dimensional velocity dispersion σ. If all, or some fixed fraction, of the baryons in the halo cool to the galactic disk, then a perfect relation between the velocity dispersion σ and the mass of the halo M would produce virtually no scatter in the Tully-Fisher relation. This follows from the most optimistic assumption that the circular velocity of the disk at large radius is simply proportional to the velocity dispersion of the halo, neglecting any complications due to triaxiality of the halo, ellipticity of the galactic disk (Franx & de Zeeuw 1992), or variations in the rotation curve due to self-gravity of the baryonic disk.

However, as illustrated by the spherical collapse model (Gunn & Gott 1972; White 1995), σ is not simply a function of M. Because σ^2 characterizes the kinetic energy of a virialized system, we can fix σ by requiring that energy be conserved through the virialization process. For the collapse of a spherical shell without a cosmological constant ($\Lambda = 0$), the shell energy per unit mass is

$dE/dM = -GM/R_{ta}$, where M is the mass enclosed within the shell and R_{ta} is the radius of the shell at turn-around. For $\Lambda = 0$ cosmologies, the spherical shell solution gives

$$R_{ta} \propto M^{1/3} t_{coll}^{2/3}, \tag{1}$$

where t_{coll} is the time of collapse. We may then integrate over all shells to find the total energy E, which yields the velocity dispersion $\sigma^2 \propto E/M$. If an object collapses at a single time, $\sigma \propto M^{1/3} t_{coll}^{-1/3}$. The constant of proportionality is unimportant for this work.

During the hierarchical process of structure formation, the formation of an object is not characterized by a single redshift of collapse. Rather, the object is assembled over time by a combination of mergers and smooth accretion. Thus, one would expect variations in these formation histories to produce a scatter in the velocity dispersion for an ensemble of objects of a given mass. To approach this problem analytically within the framework of bottom-up models whose initial density perturbations are described by Gaussian random fields, we apply the excursion set formalism (BCEK) that has been used to predict halo formation histories (LCa,b). Given a collapsed object of mass M_2 at some redshift z_2 this formalism predicts the probability distribution for the object to have a mass M_1 at some higher redshift z_1. We denote by $\sigma_M(M)$ the *rms* fluctuation amplitude $\delta M/M$ (using a k-space top-hat window function) as a function of mass M and define $\delta_c(z)$ to be the overdensity required for the collapse of a spherical perturbation to occur at redshift z; both these quantities are measured at the present epoch as extrapolated in linear theory. Then the value of M_1 is found implicitly from the relation

$$\sigma_M^2(M_1) = \sigma_M^2(M_2) + \frac{[\delta_c(z_2) - \delta_c(z_1)]^2}{x^2}, \tag{2}$$

where x is a Gaussian-distributed random number with zero mean and unit variance (see eq. 2.16 of LCb and change variables to $x^2 = [\delta_{c1} - \delta_{c2}]^2/[\sigma_1^2 - \sigma_2^2]$). Using this equation, we start from a given mass at the present epoch and step backwards in time (with steps $\Delta z = 0.05$) to find its mass at a sequence of higher redshifts. We continue this procedure until M_1 is less than 10% of the present-day mass. Note that because we start from the present-day mass of the object, we are restricting our sample to isolated galaxies. Because isolated spiral galaxies do satisfy the Tully-Fisher relation, this is an acceptable restriction.

Next, we need to convert a given mass history into a binding energy. We do so by assuming that the mass gained in an infinitesimal redshift interval was in fact accreted in the most regular fashion, i.e. in spherical symmetry and with a constant dM/dz. Any variation on these simplifying assumptions would introduce additional degrees of freedom, such as the orbital parameters or the internal structure of the merger components. For spherical accretion in an arbitrary cosmology, the infinitesimal energy added to the system is

$$dE \propto \frac{GM\,dM}{M^{1/3}} \left(\frac{5}{3g} \delta_c(z) - \Omega_R \right), \tag{3}$$

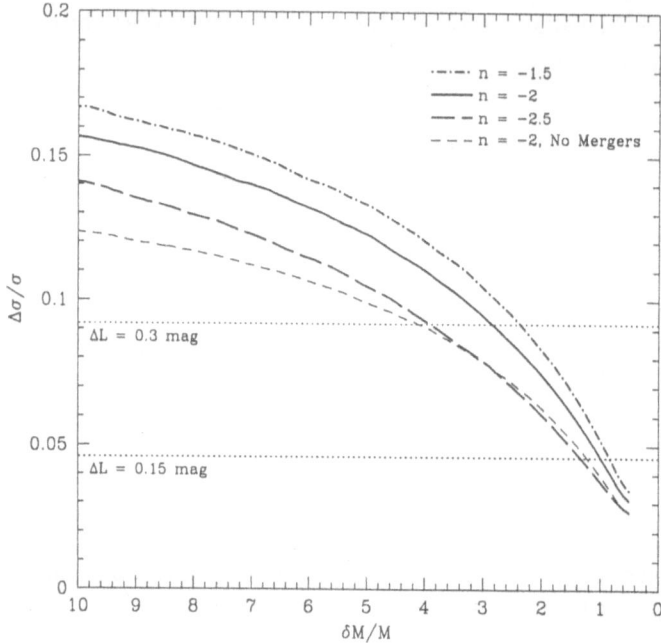

Fig. 1. Fractional scatter in the velocity dispersion σ as a function of the present-day overdensity $\delta M/M \equiv \sigma_M(M_0)$ on the mass scale of the final object M_0. The different curves correspond to scale-free power spectra, $P(k) \propto k^n$, with different values of n, in an $\Omega = 1$ cosmology. The "No Mergers" case requires that the object never accrete more than 20% of its final mass in any given time step.

where Ω_0 is the present fraction of the critical density in non-relativistic matter, Ω_Λ is the present fraction of the critical density in the cosmological constant, $\Omega_R = 1 - \Omega_0 - \Omega_\Lambda$ at the present epoch, M is the mass interior to the added shell, dM is the shell mass, and

$$g = \frac{5}{2} \int_0^1 \frac{a^{3/2}\,da}{(\Omega_\Lambda a^3 + \Omega_R a + \Omega_0)^{3/2}}. \tag{4}$$

The last factor in (3) is the generalization of the factor of $t_{coll}^{2/3}$ in (1). For finite Δz and ΔM, we integrate over ΔM assuming $dM/dz = \Delta M/\Delta z$ to find ΔE. We then sum over all the redshift steps to find the total energy, which we set proportional to the square of the velocity dispersion.

The above prescription allow the possibility that the mass accreted in a given time step (ΔM) is greater than the mass in the progenitor (M_1). This is in conflict with our goal to track the mass of the largest progenitor of an object. In the context of (3), it means that we are allowing most of the object to accrete spherically onto a small central seed in a short redshift interval. As this process seems unrealistic, we alter the rules so that if the mass M_1 found from (2) in a

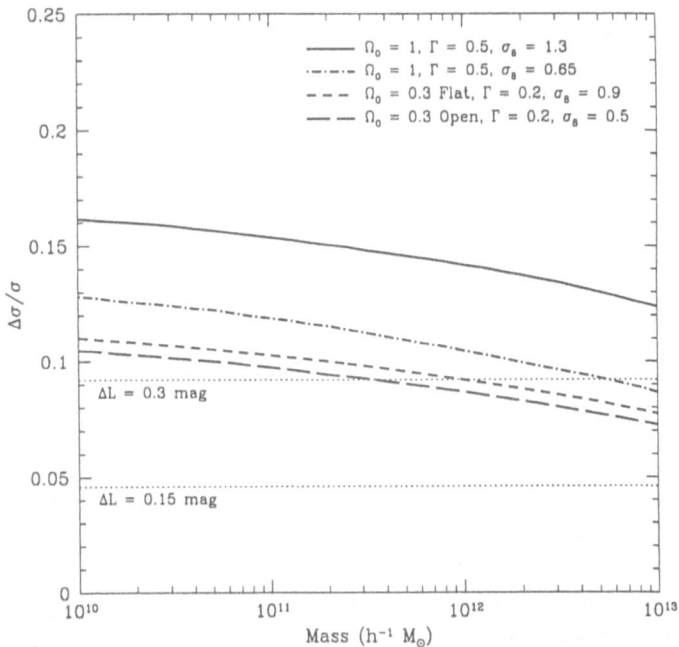

Fig. 2. Fractional scatter in the velocity dispersion σ in four different CDM models. The flat, low-Ω_0 model is the only one with a non-zero cosmological constant. The parameter $\Gamma = \Omega_0 h$ is used to fit the power spectrum (cf. Efstathiou et al. 1992).

particular redshift step is less than half of M_2, we flip the roles of the pieces and instead take the mass to be $M_2 - M_1$. Allowing this swap results in about 15% less scatter in the M–σ relation than would occur otherwise. Other methods for generating merger histories exist (Kauffmann & White 1993; Cole et al. 1994); using that of Kauffmann & White doesn't change our results.

We may now integrate over many generated mass histories to find the variation in the velocity dispersion on a given mass scale. We first consider scale-free power-spectra $P(k) \propto k^n$, for which $\sigma_M \propto M^{-(n+3)/6}$, and find $\Delta\sigma/\sigma$ as a function of $\sigma_M(M_0)$ and n, where M_0 is the present-day mass and $\Delta\sigma$ is the standard deviation of the distribution of velocity dispersions σ. The results are shown in Fig. 1 and compared to the scatter in luminosity that would result if we enforced a Tully-Fisher relation of the type $L \propto \sigma^3$ (dotted horizontal lines). The figure shows that the scatter in σ drops strongly with decreasing $\sigma_M(M_0)$ and with more negative n. Both these changes tend to bring the collapse to lower redshift so that the ensemble of objects is described by a narrower range of collapse redshifts. An open universe ($\Omega_0 < 1$, $\Lambda = 0$) results in a larger scatter at a given $\sigma_M(M_0)$ and n. A flat universe with $\Lambda \neq 0$ produces the same scatter as its flat $\Lambda = 0$ counterpart, because if $\Omega_R = 0$ then both (2) and (3) involve only δ_c and functions of mass, with Λ entering only in the constant of proportionality of the energy and therefore dropping out of $\Delta\sigma/\sigma$.

We next consider various cold dark matter (CDM) models, as parameterized by the constant Γ (Efstathiou et al. 1992). We normalize the power-spectrum to COBE using σ_8, the *rms* amplitude of fluctuations in spheres of radius $8h^{-1}\,\mathrm{Mpc}$ (Górski et al. 1995a,b; Stompor et al. 1995). Figure 2 presents the resulting scatter as a function of mass scale. We find that $\Omega = 1$, COBE-normalized CDM (solid line) produces too much scatter; this reflects the tendency of this model to have too much power on small-scales. Under-normalized $\Omega = 1$ CDM (dotted line) produces less scatter, while open and Λ-dominated universes give yet lower values. The lower value of σ_8 is the major factor in reducing the spread. However lower values of σ_8 than the ones shown in the figure are not allowed by the abundance of clusters of galaxies (White et al. 1993).

3   Discussion

We next examine the implications of the predicted scatter in $M-\sigma$ for the Tully-Fisher relation. In attempting to rule out a primordial origin for this relation, we adopt the most optimistic approach for connecting $M-\sigma$ to $L-v_c$. We assume that the circular velocity of the baryonic disk is strictly determined by the velocity dispersion of its spherical isothermal halo and that the overall luminosity of the galaxy is tightly related to the mass of its halo. The latter relation might exist if all the baryons in the halo cooled onto the central disk and were turned into stars with a universal mass-to-light relation. We then treat the scatter in σ on a given mass scale as a minimal source of intrinsic scatter in the Tully-Fisher relation.

How much scatter is allowed by the observations? The full observed scatter in the largest samples is 0.4 ± 0.02 magnitudes (Willick et al. 1995, 1996a,b), with brighter galaxies yielding smaller scatter (Giovanelli, this volume). Many different effects contribute to this error budget. Observational errors are usually estimated to be 0.15–0.30 mag (Strauss & Willick 1995). The ellipticity of the halo potential leads to error in measuring the circular velocity and estimating the inclination of the baryonic disk (Franx & de Zeeuw 1992); these effects have been measured in a sample of spirals and estimated to contribute a scatter of about 0.15 mag (Rix & Zaritsky 1995). The above effects alone reduce the remaining error budget to $\lesssim 0.3$ mag, which corresponds to about 10% scatter in the circular velocity. Indeed, Giovanelli (this volume) estimates the intrinsic scatter at 0.2–0.25 mag for all but the dimmest galaxies. It seems likely that variations in the mass-to-light ratio would lead to additional scatter, leaving less of the error budget available for intrinsic variations in σ. In summary, $\Delta\sigma/\sigma = 10\%$ provides an upper limit to the intrinsic scatter in σ at fixed M that is allowed by observations.

Naively, if each halo had a single formation redshift z_f, then the above 10% limit would require that all halos of a given mass form within an unusually narrow redshift interval, $\Delta z_f/(1 + z_f) \lesssim 20\%$. Indeed, Fig. 2 indicates that reasonable cosmologies are at best marginally consistent with the above 10% limit. Smaller masses are predicted to have a larger scatter, matching a trend in the data (Federspiel et al. 1994; Willick et al. 1996b), although various observational effects could also introduce larger scatter in dimmer galaxies.

In calculating the minimum Tully-Fisher scatter due to differences in the formation histories of dark halos, we have introduced two simplifying assumptions. First, we assumed spherical accretion in relating the mass history of an object to its final binding energy E [cf. (3)]. Non-sphericity would add new degrees of freedom to our analysis, such as the orbital parameters or the internal structure of the merger components. These additional parameters could increase the resulting M–σ scatter. Second, we have assumed that the mass and stellar luminosity of the gas that cools to form the galactic disk is a strict function of the mass of the halo. If not all of the baryons originally associated with a halo are currently in its stellar disk and if the fraction of baryons included in the disk varies from one galaxy to another, then the total luminosity of the galaxy will not be as tightly constrained by the mass of its halo and the expected Tully-Fisher scatter would increase. As for the fate of the remaining baryons, they may either be expelled from the galaxy by supernova-driven winds (e.g. Spitzer 1956; Corbelli & Salpeter 1988) or remain in the halo in the form of hot pressure-supported gas or as dark compact objects. X-ray observations exclude the existence of extended x-ray halos or cooling flows in most spiral galaxies (Fabbiano 1989). Searches for microlensing events toward the LMC indicate that a noticeable fraction of the halo mass may consist of baryonic objects (Alcock et al. 1995; Gates, Gyuk, & Turner 1994). If these objects formed in the intergalactic medium at very high redshift, then they would be essentially indistinguishable from other types of cold dark matter. However, if they formed within proto-galaxies, then their mass fraction may vary from galaxy to galaxy. Similarly, if supernovae are effective at expelling gas out of the disk, then the amount of gas expelled would likely vary from one galaxy to another due to differences in their star formation history, producing additional scatter in the relation. Finally, differences in the merger histories between individual galaxies will affect the cooling of the halo gas and its conversion into stars (Kauffmann, White, & Guiderdoni 1993), causing variations in the stellar ages and the amount of cool gas.

The Tully-Fisher relation applies to a particular morphological class of galaxies, and so one may argue that galaxies which have undergone major mergers since the formation of their stellar disks would not appear as thin spiral galaxies today (Tóth & Ostriker 1992). In order to examine the significance of this restriction, we modify our algorithm so that any object that has undergone a merger that added more than 20% of its present-day mass is rejected. This constraint rejects 70% of the systems and reduces the scatter in v_c by a factor of 20% (cf. the lowest curve in Fig. 1). Hence, while this morphological constraint goes in the right direction, it leads to a small overall effect.

Another way to implement the above morphological restriction is to argue that spiral galaxies already exist in their current form at some high redshift, z_f. If accretion does not affect the rotation curve or luminosity of these galaxies between this early redshift and the present time, then the M–σ relation present at the earlier time would be frozen in. This lowers the effective value of $\sigma_M(M_0)$ by the ratio of the growth factor at z_f to that of today, which in turn reduces the scatter in the M–σ relation, as can be seen in Figs. 1 and 2. Note that the fact that structure formation cuts off at late times in an open universe is not sufficient for this purpose. However, if galaxies form in highly overdense regions at high redshift, they will generally imbedded in much more massive systems today, leaving one to explain why the systems are unchanged by their infall into larger objects. At first glance, it seems unlikely that both the baryonic mass and the velocity dispersion of the halo center would remain unchanged as the growth factor increases by a significant factor and the corresponding accretion occurs. However, detailed consideration of these conditions raises difficult questions concerning the maximum mass of galaxies (Thoul & Weinberg 1995), the accretion of satellite galaxies (Tóth & Ostriker 1992; Navarro, Frenk, & White 1994), and the fate of dark matter cores as they fall into larger objects (Moore, Katz, & Lake 1995).

Empirical evidence that cold gaseous disks have already condensed at early times comes from the fact that the mass of neutral hydrogen (HI) in damped Lyα systems at $z \approx 2-3$ is comparable to the total mass in stars of local disk galaxies (Lanzetta et al. 1995). Such systems have column densities similar to those found in nearby galactic disks (Broeils & Van Woerden 1994) and velocity structure consistent with thick, rotating disks (Turnshek et al. 1989; Lanzetta & Bowen 1992; Prochaska & Wolfe 1997). These facts suggest that a considerable fraction of the cold gas necessary to form the stellar content of galactic disks today was already available at $z \gtrsim 2$ (Wolfe et al. 1995) and could not have been the result of a recent cooling process. The inferred metallicity of this high-redshift gas is an order of magnitude lower than solar (Pei, Fall, & Bechtold 1991; Sembach et al. 1995), implying subsequent star formation and metal enrichment. If the central halos of disk galaxies were also formed at $z \gtrsim 1$ and have not accreted mass afterwards, then the low scatter in the Tully-Fisher relation could be explained.

A different approach to the Tully-Fisher relation relies on the notion that the cooling of the gas is incomplete. Here, one assumes that after collapse the gas acquires an isothermal density profile similar to that of the dark matter. If one then considers the cooling time as a function of radius and includes in the galactic disk only the mass of the gas that has had time to cool, M_{cold}, then the reservoir of cold gas depends on the velocity dispersion and not on the mass of the halo, since the former sets the density profile of the gas. Thus, it is possible to get a tight $\sigma - M_{cold}$ relation regardless of the actual halo mass and its formation history. However, the ongoing cooling of gas presents serious problems for this scenario (White 1994). Even for a minimum steady-state accretion rate of a few $M_\odot\,\mathrm{yr}^{-1}$, necessary to accumulate a typical disk mass in a Hubble time, the dissipation of the radial-infall kinetic energy should already produce $\sim 10^{41}\,\mathrm{erg\,s}^{-1}$, well beyond the typical limits on diffuse X-ray emission from spirals, $\lesssim 5 \times 10^{39}\,\mathrm{erg\,s}^{-1}$ (Bregman & Glassgold 1982; Fabbiano 1989). The lack of X-ray halos in spiral galaxies suggests that gas accretion from the halo has essentially ended and that an isothermal gaseous halo cannot be the regulating mechanism for the Tully-Fisher relation. SPH simulations (Navarro & White 1994) support this view by showing that the halo gas tends to cool efficiently due to the clumping that characterizes hierarchical merging, so that rather little gas is left in X-ray halos at the present time. Although these simulations do not include heating from supernova-driven winds that might slow the cooling or return gas to the halo, it is unlikely that this heating will maintain the gas just at the virial temperature of the halo at all times.

Collisionless N-body simulations also provide estimates of the uniformity of dark matter halos. White (this volume; Navarro et al. 1996b) presented estimates of scatter in the correlation between the mass of a dark matter halo and the maximum of its rotation curve that are smaller than the estimates from our analytical work. On the other hand, simulations from Cole & Lacey (1996) show larger scatter, consistent with our estimates. One difficulty in comparing simulations to the analytic work has to do with the measurement of the mass of a halo. In these simulations, the mass of a halo is chosen to be the mass within a radius inside of which the overdensity achieves a particular multiple (e.g. 200) of the critical density. If we imagine an unchanging halo, its mass will be found to increase slightly with time due to the decrease of the critical density. This suggests that objects that form at earlier times will have higher σ for their 'analytic' mass *and* will also place more of this mass inside the threshold radius. While this effect is small, it is well-correlated with the residuals in the $M-\sigma$ relation and seems to reduce the scatter by 1–2% in sample calculations. This may reduce the discrepancy between our results and those of Navarro et al.

We conclude that the expected scatter in the $M-\sigma$ properties of dark matter halos is too large to result in the observed Tully-Fisher relation. We see two paths for breaking the assumptions leading to this conclusion. First, if galaxies formed at redshifts $z \gtrsim 1$ and did not accrete material subsequently (contrary to expectations from the standard hierarchical clustering picture), then the scatter in the $M-\sigma$ relation would be significantly reduced. Second, if a strong feedback process comes into play, it may regularize star formation and gas dynamics and detach the overall luminosity of spiral galaxies from details concerning the formation history of their halos. Either one of these paths would have significant consequences for the process of galaxy formation.

We thank Guiseppina Fabbiano, Tsafrir Kolatt, David Spergel, David Weinberg, Simon White, and Jeffery Willick for useful discussions. D.J.E. was supported in part by a National Science Foundation Graduate Research Fellowship and NSF grant PHY-9513835.

References

Aaronson, M., Huchra, J., & Mould, J. 1979, ApJ, 229, 1

Alcock, C., et al. 1995, PRL, 74, 2867

Bernstein, G.M., Guhathakurta, P., Raychaudhury, S., Giovanelli, R., Haynes, M.P., Herter, T., & Vogt, N.P. AJ, 107, 1962

Bond, J.R., Cole, S., Efstathiou, G., & Kaiser, N. 1991, ApJ, 379, 440 (BCEK)

Bregman, J.N., & Glassgold, A.E. 1982, ApJ, 263, 564

Broeils, A. H., van Woerden, H., 1994, A&AS, 107, 129

Cole, S., Aragon, A., Frenk, C.S., Navarro, J.F., & Zepf, S. 1994, MNRAS, 271, 781

Cole, S., & Kaiser, N. 1989, MNRAS, 237, 1127

Cole, S., & Lacey, C. 1996, MNRAS, 281, 716

Corbelli, E., & Salpeter, E.E. 1988, ApJ, 326, 551

Efstathiou, G., Bond, J.R., & White, S.D.M. 1992, MNRAS, 258, 1P

Eisenstein, D.J., & Loeb, A. 1996, ApJ, 459, 432

Fabbiano, G. 1989, ARA&A, 27, 87

Federspiel, M., Sandage, A., & Tammann, G.A. 1994, ApJ, 430, 29

Franx, M., & de Zeeuw, T. 1992, ApJ, 392, L47

Gates, E., Gyuk, G., & Turner, M.S. 1994, PRL, 74, 3724

Górski, K.M., Ratra, B., Sugiyama, N., & Banday, A.J. 1995b, ApJ, 444, L65

Górski, K.M., Stompor, R., & Banday, A.J. 1995a, ApJ, submitted, astro-ph/9502033

Gunn, J.E., & Gott, J.R. 1972, ApJ, 176, 1

Kauffmann, G. & White, S.D.M. 1993, MNRAS, 261, 921

Kauffmann, G., White, S.D.M., & Guiderdoni, B. 1993, MNRAS, 264, 201

Lacey, C.G., & Cole, S. 1993, MNRAS, 262, 627 (LCa)

Lacey, C.G., & Cole, S. 1994, MNRAS, 271, 676 (LCb)

Lanzetta, K. M., Wolfe, A. M., & Turnshek, D. A. 1995, ApJ, 440, 435

Mathewson, D.S., & Ford, V.L. 1994, ApJ, 434, L39

Moore, B., Katz, N., & Lake, G. 1995, ApJ, 457, 455

Navarro, J.F., Frenk, C.S., & White, S.D.M. 1994, MNRAS, 267, L1

Navarro, J.F., Frenk, C.S., & White, S.D.M. 1996a, ApJ, 462, 563

Navarro, J.F., Frenk, C.S., & White, S.D.M. 1996b, ApJ, submitted, astro-ph/9611107

Navarro, J.F., & White, S.D.M. 1994, MNRAS, 267, 401

Pei, Y. C., Fall, S. M., & Bechtold, J. 1991, ApJ, 378, 6

Press, W.H., & Schechter, P.L. 1974, ApJ, 187, 425

Prochaska, J.X., & Wolfe, A.M. 1997, ApJ, 474, 140

Rix, H.W., & Zaritsky, D. 1995, ApJ, 447, 82

Sembach, K. R., Steidel, C. C., Macke, R. J., & Meyer, D. M. 1995, ApJL, 445, L27

Spitzer, L. 1956, ApJ, 124, 20

Sprayberry, D., Bernstein, G.M., Impey, C.D., & Bothun, G.D. 1995 ApJ, 438, 72

Stompor, R., Górski, K.M., & Banday, A.J. 1995, ApJ, submitted, astro-ph/9502035

Strauss, M.A., & Willick, J.A. 1995, Phys. Rep., 261, 271

Thoul, A.A., & Weinberg, D.H. 1995, ApJ, 442, 480

Tóth, G., & Ostriker, J.P. 1992, ApJ, 389, 5

Tully, R.B., & Fisher, J.R. 1977, A&A, 54, 661

Turnshek, D. A., Wolfe, A. M., Lanzetta, K. M., Briggs, F. H., Cohen, R. D., Foltz, C. B., Smith, H. E., & Wilkes, B. J. 1989, ApJ, 344, 567

White, S.D.M. 1994, Les Houches Lectures, in press, MPA preprint 831

White, S.D.M., Efstathiou, G., & Frenk, C.S. 1993, MNRAS, 262, 1023

Willick, J.A., Courteau, S., Faber, S.M., Burstein, D., & Dekel, A. 1995a, ApJ, 446, 12

Willick, J.A., Courteau, S., Faber, S.M., Burstein, D., Dekel, A., & Kolatt, T. 1996a, ApJ, 457, 460

Willick, J.A., Courteau, S., Faber, S.M., Burstein, D., Dekel, A., & Kolatt, T. 1996b, ApJS, in press, astro-ph/9610202

Wolfe, A.M., Lanzetta, K.M., Foltz, C.B., & Chaffee, F.H. 1995, ApJ, 454, 698

Zwaan, M.A., van der Hulst, J.M., de Blok, W.J.G., & McGaugh, S.S. 1995, MNRAS, 273, L35

The Tully–Fisher Relation:
A Numerical Simulator's Perspective

August E. Evrard[1,2,3]

[1] Physics Department, University of Michigan, 48109-1120 USA
[2] Institut d'Astrophysique, 98bis Blvd Arago, 75014 Paris, France
[3] Max–Planck–Institut für Astrophysik, Karl–Schwarzschild–Str. 1,
  Garching bei München, Germany

Abstract. In this brief contribution, I will outline hopes of understanding the origin of galaxy scaling relations using numerical simulations as a tool to gain understanding. The case of the Tully–Fisher relation for disk galaxies is used as a working example to illustrate the modest achievements to date and the difficult tasks ahead.

1 Overview

Galaxies, like people, come in a variety of shapes and sizes. Like people, galaxies tend to have many features in common, but explaining in detail the origin of these features can be a difficult task. Understanding why the luminosity of a disk galaxy should scale as a power of its circular velocity is a bit like understanding the relation between a person's weight and his or her belt length. There are basic governing principles — bodies are (crudely speaking) similarly shaped bags of water and galaxies are (crudely speaking) similarly structured gravitational objects — but other factors creep in when you start giving it more careful thought. A skinny basketball player and a jockey might have the same waistline, but their body masses could differ by more than 50% because of their height difference. Why couldn't two galaxies lying in potential wells of the same circular speed differ in luminosity by, say, a factor 3 because of differences in their gas dynamic/stellar evolutionary histories? It appears nature does not allow this to happen; the scatter about the Tully–Fisher relation is remarkably small (see Giovanelli and others in this proceedings).

The remarkable nature of such a tight correlation in non–trivially linked physical properties is made apparent when one considers their complicated birthing process shown schematically in Figure 1. This picture was laid out theoretically in the late 1970's, and the seminal paper of White and Rees (1978) cemented the elements together within a modern, hierarchically clustering framework. In hierarchical models, gravity amplifies density perturbations on ever–increasing mass and length scales, driving an overdense, filamentary/knotty network which evolves in a nearly self–similar fashion. On mass scales roughly between 10^8 and $10^{12} M_\odot$ and in the absence of significant non–gravitational heat input, cooling via radiative processes removes thermal pressure support from the baryons. This process acts to concentrate the baryons within an assumed, dominant halo of dark matter. Once self–gravitating, star formation is ignited in a manner poorly understood from first principles (hence the "black box" in the figure) and *poof!* we end up with a disk galaxy rotating in its dominant dark halo.

Given that rotation speed is a direct measure of total mass within a fixed density contrast M_δ (see below), then a suspiciously simple interpretation of the tight, observed Tully–Fisher relation in the context of Figure 1 is that cooling and star formation are highly regular and dependent primarily on M_δ. Such a simple picture appears to be true for the structure of collisionless halos formed from hierarchical, gravitational clustering. A single characteristic function with parameters smoothly varying with mass appears to describe the density and velocity structure of collapsed objects (see White's contribution in this volume).

The situation in Figure 1 is simplistic in a number of ways. Formation of a single galaxy actually entails a network of such segments, inter-connected in a manner reflecting the particular merger history of that object. An ensemble of equal mass objects observed today will naturally arise from a variety of merger histories/inter-connections. Why doesn't this variety evidence itself as a large scatter in the Tully–Fisher relation? (Eisenstein in this volume presents a similar argument from a slightly different perspective.)

2   Looking Behind

Numerical simulations with gas dynamics are now beginning to be used to address the origin of disk galaxies and the Tully–Fisher relation (Katz & Gunn 1991; Evrard, Summers & Davis 1994, Navarro & White 1994, Steinmetz & Müller 1994; Navarro & Steinmetz 1996; Tissera, Lambas & Abadi 1996; Groom 1997). Most of these simulations ignore star formation altogether. Those in which it was included failed to form a disk of stars, forming spheroids instead. So the best we can do at the moment is analyze the gas disk properties. An idea of where we stand is shown in Figure 2, which compares the cold, gas mass in the galaxy to its circular speed. Data shown are from Navarro & White (1994; hereafter NW) and from a unpublished P3MSPH simulation by myself of a random, 16 Mpc ($H_o = 50$ km s^{-1} Mpc^{-1}) patch of a standard cold dark matter universe. The characteristics of the simulation are identical to that detailed in Evrard *et al.* (1994), with the exception of it being a random, (instead of constrained cluster) spatial region and it being evolved to the present (instead of $z = 1$). The interested reader should consult these papers for further details. The "raw" points in the figure use the peak in the measured circular speed of the gas disk, while the "corrected" points enforce centrifugal equilibrium at that point in the rotation curve. The correction is necessary because the size of the disks is within a factor of a few of the spatial resolution limit of the simulation. The dotted line in the figures has slope 2.45.

The good news from Figure 2 is that the scatter in both data sets is quite small, in fact, smaller by a factor 2 than typical observed values. The bad news is that the left hand panel of the figure is a dishonest comparison of two independent experiments. The NW rotation speed is actually $\sqrt{GM/r}$ measured at a density contrast of 200 with respect to the critical background. Going back and

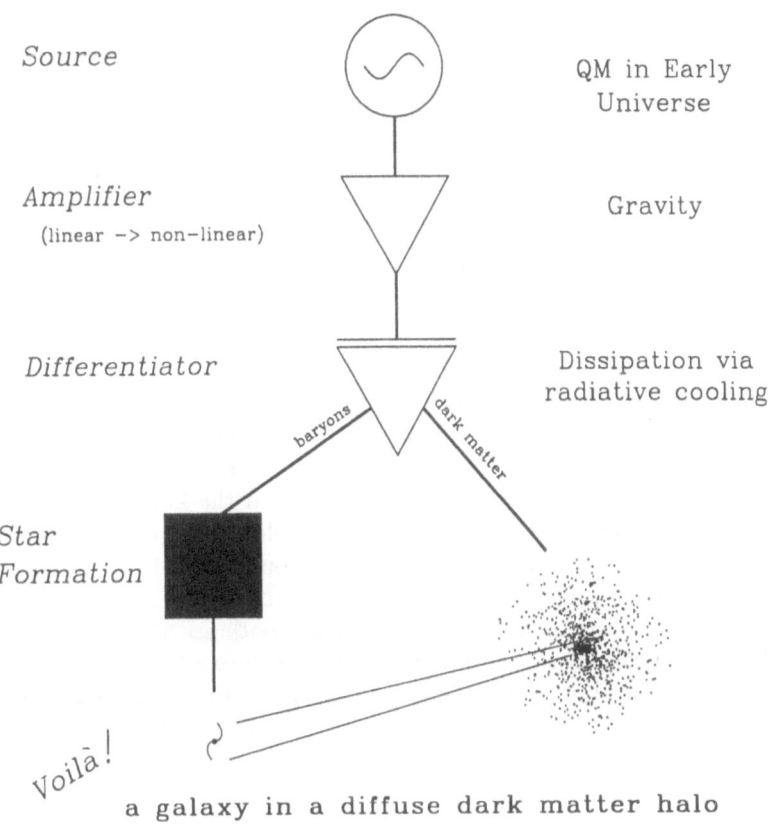

Fig. 1. Schematic illustrating the basic principles behind the White–Rees picture of galaxy formation discussed in the text.

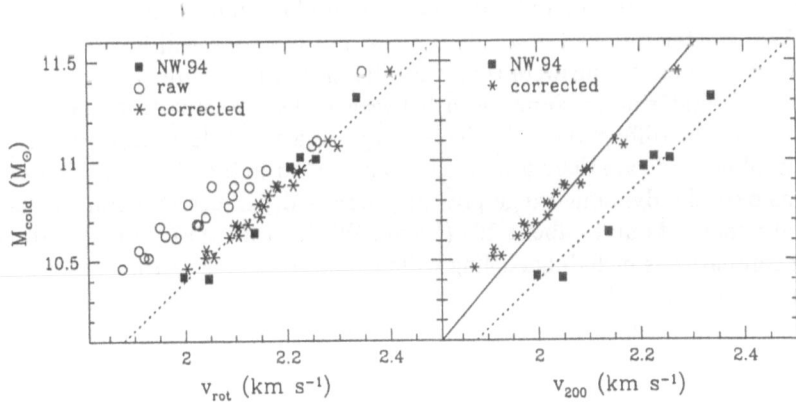

Fig. 2. Cold gas mass versus rotation speed derived from gas dynamic simulations of Navarro & White 1994 (NW) and Evrard (1995, unpublished). See text for a discussion.

measuring the same quantity for the disk galaxies in the P3MSPH simulation (the "corrected" values in the right panel) results in a systematic offset in the intercept of the two data sets. This intercept is not due to different values of Hubble constant, both use $h = 0.5$, or cosmic baryon fraction, since both assume $\Omega_b = 0.1$ and $\Omega = 1$. In this cosmology, the radius and total mass defining a density contrast of 200 at the present epoch are

$$r_{200} = \sqrt{\frac{2}{\delta}} \frac{v_{200}}{H_o} = 100 \left(\frac{v_{200}}{100 \text{ km s}^{-1}} \right) h^{-1} \text{ kpc} \tag{1}$$

$$M_{200} = 2.32 \times 10^{11} \left(\frac{v_{200}}{100 \text{ km s}^{-1}} \right)^3 h^{-1} M_\odot \tag{2}$$

If all the baryons within this density contrast cool and sink to the center of the halo, then the cold, galactic mass will simply be $\Omega_b M_{200}$. This implies a Tully–Fisher relation shown as the solid line in the right panel (using $h = 0.5$ as in the simulations). One interpretation of this panel is that the P3MSPH treatment is allowing nearly all the gas to cool in the halos while NW's treatment allows half the baryons to cool, with the remainder in a tenous, hot halo. It remains to be seen if this interpretation is correct but, at any rate, the offset between the two data sets is most likely numerical in origin, since both are attempting to model essentially identical physical situations. The silver lining here is that the small degree of scatter in the relation appears insensitive to the detailed numerical treatment.

3   Looking Ahead

The example above illustrates our current level of uncertainty in modeling just some of the physical processes associated with Figure 1. The black box of star formation is largely unexplored territory. Presumably different physical and numerical parameterizations for star formation and feedback will lead to an even larger range of possible answers than that illustrated in Figure 2.

On the bright side, a comparison between codes attempting to model the branch above the differentiator in Figure 1 (gravitational clustering without radiative cooling) indicate there is quite good agreement in the gas and dark matter solutions over the dynamic range presently accessed by such experiments, that is, density contrasts up to about 10^4 (Frenk, White *et al.* 1997, in preparation). Similar comparisons including cooling will ultimately enable sorting out of physical versus numerical effects.

In the realm of galaxy scaling relations, theorists are in the typical position of attempting to understand current observations; predictive power is tenuous at best. From the excellent new data presented at this meeting, particularly in the area of evolution in the scaling relations at moderate to high redshift (*e.g.*, the contributions of Franx, Pahre, Schade, Guzman, Dickinson, and Ziegler in this volume), it seems the observers are accelerating their pace! Modeling this wealth of data presents a formidable challenge for the foreseeable future.

This work was supported in the USA by NASA through grant NAG5–2790 and in France by the CIES and CNRS at the Institut d'Astrophysique in Paris. It is a pleasure to thank L. da Costa, A. Renzini and the rest of the organizers for putting together an excellent workshop. I am most grateful to S. White for hospitality extended during my stay at MPA, where this proceedings was written.

References

Evrard, A.E., Summers, F.J. & Davis, M. 1994, ApJ, 422, 11

Groom, W. 1997, PhD Thesis, Cambridge University.

Katz, N. & Gunn, J.E. 1991, ApJ, 377, 365

Navarro, J. & White, S.D.M. 1996, MNRAS, 267, 401

Navarro, J. & Steinmetz, M. 1996, ApJsubmitted (astro-ph/9605043)

Steinmetz, M. & Müller, E. 1994, AA, 268, 391

Tissera, P.B., Lambas, D.G, Abadi, M.G. 1996, MNRAS, submitted (astro-ph/9603153)

White, S.D.M. & Rees, M. 1978, MNRAS, 183, 341

A Universal IMF: Tilt of the Fundamental Plane

Cesare Chiosi[1], Alessandro Bressan[2]

[1] Astronomy Department, Vicolo Osservatorio 5, 35122 Padova, Italy
[2] Astronomical Observatory, Vicolo Osservatorio 5, 35122 Padova, Italy

Abstract. In this paper, basic properties of elliptical galaxies such as the integrated spectra, the chemical abundances and enhancement of α-elements, the color-magnitude relation (CMR), the line strength indices, and the mass-luminosity ratios, are examined at the light of current theoretical interpretations, and attention is called on several points of internal contradiction. Specifically, in the context of standard star formation in galactic-wind driven models that are at the base of present-day understanding of the CMR (and UV excess) it is difficult to explain the slope of the M/L_B versus M_B relation, the enhancement of the α-elements in the brightest elliptical galaxies, and the distribution of elliptical galaxies in the H_β - $< MgFe >$ plane at the same time. We suggest that the new initial mass function (IMF) of Padoan et al. (1996), which depends on the temperature, density and velocity dispersion of the medium in which stars are formed, may constitute a break-through of the difficulties in question. Models of elliptical galaxies incorporating the new IMF (varying with time and position inside a galaxy) are presented and discussed at some extent.

1 Introduction

The conventional picture of star formation in elliptical galaxies is one in which the galaxies and their stellar content formed early in the universe and have evolved quiescently ever since. This view is supported by the apparent uniformity of ellipticals in photometric and chemical properties, and the existence of scaling relations (e.g. the fundamental plane). In contrast, the close scrutiny of nearby ellipticals makes evident a large variety of morphological and kinematic peculiarities and occurrence of star formation in a recent past. All this leads to a different picture in which elliptical galaxies are formed by mergers and/or accretion of smaller units over a time scale comparable to the Hubble time. Without venturing into the question of which of the two pictures ought to be preferred, we shortly examine the pattern of main properties and their current understanding and call attention on several points of contradiction.

(1) **G-Dwarf Analog**: The near UV spectrum of E-galaxies shows that the relative percentage of old, metal-poor stars is small, thus indicating that the metallicity partition function N(Z), i.e. number of stars per metallicity bin, cannot be the one predicted by the closed box scheme, but ought to resemble those of infall or prompt enrichment (Bressan et al. 1994, BCF94; Tantalo et al. 1996 TCBF96; Greggio 1996).

(2) **Chemical Abundances**: Although passing from narrow band indices such as Mg_2 and $< Fe >$ (Carollo et al. 1993, Carollo & Danziger 1994) or integrated colors (Davies et al. 1993) to chemical abundances is not a straight

process (Bressan et al. 1997, BTC97), arguments are given to conclude that the mean $[Mg/Fe]$ ratio exceeds that of the most metal-rich stars in the solar vicinity by about 0.2-0.3 dex (enhancement of α-elements; O, Mg, Si, etc.. with respect to Fe), and the ratio $[Mg/Fe]$ increases with the galactic mass up to the this value (cf. Matteucci 1994, 1996). The enhancement of α-elements is particularly demanding as it implies that a unique source of nucleosynthesis has been contributing to chemical enrichment (type II SN). It follows from this that in the standard star formation scenario, the maximum duration of the star forming activity should be inversely proportional to the galaxy mass ($\Delta t_{SF} \propto M_G^{-1}$).

(3) **CMR**: Long ago Larson (1974) postulated that the present-day CMR, cf. Bower et al. (1992) for galaxies in the Virgo and Coma clusters and Schweizer & Seitzer (1992) for galaxies in small groups and field, is the consequence of SN driven galactic winds. In the classical scenario, massive galaxies, ejecting their gaseous content much later than the less massive ones, are able to get higher mean metallicities. This implies that $\Delta t_{SF} \propto M_G$, in contrast to what inferred from the α-enhancement problem. The tightness of the cluster CMR (in the U-V color) suggests that most galaxies are nearly coeval with age ranging from 13 to 15 Gyr (Bower et al. 1992). The CMR for field galaxies is likely compatible with more recent episodes of star formation perhaps interactions spread over several Gyr (Schweizer & Seitzer 1992). Finally, the CMR is a mass-metallicity sequence and not an age sequence (bluer galaxies being significantly younger than the massive ones). See BCF94, TCBF96, and Kodama & Arimoto (1996).

(4) **Age-Metallicity Dilemma**: The stellar content of a galaxy gets redder at increasing age and metallicity thus giving rise to the well known age-metallicity dilemma. The indices H_β and $< MgFe >$ are particularly suited to cast light on the separate effects, because H_β is a measure of the turn-off colour and luminosity, and age in turn, whereas $< MgFe >$ is more sensitive to the RGB colour and hence metallicity. The analysis of existing data (Gonzales 1993) leads to the following provisional conclusions: (i) Most galaxies seem to possess nearly identical chemical structures, i.e. high mean metallicities within a narrow range of values (Dressan et al. 1996, BCT96; Greggio 1996); (ii) Galaxies do not distribute along the locus expected from the CMR of coeval old objects. In contrast, they seem to follow a sequence of about constant metallicity and varying age (BCT96) for which recent episodes of star formation (bursts) have been suggested. The major drawback with this idea is that a sort of synchronization is required. Looking at the difference in H_β and $< MgFe >$ between the nuclear region (i.e. $Re/8$: N) and the whole galaxy (i.e. $Re/2$: W), and translating $\Delta H\beta_{NW}$ and $\Delta < MgFe >_{NW}$ into Δt_{NW} and ΔZ_{NW}, i.e. in age and metallicity difference, respectively, BCT96 pointed out that in most galaxies the nucleus is younger and more metal-rich than the external regions, and suggested that $\Delta t_{NW} \simeq \Delta t_{SF} \propto 1/\Sigma \propto 1/M_G$.

(5) **Tilt of the Fundamental Plane (FP)**: Elliptical galaxies do not populate uniformly the parameter space with coordinates the central velocity dispersion Σ_0, effective radius R_e, and surface brightness I_e. They cluster around a plane called the *Fundamental Plane*. Using the coordinate system $\kappa_{1,2,3}$ defined by Ben-

der et al. (1992) – mere rotation in the space of Σ, R_e, and I_e – the hypothesis of virialization, and the identities $L = c_1 I_e R_e^2$ and $M = c_2 \Sigma_0^2 R_e$ (c_1 and c_2 the virial coefficients), one gets $k_1 = (1/\sqrt{2})log[M/c_2]$, $k_2 = (1/\sqrt{6})log[(c_1/c_2)(M/L)Ie^3]$, and $k_3 = (1/\sqrt{3})log[(c_1/c_2)(M/L)]$. Of particular relevance is the projection of the FP onto the $k_1 - k_3$ plane, where the FP is seen edge on. Limited to the case of the Virgo elliptical galaxies to avoid distance uncertainties, the relation $k_3 = 0.15k_1 + 0.36$ with $\sigma(k_3) = 0.05$ is found to hold (cf. Ciotti et al. 1996 for details). If the virial coefficients are constant for all galaxies, the observed tilt of the FP implies a systematic increase of (M/L) with the galaxy mass (tilt of the FP): for instance for the B-band $M/L_B \propto L_B^{0.2}$. The tilt and tightness of the FP have so far eluded simple explanations (cf. Renzini & Ciotti 1993, Ciotti et al. 1996).

2   A Universal IMF

Among the various suggestions advanced to reconcile some of the above difficulties and in particular to explain the tilt of the FP, a major but suitable change of the IMF has often been invoked (cf. Matteucci 1994, 1996; Renzini & Ciotti 1996). Padoan's et al. (1996) have recently proposed a new IMF which seems to possess the desired kind of flexibility. It stems from hydrodynamical simulations of the density field emerging from randomly forced supersonic flows in star forming regions. In brief, the proposed IMF is

$$\int_0^\infty \phi(M)dM = \int_0^\infty \frac{2B^2}{(2\pi\sigma^2)^{0.5}} M^{-3} exp[-0.5(\frac{2lnM - A}{\sigma})^2]dM = 1 \quad (1)$$

The quantities A, B and σ are defined by the following relations:

$$A = 2\ln B + 0.5\sigma^2 \quad (2)$$

$$B = 1.2 \times (T/10)^{1.5} \times (n/1000)^{-0.5} \quad (3)$$

$$\sigma^2 = \ln[1 + 0.36(M_a^2 - 1)] \quad (4)$$

in which σ is the standard deviation of the number density distribution in the field with respect to the mean. M_a^2 is the Mach number $M_a^2 = (\Sigma/v_s)^2$, T is the temperature in K, n is the number density in cm^{-1}, v_s the sound velocity, and Σ the velocity dispersion of the gas (in km/s). This IMF has a long tail at high masses, an exponential cutoff at the smallest masses, a characteristic peak mass

$$M_p \simeq 0.2M_\odot \times (T/10)^2 \times (n/1000)^{-0.5} \times (\Sigma/2.5)^{-1} \quad (5)$$

and a slope continuously varying with the mass. The IMF gets flatter and M_p gets higher at increasing temperature or decreasing density and velocity dispersion.

3   Models with the New IMF

The new IMF has been implemented in a simplified version of the models of elliptical galaxies of Tantalo et al. (1997, TCB97) in which spatial gradients of density and star formation are taken into account. For the sake of simplicity and better understanding of the role played by the sole IMF, we neglect here the possible presence of dark matter, and adopt the closed-box formulation (no infall of primordial gas). The rate of star formation is the standard Schmidt law $d\rho(t)/dt = \nu\rho(t)_g^\kappa$ where $\kappa = 1$ and ν is the specific efficiency. All the results below refer to $\nu = 5$ and the central core of the galaxies ($r/R_e = 0.1$ or approximately 10% of the galaxy mass).

The IMF is calculated assuming for the density, temperature and velocity dispersion Σ the values obtained solving the energy equation

$$\frac{dE(T,\rho_g,t)}{dt} = H_r(T,\rho_g,t) + H_m(T,\rho_g,t) - \frac{\Lambda(T,\rho_g,t)}{\rho_g} \tag{6}$$

where H_r and H_m are the heating rates of radiative and mechanical origin, respectively, and Λ is the cooling rates. $H_m(T,\rho_g,t)$ as large as $10^{-4} \times H_r(T,\rho_g,t)$ turns out to be fully adequate. At each time step Δt the energy equation is solved assuming that heating-expansion and cooling-contraction of gas occur at constant pressure until equilibrium values for the temperature and density are reached. They determine the Jeans mass above which gas clouds are gravitationally unstable, and fix the velocity dispersion Σ of the gas clouds. From now on the unstable gas clouds are conceived as thermally decoupled from the surrounding medium (H_r is switched off, whereas H_m is retained) and let reach the collapse temperature and density. In any case, the gas temperature cannot be cooler than the current value of the cosmic background radiation. Depending on the competition between heating and cooling, the energy stored into the interstellar medium continuously grows thus making gas hotter and hotter. The case can be met in which the thermal energy of gas eventually exceeds the gravitational binding energy of it, or equivalently the gas temperature exceeds the virial temperature. In such a case the star formation is halted and galactic winds occur. No other details are given here for the sake of brevity.

4   Outline of the Basic Results

Inside a galaxy, the local density decreases going from the center to the periphery. Likewise, the mean density $< \rho >$ decreases passing from a low to a high mass galaxy (as a result of the underlying mass-radius relationship). Equally for the velocity dispersion Σ, which increases with the galaxy mass. Therefore, we expect a systematic variation of the IMF with the galaxy mass due to the sole scaling of $< \rho >$ and Σ, and inside a galaxy due to drop-off of density with the radial distance.

The initial temperature T_0 of the gas, perhaps determined by external conditions such as heating up of gas during the collapse phase, cosmic background radiation etc, strongly affects the very first episode of star formation by setting the peak mass of the IMF and the relative percentage of massive stars, SNII explosions, and efficiency of metal enrichment in turn. T_0 is found to bear very much on the future history of a galaxy.

T_0 in the range 30 to 100 K provides strong enough energy input to heat up the gas so that, in spite of cooling, the IMF tends to be always skewed toward the high mass range thus favoring gas heating. The virial temperature is easily reached. This typically happens in the nuclei of massive galaxies and in the most external regions of galaxies of any mass.

In contrast with $T_0 < 30$ K, the IMF tends to be skewed toward the low mass end, with little energy deposit and gas heating. The virial temperature is hardly reached so that star formation may occur for long periods of time. This typically happens in the central regions of low mass galaxies.

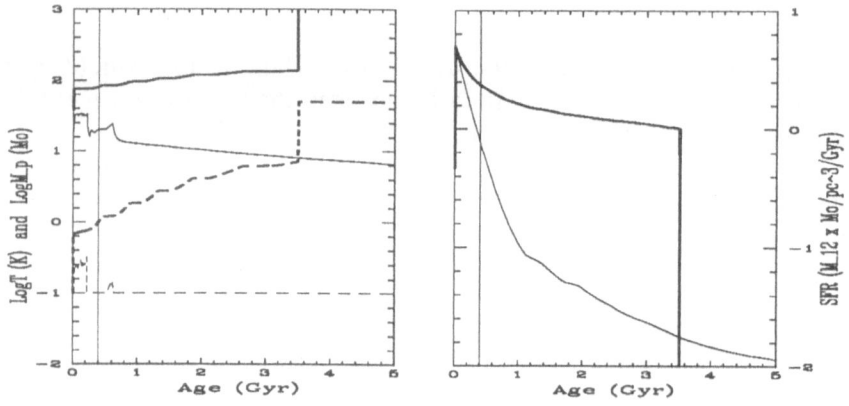

Fig. 1. Left Panel: Temperature (solid line) and M_p (dashed line) as function of the age for the $1 \times 10^9 M_\odot$ (thin lines) and $3 \times 10^{12} M_\odot$ (thick lines) galaxies. The vertical bar shows for the $3 \times 10^{12} M_\odot$ galaxy the age at which M_p gets greater than $1 \times M_\odot$. **Right Panel:** The normalized rate of star formation SFR as a function of the age for the same models. The SFR is expressed in units of $M_{12} \times M_\odot / pc^3 / Gyr$.

To illustrate the above points, in the left panel of Fig. 1 we show the time variation of the peak mass characterizing the IMF and of the gas temperature for two galaxies of different total mass, i.e. $1 \times 10^9 M_\odot$ and $3 \times 10^{12} M_\odot$. The most noticeable thing to point out is that in the case of $3 \times 10^{12} M_\odot$ galaxy after about 0.4 Gyr (vertical line in Figs. 1 and 2), M_p becomes greater than $1 \times M_\odot$, which means that fewer and fewer stars of the past generation will be visible today as their lifetime is shorter than the typical galaxy age of about 15 Gyr. These stars rapidly evolve into collapsed remnants (white dwarfs, neutron stars, black holes as appropriated to the initial value of their mass).

Star Formation History. As a result of the complex game sketched above, in a galaxy of given mass the central regions form stars for periods of time longer than the external ones. Furthermore, massive galaxies suffer from galactic winds earlier than low mass galaxies. This is shown in the right panel of Fig. 1 which displays the rate of star formation (in units of $M_{12} \times M_\odot/pc^3/Gyr$) as a function of the age. In the case of the $1 \times 10^9 M_\odot$ model actually star formation in the center never stops, even if the SFR has dropped to about 10^{-4} the initial value.

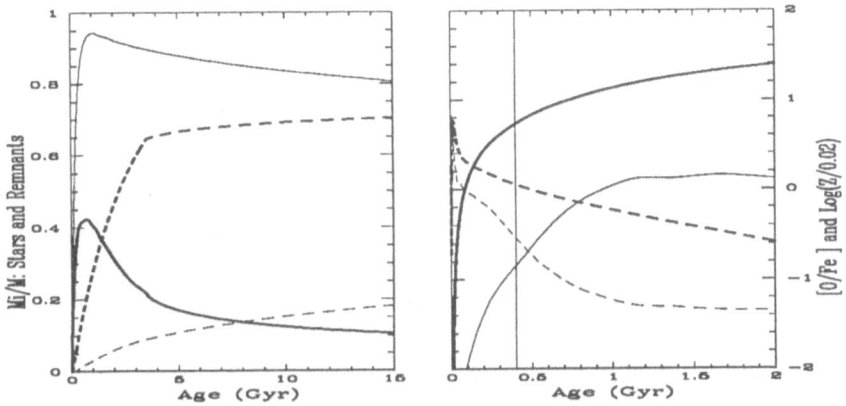

Fig. 2. Left Panel: Fractionary masses in stars (solid lines) and remnants (dashed lines) as function of the age for the $1 \times 10^9 M_\odot$ (thin lines) and $3 \times 10^{12} M_\odot$ (thick lines) galaxies. **Right Panel:** The metallicity $Z/0.020$ (solid lines) an [O/Fe] ratio (dotted lines) for the same models. The vertical line has the same meaning as in Fig. 1. See the text for more details.

Luminous versus Dark Material. Since the IMF in the course of evolution tends to progressively shift toward the high mass end, the formation of stars leaving collapsed remnants is continuously favoured. This trend is, however, driven by the detailed history of gas temperature so that differences are expected in the contents of visible stars and collapsed remnants in individual galaxies. The situation is illustrated in the left panel of Fig. 2 showing the fractionary mass of visible stars (solid line) and collapsed remnants (dashed lines) as a function of the age. In the $3 \times 10^{12} M_\odot$ model, most of the present-day mass is in remnants. **Enhancement of α-Elements.** As the IMF is more skewed toward the high mass range in massive galaxies, than in the low mass ones, the enhancement of α-elements in the former is easily met together with its dependence on the galaxy mass. This is true even in our $3 \times 10^{12} M_\odot$ model despite its rather long duration of the star forming period (3.5 Gyr). The time dependence of [O/Fe] is shown in the right panel of Fig. 2. The mean value of [O/Fe] in the stellar content of the $1 \times 10^9 M_\odot$ galaxy is much lower than solar, whereas that of the $3 \times 10^{12} M_\odot$ galaxy is about about 0.2-0.3 if one takes into account that the vast majority of the stars observable today are those generated within the first 0.4 Gyr. In the same panel we also display the gas metallicity in solar units (Z/0.02). The mean metallicity of the $1 \times 10^9 M_\odot$ galaxy is $< Z > = 0.014$, whereas that of the $3 \times 10^{12} M_\odot$ galaxy is nearly solar (once again limited to the stars visible today). Furthermore, the rapid increase of the metallicity in early stages rules out the G-Dwarf Problem.

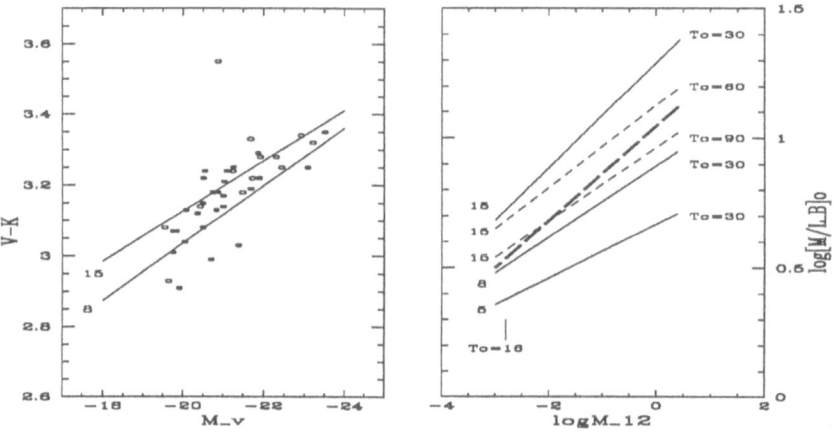

Fig. 3. Left Panel: CMR. The filled and open circles are the data for the Virgo and Coma galaxies of Bower et al. (1992). The distance modulus to Virgo is $(m - M)_0 = 31.54$ (Branch & Tammann 1992), whereas that for Coma is $(m - M)_0 = 35.12$ (Bower et al. 1992). The solid lines are the theoretical CMR for the age of 15 and 8 Gyr as indicated. **Right Panel**: The theoretical $log(M/L_B)_\odot$ ratios as a function of $logM_{12}$ for the ages of 15, 8 and 5 Gyr as indicated. The vertical bars show the initial temperature of the gas. The thick dashed line is the relation for the Virgo galaxies.

CMR. Despite the shorter duration of the star forming activity and earlier winds, the mean metallicity of massive galaxies is higher than of the low mass ones, because of the shift of the IMF toward higher stellar masses. The results are shown in the left panel of Fig. 3 for two values of the age and compared to the data of Bower et al. (1992) for the Virgo and Coma galaxies. The theoretical magnitudes have been shifted by -2.5log0.5 to take into account the contribution from the remaining regions of the galaxy not considered here. Although the agreement with the data is not perfect, yet the slope of the CMR is matched. Even in the models the CMR is a mass-metallicity sequence, however, the contradiction between CMR and α-enhancement (cf. section 1) does no longer occur.

5   Tilt of the Fundamental Plane

The question to be addressed is: can we get coincidence of the slope due to the sole variation of the IMF or changes in the virial coefficients must be still invoked? The situation is shown in the right panel of Fig.3 which displays the $log(M/L_B)_\odot$ ratio as a function of $log M_{12}$, and compares the theoretical relation with the observed one for the Virgo and Coma galaxies (cf. section 1). It is worth recalling that the theoretical slope is $\sqrt{3/2}$ times the observational one in the κ_3 - κ_1 plane, and a vertical shift of this latter is applied so that $log(M/L_B)_\odot = 0.5$ at the low mass end of the relation. Finally, we notice that while the low mass end of the theoretical relation scarcely depends on T_0 (in the range 15 to 60 K), the high mass end is more sensitive to it. It seems that agreement is possible for initial temperatures in the range $15 \leq T_0 \leq 30$ K as the galaxy mass increases from $1 \times 10^9 M_\odot$ to $3 \times 10^{12} M_\odot$. The slope of the $log(M/L_B)_\odot$ versus $log M_{12}$ relation slightly flattens at decreasing age, so that best agreement is found for old ages (say 15 Gyr). Finally, we have also looked at the $log(M/L_K)_\odot$ ratio as function of $log M_{12}$. The slope is about 0.5 the slope of the relation for the B-passband.

References

Bender R., Burstein D., & Faber S.M. 1992, ApJ, 399, 462
Bower R.G., Lucey J.R., & Ellis R.S. 1992, MNRAS, 254, 601
Branch D., & Tammann G.A. 1992, ARA&A, 30, 359
Bressan A., Chiosi C., & Fagotto F. 1994, ApJS, 94, 63, BCF94
Bressan A., Chiosi C., & Tantalo R. 1996, A&A, 311, 425, BCT96
Bressan A., Tantalo R., & Chiosi C. 1997, A&A, to be submitted, BTC97
Carollo C. M., & Danziger I.J. 1994, MNRAS, 270, 523
Carollo C.M., Danziger I.J., & Buson L. 1993, MNRAS, 265, 553
Davies R. L., Sadler E. M., & Peletier R.F. 1993, MNRAS, 262, 650
Ciotti L., Lanzoni B., & Renzini A. 1996, MNRAS 282, 1 1996
Gonzales J.J. 1993, Ph.D. Thesis, Univ. California, Santa Cruz
Greggio L. 1996, MNRAS, submitted
Greggio L., & Renzini A. 1990, ApJ, 364, 35
Kodama T., & Arimoto N. 1996, A&A in press
Larson R.B. 1974, MNRAS, 166, 585
Matteucci F. 1994, A&A, 288, 57
Matteucci F. 1996, Fundamentals of Cosmic Physics, in press
Padoan P., Nordlund A.P. & Jones B.J.T. 1996, preprint
Renzini A., & Ciotti L. 1993, ApJ, 416, L49
Schweizer F., & Seitzer P. 1992, AJ, 104, 1039
Tantalo R., Chiosi C., Bressan A., & Fagotto F. 1996, A&A, 311, 361, TCBF96
Tantalo R., Chiosi C, & Bressan A. 1997, A&A, to be submitted, TCB97

The Physical Origin of the Fundamental Plane (of Elliptical Galaxies)

Luca Ciotti

Osservatorio Astronomico di Bologna,
via Zamboni 33, 40126 Bologna, Italy

Abstract. I review the basic problems posed by the existence of the Fundamental Plane, and discuss its relations with the Virial Theorem (VT). Various possibilities are presented that can produce the observed uniform departure from *homology* (structural, dynamical) and/or from a constant *stellar* mass-to-light ratio. The role of orbital anisotropy and its relation with the FP thickness are also discussed. None of the explored solutions – albeit formally correct – are easily acceptable from a physical point of view, due to the ever-present problem of the required fine-tuning.

1 Observational Facts

Three are the main global observables of elliptical galaxies (Es): the central projected velocity dispersion σ_0, the effective radius R_e, and the mean effective surface brightness within R_e, $\langle I \rangle_e = L_B/2\pi R_e^2$. It is well known that Es do not populate uniformly this three dimensional parameter space; they are rather confined to a narrow logarithmic plane (Dressler et al. 1987; Djorgovski and Davis 1987), thus called the *Fundamental Plane* (FP). For example, for Virgo ellipticals

$$R_e \propto \sigma_0^{1.4} \langle I \rangle_e^{-0.85} . \tag{1}$$

Bender, Burstein, and Faber (1992, hereafter BBF) have introduced the k coordinate system, in which the new variables are a linear combination of the observables:

$$k_1 \equiv (2\log\sigma_0 + \log R_e)/\sqrt{2} , \quad k_2 \equiv (2\log\sigma_0 + 2\log\langle I \rangle_e - \log R_e)/\sqrt{6} , \tag{2}$$

$$k_3 \equiv (2\log\sigma_0 - \log\langle I \rangle_e - \log R_e)/\sqrt{3} . \tag{3}$$

In the new space, the k_1-k_3 plane provides an almost edge-on view of the FP, and $k_3 = 0.15k_1 + \text{const.}$ (Fig. 1 of BBF). Here I consider some of the implications of the two main properties of the FP of Virgo and Coma cluster ellipticals: 1) the FP is remarkably thin, with a 1-σ dispersion $\sigma(k_3) \simeq \pm0.05$ (Bender, private communication); and 2) this thickness is nearly constant along the FP. First, I will try to clarify some frequently misunderstood points about the relations between the FP and the VT, and the meaning of structural/dynamical non-homology.

2   The FP and its Relation with the VT

The characteristic dynamical time of Es (e.g., within R_e) is

$$T_{dyn} \simeq (G < \rho >_e)^{-1/2} \approx 10^8 \text{yrs} , \qquad (4)$$

and their collisionless relaxation time is of the same order (Lynden-Bell 1967), i.e., both are short with respect to the age of Es. As a consequence, only highly perturbed galaxies are presumably caught in a non-stationary phase. Stationarity is a *sufficient* condition for the validity of the VT, and so for Es the Virial Theorem holds. For a galaxy of total stellar mass M_* embedded in a dark matter halo of total mass M_h, the scalar VT can be written as

$$< V_*^2 >= (G\Upsilon_* L_B / R_e) \times (|\tilde{U}_{**}| + \mathcal{R}|\tilde{W}_{*h}|) , \qquad (5)$$

where $\mathcal{R} \equiv M_h / M_*$ and $\Upsilon_* = M_* / L_B$ is the *stellar* mass-to-light ratio, here defined using the galaxy total *blue* luminosity. The dimensionless functions $\tilde{U}_{**}$ and $\tilde{W}_{*h}$ are the stellar gravitational self-energy and the interaction energy between the stars and the dark matter halo. They depend only on the stellar density profile and on the relative distribution of the stellar matter with respect to the gradient of the dark matter potential, i.e., the dimensionless function on the r.h.s. of equation (5) *depends only on the galaxy structure* (see, e.g., Ciotti, Lanzoni, and Renzini 1996, CLR). Moreover, $\tilde{U}_{**}$ and $\tilde{W}_{*h}$ are known to be *weakly dependent* on the particular density profiles (see, e.g., Spitzer 1969, Ciotti 1991, Dehnen 1993). Obviously the same comments apply also to the stellar velocity virial velocity dispersion $< V_*^2 >$, that necessarily results to be *independent* of the particular internal dynamics of the galaxy (e.g., the amount of orbital anisotropy).

$< V_*^2 >$ is related to σ_0^2 through a dimensionless function that depends on the galaxy structure, its specific internal dynamics and on projection effects:

$$< V_*^2 >= C_K(structure, anisotropy, projection) \times \sigma_0^2 . \qquad (6)$$

It is important to note that – also in absence of orbital anisotropy – C_K *is very sensitive to galaxy-to-galaxy structural differences*, much more than $\tilde{U}_{**}$ and $\tilde{W}_{*h}$, because it relates a weakly structure dependent quantity $(< V_*^2 >)$ to a *local* one (σ_0^2). So, in practice structural non-homology always implies a *strong* variation in the σ^2 profile. Some authors call this phenomenon *dynamical* non-homology, but I (strongly!) suggest to call this effect *kinematical* non-homology, at least for two reasons. First, it is not very useful to call with two different terms (structural vs. dynamical non-homology) the *same* phenomenon: you cannot break structural homology without breaking also kinematical homology. Second, *dynamical* homology should be used to describe galaxies (for example) with the same relative amount of ordered rotation or orbital anisotropy. So, according to this nomenclature, kinematical non-homology can be induced by structural non-homology and/or dynamical non-homology. Moreover, globally isotropic galaxies with different density profiles are structurally non-homologous but dynamically homologous, their $R_e < V_*^2 > /G\Upsilon_* L_B$ are similar, but their C_K's can be significantly different. As a final remark, it is important to note that the strong dependence of C_K on galaxy structure is essentially due to its definition: in fact, using larger and larger aperture velocity dispersions instead of the *central* velocity dispersion σ_0^2, the projected Virial Theorem is better and better approximated, and for a spherical system without dark matter $C_K \to 3$ independently of the galaxy orbital structure (e.g., Ciotti 1994; cfr. also the results of the numerical simulations of mergers of Capelato et al. 1995).

From equations (2)-(3) and (5)-(6), defining

$$\Theta \equiv [2\pi G\,(|\tilde{U}_{**}| + \mathcal{R}|\tilde{W}_{*h}|)]/C_K \ , \tag{7}$$

one finally obtains

$$k_1 = \log(\Theta \times \Upsilon_* \times L_B/2\pi)/\sqrt{2} \ ; \quad k_3 = \log(\Theta \times \Upsilon_*)/\sqrt{3} \ . \tag{8}$$

Note that the VT does not imply any FP, in fact for fixed L_B different galaxies, all satisfying the VT, can in principle have very different Θ and Υ_*, and so be scattered everywhere in the k-space. So, the statement that *the FP deviates from the VT* – often stated because of the difference between the exponents in eq. (1) and those in the VT [eq. (5)] – is wrong: the FP deviates from *homology* (in a broad sense).

Summarizing, three ingredients are necessary for a class of hot dynamical systems to flatten about a FP: 1) to be virialized, 2) to have similar structures and internal dynamics, 3) to exhibit a small dispersion of Υ_* for any given L_B. Observations tell us that $\Theta \times \Upsilon_*$ is a very well defined function of the galaxy parameters with an intrinsic nearly constant scatter less than 12%. In particular,

$$\delta(\Theta \times \Upsilon_*)/(\Theta \times \Upsilon_*) < 0.12 \ . \tag{9}$$

As galaxies in the BBF sample span a factor ~ 200 in L_B, the tilt corresponds to a factor ~ 3 increase of $\Theta \times \Upsilon_*$ along the FP, from faint to bright galaxies. If Υ_* and Θ are not finely anticorrelated, this implies a very small dispersion, *separately* for both quantities, at any location on the FP. This sets a very severe restriction on $\Theta \times \Upsilon_*$, which translates into strong constraints on the range that each parameter can span at any location on the FP. It is evident that fine tuning is required to produce the tilt, and yet preserve the tightness of the FP. Note also how, from eqs. (8), galaxies with fixed L_B and various internal dynamics, structure, Υ_*, *move along straight lines* in the k-space, with $k_3 = \sqrt{2/3}k_1 + $ const.. The inclination of this line with respect to the FP given by BBF is equal to $\arctan(\sqrt{2/3}) - \arctan(0.15) \simeq 30\,\mathrm{deg}$: this is the reason why in numerical simulations the end-products of the merging of systems initially placed on the FP are found near the FP itself[1]. In conclusion, what is important is not the attempt to understand, perhaps by the finding of a "good" set of observational quantities, why the FP is "distant" from the VT, but, on the contrary, why galaxies are so similar in structure and dynamics, with such a small scatter.

---

[1] Moreover, this relation between k_3 and k_1 helps reduce slightly the problem of the FP thickness.

3   Exploring Various Possibilities

For simplicity the origin of the FP tilt can be sought in two *orthogonal* directions: either due to a *stellar population* effect, in which case $\Upsilon_* \propto L_B^{0.2}$ and Θ=const, or due to *structural/dynamical* effects, i.e., $\Theta \propto L_B^{0.2}$ and Υ_*=const.

3.1   A Stellar Origin: Changing the IMF

A systematic change of the stellar initial mass function (IMF) is explored in Renzini and Ciotti (1993, hereafter RC). Υ_* is obtained by convolving the present mass of the stars M with the IMF, where $M = M_i$ for the initial mass $M_i < M_{TO}$ (the turnoff mass), and $M = M_R$ (the remnant mass) for $M_i \geq M_{TO}$. For the IMF we adopt $\psi(M_i) = A L_B M_i^{-(1+x)}$, where L_B is the present day blue luminosity of the population, or a multi-slope Scalo IMF with a variable slope for $M < 0.3 M_\odot$. For details see RC.

Changing M_{inf}. In this case we assume a *decrease* of M_{inf}, the lowest stellar mass, for increasing L_B. We found that – in the case of a single-slope IMF – for no value of x small values of Υ_* (characteristic of the FP faint-end) are obtained, unless M_{inf} is unrealistically high. Reducing the slope does not help: for x below ~ 0.65 Υ_* increases again, since then the mass in remnants increases more than how much the mass in the lower main sequence stars is reduced. Only with the multi-slope Scalo IMF a low Υ_* can be realized (Fig. 1 in RC).

Changing the IMF Slope. The previous requirement of a low Υ_* at the FP faint-end forces the choice of a low x, but then Υ_* is quite insensitive to variations of M_{inf}; only for a steep IMF Υ_* is sensitive to M_{inf}. As a consequence a mere variation of M_{inf} with a constant IMF slope cannot account for the observed trend. Hence, a variation of slope is required, by an amount Δx which depends on the adopted M_{inf}. We conclude that *a major change of the IMF slope in the lower main sequence is necessary to account for the FP tilt*. There remains to consider the thickness of the FP. In RC it is shown that in order to preserve the $\sim 12\%$ upper limit on $\sigma(\Upsilon_*)$, the galaxy-to-galaxy dispersion in M_{inf} and x should be extremely small, $< \pm 10\%$ and $< \pm 0.15$, respectively. Such very small galaxy-to-galaxy dispersion, coupled to a large systematic variation of x, is a rather demanding constraint, and we conclude that fine tuning is required to produce the observed tilt of the FP, while preserving its constant thickness: the IMF should be virtually universal for a given galaxy mass, and yet exhibit a large trend with galaxy mass. A more accurate analysis of this scenario, using stellar population synthesis models, is given by Maraston (1996).

3.2   A Structural/Dynamical Origin

In this case, assuming Υ_* =const., we explore under which conditions structural/dynamical effects may cause the tilt in k_3 via a systematic increase of Θ

[RC; CLR; Ciotti and Lanzoni 1996, (CL); Lanzoni and Ciotti 1996 (LC)]. In all these investigations spherical, non rotating, two–component galaxy models are constructed, where the light profiles resemble the $R^{1/4}$ law when projected. The internal dynamics is varied using the Osipkov-Merritt formula. Here for shortness reasons only the main results are summarized.

Dark Matter Content and Distribution. In this case we assume *global isotropy*, i.e., all models are *dynamically homologous*, and we ascribe all the FP tilt to systematic variations of $\mathcal{R} = M_h/M_*$ or $\beta = r_h/r_*$ (r_h is a characteristic radius of the dark matter distribution). Obviously, the larger β, the larger the variations of $\mathcal{R}$ that are required to produce the tilt. For Hernquist+Hernquist models and Hernquist+Plummer models, for $\beta \simeq 5$ exceedingly large values of $\mathcal{R}$ are required to produce the FP tilt ($\mathcal{R} \simeq 30 - 175$). An increasing $\mathcal{R}$ may be at the origin of the observed tilt, provided that $\beta < 5$. The same problem affects all the Jaffe+Quasi-isothermal models that we have considered, for every values of β and r_t. For Jaffe+Jaffe models, $\mathcal{R}$ never becomes larger than 10 thus every value of this parameter is acceptable, for every explored value of β.

Structural Non-Homology. Systematic deviations of the Es light distribution from the standard $R^{1/4}$ profile may also possibly cause the FP tilt (Djorgovski 1995; Hjorth and Madsen 1995). We investigate this *morphological* option using isotropic $R^{1/m}$ models without dark matter, thus ascribing to a systematic variation of m the origin of the tilt. Assuming the faintest galaxies to be $R^{1/4}$ systems, in order to produce the tilt m has to increase from 4 up to ~ 10 along k_1 (CLR). If one assumes $m = 2$ for the faintest galaxies, the required variation is even larger, about a factor of 4, up to ~ 8, and its permitted variation at each FP location remains very small: a scatter of $m < 10\%$ at any location on the FP should be associated to a large variation of it with galaxy luminosities. A systematic trend of m with galaxy luminosity has been reported (Caon, Capaccioli and D'Onofrio 1993), with m increasing from ~ 1 up to ~ 15, thus spanning a much wider range than required to produce the tilt. We conclude that further observational studies are required in order to determine whether a progression of light-profile shapes along the FP really exists among cluster ellipticals.

The Role of Anisotropy. In this case we ascribed the entire tilt of the FP to a trend with L_B in the anisotropy degree of the galaxies (described in the Osipkov-Merritt formulation by the anisotropy radius r_a), assuming no dark matter. For Hernquist, Jaffe, and $R^{1/m}$ models constrained to the FP we found that, above a certain luminosity, the phase–space distribution function runs into negative values. So we conclude that anisotropy alone cannot be at the origin of the tilt, because the extreme values of r_a that would be required correspond to dynamically inconsistent models (CLR, CL, LC). A special problem with anisotropy is raised by the FP thinness: for galaxies with a positive distribution function a strong fine tuning between L_B and r_a could appear to be required. However in CL and LC we showed, using a semi–quantitative global instability indicator, that in $R^{1/m}$ models the excursion in anisotropy permitted by *stability* is much less than that given by the simple dynamical consistency, and the induced scatter on the FP is inside the observed spread in k_3. We are well aware that this result is very qualitative (where is it placed in the k–space the end-product of a radially unstable galaxy model initially on the FP?), and so we feel that this result is worth to be further studied, using N-body simulations.

4 Conclusions

Our exploration indicates that all structural/dynamical/stellar population solutions to the FP tilt are rather unappealing, though some are more so than others. This comes from the strong *fine tuning* that is required, no matter whether the driving parameter is the amount of dark matter ($\mathcal{R}$), its distribution relative to the bright matter (β), the shape of the surface brightness distribution (m), or finally the properties of the IMF. In addition to this, we have excluded a trend in the anisotropy as possible cause of the tilt, because it leads to physically inconsistent models. Finally, we showed in a semi-quantitative way that probably a fine tuning of anisotropy with the galaxy luminosity is not required for stability arguments, but deeper investigations, both analytical and numerical, are required.

References

Caon, N., Capaccioli, M., D'Onofrio, M., (1993): MNRAS, **265**, 1013

Capelato, H.V., de Carvalho, R.R., Carlberg, R.G., (1995): ApJ, **451**, 525

Bender, R., Burstein, D., Faber, S.M., (1992): ApJ, **399**, 380, (BBF)

Ciotti, L., (1991): A&A, **249**, 99

Ciotti, L., (1994): Celestial Mechanics, **60**, 401

Ciotti, L., Lanzoni, B., (1996): A&A, *in press*, (astro-ph/9610251) (CL)

Ciotti, L., Lanzoni, B., Renzini, A., (1996): MNRAS, **282**, 1 (CLR)

Dehnen, W., (1993): MNRAS, **265**, 250

Djorgovski, M., Davis, S., (1987): ApJ, **313**, 59

Dressler, A., Lynden-Bell, D., Burstein, D., Davies, R.L., Faber, S.M., Terlevich, R.J., Wegner, G., (1987): ApJ, **313**, 42

Hjorth, J., Madsen, J., (1995): ApJ, **445**, 55

Lanzoni, B., and Ciotti, L., (1996): *this workshop* (LC)

Lynden-Bell, D., (1967): MNRAS, **136**, 101

Maraston, C., (1996): *this workshop*

Renzini, A., Ciotti, L. (1993): ApJL, **416**, L49 (RC)

Spitzer, L. (1969): ApJL, **158**, L139

The Role of Photoionization During Galaxy Formation

Alain Blanchard[1], and Simon Prunet[2]

[1] Observatoire astronomique de Strasbourg, ULP, 11, rue de l'Université, 67 000 Strasbourg, France

[2] Université de Paris-Sud, Institut d'Astrophysique Spatiale, Bâtiment 121, F-91405 Orsay Cedex, France

Abstract. Baryonic dissipative physics is expected to be important during galaxy formation and to play a fundamental role in the origin of galaxies properties as well as in the mechanism controlling their distribution. The cooling criteria seems to be useful in order to distinguish halos which might form a galaxy from those which may form a cluster. However the same criteria imply the so-called "overcooling" problem, which pointed out that substantial reheating should have occurred in the past universe. Photoionization seems not to be adequate anyway. We briefly discussed an alternative possibility based on a warm IGM picture. In this picture photoionization may allow low mass galaxy formation which is inhibited otherwise.

1 Introduction

The properties of galaxies, their morphology, their spatial distribution, are challenging current theory of galaxy formation. One would like to understand the origin of some of the basic properties of galaxies: the total amount of stars present in the universe, the luminosity function of galaxies, the star formation rate, the Tully-Fisher relation, the number counts in different bands, the physical origin of the various morphologies, etc... A comprehensive understanding would also mean that we are able to follow the history of these various quantities. However the complexity of the problem, as well as the lack of information, has greatly limited theoretical investigations.

The situation has changed dramatically in the past few years, thanks to both the Keck telescope and the HST: a much wider range of information is now available, in addition to classical deep redshift surveys. In the meantime, numerical simulations in which baryonic physics is included have become very popular despite the enormous challenge that a full description of galaxy formation represents. Semi-analytical models have been elaborated and represent a first step toward a more comprehensive description.

A first attempt to specifically address the question of galaxy formation has met a first important apparent success: it has been suggested that a critical criterion in order to differentiate dark halos leading to galaxies from those leading to clusters is the cooling criterion: when gas falls in a potential it is shock-heated (and/or by adiabatic compression) up to a temperature which does allow the gas to be in hydrostatic equilibrium. This temperature is currently called the virial temperature of the gas. Numerical simulations have confirmed (Evrard, 1997) that this simple argument provides an accurate estimation of the actual temperature:

$$T_v = 5.10^5 \text{K} M_{12}^{2/3}(1+z) \qquad (1)$$

where M_{12} is the total mass of the object forming at redshift z. The typical size of the halo is:

$$R_v = \left(\frac{T_v}{10^5 \text{K}}\right)^{1/2} \frac{1.}{(1+z)^{3/2}} 45h^{-1} \text{kpc} \qquad (2)$$

When the gas reaches its virial temperature, it has a characteristic cooling time (the density within a virialized structure being, almost by definition, at least 200 times the mean density). Clusters typically represent structures for which the cooling time exceeds the age of the universe, while for galaxies this typical time scale is much shorter. It is tempting therefore to think that the cooling criterion can be used as a criterion for star formation: if the gas is able to cool, it will contract in a runaway fashion, which can end only by star formation (as there is not so much cooled gas in the universe). This argument successfully explains the order of magnitude of the luminosity of the brightest galaxies (L_*).

White and Rees (1978), in a pioneering work, took into account the cooling criterion in order to derive the luminosity functions of galaxies from the mass function of cosmic structures as given by the Press and Schechter formula. They noticed that with the simple rule they used, the low end of the luminosity function was steeper than the observed one, but argue that this was not a critical problem for their approach.

2   The Star Formation History

2.1   Some Relevant Observations

A first number one can expect from a theory of galaxy formation is the amount of observed stars. Most of the baryons which are actually observable are those seen in galaxies consisting of stars and neutral gas. With the most recent determination of the luminosity function (Zucca et al, 1997), this leads to a first evaluation of the gas which has cooled in the history of the universe:

$$\Omega_{*+gas} \approx 0.007 \qquad (3)$$

Primordial nucleosynthesis allows one to evaluate the total baryonic content of the universe. Standard nucleosynthesis leads to :

$$\Omega_{bbn} \approx 0.04 h_{50}^{-2} \qquad (4)$$

in order to reproduce the observed abundances of light elements. This allows to evaluate the integrated fraction of baryons which has the ability to cool during the cosmic history:

$$f_c \approx 0.1 - 0.2 h_{50}^{-2} \qquad (5)$$

There are two sets of observations which might obscure the previous discussion and put a significant uncertainty on the previous number: the abundance of Deuterium has been recently a matter of lively debate. The definitive conclusion is yet unclear, but it might imply that the actual value of Ω_{bbn} should be taken a factor of two smaller or higher than the previous estimate (4). The second set of observational results comes from the microlensing experiments: if the detection of Machos is real, it does strongly suggest that most of the stars in the universe have escaped direct detection up to now and the total amount of stars has therefore been substantially underestimated, being substantially larger than (3). This could mean that our parameter f_c may be somewhere in the range 0.05–0.8... Such a conclusion would severely undermine the following discussion. However, clusters are providing a nice independent argument in order to evaluate the cooled fraction f_c: we know how much gas and how many stars there are in a typical cluster like Coma (White et al, 1993):

$$f_c = \frac{M_*}{M_{gas}} \approx 0.15 h_{50}^{-3/2} \tag{6}$$

assuming the dark matter to be non-baryonic. This number is in nice agreement with the previous estimate (5), and is subject to much less uncertainty, as it is independent of the previous possible caveats.

2.2   The Overcooling Problem

The simple cooling scheme meets a fundamental problem: at high redshift the bulk of the baryons lies in small potentials with temperature in the range $10^4 - 10^6$K in which cooling is extremely efficient. Actually, most of the mass of the universe has set in such potentials at some epoch. Consequently, most of the baryons should have been cooled by now. This is in clear contradiction with the two basic facts that we have met: known stars represent only a small fraction of baryons predicted by nucleosynthesis, and most of the baryonic content of clusters is still in the gas phase. Both facts suggest that only something like 10%–20% of the primordial baryons were actually turned into stars during the cosmic history. This problem was first pointed out by Blanchard et al. (1990, 1992) and Cole (1991). The reality of this overcooling problem is not easy to test by means of numerical simulations: one has to be sure that the smaller scale which leads to halo with temperature of the order to 10^4 at high redshift is fully resolved. Typically this means, according to (2), a comoving size of the order of 5 kpc at a redshift of 10, with a typical mass of 3×10^6 M$_\odot$. To our knowledge, such high resolution simulation has not been performed. However, the high resolution of Navarro and Steinmetz (1996) is not very far from these requirements. Despite the fact that the resolution does not seem to be sufficient they found a large amount of cooling, in rather good agreement with the theoretical speculations. It is therefore clear that the problem is real and probable that its solution implies that the gas has undergone some substantial reheating.

As we have seen, within the simple spherical model it is possible to evaluate the virial temperature. The knowledge of the mass function allows to estimate

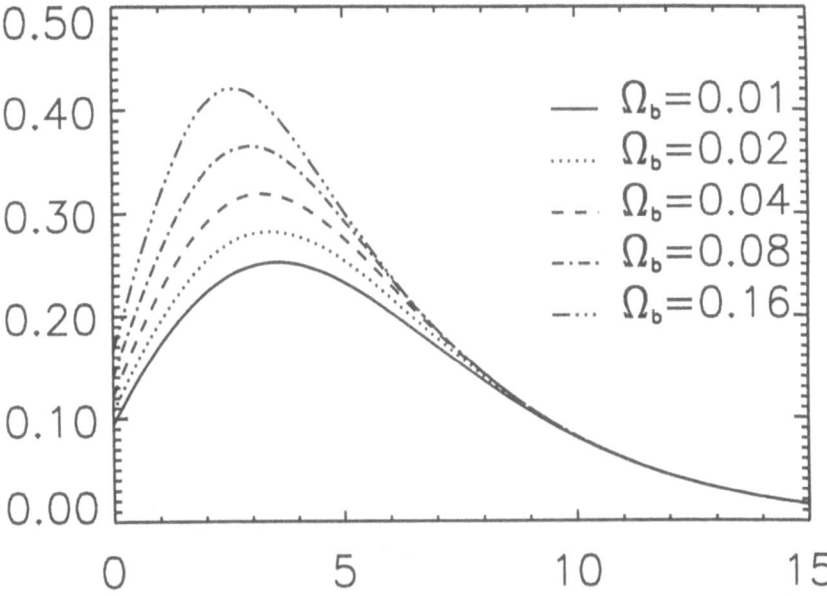

Fig. 1. Instantaneous fraction of gas able to cool at various redshifts. The different curves correspond to different values of the baryonic density parameter for the standard Cold Dark Matter scenario.

the fraction of gas able to cool at any redshift by considering the mass range in which cooling is likely to occur. The non-photoionized case is shown in figure 1. This figure represents the instantaneous fraction of gas able to cool within structures at various redshifts. The dependence on the total baryonic content of the universe is rather moderate (the instantaneous maximum fraction is changed by less than a factor of 2 when Ω_b is changed by a factor of 8). This fraction has been evaluated as in Blanchard et al. (1992) by assuming a cooling function for an IGM with no metals – inclusion of metals increased this amount by a factor smaller than 2).

The computation of the total amount of cooled gas seems to be a rather non-trivial question, as one has to evaluate in halos too hot for cooling to occur the fraction of gas which has undergone cooling in a previous structure of the hierarchy. This problem leads to the use of the hierarchical tree making any calculation rather heavy. However, a useful hint has been given by Blanchard et al (1992): it has been argued that in every halo with a temperature higher than 10^4 K, any gas element has been previously at some redshift in a structure for which the cooling criterion was satisfied. In this case the total amount of cooled gas is just:

$$F_c = \frac{1}{\rho} \int_{m_4(z)}^{+\infty} N(m)m\,dm \qquad (7)$$

where $m_4(z)$ represents the mass of halos which have a virial temperature of 10^4K at redshift z. The integrated fraction has been evaluated in the CDM picture on figure 2.

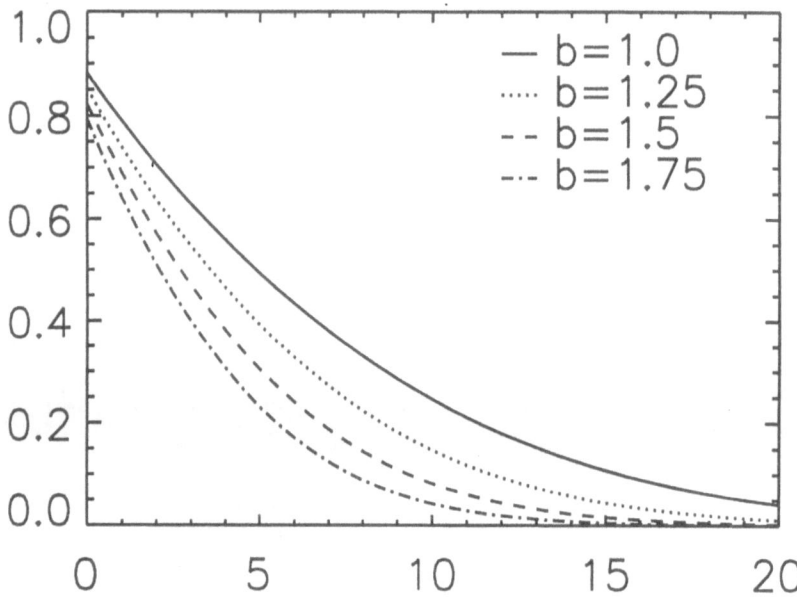

Fig. 2. Integrated fraction of gas able to cool at various redshifts. The different curves correspond to different values of the bias parameter.

3   Reheating Mechanisms

The solution of the "overcooling" problem is very likely to be a substantial reheating phase. Such reheating has been advocated in various contexts in galaxy formation scenarios. For instance, White and Frenk (1991) argued that the energy input of supernova from the first generation of stars is able to prevent the cooling of the gas that remains confined in galactic scale halos. On the contrary Blanchard et al. (1992) suggested that the first objects which form heat the IGM to a temperature high enough that most of the gas does not fall in most of the forming potentials because of its thermal energy. This introduces the idea that a key physical quantity controlling galaxy formation is the temperature of the IGM, which could be regulated by galaxy formation.

3.1   Photoionization

It has been suggested that photoionization can strongly suppress cooling in small potentials and therefore small galaxy formation (Efstathiou, 1992). Let us examine whether photoionization can act efficiently to suppress the overcooling problem. The first effect of photoionization is to heat the IGM to a temperature of the order of 10^4K and therefore this prevents the formation of small potentials with virial temperature smaller than this. This is not directly relevant to the overcooling problem, as cooling is expected to occur only in potentials with higher temperature. However, this could be an important help, as a situation in which most of the primordial gas fell in small and dense potentials, even without being cooled, will be problematic. Another important point is that the photoionization substantially alters the cooling rate of the gas (which is normally dominated by collisional excitation cooling). It is therefore interesting to examine whether this significantly reduces the fraction of gas able to cool.

In figure 3 the same quantity as in figure 1 is evaluated in the case of a fully ionized gas, which represents the most extreme case of the photoionization regime. It is clear that although the photoionization reduces the instantaneous fraction, the change is at most a factor of two in the regime where the cooling is the most efficient. Such a factor is in agreement with the conclusion of Navarro and Steinmetz (1996). It is therefore unlikely that photoionization can be a solution to the overcooling problem.

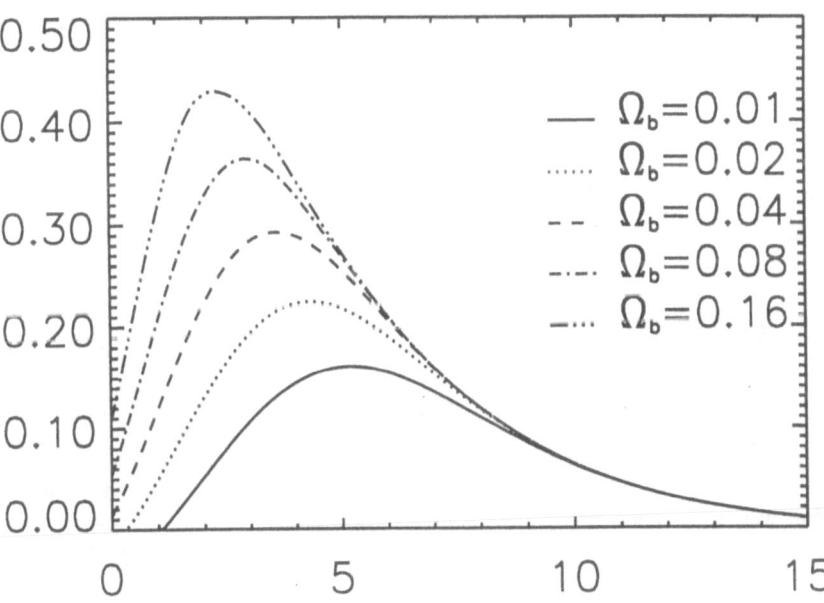

Fig. 3. Same quantity as in figure 1, but here the cooling curve is taken in the fully ionized case, which corresponds to a highly photoionized regime.

3.2   A Self Regulated IGM

In this section we will briefly discuss the scenario in which the temperature of the IGM is regulated by the galaxy formation, the basic mechanism being the energy injection from supernova during the early phase of galaxy formation. A substantial fraction of the energy produced by the supernova is assumed to be the source of the heating of the IGM. A fully efficient heating corresponds to a value for our normalized heating parameter of the order of 1–2. As one can see on figure 4, even if a small fraction of the energy of the supernova is transferred to the IGM, the suppression of the overcooling problem is quite natural. This makes this scenario quite appealing.

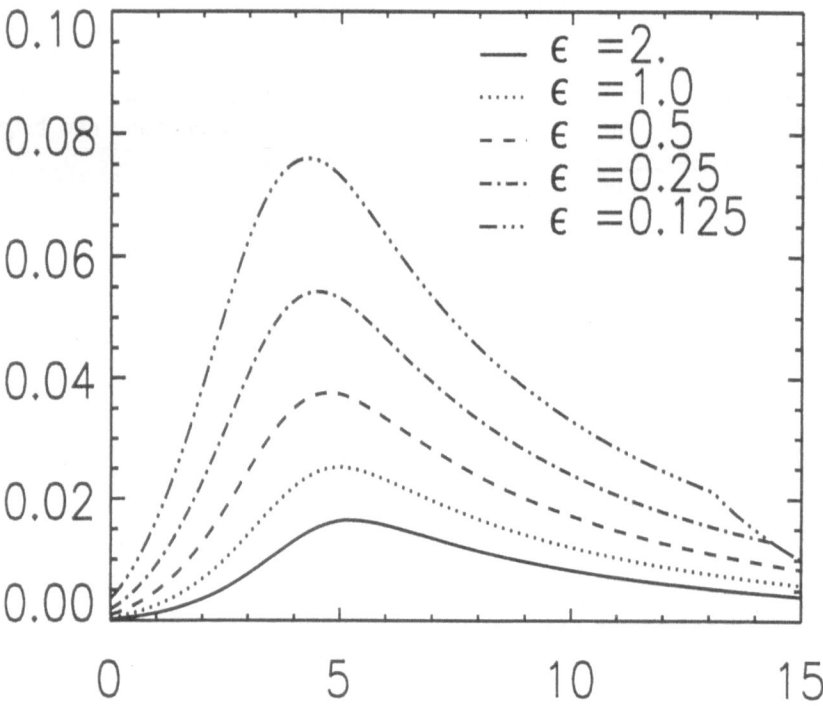

Fig. 4. Same quantity as in figure 1, but here the IGM is heated in a self-regulated way. Moreover, the cooling curve is taken in the photoionized case. The various curves correspond to different values of the efficiency of the energy injection in the IGM from supernova. The parameter ϵ is not well constrained, but has a maximum plausible value of 2.

A key hypothesis is the fact that if the IGM is at some temperature T, only halos which have a higher virial temperature have a chance to actually undergo baryon accretion and therefore to be identified as a galaxy. The amplitude of the cooled fraction is reduced as the temperature grows. The resulting integrated cooled fraction can then be of the right order of magnitude. The maximum temperature which is reached is of the order of 10^6K. At that time only L_* galaxies could form. Later as the temperature decreases, lower mass potentials can undergo cooling. In this model, galaxy formation occurs in an inverse hierarchical way: large galaxies form first and then smaller ones form later. One problem which occurs in this model is that the temperature remains high at low redshift (of the order of 10^5K), preventing the accretion of gas and therefore the formation of stars in halos with circular velocity smaller than ~ 100km/s.

The effect of photoionization in this case is to allow the temperature of the IGM to be smaller, reaching temperature of the order of 10^4K at low redshift. In such case low mass galaxy formation is not suppressed anymore.

4   Conclusion

What we know from baryons clearly indicates that most of the bulk of the baryons has undergone a strong reheating in order to avoid the overcooling catastrophe. Photoionization does not seem to be able to provide the adequate energy injection and consequently, the overcooling problem cannot be solved easily by the suppression of the collisional excitation cooling in such a way: although photoionization reduces the amplitude of the cooled fraction, the effect seems too small to offer a solution to the overcooling problem. Detailed numerical simulations lead to similar conclusions (Navarro, Steinmetz, 1996). The simplest solution which agrees quite well with present day information is the existence of a hot IGM heated by some other mechanism as suggested by Blanchard et al (1992): in such a case the high temperature of the gas prevents it from falling in most of the potentials at high redshift. If the temperature of the IGM is greater than a few 10^5K the overcooling problem can be solved.

References

Blanchard, A., Valls-Gabaud, D., Mamon, G. (1990): in *"Particle Astrophysics, the Early Universe and Cosmic Structures, Proceedings of the XXth Rencontres de Moriond in Astrophysics"*, ed. J.-M. Alimi, A. Blanchard, A. Bouquet, F. Martin de Volnay and J. Trân Thanh Vân, Editions Frontières, Gif-sur-Yvette, p. 403.
Blanchard, A., Valls-Gabaud, D., Mamon, G. (1992): A&A **264**, 365–378
Cole, S, (1991): ApJ **367**, 45
Efstathiou, G., (1992): MNRAS **456**, 43p
Navarro, J.F., Steinmetz, M. (1996): astro-ph/9605043
White, S., Rees, M. (1978) MNRAS **183**, 341–358
White, S.D.M., Frenk, C..S. (1991): ApJ **379**, 52–79
White, S.D.M., Navarro, J..F., Evrard, A.E., Frenk, C..S. (1993): Nat **366**, 429–433
Zucca, H., et al., preprint.

Semianalytic Modelling of Galaxy Evolution

C.M. Baugh[1], S. Cole[1], C.S. Frenk[1] and C.G. Lacey[2]

[1] Department of Physics, South Road, Durham, DH1 3LE, UK.
[2] Theoretical Astrophysics Center, Juliane Maries Vej 30, DK-2100 Copenhagen Ø, Denmark.

Abstract. A coherent picture of the process of galaxy formation is now being built up from a combination of new data from the Hubble Space Telescope and large ground based telescopes such as the Keck, and from the interpretation of these data using semi-analytic and numerical techniques. Here we concentrate on the semianalytic approach, outlining the basic features of this class of model. We review some of the successful predictions already made before focussing on the Tully-Fisher and colour-magnitude scaling relations, both locally and at high redshift, two areas in which VLT observations are expected to have a major impact. Finally, to illustrate the full power of the semianalytic approach, we compare our models with observations of Lyman break galaxies in ground based and HST data. These observations can be explained within the context of a cosmology in which structure formation is hierarchical and suggest that the beginning of the epoch of galaxy formation has been discovered.

1 Introduction

The problem of galaxy formation is being attacked theoretically on two main fronts: numerical simulations and semi-analytic modelling. Any potential model has to explain a range of complex phenomena operating on very different length and time scales, that can be broken down into six broad areas:

- 1 The formation of dark matter halos.
- 2 The dynamics of cooling gas inside these halos.
- 3 Star formation.
- 4 Energy feedback into the cooling gas due to supernovae and stellar winds.
- 5 Evolution of the colour and luminosity of the stars that form.
- 6 The mergers of galaxies.

Numerical simulations that follow the development under gravity of the dissipationless component of the universe are the most mature of the techniques available. We now have a good understanding of the way in which dark matter halos are built up hierarchically by accretion and mergers. Recent simulations by Navarro, Frenk & White (1996) (see also the contribution of S. White to this meeting and Lacey & Cole (1996)) have shown that dark matter halos have a universal profile, valid in a wide range of different cosmologies. With the increase in computing power, it is now possible to achieve the spatial resolution necessary to attempt to follow the evolution of the gaseous component. Although these simulations are still in their infancy, there have been some notable successes. The shock heating of gas that falls into the gravitational potential well of a dark matter halo has been demonstrated in the non-radiative limit in the Smooth Particle Hydrodynamics simulations of Evrard (1990). The gas is found to adopt a density profile that traces that of the dark matter (Navarro, Frenk & White 1995). However, one problem is that simulations of the formation of individual galaxies which include cooling, whilst producing disks and spiral arm features, tend to produce disks that are smaller than observed (Steinmetz & Muller 1995). This is a result of the transfer of angular momentum away from the disk to the halo as a consequence of mergers.

The semianalytic approach has been developed by several groups (White & Frenk (1991), Cole (1991), Lacey et al. (1993), Kauffmann, White & Guiderdoni (1993), Cole et al. (1994)) since 1991, building on the pioneering work of White & Rees (1978). The basis of the models is a Monte Carlo representation of the hierarchical formation of dark matter halos, which implements the extended Press & Schechter (1974) theory derived by Bond et al. (1991) and Bower (1991). Our current understanding of processes (2) - (5) outlined above can be encapsulated into a set of simple rules. The evolution of galaxies can then be traced from the smallest branches of the halo merger tree (*cf* figure 6 of Lacey & Cole 1993) at high redshifts, all the way through to the present day.

The numerical and semianalytic approaches are complementary. The semianalytic models can probe a wide parameter space quickly, evaluating how the particular representation of the complex processes listed above affects the galaxy population. Numerical simulations attempt to model the physics involved from first principles and give guidance about the form that parameterisations of rules should take for input to the semianalytic models.

In this review, we give a brief outline of the model of Cole et al. (1994) and list some of its successes, before discussing in more detail the Tully-Fisher and colour magnitude scaling relations. Finally, we discuss the recent observations of star forming galaxies at high redshift by Steidel et al. (1996a) and Madau et al. (1996) as a further example of the power of the semianalytic approach.

2   A Recipe for Galaxy Formation

The basic rules that govern the physical processes in our scheme are presented in detail in Cole et al. (1994). When a dark matter halo collapses, its associated gas is shock heated and settles into quasistatic equilibrium at the virial temperature of the halo. This gas cools radiatively over the lifetime of the halo and the cold gas is turned into stars at a rate proportional to the instantaneous cold gas mass. Feedback from supernovae and stellar winds returns some of the cold gas to the hot phase, strongly inhibiting star formation in low circular velocity halos. During a merger, the dark matter halos coalesce rapidly, but the galaxies within them can survive longer, eventually merging on a timescale related to the dynamical friction time.

A given cosmological model is defined by the values of the following parameters: the mean cosmic density (Ω_0), the cosmological constant (Λ_0), Hubble's constant ($H_0 = 100h\text{kms}^{-1}\text{Mpc}^{-1}$), and the mean baryon density in units of the critical density (Ω_b). Our galaxy formation prescription then requires the specification of 5 physical parameters, with the rules calibrated against the results of numerical simulations where possible: (i) a star formation timescale, (ii) a "feedback parameter," (iii) the initial mass function (IMF) of stars, (iv) a merger timescale for galaxies and (v) an overall luminosity normalisation that sets the fraction of stars that are below the hydrogen burning limit. To describe the broad morphology of a galaxy (ie. its bulge-to-disk ratio) a sixth parameter is required: the threshold mass that is to be accreted in a merger in order to turn a disk into a spheroid (see Baugh, Cole & Frenk (1996a) for further details.)

The general strategy that we have adopted is to choose a combination of parameters that gives the best match to the local luminosity function and the local morphological mix of galaxies in the field. The remaining properties of the model galaxies, for example the Tully-Fisher relation, morphology-density relation, number counts and redshift distributions, are all predictions that can be tested against other data.

Due to the strong feedback assumed, which severely restricts star formation in halos with low circular velocity, the model of Cole et al. (1994) produces a flatter faint end slope for the luminosity function than most other semianalytic models. However, the recovered slope of $\alpha \sim -1.5$ implies more faint galaxies than are observed, though local surveys may not be sensitive to low surface brightness galaxies (McGaugh 1994). We use the revised stellar population models of Bruzual and Charlot (Charlot, Worthey & Bressan (1996)). The colours of the reddest galaxies in our models now match those of observed ellipticals (see figure 2 of Frenk, Baugh & Cole 1996).

Other successes include the prediction by Cole et al. (1994) of the redshift distribution of faint galaxies limited to $B = 24$ observed by Glazebrook et al. (1995) and in particular by Cowie et al. (1996) (see figure 1 of Frenk, Baugh & Cole 1997) and the comparison of the faint number counts as a function of morphological type (Baugh, Cole & Frenk 1996b) with HST observations.

3   The Tully-Fisher Relation

The Tully & Fisher (1977) relation predicted by Cole et al. has a scatter and slope quite similar to those observed. However, it has so far proved impossible to obtain a model that matches simultaneously the zero point of this relation and the amplitude of the luminosity function. This problem is a generic feature of standard CDM, which predicts an overabundance of galactic dark halos. The discrepancy is reduced slightly in low density CDM models (Heyl et al. 1995).

The local Tully-Fisher relation in the I-band predicted by our models is shown in Fig. 1a. Note that in this and subsequent panels we plot only the central galaxies in halos. We have plotted twice the circular velocity to allow direct comparison with observed velocities inferred from linewidths. The data points

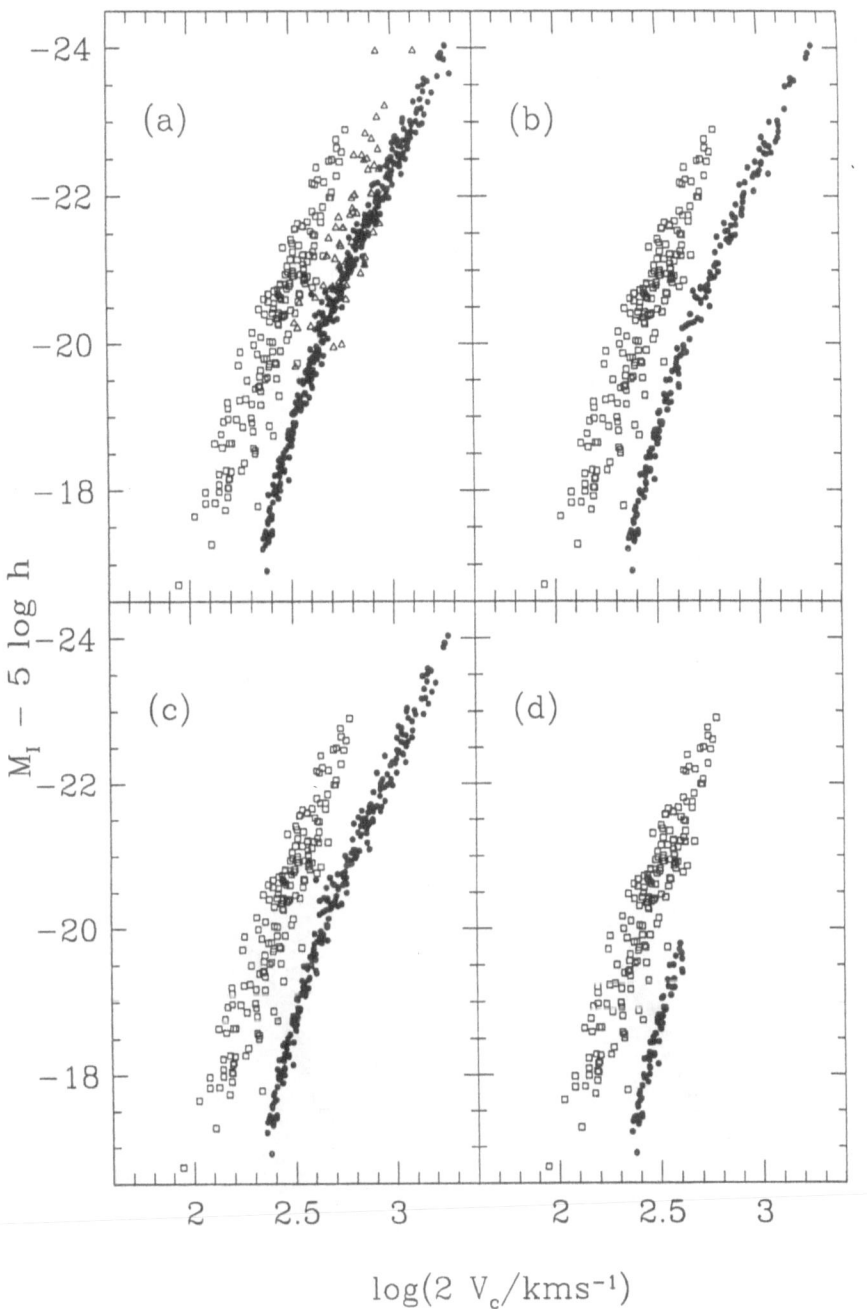

Fig. 1. The Tully-Fisher relation at redshift zero. The squares are a compilation of data taken from Young (1996). The triangles in (a) show elliptical galaxies from the Coma cluster. The filled points show the predictions of our model for different selections discussed in the text.

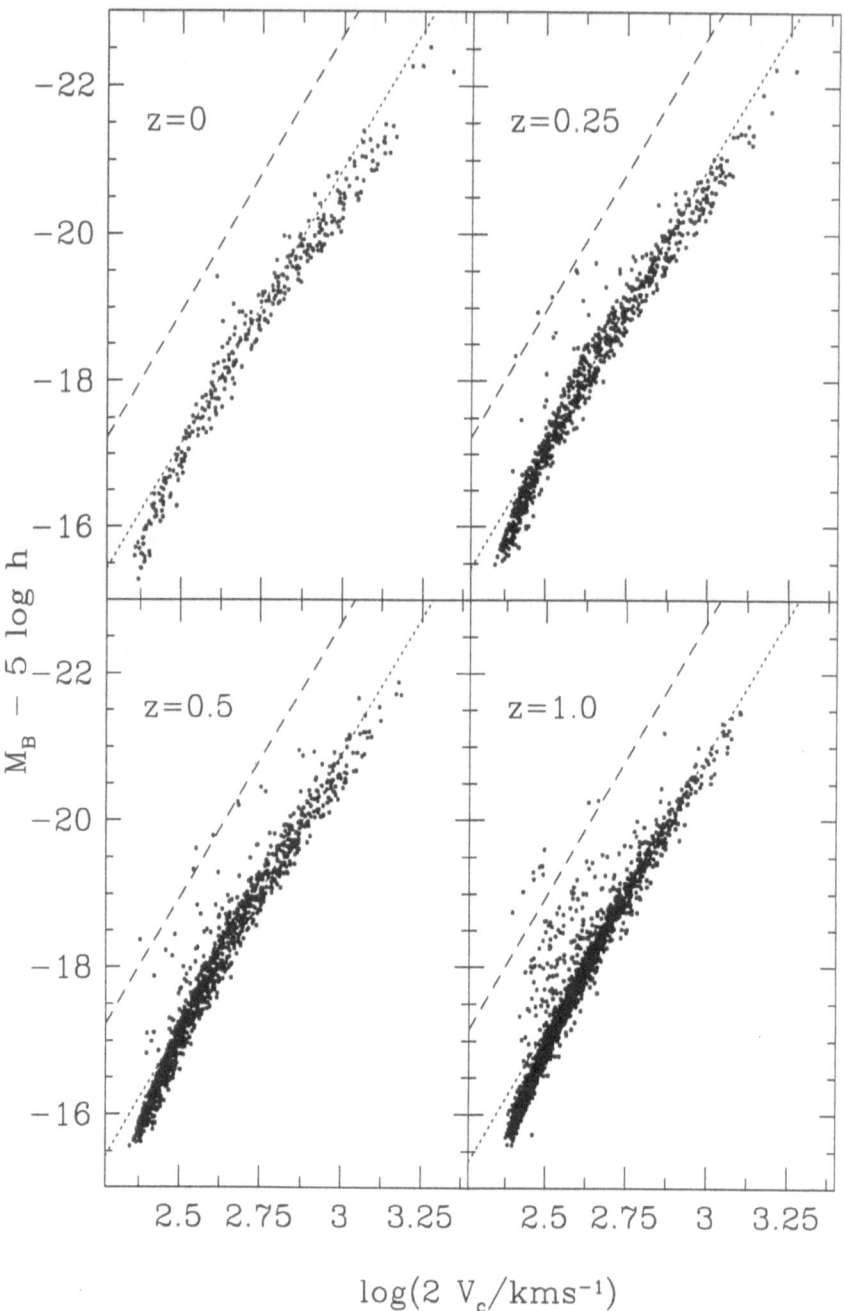

Fig. 2. The Tully-Fisher relation in the rest frame B band at different redshifts. The dashed line is a fit to the local data from Pierce & Tully (1992). The dotted line goes through the model prediction at redshift zero. These two lines are repeated in the subsequent plots.

(squares) are a compilation of new and published data from Young (1996). The triangles in Fig. 1a show a sample of elliptical galaxies from the Coma cluster from Lucey et al. (1991) that have been placed on the diagram by assuming an effective circular velocity in terms of the observed velocity dispersion $V_c = \sqrt{3}/1.1\sigma_{1D}$. The zeropoint offset can be expressed in terms of the galaxies being ~ 1.8 magnitudes too faint for the halo circular velocity or the halo V_c being $\sim 50\%$ too large for the central galaxy luminosity. In panel (b) we plot only the spiral galaxies, which are defined to have a I-band bulge to disk luminosity ratio less than 40%. Panel (c) shows galaxies bluer than $B - K = 3.5$, which is the median colour of the sample. Finally, in panel (d) we show spiral galaxies (as defined by their bulge to disk ratio) that have a cold gas fraction in excess of 5% of the total gas mass.

The evolution of the Tully-Fisher relation with redshift is controversial due in part to the small sample sizes and the different selection criteria used to define these samples (Bershady 1996). Bershady (1996) and Vogt et al. (1996) (whose observations extend to $z \sim 1$) find little evidence for a change in luminosity for a given circular velocity, whereas Rix et al. (1996) and Simard & Pritchet (1996) find a brightening of $1.5 - 2$ magnitudes by redshifts $z = 0.25 - 0.4$. The Tully-Fisher predictions of our model in the *rest*-frame B-band at different redshifts are shown in Fig. 2. The dashed line is a fit to the local data from Pierce & Tully (1992). The dotted line goes through the centre of the model prediction at redshift zero. These two fiducial lines are repeated in all the subsequent panels. Essentially no evolution in the zero point is seen in the models, though the scatter increases slightly with redshift.

4 The Colour Magnitude Relation for Cluster Ellipticals/SOs

The small scatter in the observed (U-V) colours of elliptical and S0 galaxies in cluster environments is a powerful constraint on the formation histories of these objects (Bower, Lucey & Ellis 1992). One interpretation is that no star formation has taken place within the past $8 - 12$ Gyr, which essentially means that the star formation in cluster ellipticals would have been complete as early as $z \sim 3$.

In the hierarchical picture, ellipticals are built up by a series of galaxy mergers, so one might naively expect to recover a much larger scatter than is observed in the colour-magnitude relation. Kauffmann (1996) and Baugh, Cole & Frenk (1996a) have demonstrated that this is not the case. Although elliptical galaxies tend to form more recently than $z \sim 3$ in these models, the formation process consists largely of the assembly of smaller fragments in which the star formation is mostly complete. Baugh, Cole & Frenk (1996a) show that the burst of star formation that accompanies a violent merger accounts for around only 5% of the final stellar mass of the elliptical when such a merger occurs between redshifts $0 < z < 0.5$.

Several observations have extended the colour-magnitude relation to high redshift clusters. In Fig. 3 we compare the predictions of our model with the

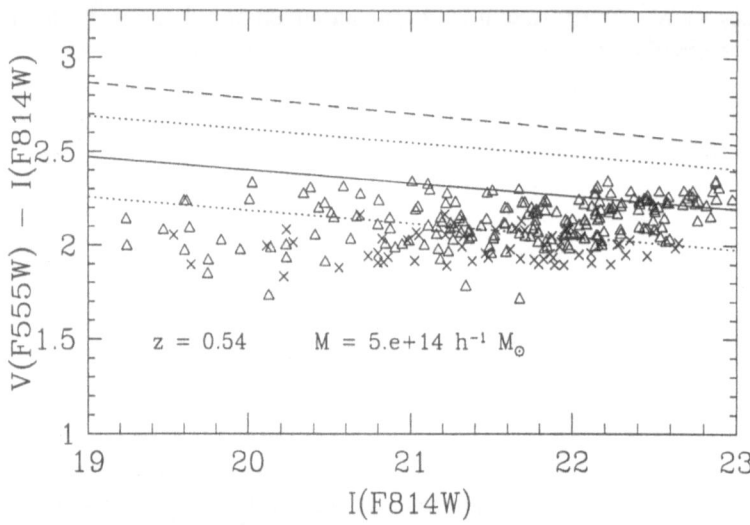

Fig. 3. The colour-magnitude relation for ellipticals and S0s in a cluster of mass $5 \times 10^{14} h^{-1} M_\odot$ that forms at $z = 0.54$. The solid line is a fit taken from Ellis et al. (1996) and the dotted lines show the 2σ errors. The dashed line is an extrapolation of the local colour-magnitude relation measured by Bower, Lucey & Ellis (1992), also taken from Ellis et al. The triangles show elliptical galaxies, defined by a bulge to disk ratio in the observer frame I band in excess of 60% and the crosses show S0 galaxies, which have a bulge to disk ratio between 40% and 60%.

observation of three high redshift clusters by Ellis et al. (1996). The zero point of the colour magnitude relation has shifted bluewards by roughly the same amount in both the data and the model. The scatter in the model colour magnitude relation is comparable to that seen in the data. We note however that the observed colour-magnitude relation has a small but significant slope, which is not seen in our model. We have ignored chemical enrichment in the model, using only a solar metallicity stellar population model.

5   The Epoch of Galaxy Formation

The discovery of Lyman break galaxies in ground based data by Steidel et al. (1996a) and in the Hubble Deep Field by Steidel et al. (1996b) and Madau et al. (1996) has revealed the existence of a population of star forming galaxies that are in place by redshifts $3 - 4.5$. Within the framework of the model discussed here, these are the first objects in which significant star formation takes place and herald the start of the epoch of galaxy formation.

Candidate high redshift galaxies are identified by deep imaging of a field in three passbands, e.g. $U(F300W)$, $B(F450W)$ and $I(F814W)$ in the case of the Hubble Deep Field. In this example, the 912Åbreak passes through the observed $U(F300W)$ passband at redshifts in the range $2.0 < z < 3.5$. The decrement in the flux in the U band is accentuated by absorption by intervening cold gas. Hence a high redshift galaxy can be identified as one for which the $U(F300W) - B(F450W)$ colour is red compared with the $B(F450W) - I(F814W)$ colour, which is blue if star formation is in progress.

The Hubble Deep Field was imaged in four passbands, so it is also possible to apply this colour selection technique to galaxies that drop out in the $B(F450W)$ band, which singles out galaxies with redshifts in the range $3.5 < z < 4.5$. The ground based observations, which use a different set of filters, select galaxies that lie at $3.0 < z < 3.5$.

We have made a detailed comparison of our model predictions for the abundance of high redshift galaxies and their properties for a range of cosmologies based on CDM (Baugh et al. 1997). We have used the same filters employed in the observations and have included the effects on the observed galaxy luminosity due to absorption by cold gas.

Here, we summarise the results for two models both of which assume a Miller-Scalo IMF; a standard CDM model ($\Omega_0 = 1, h = 0.5$) normalised to reproduce the abundance of rich clusters ($\sigma_8 = 0.67$) and a COBE normalised ' ΛCDM ' model with $\Omega_0 = 0.3, \Lambda_0 = 0.7$ and $h = 0.6$. The abundance of high redshift galaxies in these models expressed as a percentage of the total counts to the magnitude limit of the observations is compared with the data in Table 1.

Table 1. The abundance of high redshift galaxies per square degree. The final two columns give the number of galaxies from our model, for two different cosmologies, that satisfy the colour selection criteria applied to the observations, to the same magnitude limits.

redshift range	observed	standard CDM	ΛCDM
2.0 - 3.5	46000 ± 6000	82000	139000
3.0 - 3.5	1400 ± 300	1300	3600
3.5 - 4.5	9300 ± 2500	6200	14000

The star formation histories of a selection of present day galaxies that have a Lyman break progenitor at $z \sim 3.2$ are shown in Fig 4. The filled circle marks the branch of the star formation tree that corresponds to the Lyman break progenitor. The width of the shaded region represents the stellar mass of the fragments. The plots are all normalised to unit width at $z = 0$, with the actual stellar mass of the final galaxy labelled at the top of each panel. Also given is the absolute R-band magnitude. The Lyman break galaxies can end up in both elliptical and spiral galaxies at the present day, with a bias towards ellipticals.

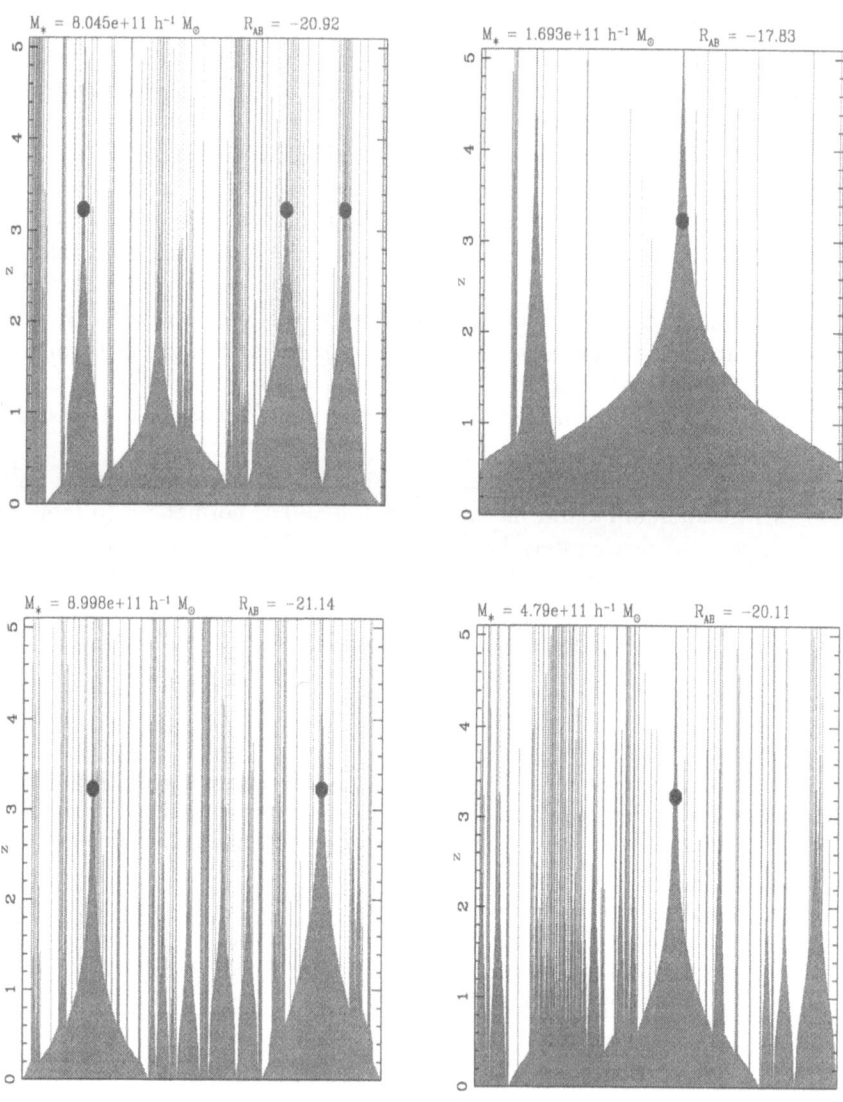

Fig. 4. The star formation history of present day galaxies that have a progenitor at $z \sim 3.2$ (shown by the circle) that matches the Steidel et al. selection criteria. The width of shaded regions is proportional to the stellar mass. The final stellar mass of each object is marked at the top of the plot along with the absolute R band magnitude.

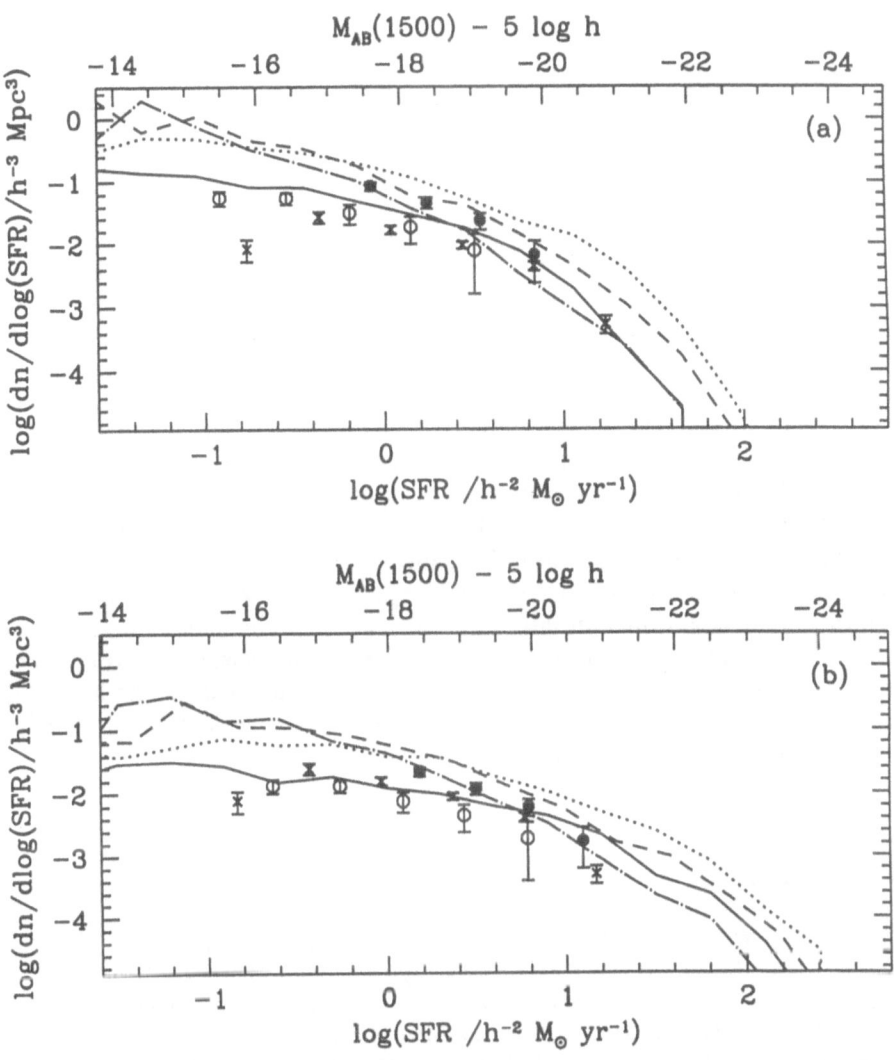

Fig. 5. The distribution of star formation rates (number of star forming systems per $\log_{10}(SFR)$ per comoving volume) at different redshifts, computed from our models and compared to observational data. Panel (a) shows the standard CDM model and panel (b) shows the ΛCDM model. The solid, dotted, dashed and dot-dashed lines show the model predictions for $z = 0, 1, 2.75$ and 4 respectively. The symbols with error bars are observational data as follows: the crosses are for $z = 0$ (Gallego et al. 1995), the filled circles for $\langle z \rangle = 2.75$ and the open circles for $\langle z \rangle = 4$ (Madau 1997). The observational data have been converted into total SFRs with the same IMF as in our models. The top scale shows L_{1500} ergs^{-1}Hz^{-1} in AB magnitude units: $M_{AB}(1500) = -2.5 \log L_{1500} + 51.60$.

In some cases, a galaxy may have two or more Lyman break progenitors that have merged.

The abundance of high redshift galaxies is very sensitive to the normalisation of the density fluctuations in the cosmological model and to the choice of stellar initial mass function (IMF). For example, in the standard CDM model discussed above, changing either the IMF from a Miller-Scalo to Scalo form or reducing the normalisation σ_8 from 0.67 to 0.50 would have the effect of reducing the number of Lyman break galaxies by an order of magnitude.

Regardless of the uncertainties just discussed, the Lyman-break galaxies of Steidel et al. correspond to the first objects in our models in which significant star formation is taking place. Only a small fraction of the final stellar component of the universe has formed by $z = 3.5$ in the models discussed here. The standard CDM and ΛCDM models have almost identical star formation histories and both have formed less than 5% of the total stellar population by $z = 3.5$. Thus, in the class of models we are considering, the redshift $z \simeq 3.5$ at which the Steidel et al. Lyman break galaxies are found is close to the onset of galaxy formation. Very few bright objects should exist beyond this redshift.

Finally, our predicted (differential) star formation rates at four different redshifts are shown in Figure 5. The upper panel gives results for the standard CDM model and the lower panel for the ΛCDM model. The lower abscissa is labelled by the actual star formation rate while the corresponding 1500 Å luminosities are given in the upper abscissa. In both cosmological models, the distribution of star formation rates has a similar shape at all times, but the rates are higher at $z = 1$ than at $z = 3$ or $z = 0$. The evolution in the star formation rate is relatively mild: over most of the range, the comoving abundance of galaxies varies by less than an order of magnitude between the peak at $z \simeq 1$ and the present.

References

Baugh, C.M., Cole, S., Frenk, C.S., (1996a): MNRAS, **283**, 1361–1378.
Baugh, C.M., Cole, S., Frenk, C.S., (1996b): MNRAS, **282**, L27–L32.
Baugh, C.M., Cole, S., Frenk, C.S., Lacey, C.G., (1997): MNRAS, submitted.
Bershady, M.A., (1996): astro-ph/9611036, to appear in *Dark Matter 1996*.
Bond, J.R., Cole, S., Efstathiou, G., Kaiser, N., (1991): ApJ, **379**, 440–460.
Bower, R.G., (1991): MNRAS, **248**, 332–352.
Bower, R.G., Lucey, J.R., Ellis, R.S., (1992): MNRAS, **254**, 601–613.
Charlot, S., Worthey, G., Bressan, (1996): ApJ, **457**, 625–644.
Cole, S. (1991): ApJ, **367**, 45–53.
Cole, S., Aragon-Salamanca, A., Frenk, C.S., Navarro, J.F., Zepf, S.E. (1994): MNRAS, **271**, 781–806.
Cowie, L.L., Songaila, A., Hu, E.M., Cohen, J.G., (1996): ApJ, **112**, 839–864.
Ellis, R.S., et al. (1996): ApJ, in press.
Evrard, A.E., (1990): ApJ, **363**, 349–366.
Frenk, C.S., Baugh, C.M., Cole, S., (1996): IAU Symposia 171, 247–254.
Frenk, C.S., Baugh, C.M., Cole, S., Lacey, C.G., (1997): astro-ph/9612109, to appear in *Dark Matter 1996* eds. M Persic & P. Salucci.

Gallego, J., et al. (1995): ApJ, **455**, L1.

Glazebrook, K., et al. (1995): MNRAS, **273**, 157–168.

Heyl, J.S., Cole, S., Frenk, C.S., Navarro, J.F., (1995): MNRAS, **274**, 755.

Lacey, C., Guiderdoni, B., Roccavolmerange, B., Silk, J. (1993): ApJ, **402**, 15–41.

Lacey, C., Cole, S., (1993): MNRAS, **262**, 627–649.

Lacey, C., Cole, S., (1996): MNRAS, **281**, 716–736.

Kauffmann, G., (1996): MNRAS, **281**, 487–492.

Kauffmann, G., White, S.D.M., Guiderdoni, B. (1993): MNRAS, **264**, 201–218.

Lucey, J.R., Guzman, R., Carter, D., Terlevich, R.J., (1991): MNRAS, **253**, 584–598.

Madau, P., et al. (1996): MNRAS, **283**, 1388–1405.

Madau, P., (1997): To appear in *Star Formation Near and Far* ; astro-ph/9612157

McGaugh, S.S., (1994): Nature, **367**, 538–541.

Navarro, J.F., Frenk, C.S., White, S.D.M. (1996): ApJ submitted, astro-ph/9611107

Navarro, J.F., Frenk, C.S., White, S.D.M. (1995): MNRAS, **275**, 56–66.

Pierce, M.J., Tully, R.B., (1992): ApJ, **387**, 47.

Press, W.H., Schechter, P., (1974): ApJ, **187**, 452.

Rix, H., Guhathakurta, P., Colless, M., Ing, K., (1996): MNRAS, in press.

Simard, L., Pritchet, C.J., (1996): ApJ, submitted, astro-ph/9606006.

Steidel, C.C., Giavalisco, M., Pettini, M., Dickinson, M., Adelberger, K.L., (1996a): ApJ, **462**, L17.

Steidel, C.C., Giavalisco, M., Dickinson, M., Adelberger, K.L., (1996b): AJ, **112**, 352.

Steinmetz, M., Muller, A., (1995): MNRAS, **276**, 549–562.

Tully, R.B., Fisher, J.R., (1977): A&A, **54**, 661.

Vogt, N.P., et al. (1996): ApJ, **465**, L15.

White, S.D.M., Frenk, C.S., (1991): ApJ, **379**, 52–79.

White, S.D.M., Rees, M.J., (1978): MNRAS, **183**, 341.

Young, P., (1996): PhD. Thesis, Univ. of Durham.

Dark Halo Scaling from Galaxy-Galaxy Lensing

Peter Schneider

Max-Planck-Institut für Astrophysik, Postfach 1523, D-85740 Garching, Germany

Abstract. The determination of the mass profile in galaxies is usually dependent on the presence of dynamical tracers like stars or neutral gas. The current view of the mass distribution in galaxies implies the existence of a dark halo which extends far beyond the radius at which dynamical tracers can be observed. Gravitational lensing by galaxies offers the possibility to trace the halo profile to radii beyond $\sim 30h^{-1}$ kpc and may become the most sensitive probe for the statistical properties of galaxy dark halos. First attempted more than 12 years ago, the first observational results are encouraging, and the upcoming wide-field cameras, super-seeing imaging instruments, and the 8-m class telescopes will provide a great opportunity for the study of the mass distribution of isolated and cluster galaxies.

1 Introduction

The potential of gravitational lensing to determine the mass and mass distribution of distant cosmic objects was realized (Zwicky 1937) long before the first gravitational lens system (Walsh, Carwell & Weymann 1979) was observed. With the currently available sample of multiply imaged QSOs and AGN (for a recent compilation, see Keeton & Kochanek 1996) it has been successfully attempted to constrain the mass distribution in galaxies, both on a case-by-case basis for those systems where a sufficiently large number of observational constraints are available (e.g., for quadruple systems, Rix, Schneider & Bahcall 1992; Keeton, Kochanek & Seljak 1996 and references therein, and for radio rings, Kochanek 1995; Wallington, Kochanek & Narayan 1996), and in a statistical sense (e.g., Kochanek 1996, and references therein). These studies have demonstrated that the mass distribution of galaxies has a small core size and that a dark halo is needed both for spiral galaxies and for early type galaxies (Maoz & Rix 1993).

The information about the mass distribution obtained from multiply imaged QSOs is restricted to scales on which the multiple images occur, which is typically less than $\sim 5h^{-1}$ kpc. In order to probe the mass distribution at larger radii, one needs to employ the weak gravitational lensing effect. This effect is now well established in the context of clusters of galaxies: from the (weak) distortion of the images of background galaxies, which can be detected (Tyson, Valdes & Wenk 1990) from a local average of galaxy ellipticities and assuming that galaxies are intrinsically oriented randomly, the mass profile of clusters can be reconstructed (Kaiser & Squires 1993; Fahlman et al. 1994; Schneider & Seitz 1995; Seitz & Schneider 1995, 1996; Squires et al. 1996; Seitz et al. 1996). As for galaxies, strong lensing effects like arcs and multiply imaged galaxies (see Fort & Mellier 1994 for a recent review, and Kneib et al. 1996 for a spectacular example) can probe the mass distribution only in the inner part of a cluster, whereas the outer parts are studied by weak lensing effects (e.g., Bonnet, Mellier & Fort 1994).

Whereas clusters are sufficiently massive to investigate them individually by weak lensing techniques, galaxies are too weak lenses for this to be possible, unless their characteristic velocity dispersion exceeds $\sim 400\,\mathrm{km/s}$. Their weak lensing effect can only be studied statistically, and the term 'galaxy-galaxy lensing' has been coined for such studies. The first attempt was carried out by Tyson et al. (1984). Whereas they used a large sample of galaxies to search for this effect, the use of photographic plates and the seeing prevented a significant detection. The discovery of galaxy-galaxy lensing was reported only recently (Brainerd, Blandford & Smail 1996; hereafter BBS), with several subsequent detections using HST data (Griffiths et al. 1996; Dell'Antonio & Tyson 1997; Hudson, this volume; R. Blandford, private communication). In this paper I will describe the basic methods for galaxy-galaxy lensing and summarize the observational results.

2   Tidal Distortion of Galaxy Images

A gravitational lens provides a map from the observer's sky to the undistorted sky, caused by the gravitational field of the deflector. The deflection angle is determined by the mass distribution of the lens, and the mapping from the source to the observable images depends in addition on the redshifts of the sources and the lens. Light bundles are not only deflected as a whole, but due to the tidal component of the gravitational potential, they are distorted. The image of a circular source will, in first order, be an ellipse. Thus, if all background galaxies were circular, the observable ellipticity of galaxy images would provide a direct means to measure the local tidal gravitational field of the lens. Since galaxies are intrinsically not circular, this simple method is impractical. However, since it can be assumed that the orientation of galaxies is random, a tidal field would cause a preferred orientation of the images. The distortions are such that galaxy images are preferentially aligned tangent to the mass center.

Denote by $\epsilon^{(s)}$ the complex intrinsic ellipticity of a galaxy (defined in terms of its second order brightness moments), and let ϵ be the corresponding complex ellipticity of the observed image. The variable ϵ is defined such that for an image with elliptical isophotes of axis ratio r and orientation φ of the major axis relative to a fixed reference direction, $\epsilon = (1 - r)/(1 + r)e^{2i\varphi}$. The locally linearized gravitational lens equation yields the transformation between $\epsilon^{(s)}$ and ϵ, which for the case of weak distortions reads

$$\epsilon = \epsilon^{(s)} + \gamma \,, \tag{1}$$

and γ is the shear or the tidal gravitational field of the lens (for definition, see, e.g., Schneider, Ehlers & Falco 1992) constructed from the trace-free part of the Hessian of the gravitational potential. The random orientation of intrinsic ellipticities then implies that the expectation value of ϵ equals γ. Since the dispersion of intrinsic galaxy ellipticities is much larger than the expected shear at $\sim 10''$ away from a typical galaxy, the shear cannot be detected around individual galaxies, and one has to refer to statistical methods. Note that for a galaxy modelled as a singular isothermal sphere, $|\gamma| = \theta_E/(2\theta)$, where θ is the angular separation from the center of the galaxy, and $\theta_E = (4\pi\sigma^2/c^2)(D_{ds}/D_s)$ is the Einstein angle of an isothermal sphere with velocity dispersion σ; D_{ds} and D_s are the (angular-diameter) distances from the lens and the observer to the source, respectively.

3   Detecting Galaxy-Galaxy Lensing

The transformation (1) implies that the images of background galaxies are preferentially oriented tangent to the line on the sky connecting the background galaxy image to its nearest foreground galaxy. Assume for a moment that only this nearest foreground galaxy acts on the light bundle from the background galaxy, and assume in addition that we can separate 'foreground' and 'background' galaxies (e.g., by apparent magnitude and/or colors). One can then define the angle ϑ between the major axis of the equivalent image ellipse (of course defined in terms of ϵ) and the line connecting foreground and background galaxy; owing to symmetry we can restrict $\vartheta \in [0, \pi/2]$. In the absence of lensing, the probability distribution $p(\vartheta)$ would be flat; including lensing, the tangential alignments predict that the distribution in skewed towards $\pi/2$. In fact,

$$p(\vartheta) = \frac{2}{\pi}\left[1 - \langle|\gamma|\rangle \left\langle \left|\epsilon^{(s)}\right|^{-1}\right\rangle \cos(2\vartheta)\right] \tag{2}$$

(BBS), where the shear average is taken over all foreground galaxies within the selected angular range where pairs are considered, and the ellipticity average implies that the effect is stronger the narrower the intrinsic ellipticity distribution is. In particular, in the presence of lensing, $\langle\vartheta\rangle > \pi/4$.

Tyson et al. (1984) attempted a measurement of $p(\vartheta)$ using a sample of 47000 background and 12000 foreground galaxies. We now know that the lensing effect is too weak to be detectable on deep images with seeing well above $1''$, and so the analysis of Tyson et al. resulted in an upper limit on the characteristic velocity dispersion σ_* of an L_* galaxy which was later revised upwards by Kovner & Milgrom (1987). From a deep CCD image of field size $\sim 9\overset{'}{.}6$ and final seeing of $0\overset{''}{.}87$, BBS selected 439 galaxies with $20 \le r \le 23$ ('foreground galaxies') and 506 ('background') galaxies with $23 \le r \le 24$. Although they have detected many more galaxies fainter than $r = 24$, the measurement of their ellipticities is not accurate enough to allow quantitative conclusions. Selecting an annulus with $5'' \le \theta \le 34''$, they obtained $p(\vartheta)$ from the resulting 3202 foreground-background pairs which is nicely fitted with the prediction (2); a uniform distribution in ϑ is excluded at the 99.9% confidence level. Griffiths et al. (1996) used data from the Hubble Space Telescope Medium Deep Survey (MDS), splitting their sample into foreground ($15 \le I \le 22$) and background ($22 \le I \le 26$) galaxies, and estimated the mean value of the angle ϑ. They obtained a clear galaxy-galaxy lensing signal from their 1600 foreground and 14000 background galaxies, and they have divided their foreground sample further into early and late-type galaxies; for both subsamples is $\langle \vartheta \rangle$ different from $\pi/4$ within ~ 10 half-light radii. Dell'Antonio & Tyson (1996) have detected galaxy-galaxy lensing in the Hubble Deep Field (HDF), using galaxies with $22 \leq I \leq 25$ as 'foreground' galaxies (110) and blue galaxies brighter than $B = 27.5$ as 'background' galaxies (645). A clear signal has been found within $5''$ of the foreground galaxies. Further detections of galaxy-galaxy lensing in the HDF have been reported by M. Hudson (thee proceedings) and R. Blandford (private communication); they find a signal on larger angular scales than reported by Dell'Antonio & Tyson (1996) which is possible due to the use of spectroscopic or photometric redshifts.

4   Quantitative Analysis of Galaxy-Galaxy Lensing

Whereas the statistically significant deviation of $p(\vartheta)$ from a flat distribution corresponds to the detection of galaxy-galaxy lensing (and BBS have run a large number of tests to ensure that this deviation is not due to any systematic effect), the interpretation of the amplitude of the cosine in (2) is not at all straight-forward. In the simplest case, where all lens galaxies are modeled as isother-mal spheres, this amplitude should decrease as θ^{-1} as the mean foreground-background separation decreases (as indeed in the case in the results of BBS). The separation into foreground and background based on magnitude cuts only cannot be sharp, i.e., some of the 'background' galaxies will lie in front of the nearest 'foreground' galaxy. If the angular separation θ is not small relative to the mean separation of 'foreground' galaxies, more than one lens will contribute to the shear γ for each background galaxy. In addition, galaxies populate a fairly broad luminosity function: at fixed apparent magnitude, a galaxy may be either a luminous galaxy at relatively high redshift (and thus a strong lens), or a relatively weak galaxy at small redshift (and thus a weak lens).

In order to make quantitative progress, BBS introduced a parametrized model for the galaxy population. The mass profile of each galaxy was modelled by a truncated isothermal sphere with velocity dispersion σ (or, equivalently, ro-tational velocity $V = \sqrt{2}\sigma$), and a truncation radius s. A scaling relation of σ and s with luminosity was assumed, i.e., $\sigma/\sigma_* = (L/L_*)^{1/\eta}$, where η (chosen to be 4) plays the role of a Tully-Fisher-Faber-Jackson index, and $(s/s_*) = (L/L_*)^{1/2}$. These scalings then predict a constant mass-to-light ratio of galaxies, indepen-dent of L. An approximate relation between apparent r-magnitude and luminos-ity was used. Most importantly, a probability distribution for the redshift as a function of r-magnitude was used,

$$p(z; r) \propto z^2 \exp\left[-\left(\frac{z}{z_0}\right)^{\beta}\right] \qquad z_0 = a + b(r - 22) , \qquad (3)$$

with $\beta = 1.5$, and z_0 being a linear function of r magnitude. This probability distribution is in good agreement with the observed redshift distribution of the brighter $20 \leq r \leq 23$ galaxy subsample.

BBS conducted Monte Carlo (MC) simulations in the following manner: for each observed galaxy with $20 \leq r \leq 24$, a galaxy was assigned a random position on the CCD, and its redshift was drawn from the distribution (3). With this redshift and apparent magnitude, the luminosity of the galaxy is obtained from the flux-luminosity relation, and the previously mentioned scaling relations then fix σ and s. The galaxies were assigned a random intrinsic ellipticity drawn from a distribution similar to the observed one, and then sheared by the tidal field of all the foreground galaxies (here, the approximation is made that the total shear distorting a galaxy image can be obtained from the sum of the shears caused by the individual deflectors which is a valid approximation in the weak lensing regime). Then for each MC realization, the distribution $p(\vartheta)$ can be calculated, and by repeating the MC simulation (~ 1000 times), a fair representation of $p(\vartheta)$ and its standard deviation has been obtained. Binning the 'foreground-background' pairs into angular bins, the observed $p(\vartheta; \theta)$ can be compared with the MC results. The resulting χ^2-function in a parameter plane spanned by s_* and σ_* exhibits a characteristic shape; the contours of χ^2 are 'open' in the direction of increasing s_*. This implies that the data used by BBS cannot be used to determine an estimate for s_*; however, a lower limit on s_* can be derived, $s_* \gtrsim 20h^{-1}$ kpc at the 1-σ level, and $\sigma_* \gtrsim 10h^{-1}$ at the 2-σ level. On the other hand, σ_* can be determined fairly accurately; the 90% confidence interval is $100 \lesssim \sigma_*/(\text{km/s}) \lesssim 210$, with the best-fitting value being $\sigma_* \approx 155$ km/s.

Whereas the results quoted depend somewhat on the various assumptions entering the analysis, such as the assumed K-correction, the extrapolation of the mean redshift to fainter magnitudes, and the Tully-Fisher index η, BBS showed that variations of these parameters within reasonable limits do not affect the conclusions seriously.

The main conclusion from the Griffiths et al. (1996) study of the MDS is the strong evidence for an extended dark halo around spiral and elliptical galaxies; the dark halo has to extend to at least 10 half-light radii, and a constant M/L de Vaucouleurs mass profile is clearly ruled out. The small-angle galaxy-galaxy lensing study of Dell'Antonio & Tyson (1996) in the HDF is of course unable to put strong constraints on the halo size ($s_* \gtrsim 15h^{-1}$ kpc), but finds for the velocity dispersion of an L_*-galaxy a value of $\sigma = 185 \pm 30$ km/s (1-σ limits), well in agreement with the value of BBS.

5   Outlook

The full potential of galaxy-galaxy lensing as a probe for the mass distribution of galaxies and their scaling relations has not been explored yet. The MC simulations of BBS have distributed their simulated galaxy images randomly on the CCD. Schneider & Rix (1997; hereafter SR) have recently proposed a modified scheme to obtain fits to the observed image distortions with a parametrized galaxy model. Their approach estimates the ellipticity probability distribution for all (faint) galaxies in a given data field, using the same parametrization for the galaxy population as used in BBS, but keeping the observed positions of the galaxy fixed. With a MC integration over the redshift probability distribution of all observed galaxies, the probability distribution for the shear acting on a given galaxy can be determined, and together with the intrinsic ellipticity distribution, the probability density for the observed ellipticity is obtained. If $p^{(s)}(\epsilon^{(s)})$ is the isotropic intrinsic ellipticity distribution, the corresponding probability density for galaxy i then becomes

$$p_i(\epsilon_i) = \frac{1}{N_{\text{MC}}} \sum_{j=1}^{N_{\text{MC}}} p^{(s)} \left(\epsilon_i - \sum_k \gamma_{ik}^j \right) , \qquad (4)$$

where N_{MC} is the number of MC realizations of the galaxy redshifts, γ_{ik}^j is the shear caused by the k-th galaxy in the j-th MC realization on galaxy i, and the sum over k extends over all galaxies with redshift smaller than the redshift of galaxy i in the j-th MC realization and which lie in an annulus with $\theta_{\min} \le \theta \le \theta_{\max}$ of the galaxy i.

From that distribution, a likelihood function is defined, $\mathcal{L} = \prod p_i(\epsilon_i)$, and maximized with respect to the model parameters. As shown in SR with synthetic data, this maximum likelihood approach yields more accurate estimates of the model parameters σ_* and s_* than the approach used in BBS. A likelihood ratio test can then be used to check whether the best model parameters are indeed compatible with the observed galaxy distortions, or whether a different (and probably more sophisticated) parametrization is requested by the data.

On the observational side, progress is 'around the corner': The availability of wide-field panoramic cameras (such as the UH8K at the CFHT) will allow to obtain data samples which are far larger than those used by BBS. In fact, it is not required that the data are taken with the special purpose of lensing in mind, but all sufficiently deep high-resolution images which cover enough 'empty area' (i.e., away from targets like nearby galaxies or clusters) can be used and combined for the statistical analysis. Several wide-field imaging surveys, such as the SDSS and the EIS, are soon to be conducted; whereas these surveys are quite shallow (compared to the BBS data), they cover such an enormous solid angle that galaxy-galaxy lensing should be easily detectable in them. The advantage of a very wide shallow survey relative to a narrower deep survey is that most of the uncertainties, like K-correction, redshift distribution etc., are negligible, and that for each potential background galaxy, the nearest foreground galaxy is farther away. Therefore, one might hope that the wide and shallow strategy will allow to obtain a more sensitive upper bound on the size scale s_*. In his recent diploma thesis, T. Erben has estimated that the formal uncertainty of σ_* as obtained from the EIS will be less than 5%, and that secure upper and lower limits on s_* can be derived. In addition he demonstrated that the assumption of a spherical matter distribution for the lensing galaxies is not critical; if the lenses are elliptical, the estimate of σ_* and s_* is basically unaffected, but the errors would be underestimated somewhat. One can also check the hypothesis that the light distribution of galaxies is a good indicator for the orientation of a supposedly flattened dark halo.

Hence, we shall soon be able to obtain data sets of several 10000 galaxies with measured shapes. As shown in SR, such samples can not only be used to obtain accurate estimates of σ_* and s_*, but also other model assumptions can be tested. For example, the likelihood can be calculated as a function of η or the parameters a and b describing the evolution of mean redshift as a function of apparent magnitude, and fairly narrow confidence intervals for these parameters can be derived with $\sim 10^4$ galaxies.

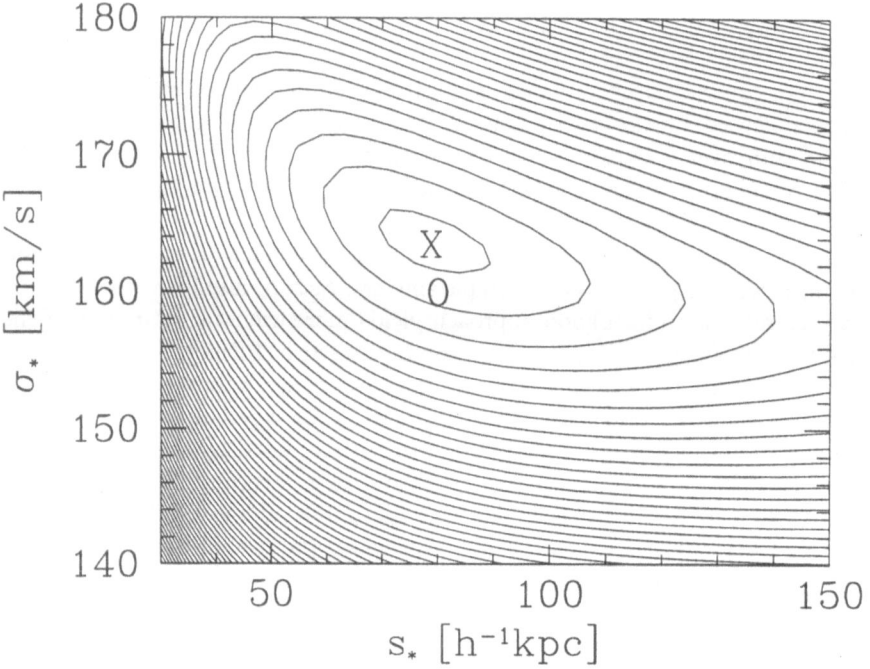

Fig. 1. Log-likelihood contours obtained from the SR method, applied to a synthetic data sample with the following specifications: A quadratic data field with 40′ sidelength, galaxies with $r \leq 24$ were 'observed' photometrically, and for those with $r \leq 23.5$, the shape can be measured. For the likelihood analysis, potential lenses in the angular range $5'' \leq \theta \leq 2'$ were included. In addition it has been assumed that the redshift of the galaxies can be estimated with a relative accuracy of 50%. Contour levels are $\Delta \ln \mathcal{L} = 0, -0.1, -0.5, -1, ...,$ O denotes the parameters used for the synthetic generation of the data, and X is the location of the maximum likelihood

Even better still, progress has recently been made in the estimate of redshifts from multi-color photometry (e.g., Connolly et al. 1995). In galaxy-galaxy lensing applications, knowing the precise value of the redshifts of galaxies in unnecessary; however, a rough estimate which allows to separate foreground from background galaxies would be particularly useful. In RS, a simulation was carried out where it was assumed that the redshifts of galaxies can be determined with a relative error of $\sim 50\%$. The corresponding log-likelihood contours are shown in Fig. 1; the galaxy sample used here contains about 10^4 galaxies with measured ellipticity. As can be seen, the assumed parameters used for the generation of a synthetic data set are very well reproduced and, in contrast to the same dataset without any redshift information, an upper limit on s_* can be obtained. Given that the only dynamical tracer of galactic potentials at distances $\gtrsim 30h^{-1}$ kpc are satellite galaxies (Zaritzky & White 1994), a size estimate from a data set as the one used in Fig. 1 would be most useful.

Galaxy-galaxy lensing is not restricted to the study of field galaxies; it can also be used to probe the mass distribution of galaxies in clusters (Natarajan & Kneib 1996; Geiger, this volume) and thus study the exciting question of whether the dark halo of cluster galaxies has been stripped off due to dynamical processes during cluster evolution.

As is true for all weak lensing application, work on galaxy-galaxy lensing requires high-quality imaging; in particular, the PSF of the frames must be well understood. The ESO telescopes can therefore play a leading role in this field, given the very high imaging quality of SUSI at the NTT and the expected imaging capability of FORS on the VLT, hopefully supplemented by another SUSI-like camera which makes use of periods of excellent seeing and which has superb intrinsic image quality. Since galaxy-galaxy lensing does not necessarily require very deep images (as, e.g., for the mass reconstruction of clusters), but depth can partly be compensated by field area (see comments made above), the new 8K camera for the 2.2m telescope can probably also be used for extensive galaxy-galaxy lensing studies. On the other hand, the development of imaging capabilities at other telescopes (such as the CFHT with its current UH8K camera, soon to be replaced by even larger arrays such as the 1 degree MEGACAM, and the wide-field camera on SUBARU) necessitates a quick move towards these programs in order to play a competitive role in that field.

This work was supported by the "Sonderforschungsbereich 375-95 für Astro–Teilchenphysik" der Deutschen Forschungsgemeinschaft.

References

Bonnet, H., Mellier, Y. & Fort, B. 1994, ApJ 427, L83.

Brainerd, T.G., Blandford, R.D. & Smail, I. 1996 ApJ 466, 623 BBS)

Connolly, A.J., Csabai, I., Szalay, A.S., Koo, D.C., Kron, R.G. & Munn, J.A. 1995, AJ 110, 2655.

Dell'Antonio, I. & Tyson, J.A. 1996, ApJ 473, L17.

Fahlman, G., Kaiser, N., Squires, G. & Woods, D. 1994, ApJ 437, 56.

Fort, B. & Mellier, Y. 1994, A&AR 5, 239.

Griffiths, R.E., Casertano, S., Im, M. & Ratnatunga, K.U. 1996, MNRAS 282, 1159.

Kaiser, N. & Squires, G. 1993, ApJ 404, 441.

Keeton, C.R. & Kochanek, C.S. 1996, IAU Symp. 173, 419.

Keeton, C.R. Kochanek, C.S. & Seljak, U. 1996, astro-ph/9610163.

Kneib, J.-P., Ellis, R.S., Smail, I., Couch, W.J. & Sharples, R.M. 1996, ApJ 471, 643.

Kochanek, C.S. 1995, ApJ 455, 559.

Kochanek, C.S. 1996, ApJ 466, 638.

Maoz, D. & Rix, H.-W. 1993, ApJ 416, 425.

Natarajan, P. & Kneib, J.-P. 1996, astro-ph/9609008.

Rix, H.-W., Schneider, D.P. & Bahcall, J.N. 1992, AJ 104, 959.

Schneider, P., Ehlers, J. & Falco, E.E. 1992, *Gravitational lenses*, Springer: New York.

Schneider, P. & Rix, H.-W. 1997, ApJ 474, 25 (SR).

Schneider, P. & Seitz, C. 1995, A&A 294, 411.

Seitz, C., Kneib, J.-P., Schneider, P. & Seitz, S. 1996, A&A 314, 707.

Seitz, C. & Schneider, P. 1995, A&A 297, 287.

Seitz, S. & Schneider, P. 1996, A&A 305, 383.

Squires, G. et al. 1996, ApJ 461, 572.

Tyson, J.A., Valdes, F., Jarvis, J.F. & Mills Jr., A.P. 1984, ApJ 281, L59.

Tyson, J.A., Valdes, F. & Wenk, R.A. 1990, ApJ 349, L1.

Wallington, S., Kochanek, C.S. & Narayan, R. 1996, ApJ 465, 64.

Walsh, D., Carswell, R.F. & Weymann, R.J. 1979, Nature 279, 381.

Zaritzky, D. & White, S.D.M. 1994, ApJ 435, 599.

Zwicky, F. 1937, Phys. Rev. 51, 290.

Constraining the Mass Distribution of Cluster Galaxies by Weak Lensing

Bernhard Geiger

Max-Planck-Institut für Astrophysik, Karl-Schwarzschild-Straße 1, Postfach 1523, D-85740 Garching bei München, Germany

Abstract. Analysing the weak lensing distortions of the images of faint background galaxies provides a means to constrain the mass distribution of cluster galaxies and potentially to test the extent of their dark matter halos as a function of the density of the environment. Here I describe simulations of observational data and present a maximum likelihood method to infer the average properties of an ensemble of cluster galaxies.

1 Introduction

Measurements of the rotation curves of spiral galaxies indicate that they are embedded in massive dark matter halos. The deflection of light rays through the gravitational action of mass concentrations, usually called gravitational lensing, provides a way to obtain information about the mass distribution of galaxies at radial distances from their centre where there are no more luminous test particles to probe the gravitational potential. The light deflection causes small distortions of the images of faint background galaxies. Recent statistical analyses (Brainerd et al. 1996, Griffiths et al. 1996) of these weak distortion effects suggest that the dark galaxy halos are indeed rather extended, as some popular theories of structure formation predict them to be. During the formation of galaxy clusters the extended halos of galaxies may be stripped off due to tidal forces of the cluster potential or during encounters with other galaxies. Ultimately the individual galaxy halos should merge and form a global cluster halo. In this contribution I discuss how this merging picture could be tested observationally by exploiting the weak lensing effects.

The distortions of the images of background galaxies produced by massive galaxy clusters are strong enough to allow a parameter-free reconstruction of the clusters' surface mass density, and several algorithms have been developed for this purpose (e.g. Kaiser and Squires 1993, Seitz and Schneider 1995, 1996). The smoothing length which has to be implemented in these techniques, however, is larger than galaxy scales, i.e., the amount of information available does not suffice to reconstruct cluster galaxies individually. Therefore, one has to superpose the effects of a large number of galaxies statistically in order to infer the average properties of an ensemble of galaxies.

Section 2 presents simulations of a galaxy cluster which are sufficiently realistic for the purposes of this work, and demonstrates how individual galaxies modify the distortion pattern of a smooth cluster mass distribution. Section 3 discusses a maximum likelihood method for constraining the mass distribution of cluster galaxies, and Sect. 4 presents results of the simulations. Finally, in Sect. 5 some suggestions for refining the simulations are mentioned, and observational prospects are discussed. A closely related work was recently published by Natarajan and Kneib (1997); in contrast to their maximum likelihood method, the mass profile of the cluster is not assumed to be known but is reconstructed from image distortions as mentioned above.

2 Simulations

2.1 Cluster and Cluster Galaxies

A galaxy cluster with a total mass of about $10^{15}h^{-1}$ M$_\odot$ located at a redshift of $z_d = 0.16$ was selected from numerical N-body simulations (Bartelmann et al. 1995). Within this paper a quadratic field of view with side length $10'$ is considered, which roughly corresponds to a physical size of $1h^{-1}$ Mpc at the cluster redshift. In order to populate the dark matter distribution of this cluster with galaxies the following requirements were specified:

1. The total mass-to-light ratio of the cluster was chosen to be $300h$ M$_\odot$/L$_\odot$.
2. Galaxy luminosities L were drawn from a Schechter function with canonical parameters (and a cutoff at $0.1\,L_*$).
3. Galaxy positions were randomly drawn from those of the N-body particles.

This procedure resulted in a rich cluster of 359 galaxies, 40 of which are brighter than L_*. For the mass distribution of the cluster galaxies, a simple truncated isothermal sphere (Brainerd et al. 1996) was used. The surface mass density Σ as a function of the projected radius ξ is given by

$$\Sigma(\xi) = \frac{\sigma^2}{2G\xi} \left(1 - \frac{\xi}{\sqrt{s^2 + \xi^2}} \right), \tag{1}$$

where the two parameters, velocity dispersion σ and cutoff radius s, were chosen as functions of the luminosity according to the following scaling relations:

$$\sigma = \sigma_* \left(\frac{L}{L_*} \right)^{1/\eta} \quad \text{and} \quad s = s_* \left(\frac{L}{L_*} \right)^{\nu}. \tag{2}$$

For the first of these relations, which is motivated by the observed Tully-Fisher and Faber-Jackson relations, a value of $\eta = 4$ was used for the scaling index and the velocity dispersion σ_* of an L_*-galaxy was fixed at 200 km/s. For simplicity, no distinction between spiral and elliptical galaxies was made. The scaling relation for the cutoff radius is more conjectural, and choosing $\nu = 0.5$ yields a mass-to-light ratio for the galaxies which is independent of luminosity. To test the method, two models were used for the cutoff radius. Choosing $s_* = 3.4h^{-1}$ kpc gives a total L_*-galaxy mass of $M_* = 10^{11}h^{-1}$ M$_\odot$, whereas an extended halo of $s_* = 34h^{-1}$ kpc results in $M_* = 10^{12}h^{-1}$ M$_\odot$. These galaxy mass models were added to the global cluster mass distribution, which was scaled such that the total mass of the system remains constant (see Fig. 1 left).

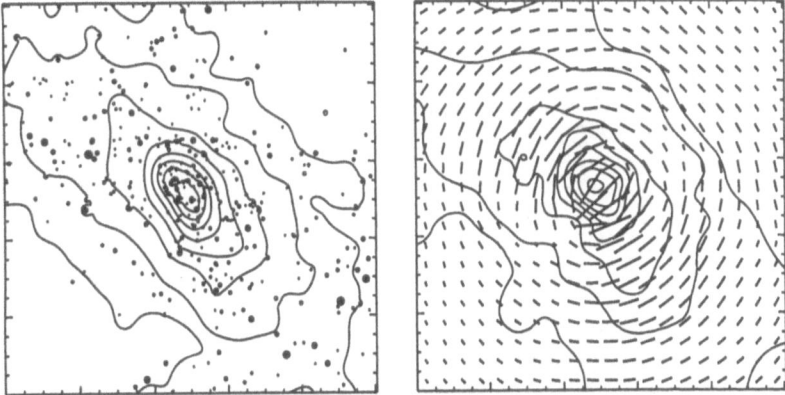

Fig. 1. Left: The mass distribution of the cluster including the cluster galaxies with $s_* = 34h^{-1}$ kpc. **Right:** The distortion pattern determined from the ellipticities of background galaxy images overlaid with the reconstructed cluster mass distribution. The field of view is $10'$ and the contours are $\kappa = 0.05, 0.1, 0.2, 0.3, 0.4, 0.5, 0.6$, and 0.7.

2.2 Distortion Effects and Background Galaxies

The lensing properties of the galaxy cluster are specified by the dimensionless surface mass density κ and the (complex) shear γ which are second derivatives of a common two-dimensional scalar potential. However, image distortions are only sensitive to the combined quantity $g = \gamma/(1 - \kappa)$. Figure 2 shows a map of $|g|$, which is a measure of the strength of the distortion effects on the images of background galaxies. In general, these distortions tend to be aligned tangentially towards the centre of mass concentrations. The figure illustrates the perturbing effects of the individual cluster galaxies. At their positions in a radial direction towards and away from the cluster centre the strength of the distortion is locally increased because the effects of the global cluster mass distribution and the cluster galaxy then act in the same direction. But in the direction tangential to the cluster centre the orientation of the galaxy contribution to the shear is perpendicular to the cluster's shear direction, and therefore these effects cancel out which leads to a reduction in the strength of the distortion effects.

Unfortunately Nature does not provide us with a continuous map of the lensing properties, but only with very noisy estimates of the parameter g at the discrete positions of background galaxy images. For these simulations, a random population of background galaxies was generated with a number density of $40/\text{arcmin}^2$, including a realistic intrinsic ellipticity distribution and a reasonable redshift distribution. Figure 1 (right) shows the gridded distortion pattern calculated from the 'observed' ellipticities of the background galaxy images by employing a suitable averaging procedure. This figure also displays the reconstruction of the mass distribution using a finite-field non-linear inversion method (Seitz and Schneider 1996).

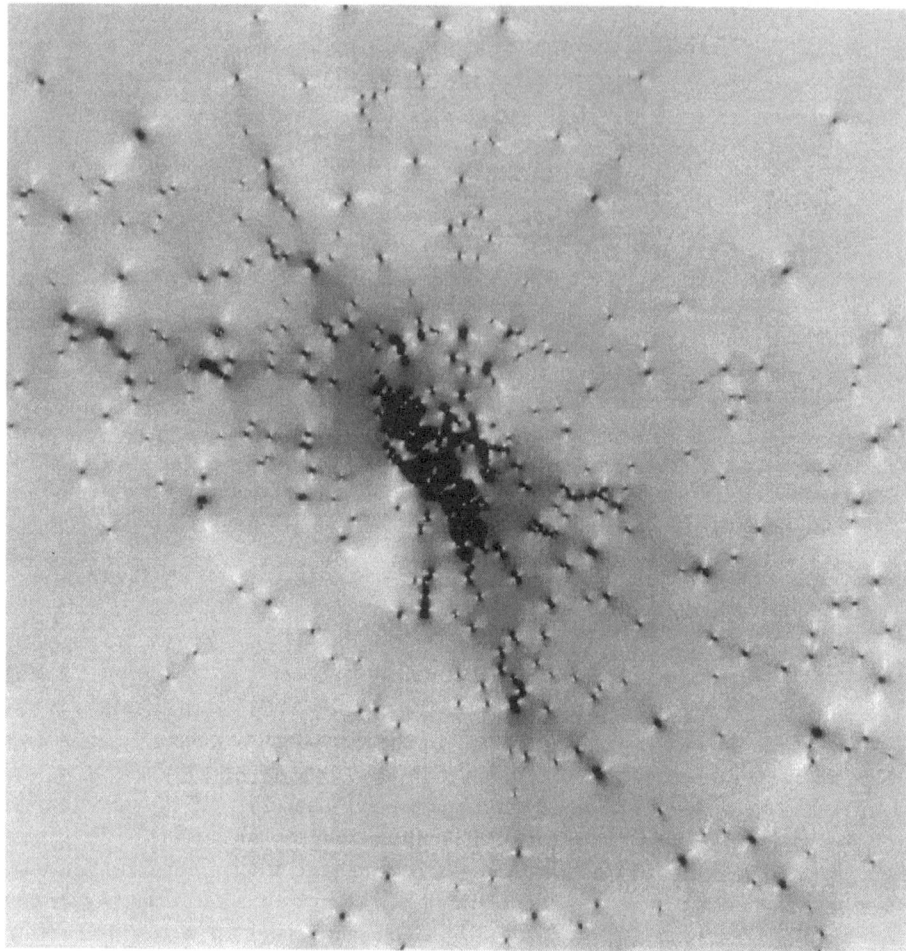

Fig. 2. The modulus $|g|$ of the reduced shear. This quantity is a measure for the strength of the distortion effects.

3   Method

In order to constrain the mass distribution of cluster galaxies a maximum likelihood method was developed which follows in part the prescription of Schneider and Rix (1997) for weak lensing by field galaxies. The image distortions are a consequence of the interplay between the effects of a global cluster potential and the perturbations due to individual galaxies. In addition to specifying a parametrized mass model for the galaxies it is, therefore, important to have an accurate description of the cluster mass distribution which is provided by the reconstruction mentioned above. As a model for the galaxy mass distribution I again used the truncated isothermal sphere (1). Of course, this model is appropriate for the simulated data used here, whereas one could argue that realistic galaxy halos in clusters might rather be flattened or completely irregular. However, this analysis is aimed at determining the average properties of an ensemble of galaxies which might still be reasonably described by a simple model with a characteristic scale and normalization as parameters. In order to add the information from galaxies with different luminosities, the scaling relations (2) were applied. Adding the mass models for each of the cluster galaxies to the cluster reconstruction then yields a model for the total mass distribution of the system as a function of the velocity dispersion $\sigma_\star$ and the cutoff radius $s_\star$ of an $L_\star$-galaxy. A complication which has to be taken into account when performing this procedure is the following. If the individual galaxies do have extended halos, the mass in galaxies constitutes a significant fraction of the total cluster mass ($\approx 30\%$ for the model with $s_\star = 34h^{-1}\,\mathrm{kpc}$). The cluster reconstruction is sensitive to the total mass and therefore it already includes the masses of the galaxies. This means that the additional mass added by the galaxy models has to be compensated in some way. This was done by simply scaling down the reconstruction appropriately or by subtracting surplus mass locally on scales of roughly 1′ at the position of cluster galaxies. The merits and limitations of this (*ad hoc*) procedure will be discussed in Sect. 4.

The total mass model constructed in this way determines the values for the lensing parameters κ and γ at the position of each background galaxy image. The strength of the lensing effect also depends on the distance of the sources, and in the following the symbols κ_∞ and γ_∞ are used to indicate the reference to (hypothetical) sources located at infinite redshift. The probability density $p_\epsilon(\epsilon\,|\,\kappa_\infty, \gamma_\infty)$ for observing the (complex) image ellipticity ϵ if the source is lensed by the specified mass model is given by

$$p_\epsilon(\epsilon\,|\,\kappa_\infty, \gamma_\infty) = \int\limits_0^\infty dz\, p_z(z)\, p_{\epsilon_s}(\epsilon_s(\epsilon\,|\,\kappa_\infty, \gamma_\infty, z)) \left|\frac{\mathrm{d}^2\epsilon_s}{\mathrm{d}^2\epsilon}\right| (\epsilon\,|\,\kappa_\infty, \gamma_\infty, z)\,. \quad (3)$$

To calculate this probability, it is necessary to know the intrinsic ellipticity distribution $p_{\epsilon_s}(\epsilon_s)$ of the sources, which can be determined from 'empty' fields, and an estimate for the redshift distribution $p_z(z)$. In addition, non-linear properties of the lens mapping have to be taken into account, and the last term under the integral is the Jacobian determinant for the transformation of image ellipticities ϵ to source ellipticities ϵ_s. The likelihood function $\mathcal{L}$ is defined as the product of the probability densities of the actually measured ellipticities ϵ_i of all the background galaxy images:

$$\mathcal{L} = \prod_i p(\epsilon_i)\,. \quad (4)$$

The best-fit galaxy mass distribution and confidence regions can then be found by maximizing this likelihood function with respect to the parameters of the model. The logarithm of the likelihood function is denoted as $l = \ln \mathcal{L}$, and in the next section contour plots of $\Delta l = l - l_{\mathrm{Max}}$ as a function of the model parameters will be presented.

4   Results

4.1   Velocity Dispersion and Cutoff Radius

Figure 3 (left) shows the result of the maximum likelihood analysis for the input galaxy model with a small cutoff radius of $s_\star = 3.4h^{-1}$ kpc. For this plot a total number of 3978 background galaxies from the entire field of view were included in the calculation of the likelihood function. The confidence region closely follows the dashed line of models with equal mass within a (projected) radius of $6h^{-1}$ kpc, which means that this is the quantity which can be determined best with this lensing method. However, the velocity dispersion and the cutoff radius individually cannot be well determined from the data. This result is not surprising because all background galaxy images which are located closer to cluster galaxies than roughly this distance of $6h^{-1}$ kpc have been excluded from the analysis as they will be outshone by the light of the cluster galaxy, and so there is no information available about the distribution of the mass inside this radius. But still it is possible to obtain tighter limits on the cutoff radius by including *a priori* knowledge on the velocity dispersion $\sigma_\star$. If we believe that the measured velocity dispersions of elliptical galaxies or the rotational velocities of spirals (divided by a factor of $\sqrt{2}$) represent the same quantity as the parameter σ of the dark matter halo model, we can include this knowledge into the likelihood function. Figure 3 (right) demonstrates the results after adding the prior information of $\sigma_\star = 200 \pm 20$ km/s. In this case very interesting limits on $s_\star$ can be achieved.

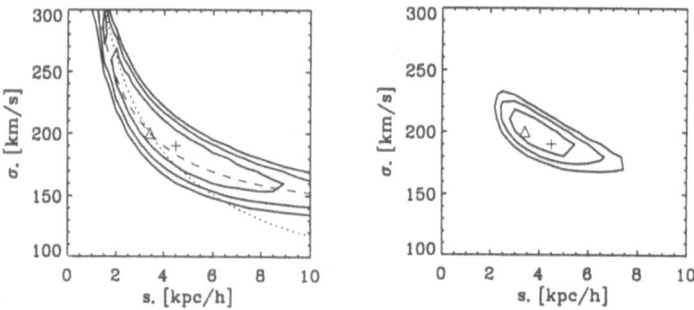

Fig. 3. The logarithm of the likelihood as a function of the velocity dispersion $\sigma_\star$ and the cutoff radius $s_\star$. The contours are $\Delta l = -1, -2, -3$. The triangles denote the input values and the crosses mark the maximum of the likelihood function. **Left:** Only including information provided by the lensing analysis. The dotted line connects models with equal total mass and along the dashed line the mass within a projected radius of $6h^{-1}$ kpc is constant. **Right:** The likelihood contours after adding the prior information of $\sigma_\star = 200 \pm 20$ km/s.

In Fig. 4 the same data is divided into two independent subsets according to the location of the background galaxy images. The left panel shows the contours of the logarithm of the likelihood function (without prior information) using all images (3714) whose distances from the cluster centre exceed $1'.5$ and in the right panel all images (264) within this limit were used. The number of cluster galaxies which are located in these areas are 265 and 94, respectively. The figure shows that the far fewer images in the centre provide almost the same amount of information as the numerous images in the outskirts of the cluster. The reasons for this are the higher cluster galaxy density in the centre and the significant enhancement of the distortion effects of individual cluster galaxies due to the underlying cluster mass distribution. Hence it is feasible to test a possible dependence of the extent of galaxy dark matter halos on the density of the environment by binning the data appropriately.

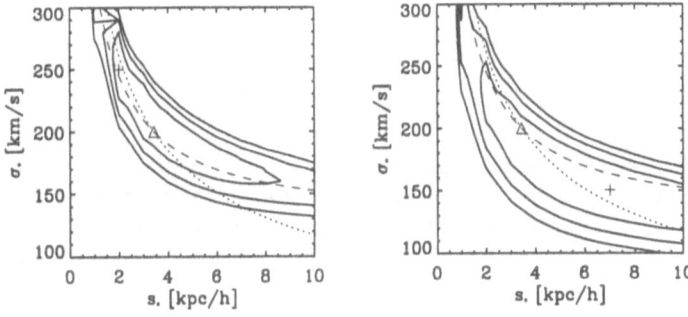

Fig. 4. The same as Fig. 3 (left) but after dividing the data set into background galaxies with a distance from the cluster centre larger than $1'.5$ (**left**) and less than $1'.5$ (**right**).

Figure 5 displays the results for the same realization of background galaxies as above – but here a large input value was used for the cutoff radius of the cluster galaxies ($s_\ast = 34h^{-1}$ kpc). The left plot, calculated from images in the outer region of the cluster, shows that in this case the velocity dispersion can be reasonably well determined, whereas the lensing effects are less sensitive to the radial extent of the mass distribution. Nevertheless a robust lower limit of about $15h^{-1}$ kpc can be set for the cutoff radius, and so this model can be distinguished with high significance from the low-$s_\ast$ model used above. Figure 5 (right) reveals the problems of the method when it is applied to the images located in the cluster centre. The input values cannot be reproduced by the likelihood analysis in this case. The reason for this is the ambiguity introduced by the mass correction procedure referred to in Sect. 3. This problem does not show up for the input model with small cutoff radius because then the mass in galaxies only amounts to a few percent of the total mass, and the mass correction is not important. In the outer regions of the cluster, the problem is less severe, because the requirements for the accuracy of the cluster mass reconstruction are less stringent in the weak lensing regime when the surface mass density is low and so the method works there even when the galaxies are massive (Fig. 5 left). In the non-linear lensing regime of the cluster centre, however, an accurate description of the cluster mass distribution is essential to obtain reliable results.

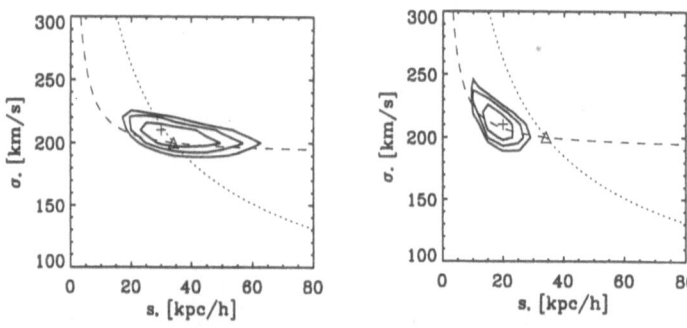

Fig. 5. The results of the likelihood analysis for a cluster galaxy input model with an extended dark matter halo. Note the change of scale on the x-axis compared to the previous figures. The binning of the data into information coming from images in the outer regions of the cluster (**left**) and the cluster centre (**right**) is the same as in Fig. 4.

In order to solve the problem becoming apparent in Fig. 5 (right) one might envisage employing a maximum likelihood reconstruction of the cluster mass distribution in the fashion of Bartelmann et al. (1996). In such a method the presence of cluster galaxies could be taken into account explicitly during the reconstruction process. For each set of parameters of the galaxy mass model, one can then determine the best representation of the underlying cluster mass distribution. Therefore, this approach would also be more satisfactory in a full maximum likelihood sense. Finally, I would like to remark that making the distinction between dark matter associated to galaxies or belonging to a global cluster mass distribution becomes somewhat artificial in the very centre of galaxy clusters when the physical distances between the galaxies become very small, and clearly the giant cD-galaxies residing in the centre of many clusters cannot be treated with the same formalism as ordinary cluster galaxies.

4.2   Scaling Parameters

In addition to $\sigma_\star$ and $s_\star$, a full description of the model for the galaxy mass distribution also requires to specify the power indices of the scaling relations (2). In the previous subsection the same values ($\eta = 4$, $\nu = 0.5$) that had been used to generate the data were taken for the likelihood analysis, but giving up that restriction and varying the scaling parameters within reasonable ranges does not affect the general conclusions drawn there.

Here the prospects for determining these scaling indices from the lensing analysis are briefly mentioned. For generating the data the galaxy model with the small cutoff radius has been adopted. Figure 6 depicts contour plots for the logarithm of the likelihood as a function of $1/\eta$ and σ_* (left) and ν and σ_* (right) including the background images from the whole field of view. Each time the two remaining parameters were fixed at the input values. The plots show that the constraints on the scaling indices are not particularly tight. In order to improve them it would be necessary to add the information from several galaxy clusters.

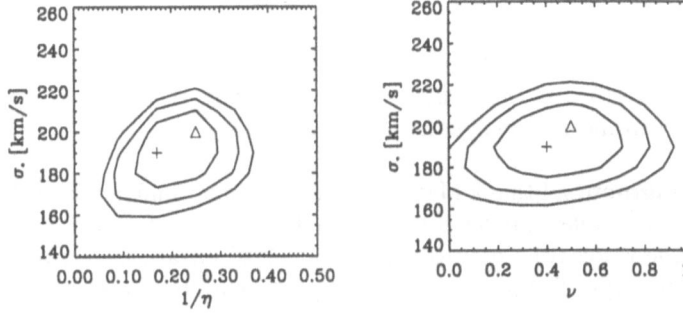

Fig. 6. The dependence of the likelihood function on the scaling indices η and ν. Again, the contours are $\Delta l = -1, -2, -3$, the triangles denote the input values and the crosses mark the maximum of the likelihood function.

5 Prospects

5.1 Simulations

There are several ways in which the simulations presented here could be refined. A distinction should be made between spiral and elliptical cluster galaxies because they require different normalizations for the velocity dispersion. An obvious thing to do is to explicitly include a dependence of the cutoff radius as a function of the (three-dimensional) density of the environment. One can then develop strategies to quantify this dependence and to assess the uncertainties introduced by projection effects. For this study it was assumed that cluster galaxies and background galaxies can be unambiguously distinguished by means of some colour criterion. The importance of this assumption can be tested by deliberately misinterpreting faint cluster galaxies as background galaxies. First investigations in this direction indicate that this is a minor problem, because the main part of the signal is contributed by the more massive cluster galaxies.

82   B. Geiger

5.2 Observations

Weak lensing is a challenging project from the observational point of view, because it necessitates measuring accurate image ellipticities for a large number of faint galaxies. To achieve the galaxy number density of $40/\text{arcmin}^2$ used in this simulations requires deep observations with a magnitude limit of about 25. The unique image quality of the (refurbished) Hubble Space Telescope allows to determine image ellipticities with a high accuracy, and in this respect it is the ideal instrument for weak lensing purposes. Its drawback, on the other hand, is the rather small field of view of its 'wide field' camera, and so time consuming mosaics of several images are required in order to completely cover a cluster which is located at a reasonable redshift. However, it has been shown in recent years that ground-based observations can be used for weak lensing studies as well, provided that they were taken in good seeing and with telescopes and instruments whose imaging properties are sufficiently well understood. Several observations – from space as well as from the ground – which are suitable for carrying out the kind of analysis described in this contribution are already available, and clearly this will be a rewarding project for the VLT-era.

Acknowledgments. I thank Matthias Bartelmann for making the N-body cluster simulation available, Peter Schneider and Stella Seitz for discussions and for providing the reconstruction algorithm, and Matthias and Peter for comments on the manuscript. This work was supported in part by the Sonderforschungsbereich 375-95 der Deutschen Forschungsgemeinschaft.

References

Bartelmann, M., Narayan, R., Seitz, S., Schneider, P. (1996), ApJ, 464, L115
Bartelmann, M., Steinmetz, M., Weiss, A. (1995), A&A, 297, 1
Brainerd, T.G., Blandford, R.D., Smail, I. (1996), ApJ, 466, 623
Griffiths, R.E., Casertano, S., Im, M., Ratnatunga, K.U. (1996), MNRAS, 282, 1159
Kaiser, N., Squires, G. (1993), ApJ, 404, 441
Natarajan, P., Kneib, J.-P. (1997), MNRAS, in press
Schneider, P., Rix, H.-W. (1997), ApJ, 474, 25
Seitz, C., Schneider, P. (1995), A&A, 297, 287
Seitz, S., Schneider, P. (1996), A&A, 305, 383

Galaxy–Galaxy Lensing in the Hubble Deep Field: the Halo Tully–Fisher Relation at High Redshift

Michael J. Hudson & Stephen Gwyn

Dept. of Physics & Astronomy, University of Victoria,
P.O. Box 3055, Victoria V8W 3P6 Canada

Abstract. We describe an analysis of galaxy–galaxy lensing in the Hubble Deep Field using photometric redshifts to determine distances and rest-frame ultraviolet luminosities of galaxies. We detect a lensing signal at the ∼ 98% confidence level from only 468 galaxies. Preliminary results suggest that the lensing galaxies in the HDF at a typical redshift of ∼ 1.2 follow a Tully–Fisher-like relation with dark matter halo circular velocities of 200 ± 50 km s^{-1}. The ultraviolet Tully–Fisher relation of the HDF galaxies has a similar slope and zero-point to that of spiral galaxies nearby. These results should be taken with caution as there remain a number of potentially important systematic effects which remain to be calibrated.

1 Introduction

Mass concentrations, such as clusters of galaxies, will distort the shapes of background galaxies by gravitational lensing. Recently, Brainerd, Blandford & Smail (1996) detected the lensing of background galaxies due to foreground galaxies using deep, ground-based, two-colour photometry. In principle, galaxy–galaxy lensing could allow the measurement of the density profiles of dark matter haloes, including their total extent and mass. In practice, however, dark matter haloes are usually assumed to be isothermal spheres (truncated and possibly with a core), and only one parameter is well constrained, namely the circular velocity V_c. Nevertheless, with redshift information this allows a measurement of a Tully–Fisher-like relation in which the circular velocity is that of the dark matter halo, not that of the stars or gas.

In this contribution, we describe ongoing work on galaxy–galaxy lensing in the *Hubble Deep Field* (HDF). The signature of galaxy–galaxy lensing in the HDF has been detected by Dell'Antonio & Tyson (1996, see also Hudson & Gwyn 1996). Dell'Antonio & Tyson found that galaxies with $22 < I < 25$ had haloes with typical velocity dispersions of 185 ± 35 km s^{-1}, equivalent to a circular velocity of 260 ± 45 km s^{-1}. In this work, we make use of the additional information provided by photometric redshifts (Gwyn & Hartwick 1996). This allows us to determine not only the redshifts of individual lens and source galaxies, but also the absolute magnitudes of the lens galaxies and hence, when combined with the circular velocities derived from the lensing analysis, their Tully–Fisher relation.

2 Data

2.1 Galaxy Shapes

The key data in any weak lensing analyses are the galaxy ellipticities. We have run the SExtractor code (Bertin & Arnouts 1996) on the HDF version 2 drizzled F814 images (chips 2-4 only) (Williams et al. 1996). SExtractor detects objects and measures their second moments (Q_{xx}, Q_{yy} and Q_{xy}) using all pixels above the detection threshold. The second moments can then converted to complex ellipticities as follows (Miralda-Escudé 1991):

$$\mathrm{Re}(\epsilon) = \frac{Q_{xx} - Q_{yy}}{Q_{xx} + Q_{yy}} \tag{1}$$

$$\mathrm{Im}(\epsilon) = \frac{2Q_{xy}}{Q_{xx} + Q_{yy}} \tag{2}$$

Thus a perfectly round object has $|\epsilon| = 0$ and highly elongated objects have $|\epsilon| = 1$. The position angle, ϕ, of a galaxy with respect to the x axis is given by $\tan^{-1}[\mathrm{Im}(\epsilon)/\mathrm{Re}(\epsilon)]/2$. For our HDF sample, the ellipticity distribution is well described by a two-dimensional Gaussian with radius $\sigma_\epsilon = 0.30$. These random galaxy shapes are the major source of noise in the analysis described below.

The top panels of Figure 1 show, for isolated galaxies, the scatter plots of the real and imaginary components of ϵ as a function of object size $\log(a)$, where the major axis a is measured in 0.04 arcsec pixels. The ellipticity ϵ appears to have roughly constant scatter for $\log(a) > 0.4$. Below this value, the HDF PSF is beginning to affect the small galaxies by making them rounder and hence reducing $|\epsilon|$. We adopt $\log(a) > 0.4$ ($a > 2.5$ pixels or 0.1 arcsec) as our selection cut. This size cut corresponds roughly to $I_{\mathrm{ST}} < 28$.

The bottom panels of Figure 1 show the difference, for isolated galaxies, between the real and imaginary components of ϵ measured on the F814 and F606 drizzled images. The scatter due to measurement errors, ~ 0.11 is considerably smaller than the total scatter found above for $\log(a) > 0.4$.

2.2 Photometric Redshifts

The photometric redshifts of galaxies in the HDF are described in Gwyn & Hartwick (1996). These photometric redshifts are based on 0.2 arcsec aperture photometry from the version 2 drizzled F300, F450, F606 and F814 filter (hereafter U_{ST}, B_{ST}, R_{ST} and I_{ST}) images. The method essentially consists of converting these magnitudes to a low-resolution spectral energy distribution (SED). A set of template spectra of all Hubble types and redshifts ranging from $z = 0$ to $z = 5$ is compiled. The redshifted template spectra are reduced to the passband averaged fluxes at the central wavelengths of the observed passbands. The SED derived from the observed magnitudes of each object is compared to each template spectrum in turn. The best matching spectrum, and hence the redshift, is then determined by minimizing χ^2. Some improvements have been made to the

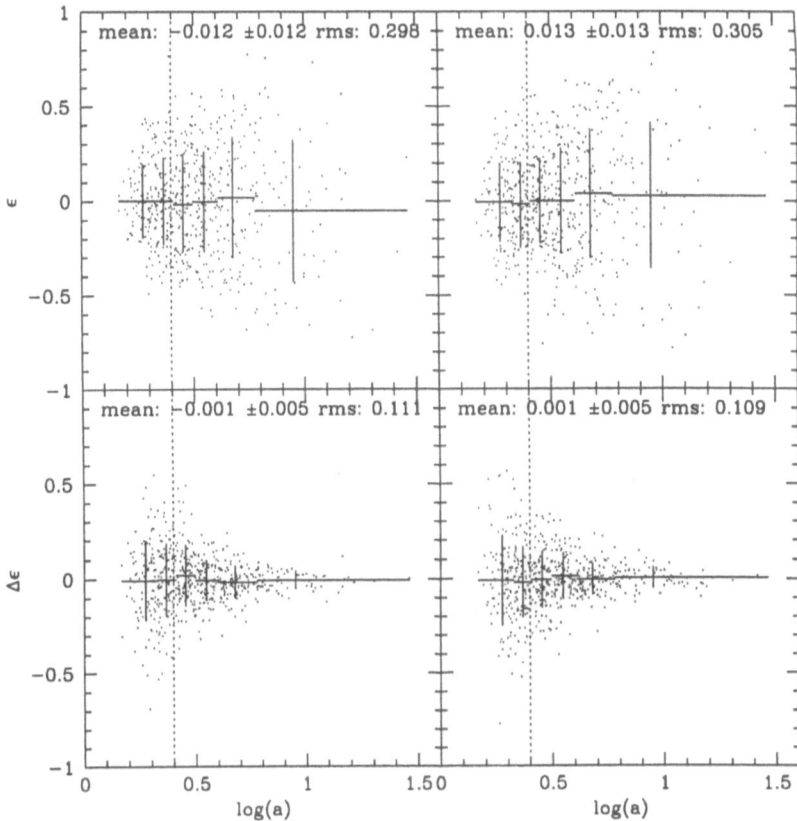

Fig. 1. Ellipticity as a function of size (a) in pixels. Top panels show the distribution of ellipticities (dots) as a function of size. The left-hand panel shows the real component and the right-hand panel shows the complex one. Crosses show the means and scatters for bins in $\log(a)$. The total means and scatters for the $\log(a) > 0.4$ sample are indicated at the top of each panel. Bottom panels show the difference in ellipticity between the F606 image and the F814 image. This is an indicator of the ellipticity measurement error. Note that only at $\log(a) \sim 0.4$ does the measurement error becomes a important contributor to the scatter in ϵ.

method of Gwyn & Hartwick (1996): the intergalactic Lyman blanketing corrections of Madau (1995) are now incorporated; and for low redshift ($z < 1.5$) the empirical SEDs of Coleman, Wu & Weedman (1980), as extended by Ferguson & McGaugh (1995), are used in preference over the evolving SEDs of Bruzual & Charlot (1993). Both of these corrections are found to improve the accuracy of the photometric redshifts when compared to spectroscopic redshifts. From such a comparison it is found that at low redshift ($z < 1.5$) the 1σ uncertainty in z is 0.16; at high redshift ($z > 1.5$) the uncertainty is 0.35. A comparison between photometric and spectroscopic redshifts is shown in Figure 2.

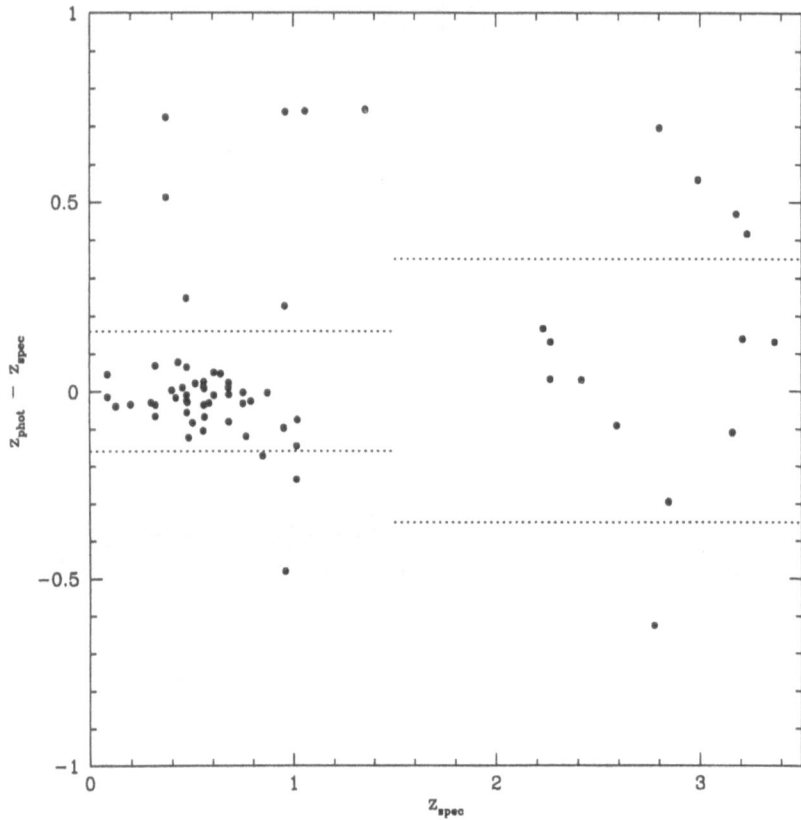

Fig. 2. Comparison between photometric redshifts and spectroscopic redshifts in the HDF. Note that two outliers with faint magnitudes $I \sim 28$ have been omitted from this plot. The dotted lines show the 1σ uncertainties discussed in the text.

We limit our sample to galaxies with 0.2 arcsec aperture magnitudes $I_{ST} < 28$ and exclude known stars. Thus our sample consists of 468 galaxies. The redshift histogram of the sample is shown in the upper panel of Figure 3.

We will show below that most of the lensing signal arises from galaxies with $0.5 \lesssim z \lesssim 2$. At $z \sim 2$, the central wavelength of the I band corresponds to the rest-frame ultraviolet. In order to minimize uncertainties in the k-corrections, rest-frame luminosities, L, are measured in a 400Å wide band centred on 2500Å. For all lens redshifts of interest, this luminosity can be obtained from a simple interpolation between the observed HDF colours.

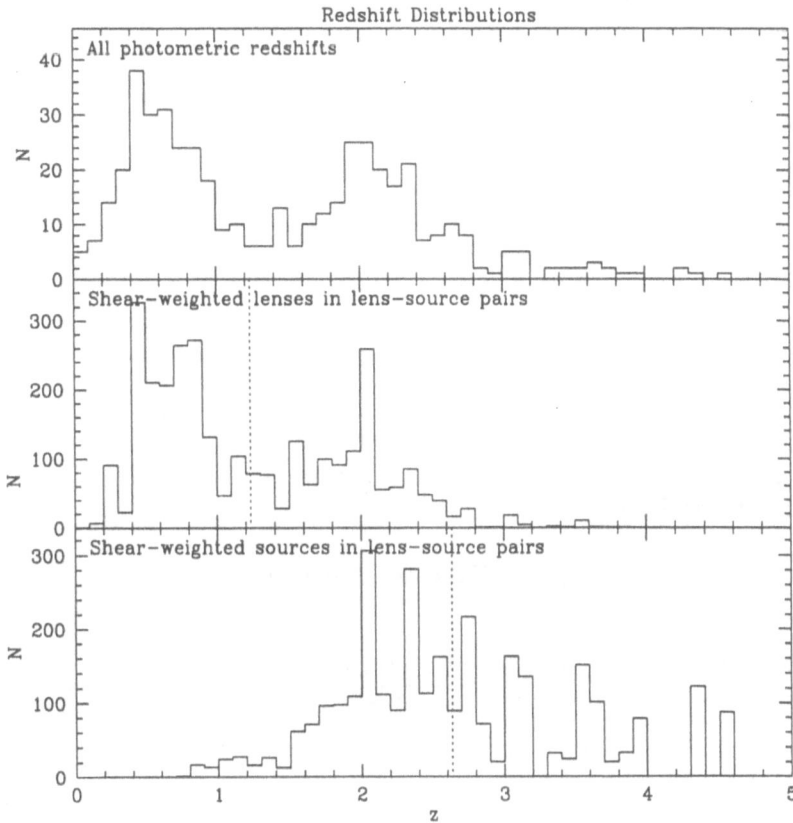

Fig. 3. Redshift histograms. The top panel shows redshift histogram of the sample. The middle and bottom panels show the shear-weighted histogram of lens and sources respectively in all lens–source pairs. For these panels, the vertical scale is arbitrary. The vertical dotted lines indicate the means of the two samples.

3   Method

In order to determine the masses of the haloes, we perform a maximum likelihood analysis similar to that of Schneider & Rix (1997). The outline of this method is as follows: for a given model of the lens galaxy mass density profiles, calculate the polarization of every source galaxies due to lensing by all galaxies in its foreground, thus obtain the true unlensed ellipticity and calculate the likelihood of this ellipticity given the ellipticity distribution determined above. A sum over all source galaxies yields a log likelihood which can be minimized with respect to the parameters of the mass model.

Our model of the dark matter haloes is as follows.

1. Galaxy haloes are truncated isothermal spheres, $\rho(r) \propto r^{-2}$ out to a truncation radius s, which have constant circular velocity V_c within s.
2. The lens galaxies follow a Tully–Fisher relation of the form

$$V_c = V_* \left(\frac{L}{L_f}\right)^\eta \tag{3}$$

where V_* is the circular velocity of a galaxy with fiducial luminosity L_f, and η is the slope of the relation. Note that L is the luminosity in the 2500Å band. For our halo Tully–Fisher relation, we set the fiducial luminosity to be $M_{2500} = -18$, which is typical of HDF galaxies at $z \sim 1.2$.
3. We scale the cutoff radius $s/s_* = (L/L_f)^{2\eta}$. However, from our fits we find that the fiducial scale s_* is not well constrained.

Given the circular velocity V_c and truncation radius, s, of galaxy, the polarization of a background galaxy with projected separation R is given by (Brainerd, Blandford & Smail 1996):

$$|p| = p_0(V_c, s, z_l, z_s)\, G(R/s) \tag{4}$$

where

$$p_0 = \frac{2\pi V_c^2 D_l D_{ls}}{s D_s c^2}, \tag{5}$$

where D_l, D_s and D_{ls} are the angular diameter distances to the lens, source and between lens and source, and

$$G(X) = \frac{(2+X)(1+X^2)^{1/2} - (2+X^2)}{X^2(1+X^2)^{1/2}}. \tag{6}$$

For $X \ll 1$, $G(X) \sim 1/X$, whereas for $X \gg 1$, $G(X) \sim 2/X^2$. It is convenient to write this as a complex quantity, $p = -|p|e^{2i\theta}$ where θ is the position angle of the line joining the lens and source galaxies. The negative sign accounts for the fact that the polarization is in the tangential direction.

The total polarization P of a source galaxy j is the linear sum over the contributions p_{ij} due to all foreground galaxies i:

$$P_j = \sum_i p(R_{ij}, V_i, s_i, z_i, z_j) \tag{7}$$

where the sum extends over all galaxies with $z_i < z_j$. In practice, we use only those pairs with a minimum redshift separation of 0.5 and projected separations at the lens in the range $10 < R < 100$ h^{-1} kpc. There are 9365 such lens-source pairs. Now to first order, the change in the ellipticity of a source galaxy due to lensing is just the polarization calculated above. We can thus obtain the true (unlensed) ellipticity by subtracting the polarization, $\epsilon_0 = \epsilon - P$. The likelihood is simply the probability of the corrected ellipticity given the measured distribution of ellipticities of galaxies in the HDF. A sum over all source galaxies yields a log likelihood

$$\log \mathcal{L} = \sum_j \left(-\frac{|\epsilon_j - P_j|^2}{2\sigma_\epsilon^2} \right) \qquad (8)$$

which is maximised by varying the free parameters V_*, s_* and η. Note that $2(\mathcal{L}_{max} - \mathcal{L})$ is distributed like χ^2 with the the number of degrees of freedom equal to the number of free model parameters.

4   Results

For each lens-source pair, we calculate a polarization p_{ij}. The middle panel of Fig. 3 shows the histogram of lens redshifts in all pairs weighted by p_{ij}. This shows that the lensing signal arises from lenses with a mean $z \sim 1.2$. The bottom panel shows the corresponding weighted histogram of source redshifts. The typical source redshift ranges from $1.8 \lesssim z \lesssim 3.5$.

Figure 4 shows the *preliminary* results of the maximum likelihood analysis expressed as contours of likelihood. The V_*-s_* plot is shown in Fig. 4a. It can be seen that s_* is not well constrained: the best fit is $s_* = 20\ h^{-1}$ kpc, but the 1σ range is from 5 to 100 h^{-1} kpc. The minimum is at $V_* = 200^{+30}_{-50}$ km s^{-1}. The V_*-η plot is shown in Fig. 4b. We find $\eta \sim 0.3 \pm 0.1$, in good agreement with canonical values of the B-band TF slope. The maximum likelihood has a $\Delta\chi^2 = -6.5$ with respect to the null hypothesis of no lensing ($V_* = 0$). For two degrees of freedom, this corresponds to a formal significance level $\sim 96\%$. If we fix $\eta = 0.3$, as observed locally, then the significance level of the one-parameter fit is $\sim 99\%$.

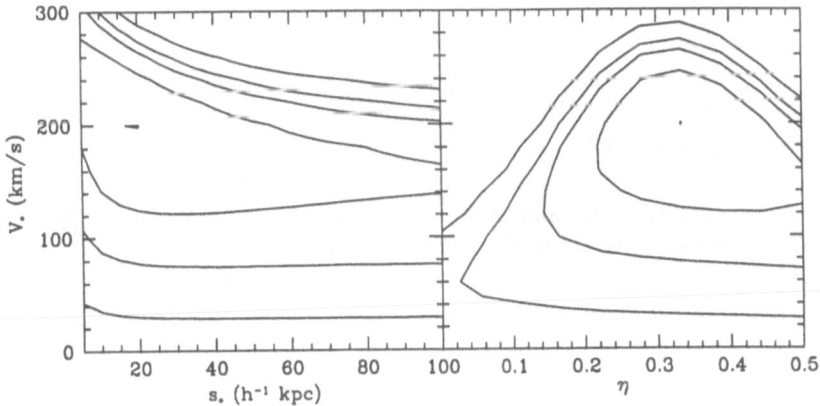

Fig. 4. Likelihood contours for V_* and s_* (right hand panel) and V_* and η (left hand panel). The smallest contour marks the location of the maximum, successive contours indicate the 68, 90, 95 and 99% confidence limits on two parameters jointly.

We have tested the significance of our results by randomizing the position angles of the source galaxies. We find, as expected, that for this control test in most cases the best-fitting model has $V_c = 0$ km s^{-1}, i.e. it assigns no mass to the lens galaxies. In $\sim 2\%$ of the control tests, we obtain a χ^2 minimum which is less than the observed χ^2, thus confirming that our result is significant at the $\sim 98\%$ confidence level.

Figure 5 compares the derived halo TF relation to the local ultraviolet TF relation of ellipticals (using velocity dispersions from Faber et al. 1989 and setting $V_c = \sqrt{2}\sigma$) and spirals (circular velocities from Aaronson et al. 1982). The 2500Å magnitudes are from Rifatto et al. (1995). Note the large spread in ultraviolet magnitudes at a given circular velocity as a function of morphological type. The hatched region shows the region occupied by $z \sim 1.2 \pm 0.7$ lenses (here we have simply equated the circular velocity of the halo with the measured maximum disk circular velocity used in the local TF relation). If we change q_0 from 0.5 to 0.05, the hatched region shifts 1.5 mag brighter. It is clear that, at a given circular velocity, the lensing objects have 2500Å luminosities similar to those of normal ($T \sim 4$, or Sbc) spirals. Note, however, that selection effects bias our $I_{ST} < 28$ sample against ellipticals; even the brightest ellipticals will drop out of our sample at $z \sim 2$ because of their low ultraviolet luminosities. In contrast, late-type spirals are ~ 4 mag brighter in the rest-frame ultraviolet and hence in the I band at $z \sim 2$. It is possible that our lens galaxies are the predecessors of present-day ellipticals undergoing a major burst of star formation so that they are brighter than present-day ellipticals by ~ 2 mag in the ultraviolet.

5   As-Yet-Uncalibrated Systematics

While the signature of lensing has been detected with high confidence, converting this signal into circular velocities is more problematic. The results of our maximum likelihood analysis are preliminary and should not at this time be given too much weight. In particular, there remain a number of systematic effects which we have not yet calibrated.

1. The effect of systematic and random errors in the photometric redshifts is currently being tested with Monte Carlo simulations.
2. The non-zero PSF of the HST will tend to circularize faint source galaxies, thus reducing the effect of the polarization.
3. Scatter in the TF relation will tend to reduce the likelihoods because it will add a random component to the polarizations and hence the corrected ellipticities. This will bias the circular velocity.

Nevertheless, the relatively small *random* errors, from only 468 galaxies in a very small field, illustrate the potential of this method as a probe of galaxy properties at high redshift.

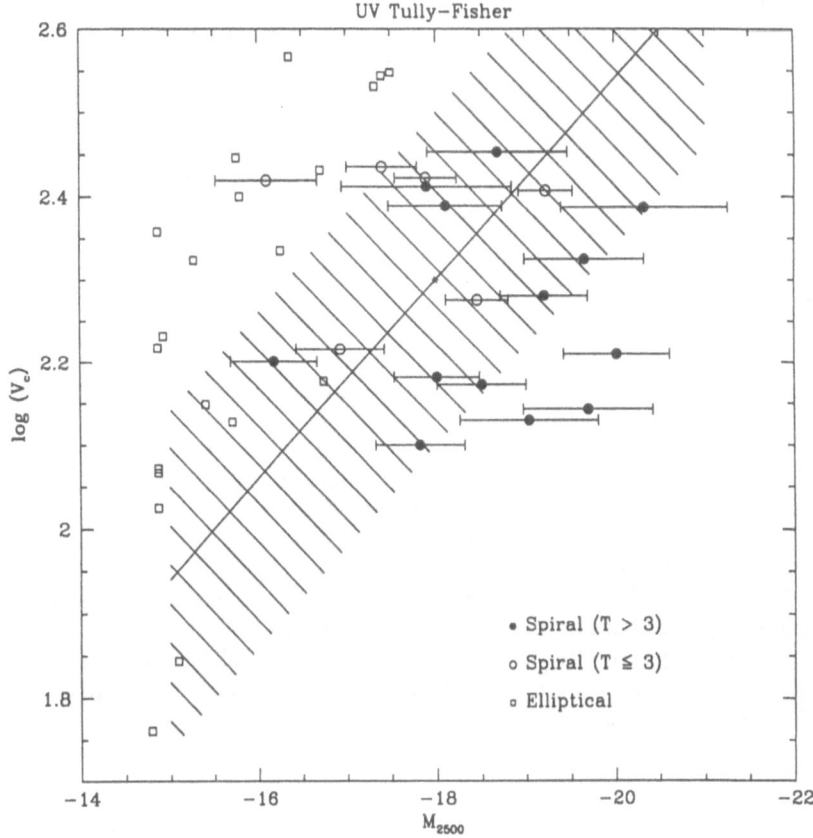

Fig. 5. The ultraviolet Tully–Fisher relation for nearby spirals (circles) and ellipticals (with $V_c = \sqrt{2}\sigma$) The magnitudes are measured in the ultraviolet at 2500Å. Note the large spread in ultraviolet magnitudes between ellipticals and late-type spirals. The mean TF relation for the lens galaxies at $z \sim 1.2$ in the HDF is indicated by the solid line. The hatched region indicates the magnitude range and an estimate of the range in circular velocity covered by the HDF lensing galaxies.

6 Summary

Our analysis of galaxy–galaxy lensing in the HDF is still in progress. Preliminary results suggest that galaxies at a typical $z \sim 1.2$ have circular velocities $V_c = 200 \pm 50$ km s^{-1} and follow an ultraviolet (2500 Å) Tully–Fisher relation similar to that of nearby spirals. These results should be taken with some caution since a number of potentially important systematics remain to be properly calibrated.

References

Aaronson, M., Tully, R. B., Fisher, J. R., Siegman, B., Huchra, J., Mould, J. R., Van Woerden, H., Goss, W. M., Chamaraux, P., Mebold, U. (1982): A catalog of infrared magnitudes and H I velocity widths for nearby galaxies. ApJS, **50**, 241–262.

Bertin, E., Arnouts, S. (1996): SExtractor: Software for source extraction. A&AS, **117**, 393–404

Brainerd, T. G., Blandford, R. D., Smail, I. (1996): Weak gravitational lensing by galaxies. ApJ, **466**, 623

Bruzual, A. G., Charlot, S. (1993): Spectral evolution of stellar populations using isochrone synthesis. ApJ, **405**, 538–553

Coleman, G. D., Wu, C.-C., Weedman, D. W. (1980): Colors and magnitudes predicted for high redshift galaxies. ApJS, **43**, 393–416

Dell'Antonio, I. P., Tyson, J. A. (1996): Galaxy dark matter: galaxy–galaxy lensing in the Hubble Deep Field. ApJ, **473**, 17L

Faber, S. M., Wegner, G,, Burstein, D., Davies, R. L., Dressler, A., Lynden-Bell, D., Terlevich, R. J. (1989): Spectroscopy and photometry of elliptical galaxies. VI - Sample selection and data summary. ApJS, **69**, 763–808

Ferguson, H. C., McGaugh, S. S. (1995): The contribution of low surface brightness galaxies to faint galaxy counts. ApJ, **440**, 470–470

Gwyn, S. D. J, Hartwick, F. D. A. (1996): The redshift distribution and luminosity functions of galaxies in the Hubble Deep Field. ApJ, **468**, L77

Hudson, M. J., Gwyn, S. (1996): Galaxy–galaxy gravitational lensing in the Hubble Deep Field. In *Proceedings of the XXXVII Herstmonceux Conference: HST and the High Redshift Universe*, in press

Madau, P. (1995): Radiative transfer in a clumpy universe: The colors of high-redshift galaxies. ApJ, **441**, 18–27

Miralda-Escudé, J. (1991): Gravitational lensing by clusters of galaxies – Constraining the mass distribution. ApJ, **370**, 1–14

Rifatto, A., Longo, G., Capaccioli, M. (1995): The UV properties of normal galaxies. III. Standard luminosity profiles and total magnitudes. A&AS, **114**, 527

Schneider, P., Rix, H.-W. (1997): Quantitative analysis of galaxy-galaxy lensing. ApJ, **474**, L25

Williams, R. E., Blacker, B., Dickinson, M., Van Dyke Dixon, W., Ferguson, H. C., Fruchter, A. S., Giavalisco, M., Gilliland, R. L., Heyer, I., Katsanis, R., Levay, Z., Lucas, R. A., Mcelroy, D. B., Petro, L., Postman, M. (1996): The Hubble Deep Field observations, data reduction, and galaxy photometry. AJ, **112**, 1335–1389

Part 2

EVOLUTION

The Fundamental-Plane and Tully-Fisher Relations Viewed in κ-Space

Ralf Bender[1], David Burstein[2] and Sandra M. Faber[3]

[1] Universitäts-Sternwarte, Scheinerstraße 1, D-81679 Munich
[2] Dept. of Physics and Astronomy, Arizona State University, Tempe, AZ 85287-1504
[3] UCO/Lick Observatory, University of California, Santa Cruz, CA 95064

1   Introduction

The structural properties of galaxies can be divided into two groups:
(a) the global scaling parameters, of which a complete set is given by a character-istic velocity (e.g. central velocity dispersion σ_o for ellipticals, rotation velocity V_{rot} for spirals), a characteristic radius, (like the effective radius r_e), and a mean density (like the mean effective surface brightness $\langle SB \rangle_e$), and
(b) the shape parameters, to which belong disk-to-bulge ratio D/B, mean ellip-ticity ε, velocity dispersion anisotropy $(v/\sigma)^*$, isophotal shape a_4/a etc.

   In this paper we analyse how galaxies are distributed within the 3-space of their basic scaling parameters and how this distribution is related to the Tully-Fisher and Fundamental-Plane relations.

2   Dynamically Hot Galaxies in κ-Space

Elliptical galaxies define a two-dimensional manifold in the 3-space of $(\sigma, r_e, \langle SB \rangle_e)$, the so-called fundamental plane (Djorgovski & Davis 1987, Dressler et al. 1987). Its defining relation is $\log r_e = 1.25 \log \sigma + 0.32 \langle SB \rangle_e + const.$ for the B-band (see Jørgensen et al. 1996). As has been shown by Bender, Burstein & Faber (1992, hereafter B²F), S0 bulges and the brighter dwarf ellipticals and dwarf spheroidals follow similar scaling relations. I.e., basically all dynamically hot galaxies (DHGs), except for the faintest dark matter dominated dwarfs lie close to the fundamental plane of ellipticals.

   It is now generally agreed that the fundamental plane is simply a consequence of the Virial theorem and the fact that most DHGs have similar mass-to-light ratios and close to homologous structure at a given luminosity (e.g. Faber et al. 1987, Djorgovski et al. 1989, B²F).

A simple orthogonal rotation in the observables-space[1]:

$$\kappa_1 = \tfrac{1}{\sqrt{2}} \log \sigma_o^2 + \tfrac{1}{\sqrt{2}} \log r_e$$

$$\kappa_2 = \tfrac{1}{\sqrt{6}} \log \sigma_o^2 - \tfrac{1}{\sqrt{6}} \log r_e + \tfrac{2}{\sqrt{6}} \log I_e$$

$$\kappa_3 = \tfrac{1}{\sqrt{3}} \log \sigma_o^2 - \tfrac{1}{\sqrt{3}} \log r_e - \tfrac{1}{\sqrt{3}} \log I_e$$

results in a coordinate system (named κ) which is especially well suited for the analysis of the properties of DHGs in general (see B^{2}F). Since DHGs are *virialized* systems, it follows that, *apart from structure constants*, κ_1 is proportional to the logarithm of the mass of the system, κ_2 proportional to $\log(I_e(M/L)^{1/3})$ (with M/L being the mass-to-light ratio within the luminous part of the object), and κ_3 proportional to $\log M/L$. As discussed in B^{2}F, the $\kappa_1 - \kappa_2$ projection is very close to a face-on projection of the fundamental plane of ellipticals and similar to a surface-brightness luminosity diagram, the $\kappa_1 - \kappa_3$ projection shows it edge-on and corresponds to a mass-to-light-ratio luminosity diagram. The edge-on view is mostly indicative of the degree of similarity between DHGs of a given luminosity, the face-on view mostly provides information about formation processes and evolution.

Figure 1 shows the distribution of DHGs in the κ-coordinate system following B^{2}F and Bender et al. (1994). Within the fundamental plane ($\kappa_1 - \kappa_2$ projection), a major sequence is defined by luminous ellipticals, bulges and most compacts, which together constitute a smooth continuum in κ-space ranging from objects like the compact companion of M 87, NGC 4486B, to central cluster galaxies, like M 87 itself. Several properties vary smoothly with mass along this sequence (though with significant scatter), including bulge-to-disk ratio, radio and X-ray properties, rotation, degree of velocity anisotropy, and peculiar kinematics. Core properties do so as well (Kormendy 1985, 1987), but a dichotomy between boxy and disky objects may exist (Nieto et al. 1991, Faber et al. 1997). These trends are consistent with the idea that the final mergers leading to larger galaxies were systematically more stellar (and less gaseous) than those in smaller galaxies (*i.e.*, with a 'gas/stellar' continuum, see B^{2}F). CDM predictions (full line in Fig. 1 shifted by a constant to include a constant collapse factor) can be made to fit the data only if dissipation *decreases* significantly with increasing mass.

A second major sequence in the $\kappa_1 - \kappa_2$ projection is comprised of dwarf ellipticals and dwarf spheroidals. These systems populate an elongated locus running at right angles to the main elliptical locus. Various evidence suggests that mass loss is a major factor in hot dwarf galaxies, but the dwarf sequence cannot be simply a mass-loss sequence as it has the wrong direction in κ-space. Hot dwarfs must have come from a range of progenitor galaxies that are not visible today as hot galaxies. The existence of a primarily one-dimensional hot

---

[1] σ is measured in km/s and r_e in kpc; for the surface brightness variable we use $\log I_e = -0.4((\mathrm{SB})_e - 27)$ in units of $L_{B,\odot}/pc^2$.

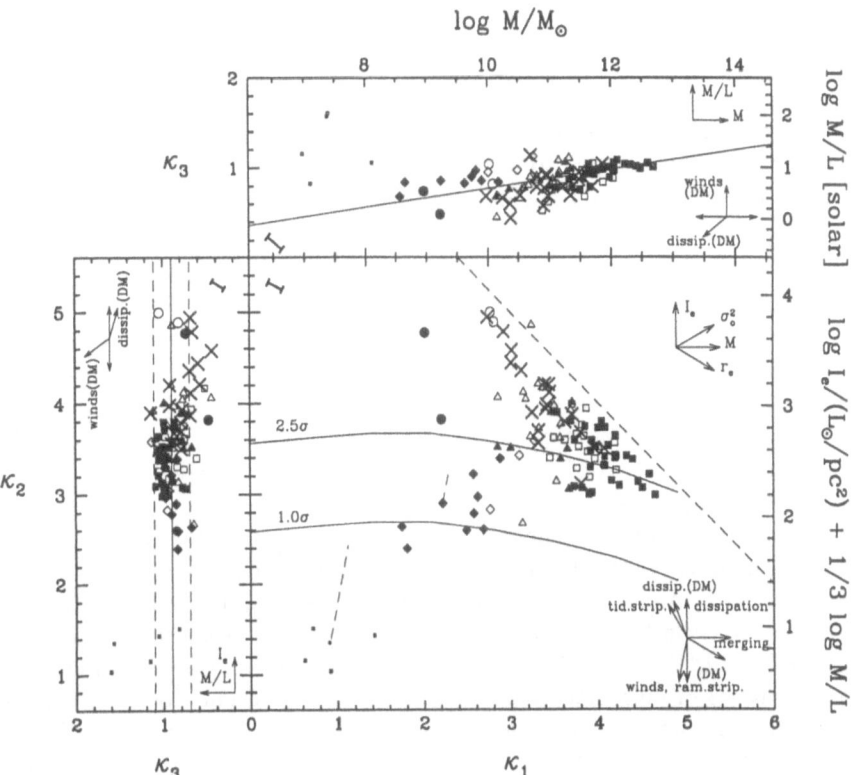

Fig. 1. The distribution of all types of Dynamically Hot Galaxies in κ-space, a special coordinate system within the 3-space of the global scaling variables (σ, r_e, $\langle SB \rangle_e$). Opposite of the κ-axes the corresponding physical parameters are given. Big squares denote giant ellipticals ($M_T < -20.5$ mag), triangles denote ellipticals of intermediate luminosity (-20.5 mag $< M_T < -18.5$ mag), circles denote compact galaxies and diamonds denote dwarf galaxies, all with known kinematics. Open big symbols are galaxies that are rotationally-flattened; filled big symbols are galaxies that have anisotropic kinematics. Bulges are represented by only one symbol (crosses), all those for which internal kinematics has been obtained appear to be rotationally flattened. The five small filled squares refer to the Fornax dwarf spheroidal (which is anisotropic) and four other dwarf spheroidal companions of the Galaxy for which no spatially-resolved kinematics are available. Error bars correspond to a distance uncertainty of $\pm 30\%$. In each panel, one set of arrows indicates the directions in which the basic global parameters of hot stellar systems increase. Another set of arrows indicates how the major processes move objects in κ space (tid. strip = tidal stripping, ram.strip = ram pressure stripping; (DM): if dark matter dominates). The range of directions for merging is approximate. The curved lines in the $\kappa_1 - \kappa_2$ plane indicate locations for galaxies of different mass but constant collapse factor arising from a CDM density fluctuation spectrum. The two lines correspond to different degrees of overdensity. The dashed lines emanating from two galaxy points show how baryon-dominated dwarfs would move due to wind-driven mass-loss. The dashed diagonal line in the $\kappa_1 - \kappa_2$ plane delineates the 'zone of exclusion', no galaxies are found above of it. The full line in the $\kappa_1 - \kappa_3$ projection represents the fundamental plane relation for elliptical galaxies ($\kappa_3 = 0.15\kappa_1 + 0.36$).

dwarf sequence is surprising, and may be at least partially an artifact of selection effects.

The most massive and the least massive hot galaxies are anisotropic, separated by a strip of galaxies of intermediate mass that are isotropic rotators. The origin of anisotropy in giants and dwarfs is probably different, with that of giants likely being due to stellar mergers, and that of dwarfs possibly being due to expansion following mass loss, or less isotropisation due to small collaps factors.

Because of distance uncertainties and uncertainties in the structure parameters of dwarfs and bulges, the $\kappa_3 - \kappa_1$ projection in Figure 1 does not show that the fundamental plane for ellipticals alone is in fact very tight. Jørgensen et al. (1996) find a typical rms-scatter of 20% in R_e. In the case of the Coma cluster, the scatter is about 10% (e.g., Saglia et al. 1993), which is quite surprising for such complex objects as ellipticals. Furthermore, the line-of-sight velocity distributions in ellipticals are found to be very similar (Bender, Saglia & Gerhard 1994). One explanation for this regularity could be the presence of at least some gas in all merging events Es may have undergone during their formation. Gas dissipates much more than stars during mergers causing it to form central density cusps which in turn lead to a de-population of box-orbits in favor of z-axis tubes (Barnes & Hernquist 1996, Merritt 1997). Therefore, gas can help to make the kinematics of ellipticals more 'boring' and bring them close to an oblate shape which is suggested by both kinematic and photometric studies (e.g. de Zeeuw & Franx 1991). If gas were not important in the formation of ellipticals, we would expect them to be predominantly prolate-triaxial with a large range of different axis ratios, like the dark matter halos formed in collisionless collapse (e.g. Frenk et al. 1988). On the other hand, if the gas fraction was always similar at a given mass then Es of similar luminosity may have similar phase space structure, despite of some residual peculiarity in velocity fields which may not be more than just 'frost on the cake'. In such a way, gas may be the origin of the tight fundamental plane.

The tilt of the fundamental plane of ellipticals relative to the simple virial relation could be due to a variety of effects; metallicity, age, dynamical structure, density profile, dark matter content etc. If the tilt is *not* due to changing structure constants and a true M/L effect, then it implies $M/L \propto L^{1/4}$. Population synthesis models (e.g. Worthey 1994) show that metallicity is unlikely to be the sole cause of the tilt. In the B-band the predicted slope is too small and in the K-band it should be negative, which is not observed (Pahre et al. 1995). Faber et al. (1995) and Worthey (1996) have argued that both line-strengths and mass-to-light ratios indicate that a systematic variation of mean ages with galaxy mass may account for the tilt, with small objects being systematically younger than big objects. This, however, seems at odds with the small scatter perpendicular to the fundamental plane — unless a metallicity-age conspiracy is at work. In the absence of a conspiracy, the small scatter constrains the variation of the ages of at least bright ellipticals to be smaller than about 15% (e.g., Bender, Burstein & Faber 1993, Renzini & Ciotti 1993). Such a small age variation is also implied by the small redshift evolution of the fundamental plane and the $Mg_b - \sigma$ relation (van Dokkum & Franx 1996, Bender, Ziegler & Bruzual 1996). Renzini & Ciotti (1993), Ciotti et al. (1996), and Graham & Colless (1997) have analysed the influence of a variation of structural parameters and dark matter on both tilt and scatter. They conclude that the tilt of the fundamental plane can be explained by the systematic variation of surface brightness profiles with luminosity.

3   Spirals and Irregulars in κ-Space

Figure 2 shows the distribution of intermediate and late-stype spiral and irregular galaxies in κ-space following Burstein et al. (1995) and Burstein et al. (1997, hereafter B^{2}FN). Early-type spirals are not shown in Figure 2 because their distribution basically falls on top of the one of elliptical galaxies. The photometric parameters refer to the B-band and are characteristic for the galaxies as a whole (no D/B decomposition was attempted). Characteristic velocities were taken to be rotation velocities divided by $\sqrt{2}$ (for sample definition, selection effects, more details on the definition of the parameters, see B^{2}FN). It is evident that the distribution of spirals and irregulars in κ-space is NOT fundamentally different from the one of DHGs. This is not completely surprising since disk galaxies are virialized systems and have mass-to-light ratios roughly similar to DHGs. In detail though, markable differences between spirals, irregulars and DHGs do exist. Clearly, the planes which are defined by intermediate and late-type galaxies are tilted in the $\kappa_2 - \kappa_3$ projection while the plane of DHGs isn't. This difference in the tilts causes the tightest scaling relations for DHGs and spirals to be different. In one case, it is the fundamental plane, in the other it's the Tully Fisher relation (or one very close to it).

A generalized Tully-Fisher relation can be expressed as $\log L = A_C \log V_{rot} +$ const. with L being the velocity and V_{rot} the maximum rotation velocity. We can re-formulate this in terms of the κ-parameters as:

$$\frac{4 - A_C}{\sqrt{2}} \kappa_1 - \frac{A_C}{\sqrt{6}} \kappa_2 - \frac{A_C + 6}{\sqrt{3}} \kappa_3 = const.,$$

which is the equation of a plane in κ-space. If $A_C = 4$, then

$$L \propto V_{rot}^4 \quad \Rightarrow \quad \kappa_3 = -\frac{1}{2.5\sqrt{2}} \kappa_2 + const.$$

while for $A_C = 3$ (the value being most appropriate for our sample of spirals), the plane is tilted in a way that in no κ-projection the scatter is minimized (which is what we also see in Figure 2).

The origin of the modest differences between the best-fitting planes of the different Hubble types is likely due to differences in disk-to-bulge ratio, star formation rate, dark matter content within the visible confines, and structure constants. At present, none of these parameters could be reliably identified as a dominant source for the differences.

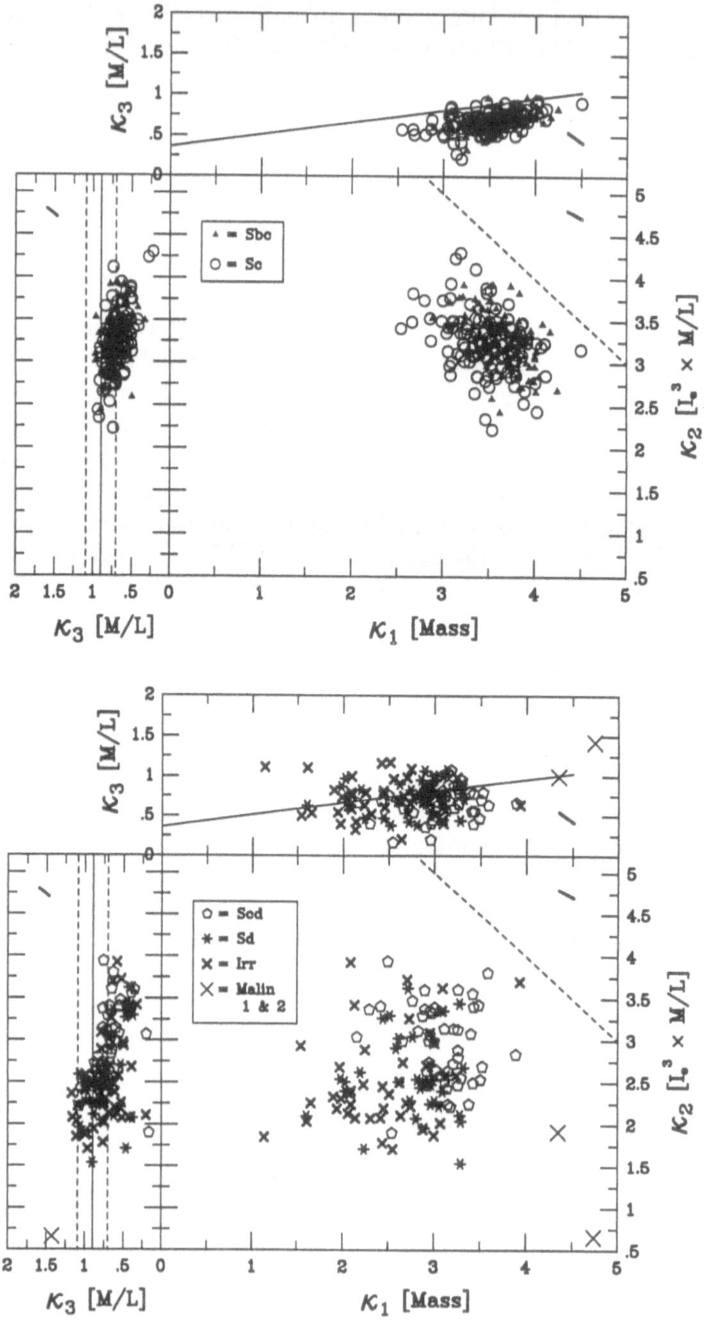

Fig. 2. The distribution of Hubble types Sbc to Sc (top) and Scd to Irr (bottom) in κ-space. The full and dashed lines correspond to those in Figure 1.

Looking at the close to face-on distribution of disk galaxies in the $\kappa_1 - \kappa_2$-projection shows that later types are in the mean located at larger and larger distances from the dashed diagonal line in Figure 2. This line defines the so-called 'zone-of-exclusion', above of which neither galaxies nor groups nor clusters are found (see B^2FN). It defines the highest possible density that can be reached at a given mass. The distance to zone-of-exclusion line is evidently correlated with disk-to-bulge ratio. Figure 2 also shows that late-type spirals and irregulars are *on average* less massive and have lower surface brightness than early-type spirals and ellipticals, a well known observational fact. The correlation between mass and surface brightness may show, as in the case of dwarf ellipticals and dwarf spheroidals, the increasing influence of galactic winds with decreasing mass. However, within each separate galaxy type, be it spirals, S0s or Es, we observe the opposite behaviour, low mass objects are more compact than high mass objects.

Further results and discussions on irregular and spirals in κ-space, as well as on galaxy groups and clusters, can be found in B^2FN.

4   Conclusions

We analysed the distribution of all types of galaxies in the 3-space of their basic scaling parameters (characteristic radius, density, and velocity). Because all galaxies are virialized systems and, except for the faintest dwarfs, have similar mass-to-light ratios, they define a highly flattened distribution within this 3-space. Different galaxy types populate almost parallel planes, with only slight offsets in mass-to-light ratio and small differences in the slopes over surface brightness and mass. The consequence of these differences is that the best-fitting plane for spirals and irregulars corresponds to the Tully-Fisher relation, the best-fitting plane for ellipticals to the fundamental plane or $D_n - \sigma$ relation. All Hubble types join smoothly within the common face-on view of the different planes. Disk-to-bulge ratio increases continuously with distance from the zone-of-exclusion which is defined by the highest possible density at a given mass and is delineated by ellipticals.

Acknowledgements. Thanks go to the many astronomers who obtained the data used here in a collective effort over many years. RB was supported by the Deutsche Forschungsgemeinschaft under SFB 375 and by the Max-Planck-Gesellschaft. DB acknowledges partial support from NSF AST90-16930, SMF from NSF AST95-29008 and from NAS-5-1661 to the WFPC1 IDT.

References

Barnes, J.E., Hernquist, L., 1996, ApJ 471, 115
Bender, R., Burstein, D., Faber, S., 1992, ApJ 399, 462
Bender, R., Burstein, D., Faber, S.M., 1994, in *Panchromatic view of galaxies*, eds. G. Hensler et al., Editions Frontieres, Gif-sur-Yvette

Bender, R., Saglia, R., Gerhard, O., 1994, MNRAS 269, 785

Bender, R., Ziegler, B., Bruzual, G., 1996, ApJ 463, L51

Burstein, D., Bender, R., Faber, S.M., Nolthenius, R., 1997, in *The World of galaxies II*, eds. G. Paturel & C. Petit, Gordon and breach, London

Burstein, D., Bender, R., Faber, S.M., Nolthenius, R., 1997, AJ submitted

Ciotti, L., Lanzoni, B., Renzini, A., 1996, MNRAS 282, 1

Djorgovski, S., de Carvalho, R.R., Han, M.S., 1989, ASP. Conf. Ser. 4, 329

Djorgovski, S., Davis, M., 1987, ApJ, 313, 59

van Dokkum, P., Franx, M., 1996, MNRAS 281, 985

Dressler, A., Faber, S.M., Burstein, D., Davies, R.L., Lynden-Bell, D., Terlevich, R.J., Wegner, G., 1987, ApJ 313, L37

Faber, S.M., Dressler, A., Davies, R.L., Burstein.D., Lynden-Bell, D., Terlevich, R. & Wegner, G., 1987, in *Nearly Normal Galaxies, From the Planck Time to the Present*, ed. S.M. Faber, p. 175, (Springer Verlag, New York)

Faber, S.M., Trager, S., Gonzalez, J., Worthey, G., 1995, in *Stellar Populations*, IAU Symp. 164, eds. P.C. van der Kruit & G. Gilmore, Kluwer Dordrecht

Faber, S.M., Tremaine, S., Ajhar, E., Byun, Y.-I., Dressler, A., Gebhardt, K., Grillmair, C., Kormendy, J., Lauer, T., Richstone, D., 1997, AJ, in press

Frenk, C., White, S.D.M., Davis, M., Efstathiou, G., 1988, ApJ 327, 507

Graham, A., Colless, M., 1997, MNRAS in press

Jørgensen, I., Franx, M., Kjærgaard, P., 1996, MNRAS, 280, 167

Kormendy, J., 1985, ApJ 292, L9

Kormendy, J., 1987, in *Structure and Dynamics of Elliptical Galaxies*, IAU Symp. 127, ed. T. de Zeeuw, Reidel, Dordrecht

Merritt, D., 1997, in *2nd Stromlo Conference: The Nature of Elliptical Galaxies*, eds. M. Arnaboldi et al., ASP conf. ser., in press

Nieto, J.-L., Bender, R., Surma, P., 1991, A&A, 244, L37

Pahre, M.A., Djorgovski, S.G., de Carvalho, R.R. 1995, ApJ 453, L17

Renzini, A., Ciotti, 1993, ApJ 223, 707

Saglia, R. P., Bender, R. & Dressler, A. 1993, A&A, 279, 75

Worthey, G.,, 1994, ApJS 95, 107

Worthey, G.,, 1996, in *New Light on Galaxy Evolution*, IAU Symp. 171, eds. R. Bender, R.L. Davies, Kluwer, Dordrecht

de Zeeuw, T., Franx, M., 1991, ARAA 29, 239

Photometric Scaling Relations of Dwarf Galaxies

Bruno Binggeli[1] and Helmut Jerjen[2]

[1] Astronomical Institute of the University of Basel, Venusstrasse 7,
CH-4102 Binningen, Switzerland
[2] Mount Stromlo and Siding Spring Observatories, Private Bag, Weston Creek PO,
ACT 2611, Canberra, Australia

Abstract. We briefly review the photometric scaling relations of dwarf galaxies. Restricting our discussion to dwarf *ellipticals*, where any photometric relation has less scatter than with dwarf irregulars for lack of star formation activity, we show that outside the central region ($r > 300$ pc) the radial surface brightness profile varies *continuously* between dwarf ellipticals and giant ellipticals (however excluding the rare compact, low-luminosity ellipticals). This is expressed by the continuity in all three parameters of Sérsic's generalized de Vaucouleurs/exponential law, which can be fitted to all elliptical profiles. The break, or discontinuity in the surface brightness - luminosity relation around $M_B = -20$, which fostered the notion of a physical dichotomy between normal and dwarf ellipticals, is really restricted to the very central part of the galaxies. This seems to favour again a common origin of dwarf and normal ellipticals.

Recently raised hopes that the photometric scaling relations of dwarf ellipticals could be used as distance indicators are turned down. The observed scatter in any of these relations is larger than 0.6 mag – too large for a competition with the kinematic (Tully-Fisher and $D_n - \sigma$) scaling relations.

1 Introduction

By "scaling relation" we mean the relation between any observational parameter of a galaxy and its total luminosity (total magnitude) or radius (as defined in a certain way). By *photometric* scaling relation we mean the relation between any photometric parameter and M or R. Since we are not concerned here with the colors of the galaxies, this parameter is basically the surface brightness (μ, as defined in a certain way), but more generally it can be any characteristic measure of the (monochromatic) radial surface brightness profile. This contrasts with the *kinematic* scaling relations where a kinematic parameter such as rotational velocity or velocity dispersion enters, leading to the Tully-Fisher or Faber-Jackson relations. The third fundamental relation would connect a kinematic quantity with a density (surface brightness), which is the "cooling diagram". All of this is nicely illustrated in Djorgovski's (1992) beautiful schematic of the galaxy parameter space (his Fig. 2). In this space galaxies tend to be confined to a fairly thin plane, which is of course the "Fundamental Plane" (FP, see Bender in this volume and references therein). The FP happens to be nearly orthogonal to the kinematic, or T-F/F-J plane, which is why the kinematic relations are very well defined and are useful distance indicators (and which is of course why this conference is dominated by the T-F and FP topics). On the other hand, the FP is nearly parallel to the photometric, or $\mu - M$ plane (e.g. Guzmán et al. 1993), i.e. *the $\mu - M$ plane is essentially a face-on view of the Fundamental Plane.*

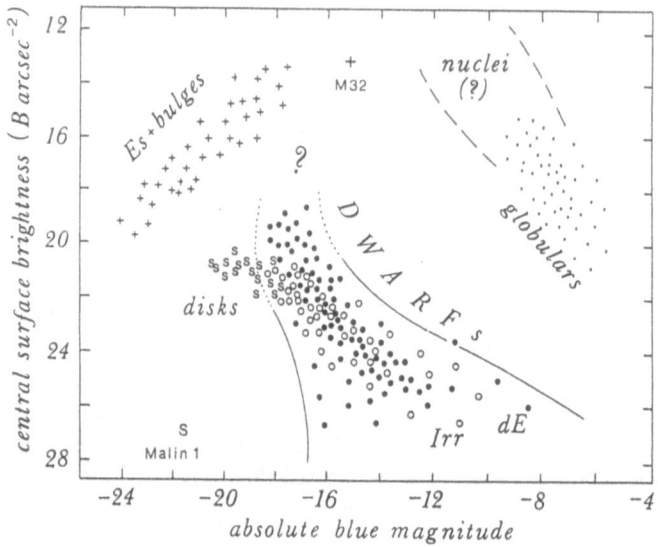

Fig. 1. The $\mu - M$ plane of stellar systems and subsystems. Dwarf galaxies follow a common sequence of decreasing surface brightness with decreasing luminosity, which is the most fundamental photometric scaling relation of galaxies. Figure taken from Binggeli (1994).

What photometric scaling relations do appear in the $\mu - M$ plane? This is shown in Fig. 1, which is a schematic taken from Binggeli (1994). A similar plot with real data can be found in Patterson & Thuan (1996, Fig. 9), and of course in Kormendy (1985) who pioneered this diagram. The surface brightness in Fig. 1 is meant to be the model-free central surface brightness of the de-nucleated stellar system. This quantity is certainly ill-defined for bright, more compact ellipticals where any nucleus cannot be easily separated from the rest of the galaxy, and where the "central" surface brightness generally depends on the resolution (e.g. Kormendy et al. 1994; Faber et al. 1997, and references therein). However, the general trend of increasing surface brightness with decreasing luminosity for normal ellipticals remains qualitatively true.

The four basic sequences of stellar systems that occupy the $\mu - M$ plane, viz. normal E's plus bulges; dwarf E's; spiral disks plus irregulars; and globular clusters are described in Binggeli (1994). The gap (?) between faint normal E's and bright dwarf E's will be referred to later (Sect. 2). The dE and S+Irr sequences are clearly distinct from each other at the bright end. Faintward of $M_B \approx -16$, however, they seem to merge into *one* sequence of dwarf galaxies, which without doubt must be viewed as the *main sequence of galaxies.* The evolutionary connection between the dE and S+Irr sequences is still unclear (Thuan

1992, Binggeli 1994, Ferguson & Binggeli 1994; for the most recent discussions see Papaderos et al. 1996 and Patterson & Thuan 1996). No matter what this connection is, it seems clear that the basic photometric relation *shared by all types of dwarf galaxies*, viz. that of decreasing surface brightness with decreasing luminosity, must have a unique physical origin. The most popular scenarios invoked here are inefficient star formation and mass loss due to galactic winds below a certain critical mass (cf. Ferguson & Binggeli 1994 for a review).

In the following we restrict our discussion to dwarf *ellipticals* which show much less scatter than dwarf irregulars in any photometric relation. The reason for this fact is simply that star formation activity changes the mass-to-light ratio, rendering surface brightness (at least in the optical) an unreliable tracer of mass density. Hence star formation activity adds noise to the data. Dwarf ellipticals are moreover of special interest with respect to their connection with normal, or giant E's. Finally, the well-defined photometric relations of dE's might be useful for distance measurements.

2   Photometric Relations of Dwarf Ellipticals

The most fundamental, model-free surface brightness measure of galaxies is certainly the "mean effective surface brightness", $\langle\mu\rangle_{\mathrm{eff}}$, which is the mean surface brightness within the radius that contains half of the total light. The dwarf sequence described above was indeed first discovered in terms of increasing $\langle\mu\rangle_{\mathrm{eff}}$ with increasing M (Caldwell et al. 1983, Binggeli et al. 1984). With this SB measure one finds a continuity between giant and dwarf ellipticals. However, there is a distinct break around $M_{\mathrm{B}} = -20$ (cf. Sandage & Perelmuter 1990, Capaccioli et al. 1993).

Fig. 2 shows the $\langle\mu\rangle_{\mathrm{eff}} - M$ relation for a large and homogeneous sample of Virgo cluster dE's. The best-fitting (regression) line is

$$\langle\mu\rangle_{\mathrm{eff}} = 11.38(\pm1.35) + 0.76(\pm0.08)B_{\mathrm{T}} \ . \tag{1}$$

The dispersion (rms scatter) about this mean regression line is considerably large with 0.69 mag. If we account for a Virgo cluster depth of $\sigma_{\mathrm{Virgo}} = 0.2$ mag and a (pessimistic) photometric error of $\sigma_{\mathrm{error}} = 0.3$ mag, we arrive at a (optimistic) intrinsic, or cosmic scatter of $\sigma_{\mathrm{cosmic}} \approx 0.6$ mag. This foreshadows the bad news about dE's as distance indicators (Sect. 3). Almost identical $\langle\mu\rangle_{\mathrm{eff}} - M$ relations are found for dwarf ellipticals in the Fornax (Ferguson & Sandage 1988) and Centaurus clusters (Jerjen 1995).

The "effective" parameters provide only a coarse measure of the radial surface brightness profile of a galaxy. A refinement is achieved by model fitting. The most popular model profiles used for dE's until recently are the exponential law and the non-analytical curves based on King's (1966) dynamical model. Both embody a *central* surface brightness, μ_0, which scales with M much like $\langle\mu\rangle_{\mathrm{eff}}$, though with a steeper slope (Binggeli & Cameron 1991, see also Fig. 1). Since normal (giant) ellipticals cannot be fitted by exponentials, or, if fitted by King models,

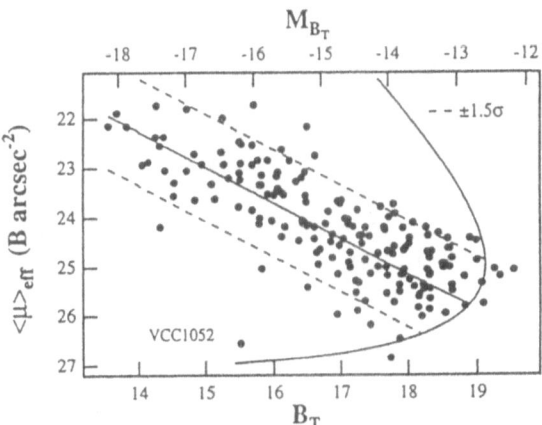

Fig. 2. The effective SB - total magnitude relation for Virgo cluster dE's (data from Binggeli & Cameron 1993). The best-fitting line is given in the text. Also shown are the ± 1.5 σ boundaries to guide the eye. The curve indicates the completeness limit for the sample. The absolute magnitude scale on top is based on a Virgo cluster distance modulus of 31.7. Figure taken from Jerjen (1995).

have much higher surface brightness than the dwarfs, the E - dE dichotomy seems perfect (see again Fig. 1).

However, both the exponential and the King model fail to represent the observed cuspyness of bright ($M_B < -16$) dE's, which in some cases can mimic the profile of a normal E (Binggeli & Cameron 1991). Young & Currie (1994), and in an earlier, ill-received paper, Davies et al. (1988) have shown that this cuspyness or *curvature* of the dE profile is perfectly explained by a third parameter, the exponent n, in a most natural generalization of the exponential law:

$$I(r) = I_0(r)e^{-(r/r_0)^n} , \qquad (2)$$

with central surface brightness $I(r)$ and scale length r_0. For $n = 1$ this reduces of course to the pure exponential. Moreover, for $n = 1/4$ one gets the classical $r^{1/4}$ law for giant ellipticals. Hence (2) is also a generalization of de Vaucouleurs' law and as such it was proposed long ago by Sérsic (1968). Sérsic's law (as we call it herefrom) not only accounts for the deviations of normal E profiles from the $r^{1/4}$ law (Caon et al. 1993, D'Onofrio et al. 1994) and for those of dE profiles from the exponential (Young & Currie 1994, Jerjen 1995). It provides a means of *re-unification of the E and dE classes*, as we now show.

We have fitted Sérsic profiles to 120 dE's and dS0's (a slight variant of the dE's), for which mean radial light profiles are provided by Binggeli & Cameron (1993). The innermost 3", corresponding to ≈ 300 pc with $(m - M)_{\rm Virgo} = 31.5$, have been left out with the fitting to circumvent the central quasi-stellar nuclei of the "dE,N". These nuclei should not be confused with the "cuspyness" mentioned before (see Binggeli & Cameron 1991 for a clear distinction). Fig. 3 shows

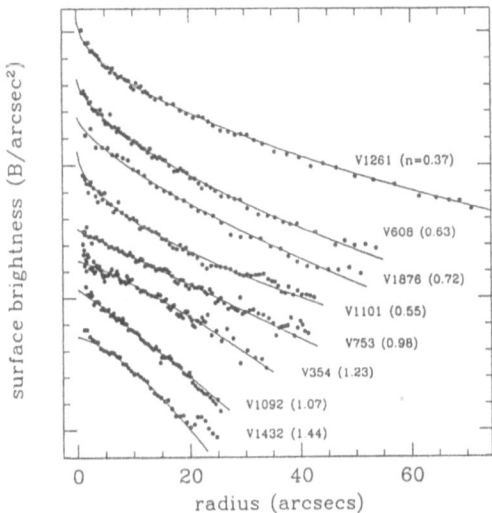

Fig. 3. Mean radial surface brightness profiles of 8 Virgo cluster dE's from the sample of Binggeli & Cameron (1991, 1993). The profiles are ordered by total magnitude but are shifted by arbitrary amounts along the ordinate. The observed profiles are given by the dots. The lines are best-fitting Sérsic profiles whose shape parameter n is given in parenthesis behind the name of the galaxy.

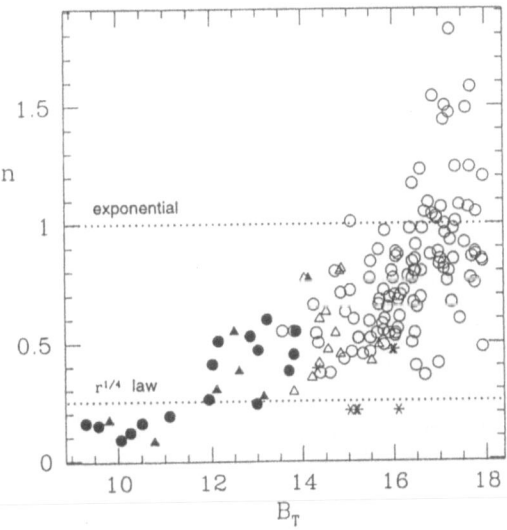

Fig. 4. Sérsic's shape parameter n versus total blue magnitude for early-type galaxies in the Virgo cluster. Dwarf galaxies (open circles: dE, open triangles: dS0) are based on the photometry of Binggeli & Cameron (1993). Data for E's (filled circles) and S0's (filled triangles) are taken from Caon et al. (1993). A few compact, M32-type E's (also from Binggeli & Cameron 1993) are added as asterisks. Note the continuity of n as opposed to the traditional bimodality, viz. exponential versus $r^{1/4}$ law, which is indicated by the dotted lines.

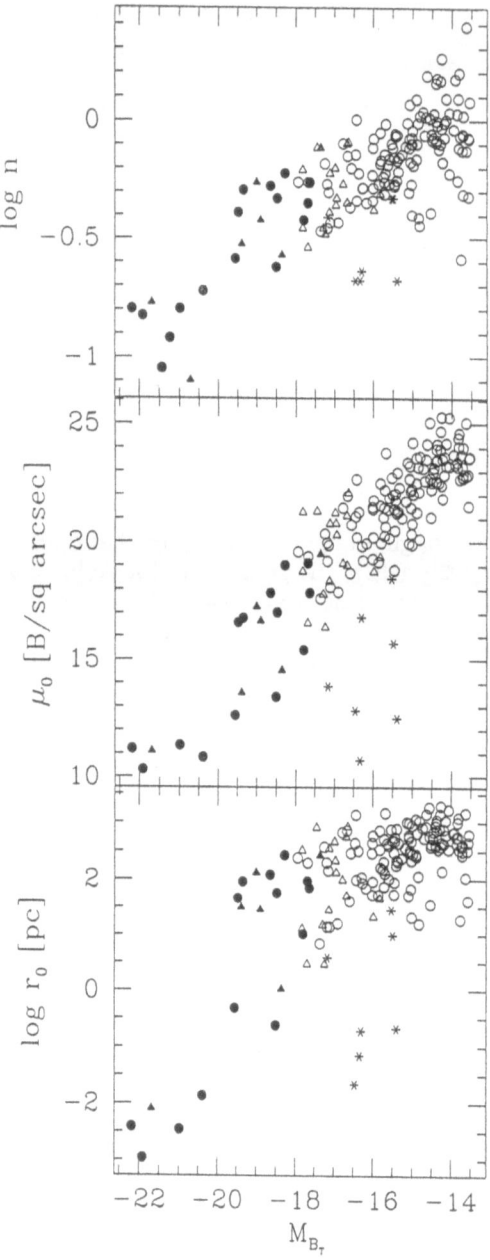

Fig. 5. The best-fitting Sérsic profile parameters for early-type galaxies in the Virgo cluster. The three parameters, as defined by (2), are shape parameter n, central surface brightness μ_0, and scale length r_0. The absolute scale is based on $(m - M)_{\mathrm{Virgo}} = 31.5$. Sample and symbols are the same as in Fig. 4 (filled circles: E, filled triangles: S0, open circles: dE, open triangles: dS0, asterisks: compact, M32-type E).

a representative sample of our dwarf profile fits. The trend of increasing cuspy-ness with increasing luminosity is striking. In terms of photometric parameters this leads to a strong correlation between Sérsic's shape parameter n and total magnitude, as shown in Fig. 4. By adding now the data of Caon et al. (1993) for giant E and S0 galaxies we get the surprising result that the "giants" (ex-cluding the compact E's) smoothly extend the "dwarf" sequence towards higher luminosities. In other words: there is a *continuity in the profile shape between E and dE*. This continuity holds in fact for all three Sérsic parameters. The corresponding relations are given in Fig. 5.

Obviously, then, the "dichotomy" between E and dE (Fig. 1) is restricted to the very central part of the galaxies (with galactocentric radius $r < 300$ pc). A similar dichotomy in central structure has recently appeared among *normal* ellipticals – that between "power law E's" and "core E's" (e.g. Kormendy et al. 1994, Faber et al. 1996). The central region is certainly a singular place, with special conditions in terms of dynamics (e.g. in the presence of a black hole) and dissipation. It simply cannot represent the *global* structure of a galaxy. In contrast, the light profile as described by the Sérsic parameters *outside the core region* c a n be regarded as a global characteristic. And here one finds a beautiful continuum that re-unites normal and dwarf ellipticals into *one* sequence. Even the brightest cluster galaxies (mostly of supergiant type cD) seem to fit into this sequence (Graham et al. 1996); likewise the bulges of S0 galaxies (Andredakis et al. 1995). This strongly suggests very similar formation mechanisms for all of these stellar systems. Theoretical modelling has shown that the shape of the radial light profile depends on the (central) potential of the galaxy (Hjorth & Madsen). Hence at the heart of it all may be a simple luminosity - potential relation for hot stellar systems.

There is one apparent exception: the compact, low-luminosity (M32-type) ellipticals. They clearly fall off the mean relations in Fig. 5. However, it is con-ceivable that these compacts were affected by a massive neighbour, either at the time of formation (Burkert 1994), or later on via tidal stripping (Faber 1973). This view is supported by the observation that compacts with a massive neigh-bour tend to fall off more strongly (are more compact) than those that haven't.

With respect to the nature of dwarf elliptical galaxies, the new continuity with normal ellipticals would suggest that dE's are not an entirely different species but are genuine ellipticals, after all. *Bright* dE's have many other char-acteristics in common with normal E's, as discussed elsewhere (Jerjen & Binggeli 1997a), while *faint* dE's (the "dwarf spheroidals") may indeed be more closely linked to dwarf spirals and irregulars.

3   Dwarf Ellipticals as Distance Indicators?

The $\mu - M$ relation of dwarf ellipticals suggests itself as a tool for distance measurements. But can it be used in practice? We have already seen (Sect. 2) that the rms scatter of the $\langle\mu\rangle_{\text{eff}} - M$ relation for Virgo cluster dE's is around 0.7 mag. The dispersion in the $\mu_0 - M$ relation is certainly not smaller (see Binggeli

& Cameron 1991). Given these uncertainties, the $\mu - M$ relation has rarely been applied for distance determinations. If it is to be of any use, it is crucial to work with very deep and complete dwarf samples (Jerjen & Binggeli 1997b), which are of course difficult to get. A singular attempt by Bothun et al. (1989) to use the (apparently!) nearly constant scale length of dE's as distance indicator remains unconvincing (but see Jerjen 1995, who used the *distribution* of scale lengths).

However, high hopes have recently been raised by Young & Currie (1994, 1995) that dE's could, after all, be used to get reliable distances. These authors propose Sérsic's profile curvature parameter n as luminosity, and hence distance indicator. The shape of the profile is of course distance-independent just as a surface brightness. Working with a small sample of Fornax cluster dE's, Young & Currie (1994) found a surprisingly tight $n - m_T$ relation, with an rms scatter of only 0.47 mag (after the exclusion of a few "outliers", however) and they went on to derive a distance modulus for the Fornax cluster based on a handful of local, calibrating dE's. A dispersion of 0.47 mag would indeed put this method nearly on a par with Tully-Fisher and $D_n - \sigma$.

In a second paper, Young & Currie (1995) applied a variant of their new distance indicator, viz. the relation between n and $\log r_0$ (Sérsic's length scale), to a sample of Virgo cluster dE's for which they had derived profiles from low-resolution Schmidt plates. The scatter of the n - $\log r_0$ relation turned out to be much larger for these Virgo cluster dE's than for an external sample of Local Group and Fornax cluster dwarfs. Young & Currie (1995) argued now, by comparing the distances from the (apparently!) independent $n - M$ and n - $\log r_0$ relations, that the *intrinsic* scatter for Virgo cluster dE's is equally small, in which case the large observed scatter for Virgo dE's must be attributed to the *depth* of the Virgo cluster. The resulting cigar-shaped cloud of dE's, stretching from ca. 8 to 20 Mpc in Young & Currie's distance scale and pointing towards us (!), seems in perfect accord with the filament of spiral galaxies advocated by Fukugita et al. (1993) and Yasuda et al. (1997) based on Tully-Fisher distances.

At this point it becomes clear that Young & Currie's method does not work. It is true that the scatter of the $n - M$ (or $\log n - M$) relation is large for the Virgo cluster (see Fig. 5, where the rms scatter for the dE's alone amounts to $\approx$ 0.9 mag). But there is absolutely no indication that this scatter is *not* intrinsic to the dwarfs. Young & Currie's small intrinsic Virgo dispersion is likely an artefact that stems from the modelling of their low-resolution Virgo dwarf profiles with high-resolution Fornax cluster and LG data. *If* the large dispersion for Virgo dwarfs were due to the depth of the cluster (or rather "cloud" in this case), one would expect to see a strong correlation between the "distance" of a dE – naively derived from its residual to the mean $n - M$ or n - $\log r_0$ relation – and its radial velocity. We have repeated Young & Currie's analysis with our large sample of highly resolved Virgo dE profiles (cf. above) and looked for such a correlation. With more than 40 known dwarf redshift there is absolutely *no* correlation between "distance" and velocity. In particular, there are a number of highly negative velocities (up to $-730\ \mathrm{km\,s^{-1}}$!). Where should these dwarfs lie if not in the very center of the cluster? Like normal E's *and unlike spirals*, dwarf ellipticals do reside only in dense environments (cf., e.g., Ferguson & Binggeli 1994). A detailed critique of Young & Currie (1995) is presented in a forthcoming paper (Binggeli & Jerjen 1997).

Given the large cosmic dispersion of the photometric structure of dE's in the Virgo cluster (and probably elsewhere), we must finally conclude that *the photometric scaling relations of dwarf elliptical galaxies are poor distance indicators.*

Acknowledgements. We thank the Swiss National Science Foundation for financial support.

References

Andredakis, Y.C., Peletier, R.F., Balcells, M. (1995): MNRAS **275**, 874

Binggeli, B. (1994): in *Panchromatic View of Galaxies – Their Evolutionary Puzzle*, eds. G. Hensler et al. (Editions Frontières, Gif-sur-Yvette), p.173

Binggeli, B. (1994): in *Dwarf Galaxies*, ESO/OHP Workshop, ed. G. Meylan & Ph. Prugniel (ESO, Garching), p. 13

Binggeli, B., Cameron, L.M. (1991): A&A **252**, 27

Binggeli, B., Cameron, L.M. (1993): A&AS **98**, 297

Binggeli, B., Jerjen, H. (1997), in preparation

Binggeli, B., Sandage, A., Tarenghi, M. (1984): AJ **89**, 64

Bothun, G.D., Caldwell, N., Schombert, J.M. (1989): AJ **98**, 1542

Burkert, A. (1994): MNRAS **266**, 877

Caldwell, N. (1983): AJ **88**, 804

Caon, N., Capaccioli, M., D'Onofrio, M. (1993): MNRAS **265**, 1013

Capaccioli, M., Caon, N., D'Onofrio, M. (1993): in *Structure, Dynamics, and Chemical Evolution of Elliptical Galaxies*, ESO/EIPC Workshop, eds. I.J. Danziger et al. (ESO, Garching), p. 43

Davies, J.I., Phillipps, S., Cawson, M.G.M., Disney, M.J., Kibblewhite, E.J. (1988): MNRAS **232**, 239

Djorgovski, S. (1992): in *Morphological and Physical Classification of Galaxies*, ed. G. Longo et al. (Kluwer, Dordrecht), p. 337

D'Onofrio, M., Capaccioli, M., Caon, N. (1994): MNRAS **271**, 523

Faber, S.M. (1973): ApJ **179**, 423

Faber, S.M., Tremaine, S., Ajhar, E.A., Byun, Y.-I., Dressler, A., Gebhardt, K., Grillmair, C., Kormendy, J., Lauer, T., Richstone, D. (1996): AJ, in press

Ferguson, H., Binggeli, B. (1994): A&AR **6**, 67

Ferguson, H., Sandage, A. (1988): AJ **96**, 1520

Fukugita, M., Okamura, S., Yasuda, N. (1993): ApJ **412**, L13

Graham, A., Lauer, T.R., Colless, M., Postman, M. (1996): ApJ **465**, 534

Guzmán, R., Lucey, J.R., Bower, R.G. (1993): MNRAS **265**, 731

Hjorth, J., Madsen, J. (1995): ApJ **445**, 55

Jerjen, H. (1995): PhD thesis, University of Basel

Jerjen, H., Binggeli, B. (1997a): in *The Nature of Elliptical Galaxies* (invited contribution), Proceedings of the Second Stromlo Symposium, Canberra, in press

Jerjen, H., Binggeli, B. (1997b): in *The Nature of Elliptical Galaxies* (poster paper), Proceedings of the Second Stromlo Symposium, Canberra, in press

King, I.R. (1966): AJ **71**, 64

Kormendy, J. (1985): ApJ **295**, 73

Kormendy, J., Dressler, A., Byun, Y.-I., Faber, S.M., Grillmair, C., Lauer, T., Richstone, D., Tremaine, S. (1994): in *Dwarf Galaxies*, ESO/OHP Workshop, ed. G. Meylan & Ph. Prugniel (ESO, Garching), p. 147

Papaderos, P., H.-H. Loose, K.J. Fricke, Thuan, T.X. (1996): A&A **314**, 59

Patterson, R., Thuan, T.X. (1996): ApJS **107**, 103

Sandage, A., Perelmuter, J.-M. (1990): ApJ **350**, 481

Sérsic, J.-L. (1968): *Atlas de galaxias australes*, Observatorio Astronomica, Cordoba

Thuan, T.X. (1992): in *The Physics of Nearby Galaxies: Nature or Nurture?*, eds. T.X. Thuan et al. (Editions Frontières, Gif-sur-Yvette), p. 225

Yasuda, N., Fukugita, M., Okamura, S. (1997): ApJS, in press

Young, C.K., Currie, M.J. (1994): MNRAS **268**, L11

Young, C.K., Currie, M.J. (1995): MNRAS **273**, 1141

The Scaling Laws of Dwarf Elliptical Galaxies

Enrico V. Held[1], Jeremy R. Mould[2], and Kenneth C. Freeman[2]

[1] Osservatorio Astronomico, Vicolo dell'Osservatorio 5, I-35122 Padova, Italy
[2] Mount Stromlo and Siding Spring Observatories, Canberra, A.C.T. 2611, Australia

Abstract. Kinematic data from our recent high-resolution spectroscopic observations of dwarf ellipticals in the Fornax cluster are used to illustrate the distribution of low-mass early-type galaxies in the space of observable parameters. The changes in mass-to-light ratio from the faintest Local Group dwarf spheroidals to bright dwarf ellipticals, and the dark-matter content of early-type dwarfs, are also briefly discussed.

1 Introduction

The scaling laws of dwarf elliptical (dE) and Local Group dwarf spheroidal (dSph) galaxies, and their comparison with more luminous galaxies, are a primary tool for understanding their relationship to normal ellipticals, and whether they lie upon the same path of galaxy formation. The photometric scaling relations for dE's are generally different from those of more massive galaxies (see Binggeli 1997, this conference). Photometric data, however, seem to indicate a gradual transition between bright dE's and low-luminosity ellipticals (Bender & Nieto 1990; Prugniel et al. 1992; Kormendy & Bender 1994; Jerjen & Binggeli 1997). Recent photometric studies show a continuous change in the shape of luminosity profiles from giant ellipticals to dE's which is correlated with galaxy luminosity (Capaccioli et al. 1993; Jerjen & Binggeli 1997). Also, the most luminous dE's approach the E sequence in the core parameter correlations (Kormendy & Bender 1994).

A comparison between ellipticals and dwarfs in the three-dimension space of physical parameters is possible only for relatively few dE's with measured velocity dispersions. Kinematic data are the most difficult to obtain, because of the low surface brightnesses of dE's and the high spectral resolution needed to measure the low velocity dispersions of low-mass galaxies. Nieto et al. (1990) gave an early discussion of the relationship of diffuse dwarfs to the (edge-on) Fundamental Plane (FP) of elliptical galaxies, mainly concluding that dE/dSph galaxies are near the low-mass extension of the FP, except for Draco, Ursa Minor, and possibly Carina. Bender et al. (1992) have discussed the distribution of the different families of hot stellar systems (including dwarf galaxies) in the space of parameters. Dwarf ellipticals in their sample appear to be slightly offset from ellipticals in an edge-on projection of the FP. Examination of the effects of various mechanisms affecting galaxy evolution (e.g., galactic winds or tidal stripping) led to the conclusion that diffuse dwarfs form a separate sequence in the FP and cannot derive from present-time ellipticals.

Special projections of the FP may also contain considerable information. An example is the Faber-Jackson relation discussed by Held et al. (1992) and Peterson & Caldwell (1993). While Held et al. (1992) suggest that dE's follow a trend of L vs. σ appropriate for self-gravitating dwarf galaxies (e.g., Yoshii & Arimoto 1987), Peterson & Caldwell (1993) conclude that data are well fit by the relation $L \propto \sigma^{5.3}$ predicted in the case of dwarfs embedded in a dominant dark-matter halo (Dekel & Silk 1986).

Observational uncertainties on the scaling laws of dE/dSph galaxies represent a major problem in modeling the manifold of hot stellar systems. These uncertainties and the paucity of kinematic data have motivated a new study of the internal dynamics of dE's and low-luminosity ellipticals, focusing on galaxies in the overlap range between the two classes. The location of low-mass early-type galaxies in the parameter space will be useful to answer two fundamental questions: Do E's and dE's form distinct sequences in the space of physical parameters? If so, does this imply a different origin?

2   Fundamental Plane Correlations

In this paper we present the first results of our recent study of internal dynamics of dE's and low-luminosity ellipticals, mostly located in the Fornax cluster. The objects were observed in two runs at the AAT and 2.3m ANU telescopes at Siding Springs, with instrument configurations yielding spectral resolutions of 0.6 and 0.9 Å, respectively. Each galaxy was observed for 1 to 3 hours along a single position angle (usually the photometric major axis). The results of a preliminary analysis are central velocity dispersions σ_0 (approximately within 0.5 R_e) and estimates of the inner mass-to-light ratio for 6 dE's observed at the AAT. Combined with all existing photometric and kinematic data for dwarf and low-luminosity ellipticals, our new data allow us to compare the different classes of early-type systems with respect to the Fundamental Plane of ellipticals. Figure 1 shows the distribution of hot stellar systems in the space of observable parameters (surface brightness, radius, and velocity dispersion). Most of the data are from the Bender's et al. (1992) compilation, with the addition of kinematic data for dE's from Peterson & Caldwell (1993) and our study. The three-dimensional display in Fig. 1 represents an effective way to provide an unbiased (although qualitative) view of the distribution of the different families. The viewpoint of Fig. 1 nearly corresponds to a face-on view of the Fundamental Plane. This plot reflects the well-known trends of surface brightness against radius for dE and E galaxies (e.g., Binggeli & Cameron 1991). Adding velocity dispersions to photometric parameters confirms that there is no discontinuity between the two sequences, the link being represented by a few "transition" low-luminosity ellipticals. Compact ellipticals of the M32 type, often considered the "true extension" of the E sequence (e.g., Kormendy & Bender 1994), are intrinsically rare, and probably reflect a specific formation history, e.g. as in Burkert's (1994) models. The sequence of ellipsoidal galaxies in Fig. 1 is probably the result of the different roles played by formation and evolution processes such as dissipation

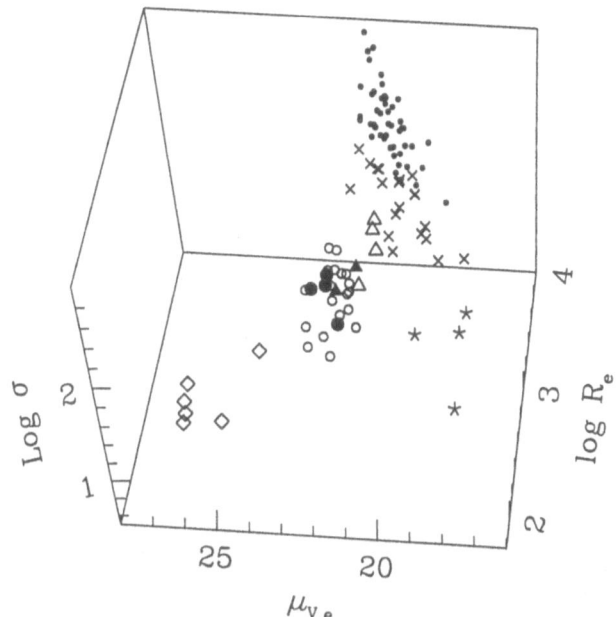

Fig. 1. The three-dimensional distribution of ellipsoidal systems in the space of observable physical properties. Dots denote giant ellipticals, stars indicate compact E's of the M32 type, crosses represent E galaxies of intermediate luminosity, while triangles indicate "transition" low-luminosity ellipticals (anisotropic). circles are diffuse dwarfs, and diamonds denote Local Group dwarf spheroidals. Filled symbols refer to our new data. This plot nearly corresponds to a face-on view of the Fundamental Plane.

and supernova-driven galactic winds (e.g., Vader 1986; Yoshii & Arimoto 1987). However, simple two-component equilibrium models such as those proposed by Kritsuk (1996) represent a first attempt to model hot stellar systems in a unified scenario. Kritsuk's (1996) models reflect a smooth transition from (centrally) dark-matter dominated dwarf spheroidals to luminous-matter dominated giant ellipticals.

A nearly edge-on projection of the FP is shown in Fig. 2. Except for the extreme dSph's in the Local Group, the diffuse dwarfs are well aligned with an extension of the FP, although they show a larger scatter than the more luminous galaxies. It has been noticed that the σ's measured so far for many dE's may represent *nuclear* rather than *halo* velocity dispersions, which may actually contribute to the scatter across the FP. Variations in the M/L ratios are the most likely origin for the widely discrepant data for the faintest Local

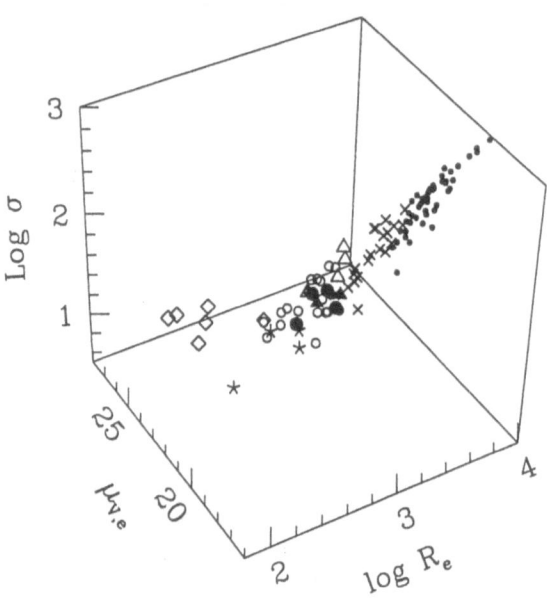

Fig. 2. The distribution of ellipsoidal systems in the parameter space. Here the FP is seen nearly edge-on. Symbols as in Figure 1.

Group dwarf spheroidals. However, since galaxy structures change continuously from normal ellipticals to dE's, non-homology is likely to be the primary origin of deviations for low surface brightness galaxies (cf. Nieto et al. 1990).

3   Mass-to-Light Ratios

One of the most interesting questions about early-type galaxies is how their dark matter halos scale as a function of mass, luminosity, and morphological type.

In Figure 3 we have collected the most recent results on central mass-to-light ratios $(M/L)_V$ for Local Group dSph's (from Mateo 1996) and dwarf ellipticals (from Peterson & Caldwell 1993 and this study). Core $(M/L)_V$ values for galaxies in our study were calculated as in Peterson & Caldwell (1993). Our data confirm the low M/L values found by these authors. We note that 3 (out of 6) dwarfs in our present sample are not really ""nucleated", so their velocity dispersions are luminosity-weighted mean values over $\sim 0.5\ R_e$ rather than nuclear dispersions.

While high central dark matter densities have been inferred for many dSph's

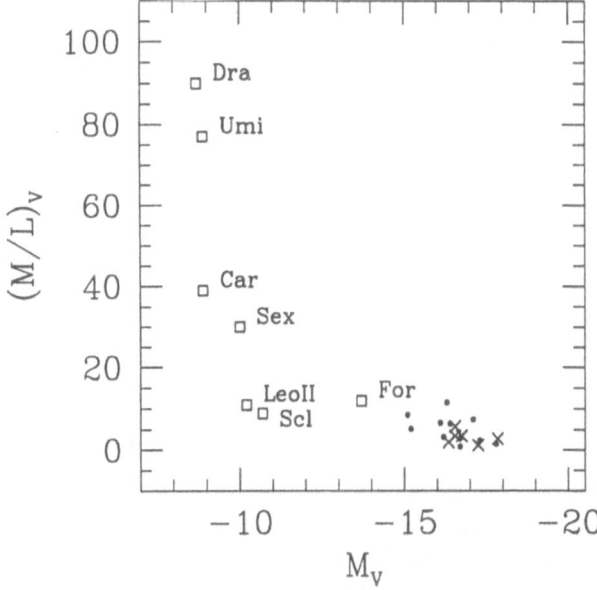

Fig. 3. The central mass-to-light ratio of dwarf spheroidals (*squares*) and dwarf ellipticals from Peterson & Caldwell (1993) (*dots*) and our study (*crosses*). Data points for dSph are from Mateo (1996).

from velocity dispersion measurements of individual member stars (e.g., Mateo 1994), the central dark-matter densities of bright dE's are similar to those found in ellipticals, as noticed earlier by Held et al. (1992).

The observed trend of $(M/L)_V$ for dwarf spheroidals appears to be consistent with the presence of a dark halo of constant mass $M_D \approx 1 - 5 \times 10^7~M_\odot$ (Mateo 1996). Thus only the faintest dSph are centrally dark-matter dominated. A transition between the dark- and luminous-matter dominated regimes is found approximately at $M_V = -15$ (cf. Held et al. 1992; Djorgovski 1993).

4 Conclusions and Future Plans

With respect to intermediate-luminosity (rotating, "disky") and bright (pressure-supported, "boxy") ellipticals, diffuse dwarfs are thought to represent a third, distinct class (e.g., Kormendy & Bender 1994). A collection of literature data and preliminary results from our kinematic study suggest that bright dE's are connected to low-luminosity ellipticals in the parameter space. A quantitative analysis of the distribution of low-mass early-type galaxies will be the subject of a forthcoming paper.

Clearly the existing kinematic sample is biased towards relatively bright dE's. Extending kinematic measurements to fainter objects will require an 8m class telescope and new generation instruments. The advent of VLT will open a new era in our ability to obtain accurate velocity dispersions for low surface brightness objects. In particular, the fibre-bundle mode of the VLT fibre spectrograph (FUEGOS) will allow us to extend our study to high-resolution spectra of faint dwarf ellipticals in nearby clusters. Off-nuclear spectra of dE's will allow us to further investigate the role of anisotropy *vs.* rotation in providing dynamical support to low-mass galaxies.

References

Bender, R., Nieto, J.-L. (1990). A&A 239, 97

Bender, R., Burstein, D., Faber, S. M. (1992). ApJ 399, 462

Binggeli, B. (1997). In ESO Workshop on *Galaxy Scaling Relations: Origins, Evolution and Applications*, ed. L. da Costa (ESO, Garching), this conference.

Binggeli, B., Cameron, L. M. (1991). A&A 252, 27

Burkert, A. (1994). MNRAS 266, 877

Capaccioli, M., Caon, N., D'Onofrio, M. (1993). In ESO/EIPC Workshop on *Structure, Dynamics and Chemical Evolution of Elliptical Galaxies*, eds. Danziger I. J., Zeilinger W. W., Kjär K. (ESO, Garching), p. 43

Dekel, A., Silk, J. (1986). ApJ 303, 39

Djorgovski, S. G. (1993). In A.S.P. Conf. Ser. 48 *The Globular Cluster – Galaxy Connection*, eds. Brodie J., Smith G. (A.S.P., San Francisco), p. 496

Held, E. V., de Zeeuw, T., Mould, J. R., Picard, A. (1992). AJ 103, 851

Jerjen, H., Binggeli, B. (1997). In 2nd Stromlo Symposium *The Nature of Elliptical Galaxies*, eds. Arnaboldi M., Da Costa G. S., Saha P. (A.S.P., San Francisco), in press

Kormendy, J., Bender, R. (1994). In ESO/OHP Workshop on *Dwarf Galaxies*, eds. Meylan G., Prugniel P. (ESO, Garching), p. 161

Kritsuk, A. G. (1996). Preprint

Mateo, M. (1994). In ESO/OHP Workshop on *Dwarf Galaxies*, eds. Meylan G., Prugniel P. (ESO, Garching), p. 309

Mateo, M. (1996). In A.S.P. Conf. Ser. 92 *Formation of the Galactic Halo ... Inside and Out*, eds. Morrison H., Sarajedini A. (A.S.P., San Francisco), p. 434

Nieto, J.-L., Bender, R., Davoust, E., Prugniel, P. (1990). A&A 230, L17

Peterson, R. C., Caldwell, N. (1993). AJ 105, 1411

Prugniel, P., Bica, E., Alloin, D. (1992). In *Morphological and Physical Classification of Galaxies*, eds. Longo G., Capaccioli M., Busarello G. (Kluwer, Dordrecht), p. 26

Vader, J. P. (1986). ApJ 305, 669

Yoshii, Y., Arimoto, N. (1987). A&A 188, 13

Bimodality of Freeman's Law

R. Brent Tully[1], and Marc A.W. Verheijen[2]

[1] Institute for Astronomy, University of Hawaii
   2680 Woodlawn Drive, Honolulu, Hawaii 96822, USA
[2] Kapteyn Astronomical Institute
   Postbus 800, NL-9700 AV Groningen, The Netherlands

Abstract. A cluster sample of 62 galaxies complete to $M_B = -16.5^m$ has been observed at B, R, I, K' bands with imaging detectors. The distribution of exponential disk central surface brightnesses is found to be bimodal. The bimodality is particularly significant at K' because obscuration is not a problem and because the high surface brightness galaxies are redder than the low surface brightness galaxies so the bifurcation is greater. The bimodality signal is especially clear when galaxies with close companions are excluded from consideration. High and low surface brightness pairs with essentially identical luminosities and maximum rotation characteristics are compared. It is suggested that the high surface brightness galaxies have self-gravitating disks while the low surface brightness galaxies are halo dominated at all radii. Evidently the intermediate surface brightness regime is unstable. If a disk has sufficiently low angular momentum and it shrinks enough that the disk potential begins to dominate the halo potential locally, then the disk must secularly evolve to the high surface brightness state characterized by a flat rotation curve.

1 Introduction

There has been a debate about whether the disks of galaxies have a quantized central surface brightness (Freeman 1970; van der Kruit 1987) or, instead, there is a broad distribution of disk central surface brightnesses (McGaugh, Bothun, and Schombert 1995; de Jong 1996). Disney (1976) and others have discussed the selection effects that could distort the observed surface brightness distribution. Surveys of the brightest galaxies tend to be dominated by the highest surface brightness objects. Special efforts are required to find low surface brightness galaxies (van den Bergh 1966; Impey, Bothun, and Malin 1988; Davies et al. 1994). It is difficult to have a proper normalization of the relative occurrence rates in different surface brightness intervals.

A large and complete sample of galaxies are considered in this study. The galaxies are drawn from the Ursa Major Cluster and the cluster assignment requires an appropriate redshift. The cluster is dynamically young (long crossing time, no central concentration, no ellipticals, many HI-rich spirals) and the galaxies may be representative of those in low density environments. There are 62 galaxies in the complete sample with $M_B < -16.5^m$. The data are presented in Tully et al. (1996).

CCD photometry at B, R, I bands has been acquired for all of the complete sample and K' HgCdTe images have been obtained for 60 of 62. This K' data is a key to the discussion. First, the K' material is not affected significantly by obscuration so surface brightnesses are only modified by geometric effects associated with viewing perspectives. Second, higher surface brightness galaxies tend to be redder than lower surface brightness galaxies so the distinctions between the different classes that will be discussed are clearer as one progresses to the red.

2   Bimodality of Central Surface Brightnesses

There is a more complete presentation by Tully and Verheijen (1997), where the situation in all available passbands is discussed. Only the best evidence will be given here. The bimodality is found in all the observed bands but is the least ambiguous at K'. In addition, it turns out that the few intermediate surface brightness cases in our sample all have close companions. If the galaxies with substantial companions closer than 80 kpc in projection are excluded, the distribution of surface brightnesses of the remaining 36 of 60 galaxies with K' photometry is that seen in Figure 1.

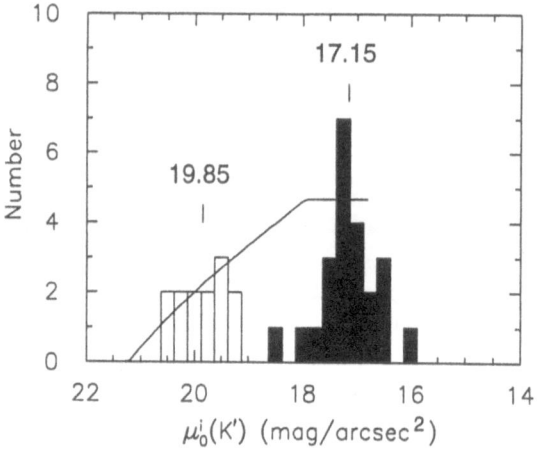

Fig. 1. Histogram of exponential disk central surface brightnesses at K'. Inclination corrected assuming no obscuration. Isolated galaxies only; galaxies with companions within 80 kpc projected are excluded. Filled histogram: HSB galaxies; open histogram: LSB galaxies. Solid curve: completeness expectation if there are equal numbers of galaxies in all surface brightness bins > 17 mag/sq. arcsec. Means of the HSB and LSB distributions are indicated.

    There is a clear separation into two peaks. The filled part of the histogram illustrates the K' equivalent of 'Freeman's law'. The galaxies that contribute belong to a *high surface brightness* [HSB] family. The open part of the histogram is made up of constituents of a *low surface brightness* [LSB] family. The superimposed solid curve illustrates a completeness expectation assuming there is uniform population of the surface brightness–exponential scale length domain for scale lengths between 500 pc and 3 kpc and for all surface brightnesses faintward of 17 mag/sq. arcsec. It can be seen that the LSB peak has probably been reduced by incompletion and may well have been truncated on the faint side. However, the critical point is that the visibility is quite adequate for the intermediate surface brightness regime between the HSB and LSB peaks and the paucity of objects in that region is highly significant.

3   Luminosity–Line Width and Surface Brightness–Scale Length Diagrams

Galaxies with essentially identical luminosities and rotation maxima can have very different central surface brightnesses and exponential scale lengths. These properties are demonstrated in Figure 2. The left panel shows the tight relation between K' luminosities and HI line profile widths. The line profiles have been transformed to approximate rotation maxima. HSB galaxies are represented by filled symbols and LSB galaxies, by open symbols. Three pairs are identified by numbers. The triplet identified by the letter a is two HSB galaxies, one with, and one without, a bulge and an LSB. The distinctions between bulge and non-bulge systems are discussed by Tully and Verheijen (1997).

The right panel illustrates the relation between K' central disk surface brightness and exponential scale lengths. Surface brightnesses have been corrected for inclination projection effects assuming no obscuration. Filled and open symbols have the same meaning as in the other panel. Crosses locate galaxies too faint to be in the complete sample. There are fewer data points in the left panel because only objects with satisfactory types, HI information, and inclinations could be plotted.

It is immediately seen how the HSB and LSB systems that were paired in the left panel have moved to separate domains in the right panel. Galaxies with a given total amount of light can be bright and compressed or dim and extended, yet generate the same rotation speeds. Galaxies apparently avoid an intermediate region in the parameter space of the right panel. The one-dimensional projection of the right panel of Fig. 2 onto the surface brightness axis gives the equivalent of Fig. 1 for the entire sample. The few intermediate surface brightness cases in Fig. 2 are filtered out of Fig. 1 because they are in close proximity to another galaxy.

4   Rotation Curve Disk–Halo Decompositions

Velocity field information will be made available for a large fraction of the Ursa Major sample (Verheijen 1997) but that work is still in progress. An HSB–LSB luminosity–rotation maximum pair that is not part of the Ursa Major sample has

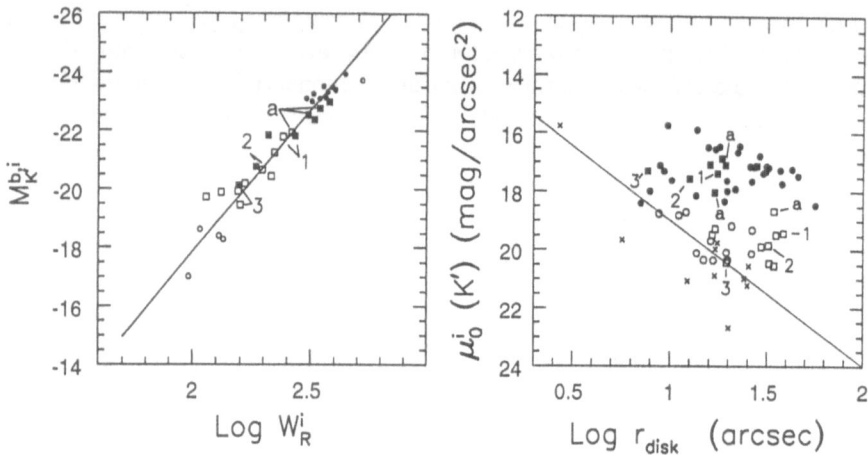

Fig. 2. *Left panel:* Luminosities at K' vs. HI profile line width; inclination corrected. Filled symbols: HSB; open symbols: LSB. HSB–LSB pairs with similar luminosities and linewidths are labeled 1,2,3. A bulge–nonbulge HSB pair and an LSB of comparable luminosity are labeled a. The solid line is the regression with uncertainties in line widths. *Right panel:* K' disk central surface brightness vs. exponential scale length. Symbols as in left panel with the addition that crosses are galaxies fainter than the completion limit. The same pairs and triplet are labeled. The solid line is the approximate completion limit.

been discussed in the literature. De Blok and McGaugh (1996) have decomposed the rotation curves of the HSB galaxy NGC 2403 and the LSB galaxy UGC 128. Their results are consistent with the following discussion.

At present, we consider only the three galaxies identified by the labels a in Fig. 2. One of these, NGC 3917, is an example of the LSB family. The other two are HSB examples: NGC 3949 has an exponential disk but no substantial bulge while UGC 6973 has a central concentration of light in addition to an exponential disk. Figure 3 provides a graphic summary of much of our information about these systems. There are B band images, I band surface brightness profiles, HI position-velocity maps, and rotation curve decompositions.

The rotation curve decompositions have three components. One is not important in any of these examples: the dotted curves indicate the contributions to rotational velocities from the interstellar gas. The dot-dash curves indicate the contributions due to the stellar distribution with an assumed fixed M/L value. The dashed curves are the residual contributions associated with a dark halo.

In each case, the same value of $M/L_{K'} = 0.4 M_\odot/L_\odot$ is assumed. This value is required for a 'maximum disk' fit (van Albada and Sancisi 1986) of the stellar components for the two HSB examples. Much larger $M/L_{K'}$ values could be accommodated for the LSB case but still a dominant dark halo would be required. A parallel situation is found at B, R, I. It is expected that M/L values for LSBs would be lower, if anything, not higher than for the HSBs.

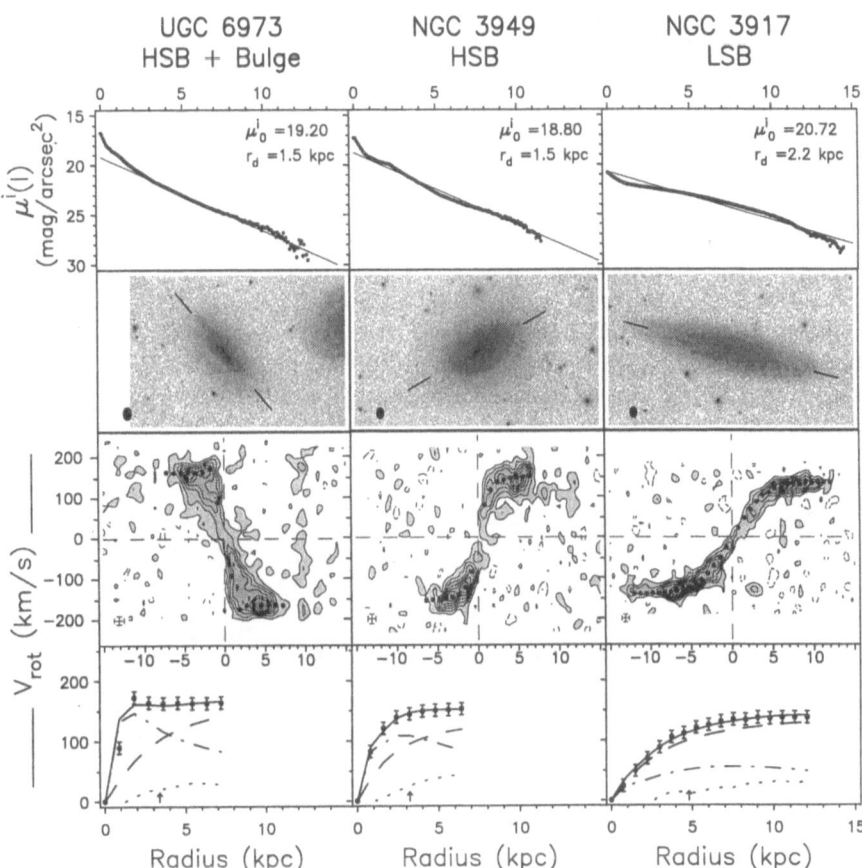

Fig. 3. Rotation curve decompositions. Examples of the three distinct classes of disk galaxies are presented in each of the vertical groups. On the left is UGC 6973, a system with a HSB exponential disk and a central bulge. In the middle is NGC 3949, a system with a HSB disk but no appreciable bulge. On the right is NGC 3917, a system with a LSB disk and no bulge. In each case, the horizontal axes are position in kpc. In the top and bottom panels the origin with respect to the nucleus is at the left axis, while in the middle panels the origin with respect to the nucleus is at the center of the plots. Surface brightnesses at I, corrected for inclination, are shown in the top panels. Images at B are shown in the second row. The major axes are indicated, as well as the FWHM beam of the HI observations. The velocity-position decomposition of the HI observations is seen in the panels of the third row. Velocities averaged over annuli are given as dots. The rotation curve decompositions are provided in the bottom panels. The dots with error bars illustrate the observed rotation curves. Dot–dashed curves illustrate the amplitude of rotation expected from the observed distributions of light and $M/L = 0.4$ at K'. The dotted curves illustrate the contribution expected from the gas component. The dashed curves demonstrate the contribution attributed to a dark matter halo modelled by an isothermal sphere. The solid lines indicate the sum of these three components.

In the LSB case, there is the same total luminosity as in the HSB cases, but the light is more diffuse. If M/L is fixed, then the mass associated with this light cannot generate such high velocities in the LSB galaxy. Yet comparable rotation velocities are observed in the LSB and HSB examples. For the HSB systems, reasonable masses associated with the observed light can explain the observed rotation in the inner regions. In the LSB case, the rotation cannot be explained by the stellar mass and must be associated with an invisible component.

Tully & Verheijen (1997) describe how a location in the μ_0, r_d plane of the right panel of Fig. 2 maps into a specified ratio of the peak rotation velocity ascribed to the luminous disk, V_{max}^{disk}, to the maximum observed rotation velocity, $W_R^i/2$, assuming M/L =constant. Loci of constant values of $V_{max}^{disk}/0.5W_R^i$ are almost horizontal lines in the right panel of Fig. 2. Hence the surface brightness histograms of Fig. 1 can be transformed into the histogram of this dynamical ratio shown in Figure 4.

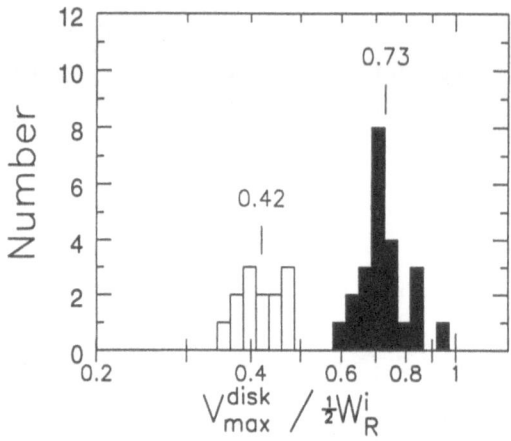

Fig. 4. Histogram of the ratio $V_{max}^{disk}/0.5W_R^i$ based on photometric properties. Isolated sub-sample only (nearest important neighbor > 80 kpc in projection). Values of V_{max}^{disk} follow from the properties of the exponential disk and $M/L_{K'} = 0.4M_\odot/L_\odot$. The relation between W_R^i and photometric parameters is given by the luminosity–line width correlation. Filled histogram: HSB systems; open histogram: LSB systems. Mean values for each family are indicated.

5   A Schematic Model

The speculation is offered that this difference between HSB and LSB systems is the rule. In LSB galaxies, the luminous material has settled into a disk that remains dominated by the halo potential. In HSB galaxies, the luminous material has become self-gravitating within the inner 2–3 scale lengths. Probably, the difference is a question of angular momentum content or exchange. High angular momentum systems may come into rotational equilibrium in the LSB state. Low angular momentum systems, or parts of systems, may shrink until the disk becomes self-gravitating.

The speculation continues that once the disk becomes self-gravitating it must secularly evolve to find a stable configuration. Evidently, the intermediate surface brightness regime is disfavored. The gap is a factor of 10 in surface density. The stable regime seems to abide by the 'disk–halo conspiracy' (van Albada and Sancisi 1986) which produces a flat rotation curve. This situation is reminiscent of the arguments of Mestel (1963) for discrete states of radial stability.

6   Acknowledgements

Mike Pierce, Jia-Sheng Huang, and Richard Wainscoat participated in the collection of data. We thank Renzo Sancisi, Erwin de Blok, and Stacy McGaugh for constructive comments. This research has been supported by NATO Collaborative Research Grant 940271.

References

Davies, J.I., Disney, M.J., Phillipps, S., Boyle, B.J., Couch, W.J. 1994, *MNRAS*, **269**, 349.

de Jong, R.S. 1996, *A&A*, **313**, 45.

de Blok, W.J.G., McGaugh, S.S. 1996, *ApJL*, accepted.

Disney, M.J. 1976, *Nat*, **263**, 573.

Freeman, K.C. 1970, *ApJ*, **160**, 811.

Impey, C.D., Bothun, G.D., Malin, D. 1988, *ApJ*, **330**, 634.

McGaugh, S.S., Bothun, G.D., Schombert, J.M. 1995, *AJ*, **110**, 573.

Mestel, L. 1963, *MNRAS*, **126**, 553.

Tully, R.B., Verheijen, M.A.W., Pierce, M.J., Huang, J.S., Wainscoat, R.J. 1996, *AJ* **112**, 2471.

Tully, R.B., Verheijen, M.A.W. 1997, *ApJ*, submitted.

van Albada, T.S., Sancisi, R. 1986, *Phil. Trans. R. Soc. Lond. Ser. A*, **320**, 447.

van den Bergh, S. 1966, *AJ*, **71**, 922.

van der Kruit, P.C. 1987, *A&A*, **173**, 59.

Verheijen, M.A.W. 1997, Ph.D. Thesis, University of Groningen, in preparation.

On the Phenomenology of Disk Galaxies

Giuseppe Gavazzi[1,2]

[1] Università degli Studi di Milano, Via Celoria 16, 20133 Milano, Italy
[2] Osservatorio Astronomico di Brera, Via Brera 28, 20121 Milano, Italy

Abstract. Using multifrequency observations, from the radio (21 cm line) to the UV (λ 2000 Å) of 928 nearby (z<0.03) late-type (giant+dwarf-Irr) galaxies, the phenomenology of late-type galaxies in the local Universe is revised as follows: 1) M/L in the NIR is constant; 2) population I indicators, namely: the atomic hydrogen content, the current star formation rate, the luminosity and the surface brightness of the young stellar population, traced by the blue and UV light are anti-correlated with the mass of galaxies; 3) conversely, the luminosity and the surface brightness of the old stellar population, traced by NIR light, increase with mass; 4) the fraction of centrally peaked NIR structures (bulges, nuclei) and their surface brightness increase non-linearly with mass. These evidences provide observational constraints to models of disk-galaxy evolution. They find a natural interpretation if it is assumed that the time scale τ of protogalaxy collapse is governed primarily by the system initial mass as $\tau = M^{-3}$.

1 Introduction

Cosmological models and numerical simulations in the CDM (or mixed DM) scenario predict successfully a number of observable properties of galaxies (see White; Baugh, this conference), such as light profiles in agreement with the observed mixtures of de Vaucouleurs & exponential laws, scale laws similar to the observed ones, both for spirals (the Tully-Fisher relation) and for Elliptical galaxies (the fundamental plane). Since the key idea in hierarchical models is that small-scale initial perturbation grow at relatively higher redshift, giving birth to low-mass galaxies first, and that massive galaxies form subsequently from the coalescence of smaller entities, they fail to predict the color- magnitude relations, which are however well established, indicating that large mass galaxies are redder than smaller systems. In the case of Elliptical galaxies the color- magnitude relation is currently interpreted as a metallicity sequence (see Arimoto, this conference). For spiral (disk) galaxies, however, there are several lines of evidence supporting that the color-magnitude relation is a genuine population sequence, as we will try to illustrate in this paper.

2 Sample Selection

The sample analyzed in this work is extracted from the multifrequency survey of nearby galaxies presented in Gavazzi & Boselli (1996). It comprises 1514 optically selected, $m_p \leq 15.7$ CGCG galaxies (Zwicky et al., 1961-68) in 8 nearby, rich clusters (A262, Cancer, A1367, Coma, A2147, A2151, A2197, A2199) and in the region of the Coma supercluster containing the bridge between A1367 and Coma ("Great Wall"). The catalogue has been subsequently extended to the Virgo cluster (Boselli et al, 1997). In total the sample includes 928 spiral galaxies. In the last 10 years extensive campaigns were carried out to observe this sample at various frequencies, in particular with Nicmos3 arrays in the Near Infrared (NIR) (Gavazzi et al, 1996a,b, Boselli et al, 1997) and with Hα imaging measurements recently obtained for 68 Coma supercluster objects (Gavazzi et al, in preparation). More details can be found in Gavazzi, Pierini & Boselli (1996) (GPB).

3   Mass – Luminosity Relation

This Section is aimed at showing that a simple photometric quantity, such as the luminosity taken in the NIR, is a reliable tracer of the dynamical mass of galaxies.

For the subsample of galaxies with type later than Sa, proper inclination and with 21cm line width accurately determined we construct the luminosity in various bands and the hybrid dynamical mass, within the optical radius, computed using the centrifugal equilibrium law: $M_{hyb} = W_c^2 R/G$; where: W_c is half the maximum rotational velocity (corrected for inclination) derived from 21cm line measurements; R is the optical radius (determined at the 25th mag arcsec^{-2} B band isophote). Fig. 1 reproduces the luminosity-M_{hyb} relation in four bands: U, B, V and H. The slope of the relations gradually increases from 0.7 (U) to 1.0 (H). We conclude, in agreement with Gavazzi (1993), that the NIR H band luminosity is an accurate tracer of the hybrid dynamical mass (within the optical radius) of galaxies. From now, throughout the rest of this paper, we will use L_H as a mass estimate of rotationally supported systems. At wavelengths shorter than NIR M/L increases with mass, in agreement with previous determinations (see e.g. M/L_B in Burstein et al., 1995).

4   Photometric Properties

Here we wish to discuss how various indicators of galaxy population (gas content, H_α E.W. and colors) scale with the dynamical mass (traced by the H luminosity)(Fig. 2, left panels). The right panels of Fig. 2 show the relations between the same quantities and the morphological type. In spite of the fact that our sample is dominated by small angular size objects (1-2 arcmin), thus difficult to classify accurately, we argue that the lack of correlation with the morphological type is genuine. On the contrary the mass offers a much more useful parametrization of galaxy properties.

The fractional HI content, normalized to L_H, anticorrelates with L_H. In other words we find that the the HI mass per unit dynamical mass is maximum in low luminosity galaxies. Dwarf galaxies have the ratio of HI gas to luminosity up to 100 times larger than giant galaxies. For these systems the majority of the "visible matter" is under the form of HI gas.

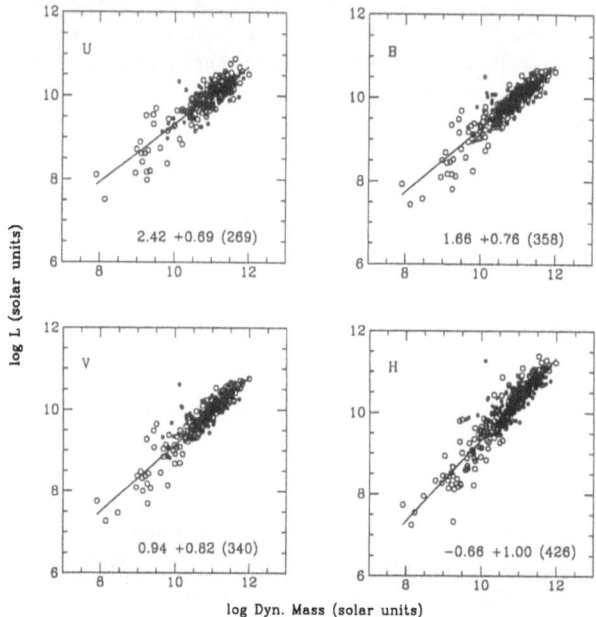

Fig. 1. The relationship between the hybrid dynamical mass and the luminosity in 4 bands: U, B, V and H. Filled symbols indicate isolated galaxies (in the Great Wall); cluster members are given with empty symbols. The parameters of the linear regression analysis (intercept+slope) and the number of points in each panel are given.

The relation between the U-B color index and H luminosity argues for the existence of a universal relationship between colors and H luminosity (mass). Gavazzi and Scodeggio (1996) have modeled the dependence on mass of various color indices using the synthesis population model of Bruzual and Charlot (1993). The model reproduces the observed colors, as a function of H luminosity if it is assumed that the initial star formation rate follows an exponential decay with a time constant τ proportional to $L_H^{-3.1}$.

The equivalent width (E.W.) of the H_α line emission integrated over the entire galaxy disks is a reliable indicator of the current ($< 10^7$ years), massive star formation rate in galaxies (see Kennicutt, 1983). Fig. 2 shows that there is a clear tendency for the H_α E.W. to increase steeply with decreasing L_H. This relation can plausibly be interpreted with a progressive change of stellar population with mass. The dotted line in Fig. 2 represents the model (assuming a Salpeter IMF and a solar metallicity) computed by Kennicutt et al (1994) as a function of τ (see their Table 1). Using the relation between τ and L_H suggested by Gavazzi and Scodeggio (1996) ($\tau = L_H^{-3.1}$), the model and the observed H_α E.W. are in satisfactory agreement.

The bottom panel of Fig. 2 shows the dependence of the NIR concentration

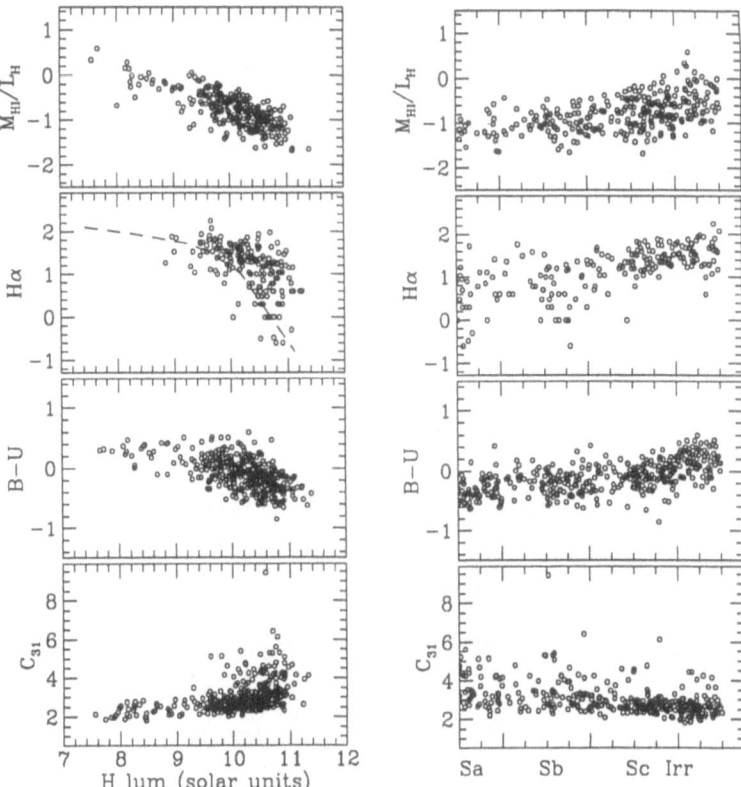

Fig. 2. Left panels: the relationship between the H band luminosity and the fractional HI gas content, U-B color index, the present star formation rate as given by the H_α E.W., and the H band concentration index C_{31}. The right panels give the dependence of the same quantities on Hubble type.

index C_{31} on galaxy luminosity. C_{31} is the ratio of the radii which contain 25% and 75% of the total light. $C_{31} < 3$ are found in pure disk galaxies. Departures from exponential light profiles due to nuclear structures and/or bulges occur at $C_{31} > 3$. The dependence of C_{31} on luminosity is well defined and strongly non-linear. Galaxies below $L_H = 10^{10}$ are pure disks (with few exceptions). Above this value the frequency of bulges increases rapidly.

4.1 Mean Surface-Brightness – Mass

Here we discuss the dependence of galaxy average surface brightness (S.B.) taken in the various bands) on H luminosity. Average S.B. are computed from the total magnitude (corrected for internal extinction as prescribed in Gavazzi and Boselli, 1996) divided by $\pi \times (a/2)^2$, where a is the galaxy major axis. For the first time the S.B. in the H band is derived using the infrared diameter $a_{21.5(H)}$, which was previously unavailable. For the visible bands we adopt $a_{25(B)}$. These dependences are given in Fig. 3 for 4 bands (open symbols). The H band data are reproduced in all plots (filled dots). The behavior of S.B. with L_H at the various frequencies is striking: the slope of S.B. with L_H goes monotonically from negative to positive with increasing wavelength from UV to H!

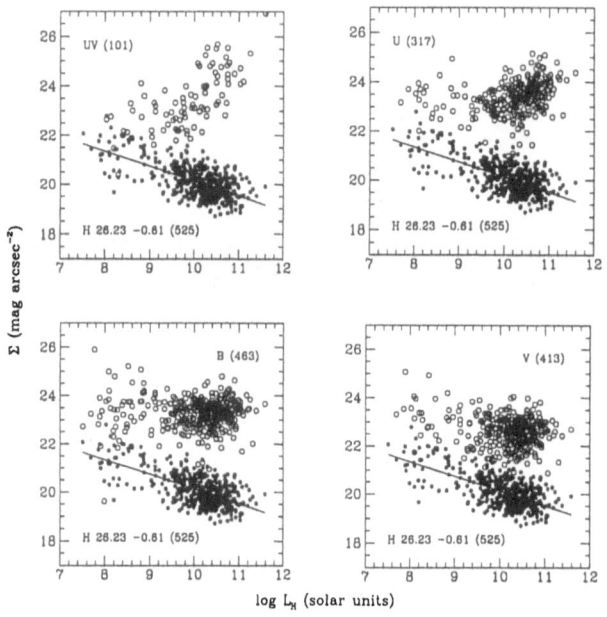

Fig. 3. The relationship between the H band luminosity and the mean surface brightness derived in 5 bands. UV, U, B and V are indicated with open symbols. The H band surface brightness (filled dots) is given in all panels for comparison (with the parameters of the linear regression analysis).

5   Discussion and Conclusion

The conclusions of the present investigation are as follows: 1) the NIR luminosity L_H is a good tracer of the (dynamical) visible matter in galaxies. 2) Population I indicators, i.e.: the fractional atomic gas content, the star formation rate, the fraction of young (blue) stars in the population, the surface brightness of the blue stars anti-correlate with mass. 3) On the contrary, Population II indicators, e.g.: the NIR light concentration index (or the bulge to disk ratio), the surface brightness of the red stars correlate with mass. These observational evidences form a coherent scenario if it is assumed that the mass is the principal parameter governing the efficiency, thus the collapse time scale of progenitors of late-type galaxies, as proposed by Sandage (1986). Systems with large initial mass underwent a prodigious initial star-burst which resulted in the old population concentrated in the bulge and in the inner disk. This event, which produced the almost complete exhaustion of the available primeval gas, must have been sufficiently fast that a large fraction of the gas is turned into stars before the collapse is completed, resulting in a galaxy with a relevant spheroidal component. Systems of smaller mass must have had a lower initial star formation efficiency, thus they consumed their gas at a much lower rate. Most of the gas was settled into a disk before forming stars (Larson, 1993). Such systems still contain enough gas to sustain a considerable star formation rate at $z=0$. Dwarf Irr galaxies can be understood as not yet fully formed systems. The time scale τ of the initial collapse is governed primarily by the galaxy mass (L_H) via $\tau = L_H^{-3.1}$ In conclusion we argue that to the first order the process of galaxy evolution can be described with a single parameter: the initial mass. The time dependence of the star formation rate SFR(t) is peaked at early epoch for massive galaxies and is shallower for smaller masses.

Acknowledgements I wish to thank A. Boselli, B. Catinella, D. Pierini and M. Scodeggio for their contribution to this work.

References

Binggeli, B., Sandage, A., & Tarenghi, M., 1984, AJ, 89, 64.

Boselli, A., Tuffs, R., Gavazzi, G., Hippelein, H. & Pierini, D., 1997, A&AS, (in press).

Bruzual, A., & Charlot, S., 1993, ApJ, 405, 538.

Burstein, D., Bender, R., Faber, S., & Nolthenius R., 1995, Ap. Lett. & Comm., 31, 95.

Gavazzi, G., 1993, ApJ, 419, 469.

Gavazzi, G., Pierini, D., Baffa, C., Hunt, L., Lisi, F., & Boselli, A., 1996a, A&AS, 120, 489.

Gavazzi, G., Pierini, D., Boselli, A. & Tuffs, R., 1996b, A&AS, 120, 521.

Gavazzi, G., & Boselli, A., 1996, Ap. Lett. & Comm., 35, 1.

Gavazzi, G., Pierini, D., & Boselli, A., 1996, A&A, 312, 397.

Gavazzi, G., & Scodeggio, M., 1996, A&A, 312, L29.

Kennicutt, R., 1983, ApJ, 272, 54.

Kennicutt, R., Tamblyn, P. & Congdon, C., 1994, ApJ, 435, 22.

Larson, R., 1993, in Physics of Nearby Galaxies: Nature or Norture?. Eds. T. Thuan, C. Balkowski & J. Thanh Van. (Gif sur Yvette: Editions Frontieres).

Sandage, A., 1986, A&A, 161, 89.

Zwicky, F., et al., 1961-1968, Catalogue of Galaxies and Clusters of Galaxies (Pasadena: Caltech).

Origin of the Colour-Magnitude Relation of Elliptical Galaxies

Nobuo Arimoto[1,2] and Tadayuki Kodama[1]

[1] Institute of Astronomy, University of Tokyo, Mitaka, Tokyo 181, Japan
[2] Institute of Astronomy, University of Cambridge, Madingley Road, Cambridge, CB3 0HA, UK

Abstract. Evolutionary models for elliptical galaxies are constructed by using a new population synthesis code. The dissipative collapse picture by Larson (1974) is adopted and model parameters are adjusted to reproduce the CM relation of Coma ellipticals. Two evolutionary sequences are calculated under the contexts of *metallicity hypothesis* and *age hypothesis*, both can equally explain the CM relation at the present epoch. The confrontation with the observational CM diagrams of E/S0 galaxies in the two distant clusters Abell 2390 ($z = 0.228$) and Abell 851 ($z = 0.407$) rejects the *age hypothesis*, and strongly suggests that the bulk of stars were formed early in elliptical galaxies, and give a reassuring confirmation of many previous contentions that the CM relation takes its origin at early times probably from galactic wind feedback. This conclusion is independent of the IMF, the SFR and the cosmological parameters.

1 Introduction

The integrated colours of elliptical galaxies become progressively bluer towards fainter magnitudes (Faber 1973; Visvanathan & Sandage 1977; Frogel et al. 1978; Persson, Frogel, & Aaronson 1979; Bower, Lucey, & Ellis 1992a; 1992b). With an accurate photometry of a large sample of elliptical galaxies in Virgo and Coma clusters of galaxies, Bower et al. (1992a, b) have shown that the colour-magnitude (CM) relations of two clusters are identical within observational uncertainties, suggesting that the CM relation is universal for cluster ellipticals. The rms scatter about the mean CM relation is typically ~ 0.04 mag, a comparable size to observational errors, and implies a virtually negligible intrinsic scatter. If the scatter is due to age dispersion among cluster ellipticals, the photometric data by Bower et al. (1992a, b) suggest that cluster ellipticals are unlikely to have formed below a redshift of 2 ($q_0 = 0.5$), which sets an upper limit of ~ 2 Gyrs for the age dispersion of the bulk stellar populations of cluster ellipticals. An identical upper limit to the age dispersion of such galaxies is also suggested by their small dispersion about the so-called fundamental plane (Renzini & Ciotti 1993). These findings are in excellent agreement with a scenario for the formation of elliptical galaxies dominated by protogalactic collapse and dissipational star formation early in the evolution of the universe (Larson 1974; Arimoto & Yoshii 1986, 1987, hereafter AY87; Matteucci & Tornambé 1987; Yoshii & Arimoto 1987; Bressan, Chiosi, & Fagotto, 1994; Tantalo et al. 1996). The CM relation itself is also quite naturally established in collapse/wind model of galaxy formation; a galactic wind is induced progressively later in more massive galaxies owing to deeper potential and stellar populations in brighter galaxies are much more enhanced in heavy elements and as a result redder in colours.

This *conventional* interpretation of the CM relation of ellipticals, however, has been questioned by Worthey (1996) who points out, based on his population synthesis model (Worthey 1994), that the sequence of colours and line strengths among ellipticals can be almost equally well explained by a progressive decrease of either a mean stellar metallicity or an effective stellar age. Recently, Bressan, Chiosi, & Tantalo (1996) have performed a detailed analysis of the CM relation of elliptical galaxies. In addition to broad band colours, the authors have considered absorption line strengths and their radial gradients, abundance ratios, spectral energy distribution in UV-optical region, and stellar velocity dispersion, all of which might be tightly coupled with the CM relation itself, and have suggested that the history of star formation in elliptical galaxies has probably been more complicated and heterogeneous than the galactic wind scenario mentioned above.

All in all, the previous interpretation that the CM relation is primarily a metallicity effect looks less secured today. It is not clear anymore whether fainter ellipticals are bluer because stellar populations are younger or because stars are more metal-deficient in average. Indeed an alternative scenario for the formation of elliptical galaxies suggests that ellipticals result from the merging of lesser stellar systems taking place mostly at later times and involving substantial star formation (Toomre & Toomre 1972; Schweizer & Seitzer 1992). This scenario is in good agreement with fine structures and kinematic anomalies of some elliptical galaxies (eg., Kormendy & Djorgovski 1989). In addition to this, the optimizing synthesis tend to pick up finite fraction of light from intermediate age populations, which also supports recent episodes of star formation in ellipticals (O'Connell 1976; Pickles 1985; Rose 1985; O'Connell 1986; Bica 1988; see also Arimoto 1996).

Although several attempts have been made to disentangle the age and metallicity effects on the line strengths as well as on the broad band colours (Buzzoni, Gariboldi, & Mantegazza 1992; Buzzoni, Chincarini, & Molinari 1993; Gonzales 1993; Worthey 1994; Worthey et al. 1994), it is still very difficult to break the age-metallicity degeneracy of elliptical galaxies so far as one sticks to photo-spectroscopic data of galaxies at the present epoch. However, it is comparatively easier to show that the CM relation is indeed primarily due to the stellar metallicity effect if one studies evolution of the CM relation as a function of look-back time. If the CM relation is an age sequence, it should evolve rapidly and should disappear beyond certain redshift, because fainter galaxies are approaching to their formation epoch. Contrary, if the CM relation is a metallicity sequence and ellipticals are essentially old, it should evolve *passively* and should be still traced at significantly high redshifts. In this paper, we compare the theoretical evolution of the CM relation with the CM relations of cluster ellipticals at cosmological distances ($z \simeq 0.2 - 0.4$) and show that the bulk of stars were probably formed early in elliptical galaxies, and that the CM relation takes its origin at early times from galactic wind feedback, thus reinsuring confirmation of previous contentions.

2   Model

2.1   Stellar Population Synthesis Code

The basic structure of our new code follows Arimoto & Yoshii's (1986) stellar population synthesis prescription that takes into account the effects of stellar metallicity on integrated colours of galaxies for the first time. New stellar evolutionary tracks are incorporated comprehensively, and late stellar evolutionary stages are fully taken into account. The code uses the so-called *isochrone synthesis* technique (cf. Charlot & Bruzual 1991) and gives the evolution of synthesized spectra of galaxies in a consistent manner with galaxy chemical evolution. The details of the code will appear in Kodama (1997).

2.2   Evolution Model of Elliptical Galaxies

A formation picture of elliptical galaxies by Larson (1974) and AY87 assumes that initial bursts of star formation occurred during rapidly infalling stage of proto-galactic gas clouds and that the gas was expelled when the cumulative thermal energy from supernova explosions exceeded the binding energy of the remaining gas. No stars formed afterwards and a galaxy has evolved passively since then.

Following AY87, we assume that star formation abruptly terminates at the epoch of galactic wind $t = t_{gw}$. A precise value of t_{gw} is rather difficult to evaluate, because it would depend strongly on how much dark matter is involved in defining a gravitational potential as well as on the change of galactic radius during the gravitationally collapsing phase. Instead, we determine t_{gw} empirically in such a way that models can reproduce $V - K$ of nearby elliptical galaxies in Coma cluster.

Model parameters are chosen as $x = 1.20$, $m_l = 0.1M_\odot$, $m_u = 60M_\odot$, $\tau = 0.1$ Gyrs, and $\tau_{in} = 0.1$ Gyrs. $m_l = 0.1M_\odot$ is chosen to give a mass-to-light ratio $M/L_B \simeq 8$ for giant ellipticals (Faber & Jackson 1976; Michard 1980; Schechter 1980). Although the choices of x, τ, and τ_{in} are somewhat arbitrary, we note that our analysis is rather insensitive to these parameters.

(a) *metallicity sequence* : The colour change along the CM relation in $V - K$ is explained simply by a difference of mean stellar metallicities of galaxies (AY87). Following this interpretation, we construct a sequence of models with different mean metallicities. All ellipticals are assumed to have an equivalent age $T_G = 15$ Gyrs. In galactic wind model, the metallicity difference can be produced by assigning different t_{gw}. A standard CM relation in $V - K$ is, then, reproduced with $t_{gw} = 0.52$ Gyrs at $M_V = -23$ mag and $t_{gw} = 0.10$ Gyrs at $M_V = -17$ mag. In total 13 models are calculated for 0.5 magnitude interval in M_V and properties of some models at the present epoch are summarized in the upper part of Table 1: M/L_B is given in solar units, galaxy mass M_G is calculated from M_V and M/L_V.

(b) *age sequence* : The CM relation can be explained alternatively by a progressive decrease of galaxy age towards fainter ellipticals (Worthey 1996). This can be represented by a sequence of models with fixed t_{gw} and various galaxy ages. A wind time is fixed to be the same as the brightest model of the metallicity sequence; ie., $t_{gw} = 0.52$ Gyrs. A standard CM relation in $V - K$ is, then, reproduced with $T_G = 15$ Gyrs at $M_V = -23$ mag and $T_G = 2.3$ Gyrs at $M_V = -17$ mag. In total 25 models are calculated for 0.25 magnitude interval in M_V and properties of some models at the present epoch are summarized in the lower part of Table 1. All models have $< \log Z/Z_\odot > \simeq 0$ in this sequence.

Table 1. Model sequences of elliptical galaxies at $z = 0$.

	M_V (mag)	-23.00	-21.98	-20.98	-20.04	-18.99	-17.97	-16.96
	$M_G(10^9 M_\odot)$	812	291	108	42.0	14.8	5.35	1.98
	T_G (Gyr)	15.00	15.00	15.00	15.00	15.00	15.00	15.00
metallicity	t_{gw} (Gyr)	0.515	0.320	0.235	0.185	0.147	0.120	0.100
sequence	$< \log Z/Z_\odot >$	-0.014	-0.116	-0.213	-0.304	-0.404	-0.502	-0.596
	$\log < Z/Z_\odot >$	0.147	0.025	-0.079	-0.173	-0.273	-0.368	-0.458
	M/L_B	8.406	7.568	6.846	6.240	5.652	5.144	4.700
	$U - V$	1.671	1.596	1.521	1.452	1.378	1.308	1.242
	$V - K$	3.355	3.273	3.192	3.116	3.032	2.949	2.868
	M_V (mag)	-23.00	-22.00	-21.00	-20.00	-19.00	-18.00	-17.00
	$M_G(10^9 M_\odot)$	812	232	68.5	22.2	6.20	1.86	0.693
	T_G (Gyr)	15.00	9.97	6.88	5.57	3.68	2.54	2.30
age	t_{gw} (Gyr)	0.515	0.515	0.515	0.515	0.515	0.515	0.515
sequence	$< \log Z/Z_\odot >$	-0.014	0.002	0.013	0.004	0.013	0.060	0.039
	$\log < Z/Z_\odot >$	0.147	0.160	0.176	0.160	0.166	0.207	0.189
	M/L_B	8.406	5.856	4.190	3.334	2.266	1.635	1.483
	$U - V$	1.671	1.571	1.465	1.408	1.326	1.211	1.146
	$V - K$	3.355	3.274	3.194	3.113	3.033	2.952	2.871

3   Comparison with Observations

3.1   Distant Clusters

Evolutions of the CM relation in $g-r$ vs. M_r diagram are presented in Figs. 1 and 2 for the metallicity sequence and the age sequence, respectively. Cosmological parameters are again $H_0 = 50$ km s^{-1} Mpc^{-1} and $q_0 = 0.1$. The age of the universe is then about 16.5 Gyrs, and a galactic age $T_G = 15$ Gyrs corresponds to the formation redshift $z_f \simeq 5.4$, and $T_G = 2.3$ Gyrs to $z_f \simeq 0.14$. Crosses on each line indicate the position of each model defined at $z = 0$. Galaxy masses at $z = 0$ are indicated for some models.

Observational data of E/S0 galaxies in two distant clusters Abell 2390 ($z = 0.228$; triangles) and Abell 851 ($z = 0.407$; circles) are superposed in Fig.1. For Abell 2390, data are taken from Yee et al. (1996). The membership is confirmed by redshift measurement and E/S0 morphologies are classified spectroscopically. Galaxies in the central field of the cluster ($7.3' \times 9.1'$) are represented by filled triangles. For Abell 851, galaxies classified as E/S0 with HST images are taken from Dressler et al. (1994). Filled circles indicate cluster members confirmed spectroscopically. Absolute magnitudes M_r are calculated from the redshift of each cluster with a help of the adopted cosmological parameters.

Both clusters Abell 2390 and Abell 852 show well-defined CM relations of E/S0 galaxies over $3 \sim 4$ magnitude range, and that these CM relations are almost in parallel to the CM relation at $z = 0$ which is calculated from model sequences calibrated with the CM relation of Coma in $V - K$ vs. M_V diagram. This trend is very well reproduced by the metallicity sequence (Fig.1). In the observer's frame, the CM relation evolves almost in parallel towards redder colours and is well defined even at $z = 1$. Indeed, Stanford, Eisenhardt and Dickinson (1995) recently analyzed the optical$-K$ CM relation of elliptical galaxies in Abell 370 ($z = 0.374$) and Abell 851 ($z = 0.407$) and showed that these two clusters have almost identical slopes and dispersions of the CM relations to those of Coma ellipticals. They therefore concluded that the CM relations of the two clusters are fully consistent with a picture of old passive evolution with age more than 10 Gyrs.

For the age sequence, on the contrary, the CM relation changes drastically (Fig.2). Galaxies *finally* on the CM relation at $z = 0$ become rapidly brighter and bluer as a function of lookback time, because smaller galaxies become considerably young as they approach to their formation epoch. For example, galaxies smaller than $2.2 \times 10^{10} M_\odot$ at $z = 0$ no longer exist in the universe at $z = 0.4$. As a result, at $z = 0.2$, the CM relation holds only for 2.4 mag from the brightest end, and virtually disappears at $z \geq 0.6$. We note that the slope of the CM relation becomes progressively steeper towards higher redshifts, which is in high contrast to the metallicity sequence whose CM relation keeps its slope nearly constant from $z = 0$ to 1.

The CM relations of the two clusters Abell 2390 and Abell 851 extend to much fainter magnitudes than the theoretical loci predicted by the age sequence models. This is a clear evidence that the CM relation cannot be simply attributed to the age difference alone. Thus, it can be concluded that the metallicity is the key factor that accounts for the CM relation of elliptical galaxies.

Our conclusion does not depend on the cosmological parameters chosen.

4 Discussion

Galaxies defining the CM relation at high redshifts may not be simply the progenitors of those defining the CM relation at $z \simeq 0$, since galaxies undergoing star formation at moderate to high redshifts could well enter the CM relation at lower redshifts, or alternatively, since galaxies in the CM relation at high redshift may

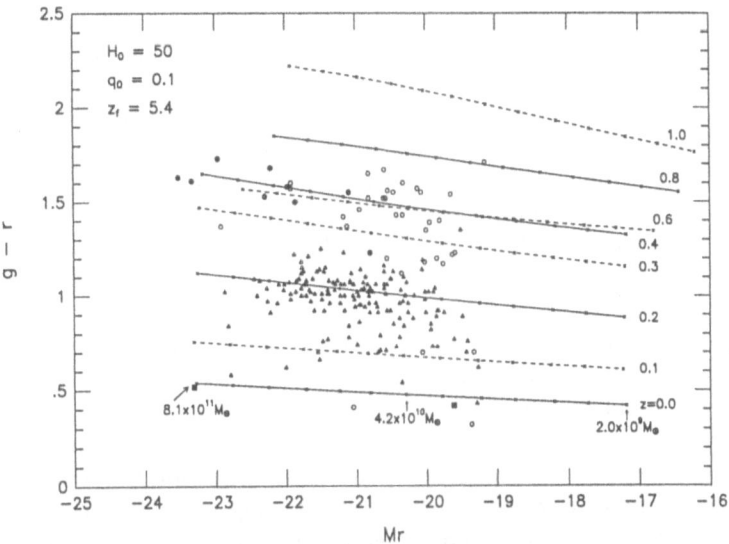

Fig. 1. Evolution of the CM relation for the metallicity sequence in the observer's frame (solid and dashed lines). Triangles and circles indicate E/S0 galaxies in Abell 2390 ($z = 0.228$) and Abell 851 ($z = 0.407$) clusters, respectively.

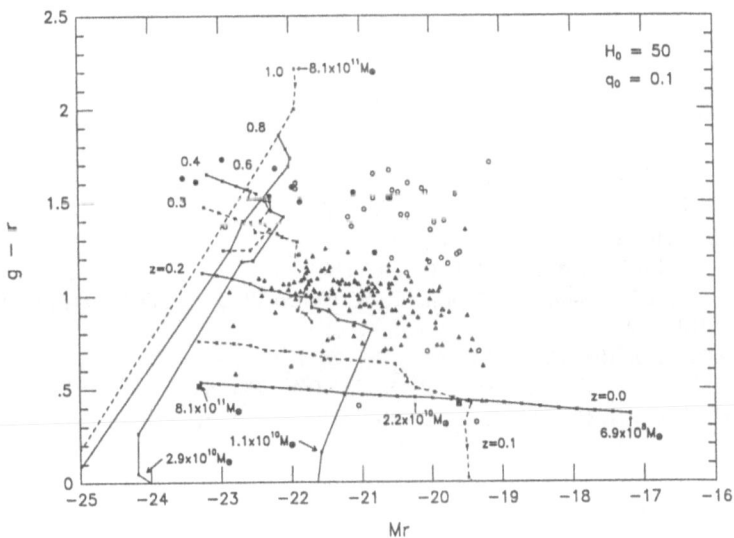

Fig. 2. The same as Fig.1, but for the age sequence.

leave it if they undergo new episode of star formation (e.g., Barger et al. 1996). However, if galaxies suffered star formation at moderate and high redshifts and if the light of the intermediate-age stellar populations dominate the luminosities of the present day ellipticals, these galaxies should deviate considerably from the CM relation, unless the star formation episodes occurred beyond a redshift of 2 (Bower et al. 1992b). In other words, the light from the intermediate-age stars cannot be a primary cause of the CM relation of cluster ellipticals today. Figure 1 shows that the CM relations were already established in Abell 2390 at $z = 0.228$ (in particular, E/S0 in the central field) and Abell 851 at $z = 0.407$. This implies that these galaxies had evolved *passively* down to these redshifts, and if the CM relations at these redshifts are due to the age effect, the epoch of galaxy formation must have been very precisely tuned for each specified galaxy mass. We believe this is very unlikely.

Finally we should comment on our definition of the epoch of galaxy formation. Quite often the epoch of galaxy formation is defined to be the time when the protogalaxy decouples from the Hubble expansion to the formation of the dark halo of the galaxy. However, this process is virtually invisible and difficult to observe. Instead, as is often done by observational cosmologists, we identify the epoch of galaxy formation with observable events, or more precisely, the time when protogalaxies (normal massive ellipticals and bulges of spiral galaxies) are undergoing their first major episode of star formation. Then, the arguments given in this paper suggest that stars formed during this episode, at $z \geq 2$, dominate the light of the elliptical galaxy and the CM relation was established at that epoch. Since only the galactic wind model has so far been successful in explaining the observed CM relation, we are led to reach a conclusion that the CM relation takes its origin at early times from galactic wind feedback.

References

Arimoto N., 1996, in: From Stars to Galaxies, eds. C. Leitherer, U. Fritz-von Alvensleben, J. Huchra (ASP Conf. Ser. Vol. 98), p.287

Arimoto N., Yoshii Y., 1986, A&A 164, 260

Arimoto N., Yoshii Y., 1987, A&A 173, 23 (AY87)

Barger A. J., Aragón-Salamanca A., Ellis R. S., Couch W. J., Smail I., Sharples R. M., 1996, MNRAS 279, 1

Bica E., 1988, A&A 195, 76

Bower R. G., Lucey J. R., Ellis R. S., 1992a, MNRAS 254, 589

Bower R. G., Lucey J. R., Ellis R. S., 1992b, MNRAS 254, 601

Bressan A., Chiosi C., Fagotto F., 1994, ApJ 94, 63

Bressan A., Chiosi C., Tantalo R., 1996, A&A 311, 425

Buzzoni A., Chincarini G., Molinari E., 1993, ApJ 410, 499

Buzzoni A., Gariboldi G., Mantegazza L. 1992, AJ 103, 1814

Charlot S., Bruzual A. G., 1991, ApJ 367, 126

Dressler A., Oemler A., Butcher H. R., Gunn J. E., 1994, ApJ 430, 107

Faber S. M. 1973, ApJ 179, 731

Faber S. M., Jackson R. E., 1976, ApJ 204, 668

Frogel J. A., Persson S. E., Aaronson M., Matthews K., 1978, ApJ 220, 75

Gonzales J. J., 1993, Ph.D. Thesis, Univ. California, Santa Cruz

dama T., 1997, Ph.D. Thesis, University of Tokyo, in preparation

Kormendy J., Djorgovski S., 1989, ARA&A 27, 235

Larson R. B., 1974, MNRAS 166, 686

Matteucci F., Tornambè F., 1987, A&A 185, 51

Michard R., 1980, A&A 91, 122

O'Connell R. W., 1976, ApJ 206, 370

O'Connell R. W., 1986, in: Spectral Evolution of Galaxies, eds. C.Chiosi, A.Renzini,
    (Dordrecht: Reidel), p.195

Persson S. E., Frogel J. A., Aaronson M., 1979, ApJS 39, 61

Pickles A. J., 1985, ApJ 296, 340

Renzini A., Ciotti, L., 1993, ApJ 416, L49

Schechter P., 1980, AJ 85, 801

Schweizer F., Seitzer P., 1992, AJ 104, 1039

Stanford S. A., Eisenhardt P. R. M., Dickinson M., 1995 ApJ, 450, 512

Tantalo R., Chiosi C., Bressan A., Fagotto F., 1996, A&A 311, 361

Toomre A., Toomre J., 1972, ApJ 178, 623

Visvanathan N., Sandage A., 1977, ApJ 216, 214

Worthey G., 1994, ApJS 95, 107

Worthey G., Faber S. M., Gonzalez J. J., Berstein D., 1994, ApJS 94, 687

Worthey G., 1996, in: Fresh Views of Elliptical Galaxies, eds. A. Buzzoni, A. Renzini,
    A. Serrano (ASP Conf. Ser. Vol. 86), p.203

Yee H. K. C., Abraham R. G., Gravel G., Carlberg R. G., Smecker-Hane T. A., Schade
    D., Rigler M, 1996, ApJS 102, 289

Yoshii Y., Arimoto N., 1987, A&A 188, 13

Stellar Populations in Bulges of Spiral Galaxies

Pascale Jablonka

DAEC-URA173, Observatoire de Paris-Meudon, France

Abstract. We discuss the integrated properties of the stellar population in bulges along the Hubble sequence and new HST data for individual stars in the bulge of M31.

1 Bulges Along the Hubble Sequence

Bulge stellar populations are still poorly investigated. It remains unclear whether bulges are formed in the very early stages of galaxy formation in synchronicity with the halo, or on the contrary on longer time scales and from disk material. The effective role of bars in bulge growth and star formation still remains unclear. The bulge-to-disk light ratio varies along the Hubble sequence, and it is important to understand whether this trend translates into a mass scale only or into differences of star formation history as well. For a long time, the bulge of our Galaxy has been considered as the prototype of bulges in general. However, the study of bulges is complicated by its contamination by disk light.

Taking advantage of Kent (1985) work on decomposition between disk and bulge lights, we show in Fig. 1, for a few face-on spiral galaxies, the radial variation of their bulge-to-disk light ratio, L_{Bulge}/L_{Disk}. It is conspicuous that the integration of light in a constant aperture for all these galaxies would gather different amount of disk light. This would severely affect any search for trends of bulge properties along the Hubble sequence. Also, this disk light contamination can be present even in the central regions of galaxies. Therefore, observations of bulges require the acceptance of some disk light contamination, yet this contamination must be kept as small as possible and in equal proportion for all galaxies; this can be done by having a different aperture for each galaxy.

The integrated-light spectra of 28 spiral galaxies were obtained in 1993, using an automatic drift scanning procedure available at the Steward Observatory 2.3-m telescope equipped with a Boller & Chivens CCD spectrograph. For each galaxy, structural parameters, such as effective radius r_e, scale height h, surface brightness values μ_e at r_e and μ_0 at the center were available (Kent 1985). Consequently, we were able to calculate the radius corresponding to a fixed value of the bulge-to-disk luminosity ratio, a ratio equal to 6 for all sample galaxies.

Fig. 2 presents the variation of the bulge Mg_2 index with its absolute magnitude in r-Gunn photometric band. The Hubble type of each galaxy is indicated by its corresponding number. There is clearly delineated a metallicity-luminosity relation for bulges along the Hubble sequence: the bulge of our Galaxy can no longer be considered as a template for all bulges. If bulges tend to be fainter in late-type spirals than in early-type ones, this is however the only segrega-

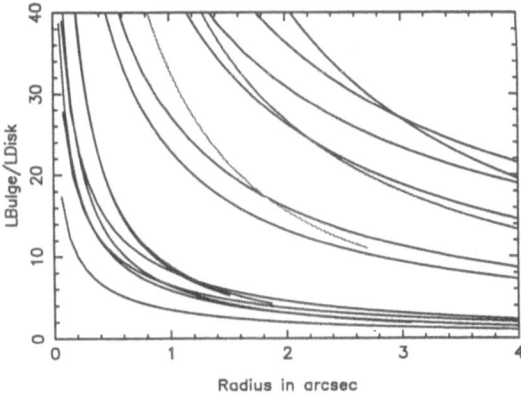

Fig. 1. The radial variation of the bulge-to-disk luminosity ratio in 28 spiral galaxies.

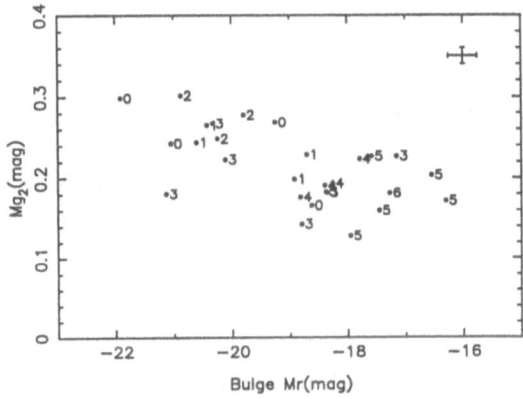

Fig. 2. The relation between the bulge total r-Gunn luminosity and Mg_2 index. Numbers represent the galaxies Hubble types (T parameter). Typical maximum errors are shown.

tion seen in Fig. 2. Besides this tendency, all Hubble types are mixed; the bulge luminosity is certainly the dominant factor in the correlation.

Fig. 3 displays the relation existing between the bulge Mg_2 index and the galaxy central velocity dispersion, when this information is available. This is the case for about half the galaxies in our sample. Our own Galaxy is indicated by an asterisk and the mean relation obtained for elliptical galaxies (Gorgas & Gonzàlez, private communication) is superimposed with its 1-σ dispersion. This figure underlines how bulges and elliptical galaxies are closely related systems and suggest common processes of formation.

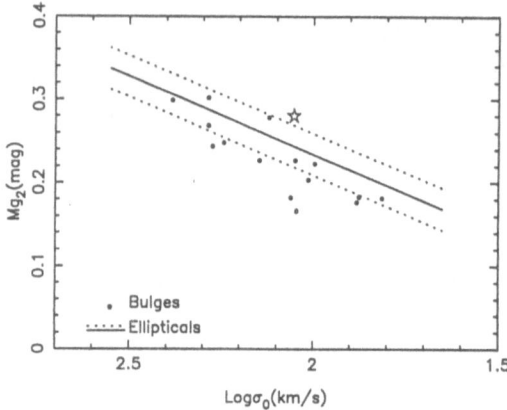

Fig. 3. The variation of the bulge Mg_2 index with the central velocity dispersion σ_0. Plain and doted lines represent the mean relation and 1-σ dispersion known for ellipticals (Gorgas & Gonzàlez, private communication). The open asterisk indicates the bulge of our Galaxy.

2   The Bulge of M31

The high inclination between the disk of M31 and its line-of-sight allows the study of the stellar populations of its bulge. The view we have of it is global and essentially free of pollution by disk stars, unlike the case for the bulge of our Galaxy.

We obtained, with the WFPC2 camera on the Hubble Space Telescope, Cycle 5 and 6 high spatial resolution images, in filters F555W and F814W, of a few fields centered on super-metal rich star clusters in the bulge of M31, clusters for which we have ground-based spectrophotometric data (Jablonka et al. 1992). Two clusters, G170 and G177, are located SW along the major axis of M31, respectively at 6.1 and 3.2 arcmin of the galaxy nucleus; another cluster, G198, is located NE along the major axis at 3.7 arcmin (Huchra et al. 1991). Adopting 1 arcmin = 250 pc from Rich & Mighell (1995), these separations correspond to projected distances of about 1.55, 0.80, and 0.92 kpc, respectively. In addition to the cluster stellar populations, these HST data give us the opportunity to study the stellar populations in the surrounding bulge fields. We present hereafter some of the results on the latter point, which are part of an extensive work to be published elsewhere (Jablonka et al. 1997).

Fig. 4 displays the cluster G177 and its surrounding bulge stellar field, as observed in the PC frame. The field is 36 arcsec by 36 arcsec in size. It is a composite image from the F814W and F555W frames. The total exposure time is 6500s in the filter F555W and 6000s in the filter F814W. This figure illustrates the compactness of the cluster G177 and the richness of its surrounding field.

We used the DAOPHOT/ALLSTAR/ALLFRAME software package for crowded-field photometry (Stetson 1994), along with the PSFs kindly provided

Fig. 4. The 36″ × 36″ PC image centered on the cluster G177

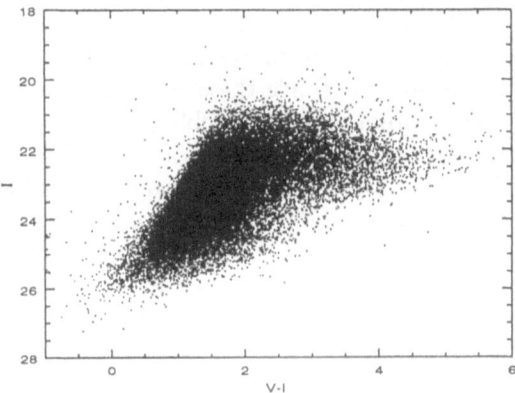

Fig. 5. The color-magnitude diagram for 65,000 stars in the field around the cluster G177, located 3.2 arcmin SW of M31 nucleus

to us by the Cepheid-Distance-Scale HST key project. The F555W and F814W instrumental magnitudes are *in fine* converted to Johnson V and Cousins I magnitudes. Fig. 5 displays the resulting (I,V–I) color-magnitude diagram obtained when the entire PC field around G177 is analyzed. About 65,000 stars are detected.

The very high density of stars in this field appears to be the main limiting factor in the detection of faint-magnitude stars, a problem which a longer exposure time might not straightforwardly solve. Our photometry reaches the horizontal-branch (HB) level, but prevents us to see its morphology. The red-giant branch (RGB) is however clearly observed, indicating a very metal-rich old population.

We are not the first ones to be attracted by the rich mine of information provided by M31. In particular, Mould (1986), Mould & Kristian (1986), and

Rich & Mighell (1995) gathered ground-based and WFPC (aberrated HST) data for various stellar fields located between 0.2 and 12 kpc from the galaxy center.

Rich & Mighell (1995), synthesizing their own and previous works, discuss an apparent brightening of the RGB tip, when moving along a sequence of stellar fields at distances decreasing from the halo to the galaxy center: at 7 kpc, the RGB tip of stars in the field appears at the same luminosity as the one observed for globular clusters (I $\sim$ 20.5 mag), while it becomes 1 mag brighter when closer to the M31 center (I $\sim$ 19.5 mag). Rich & Mighell (1995) mention the contradiction between these observations and what can be expected from a luminosity variation due to metallicity, since metal-rich globular clusters have fainter RGBs (Bica et al. 1991). They mention the hypotheses of an M31 bulge younger by 5 to 7 Gyr than the extreme Galactic halo and/or of the presence of a rare stellar evolutionary phase.

However, we do not confirm the detection of these very luminous stars, and our data are in perfect agreement with an increase of metallicity towards the center of the galaxy. The refurbished Hubble Space Telescope brings a new light on this problem: 65,000 stars detected in a $36'' \times 36''$ area, is equivalent to about 50 stars per 1 arcsec^2. Only the refurbished HST with its high spatial resolution capability is able to resolve these stars. It seems very possible that previous studies suffered from crowding problems and detected blends instead of individual stars, thus artificially overestimating their luminosities.

The bulge stellar populations around the two other M31 clusters G198 and G170 present the same characteristics of those observed for G177 and shown in Fig. 5. At this stage of the analysis, it is already clear that M31 bulge stars share the same mean locus in their color-magnitude diagrams as the two Galactic globular clusters NGC 6528 and NGC 6553 (Ortolani et al. 1995). These results favor a rapid formation of bulges during the early history of galaxies.

3   Acknowledgements

I wish to thank the conference organizers for such an interesting and lively meeting. I also wish to thank my collaborators, T. Bridges, A. Maeder, G. Meylan, G. Meynet, A. Sarajedini, for allowing me to quote some preliminary results before publication.

References

Bica E., Barbuy B., Ortolani S., 1991, PASP, 382, L15
Huchra J., Brodie J., Kent S., 1991, ApJ, 370, 495
Jablonka P., Alloin D., Bica E., 1992, A&A, 260, 97
Jablonka P., Martin P., Arimoto N., 1996, AJ, 112, 1415
Jablonka P., Bridges T., Maeder A., Meylan G., Meynet G., Sarajedini A., 1997, in
    preparation
Jablonka P., Martin P., Arimoto N., 1996, AJ, 112, 1415
Kent S.M., 1985, ApJS, 59, 115

Ortolani S. et al., 1995, Nature, 377, 701
Mould J., Kristian J., 1986, ApJ, 305, 591
Mould J., 1986 in Stellar Populations, Ed. A. Renzini, p. 9
Rich M., Mighell K., 1995, ApJ, 439, 145
Stetson P., 1994, PASP, 106, 250

The Luminosity–Width Relation of Spiral Galaxies

Riccardo Giovanelli

Department of Astronomy
Cornell University
Ithaca, NY 14853, USA

Abstract. Several technical aspects of the Luminosity–Width relation of spiral galaxies are discussed, such as the characteristics and sources of its scatter, its dependence on morphology and environment and issues related to the opacity of disks. Alternative formulations of the relationship are also considered, such as those proposed by Karachentsev (1989) and Chiba & Yoshii (1996). Results of various applications of the relation, as they apply to the determination of the peculiar velocity field, the value of H_o and the evolution of M/L are finally discussed.

1 Introduction

Twenty–years after the seminal contribution by R.B. Tully and J.R. Fisher (1977, hereafter TF), the relation between luminosity and velocity width

$$L \propto V_{rot}^{\alpha} \qquad (1)$$

is one of the most extensively applied scaling laws in extragalactic astronomy. Its applications range from the study of the internal dynamics of galaxies to cosmology, notably the determination of the peculiar velocity field, the mean matter density and the density fluctuation field, the measurement of the Hubble constant H_o and, more recently, the exploration of the evolution of M/L as a function of z.

The simplest physical justification for (1) can be found in Eisenstein & Loeb (1996). Consider, they argue, a spherical cloud of mass M collapsing at the epoch t_{coll}; the turnaround radius R_{ta} at that epoch is $R_{ta} \propto M^{1/3} t_{coll}^{2/3}$. The total energy of the cloud, after virialization, is $E \propto M\sigma^2 \propto GM^2/R_{ta}$, where σ is the velocity dispersion of its internal motions. Thus, assuming that a galaxy results from collapse at a single epoch and not from a succession of mergers, $\sigma \propto (M/t_{coll})^{1/3}$, so if all galaxies collapse in single events at the same epoch (no further merger), and if M/L ratios don't vary significantly as galaxies evolve, $L \propto \sigma^3$. Because variations in the formation histories among galaxy systems should be expected, scatter in the luminosity–width relation should result (see Eisenstein's contribution to these proceedings).

A more frequently invoked physical "syntax" behind relation (1) relies on the combination of scaling arguments and other empirical laws. Consider a pure exponential disk of central disk surface brightness $I(0)$ and scale length r_d ; its total luminosity is $L_d \propto r_d^2 I(0)$. The mass internal to radius R is $M(R) \propto RV^2$; if the rotation curve flattens in the outer regions of the disk, as is usually the case for spiral galaxies, the total mass is $M_{tot} \propto r_d V_{max}^2$. We can then write $L_d \propto (M_{tot}/L_d)^{-2} V_{max}^4 / I(0)$. If a dark matter halo is present so that the disk mass is $M_d = \Gamma M_{tot}$, then

$$L_d \propto (M_d/L_d)^{-2} \Gamma^2 V_{max}^4 / I(0) \qquad (2)$$

When a number of "standard" assumptions are made, i.e. that $\Gamma \sim$ const, $M_d/L_d \sim$ const and $I(0) \sim$ const (Freeman's law, 1970), $L_d \propto V_{max}^4$, which resembles eqn. (1). In practice, none of the assumptions of constancy for M_d/L_d, Γ and $I(0)$ apply; all those parameters exhibit mild dependencies on V_{max} (or L_d), reducing the exponent to values $\alpha < 4$, in a measure that depends on the adopted photometric band. Some workers (e.g. Aaronson et al. 1986; Willick 1990; Pierce & Tully 1996) have found significant departures from a single power law behavior, and quadratic or bilinear TF fits have sometimes been adopted. It has been argued that inappropriate extinction corrections (Giovanelli et al. 1995) or samples including a mixture of morphological types (Giovanelli et al. 1997a) may result in TF departures from linearity. There is however no a priori reason to expect that the TF relation be strictly linear. Recent N–body simulations (Navarro 1996; Navarro et al. 1996) provide further insights in the understanding of the TF relation.

In this contribution, we first review possible alternative formulations to the form (1), as ventilated in the recent literature (Section 2). We then proceed to discuss some of its basic properties, namely the characteristics of the observed scatter (Section 3), its dependence on morphology, environment and the uncertainty introduced by the adopted internal extinction relations (Section 4), the accuracy of empirically determined template relations (Section 5). Finally, in Section 6 we review briefly some of the results of applications of the TF relation, namely those regarding the study of the peculiar velocity field, the derivation of the Hubble constant and of investigations at high z. Many aspects of the work discussed here are based on results obtained in collaboration with P. Chamaraux, L. da Costa, W. Freudling, M. Haynes, J. Salzer and G. Wegner.

2   A Better Relation?

The desire to obtain a relation with as small a figure of scatter as possible, which would have obvious advantages in its predictive capacity, has prompted searches for formulations of the TF relation involving a third parameter or an alternative choice of primary parameters. The significant improvement represented by the Fundamental Plane (FP) relation between velocity dispersion, effective radius and effective surface brightness for spheroidal systems (Djorgovski & Davis 1987; Dressler et al. 1987), with respect to either the Faber–Jackson (1976) or the Kormendy (1977) relations, certainly constitutes a strong enticement to search for an analogous relation for spirals. This possibility has recently been explored by Karachentsev (1989) and, independently, by Chiba & Yoshii (1996). Here we review their work and explore the potential of their proposal.

An analytical solution for the rotation curve of an infinitely thin exponential disk was first obtained by Freeman (1970). Such form has a peak

$$V_{max} = 0.62\sqrt{2\pi G\Sigma(0)R_d} \tag{3}$$

where $\Sigma(0)$ and R_d are respectively the central mass surface density and scale-length of the disk. That peak occurs at a radial distance $R = 2.12R_d$. Relation (3) prompted Karachentsev (1989) to investigate the relationship between the rotational width, the disk scale length and, on the assumption of constant mass–to–light ratio, the disk central surface brightness $I(0)$ of systems that, at least photometrically, appeared as pure disks. Using a sample of 15 galaxies with available velocity widths and surface photometry, he found a relatively tight relation, with a dispersion on the distance as low as 12%. More recently, Chiba and Yoshii (1995) have reconsidered this approach and, apparently unaware of the earlier effort of Karachentsev, calibrated the relationship

$$\log R_d = a\log[V_{2.1}^2/I(0)] + b \tag{4}$$

using 14 nearby galaxies with Cepheid distances. In eqn. (4), $V_{2.1}$ is the measured rotational velocity at $R = 2.12R_d$. Chiba and Yoshii estimate $a \simeq 1.04$, in agreement with eqn. (3), while Karachentsev derives $a \simeq 1.4$. Both studies advocate the use of eqn. (4) as a valid, attractive, and potentially more accurate alternative to the TF relation.

Here we test relation (4) with a subset of 153 galaxies of the data presented in Giovanelli et al. (1997a), with I–band photometry. For those objects, measurements of R_d, $I(0)$ (in the form of a central disk surface magnitude $\mu(0)$) and $V_{2.1}$ are available, the latter parameter from long–slit rotation curves obtained with the 5 m Hale telescope. In the top panel of Fig. 1 we plot the data in the form suggested by eqn. (4), with R_d in kpc. The relation has high dispersion, about 0.25 in $\log R_d$ or 0.75 in the relative error on the predicted distance σ_d/d. Also shown in the same figure are the relationship between the full velocity width W and the isophotal radius $R_{23.5}$, in kpc, measured at the 23.5 mag arcsec^{-2} I band isophote, and the standard TF relation for the same set of objects, where M_I is the total I–band magnitude. The scatter is different among the three relations, and significantly smaller in this sample for the standard TF relation. What are the reasons for this substantial difference? The first is that scale lengths and central surface brightnesses are affected by notorious measurement errors. The photometric profiles of spiral galaxies are *approximately* exponential, provided that one steers clear of bulges and bars; yet even, in the outer regions of disks uncertainties arise from poor azimuthal averaging of spiral features, the inhomogeneities associated with star forming regions, etc. Knapen & van der Kruit (1992) pointed out that scale lengths obtained from exponential fits are so erratic that different authors may produce estimates of R_d of the same galaxy that vary by as much as a factor of 2. The uncertainties in the measurement of R_d and $\mu(0)$ are obviously coupled. Moreover, the aspect at which the disks are

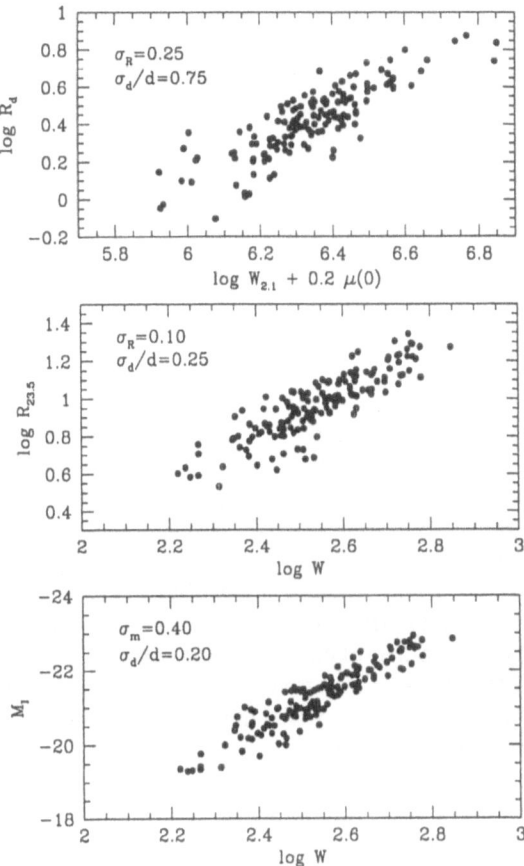

Fig. 1. Three variants on the TF relation: relation (4) is in the top panel, full velocity width vs. isophotal radius in the middle panel and the "standard" TF relation in the bottom. The scales of the three panels are set in such a way that the scale of the dispersion in *distance* is visually the same in the three cases.

seen will have a very significant effect on the observed values of R_d and $\mu(0)$: e.g. in the absence of internal extinction, $\mu(0)$ would vary by as much as 2.5 mag between the edge–on and the face–on aspect. In general, extinction will play an important role, and the recovery of the face–on values of R_d and $\mu(0)$ is a difficult task (Giovanelli *et al.* 1995). The second reason is that relation (3) is valid for pure, infinitely thin disks (Casertano 1983 has generalized the study to truncated, exponential disks of nonzero thickness). Chiba & Yoshii (1995) argue that the observed $V_{2.1}$ should approximate well the value expected at that radius from a pure exponential disk, because at that radial distance bulges become unimportant and R is still well inside the haloes. This assumption may not be correct, and especially for less luminous objects, the importance of a dark matter halo may be quite important even at short radii (e.g. Broeils 1992). The third reason is that the expectation that eqn. (3) maps to eqn. (4) with $a \sim 1$ relies on the assumption that the disk mass–to–light ratio is a constant across the spiral population, which is well known not be true (Broeils 1992).

While fig. 1 suggests that the standard form of the TF relation may be hard to beat, it is possible that eqn. (4) is better satisfied by bulge–free systems than by the morphologically mixed sample used here, as proposed by Karachentsev (1989). With his collaborators, he has produced a catalog of pure disk, edge–on spiral systems ("diskoteka": Karachentsev *et al.* 1993), with which eqn. (3) is being tested more exhaustively (see also Giovanelli *et al.* 1997c).

3   Scatter Properties

The scatter in the TF relation arises from several sources: errors in the *measurement* of TF parameters and uncertainties associated with the *corrections* applied to them combine with *variance in the galactic properties* produced by different formation and evolutive histories, characteristic of each object. The latter is often referred to as the *intrinsic* contribution to scatter, which may include not only the effects discussed by Eisenstein & Loeb (1996) but also velocity field distortions, deviations from disk planarity, other gravitational and disk asymmetries which may result from environmental circumstances. Several misconceptions regarding the nature of the TF scatter often appear in the literature: that it is well represented by a single number; that the measurement and correction errors fully account for the observed dispersion; that only errors on the velocity widths are important. Giovanelli *et al.* (1997a: G97a) have made a detailed appraisal of the sources of I band TF scatter. They find that: (a) the total TF scatter cannot be represented by a single value, but it varies monotonically with velocity width; while for low widths ($2 \times V_{rot} < 120$ km s^{-1}) it exceeds 0.5 mag, for high widths ($2 \times V_{rot} > 400$ km s^{-1}) it is about half that value; (b) measurement and processing errors are important contributors to, but they cannot fully account for the amplitude of the total error; the intrinsic scatter contribution varies between ~ 0.4 mag for low width objects and ~ 0.2 mag for high width ones; (c) while errors on the width tend to be very important, measurement and processing errors on the magnitude can also be important drivers of the scatter, especially for luminous, highly inclined galaxies.

Franx and de Zeeuw (1992) have found that the TF scatter poses strong constraints on the elongation of the gravitational potential in the disk plane of spirals. Their conclusion, that the average ellipticity of the potential in the plane of the disk must be smaller than about 0.06, is reinforced by the results in G97a. See Eisenstein's contribution in these proceedings for the constraints imposed on the merger history of spirals, by the amplitude of the intrinsic component of the scatter.

Sandage (1994 and refs. therein) and collaborators advocate for a large value of the TF scatter, near or larger than 0.7 mag, as an explanation for the high values of the Hubble constant resulting from the use of the TF relation. In the I band at least, the average scatter is about half that much. As a result, incompleteness bias corrections (see section 5) are not as large as advocated by those workers, and the impact of the scatter on the inferred value of H_o is not as severe (see section 6.4).

It has been advocated that the use of an *inverse* fit for the TF relation — one where the "independent" variable is the magnitude rather than the velocity width — does away with the need to correct for incompleteness bias (e.g. Schechter 1980). The nature of the TF scatter, especially the fact that velocity width errors can be overshadowed by other sources, weakens the case for a bias–less inverse TF relation.

4   Morphology, Environment and Extinction Laws

Early work by Roberts (1978) and others showed that in the blue part of the visible spectrum there are significant differences in TF behavior among galaxies of different morphological types. Aaronson & Mould (1983) found such differences to be imperceptible when H band photometry is used. At I band, G97a find a weak but significant difference in the zero point between early– and late–type spirals, Sa and Sab galaxies being less luminous, at a given width, than Sbc and Sc galaxies. Because cluster samples often include a sizable fraction of early–type spirals, an adjustment for the mixing needs to be made.

Concerns have often surfaced regarding the possibility that environmental circumstances might affect the photometric and kinematical properties of galaxies, thus altering their location in the TF plane. While such an effect may be discernible when blue photometry is involved (Schroeder 1996), the analysis in G97a indicate that in the red and infrared bands the effect is negligible.

An accurate determination of the inclination of the disk to the line of sight is necessary for two important reasons: it is used to estimate a correction factor to the observed velocity width, which increases in amplitude and uncertainty as the disk appears more face–on, in order to obtain a true rotational width, and it helps to estimate the amount of extinction suffered by the stellar light in propagating through the interstellar medium of the disk, which increases as the disk appears more edge–on; the amplitude of this correction is not well agreed on (see Giovanelli *et al.* 1995 and refs. therein), but it is likely that in early TF work it may have been underestimated. It also appears that the amplitude of the correction is a function of the luminosity of a galaxy. The use of inadequate extinction laws has several important consequences, besides producing inclination–dependent TF residuals: e.g. it may introduce artificial departures of (1) from a pure power law and a spurious peculiar velocity field (Giovanelli *et al.* 1994).

5   Constructing a Template Relation

Clusters of galaxies are favorite targets for TF applications for two important reasons. First, a cluster provides a large number of objects located at the same distance, thereby exempting the shape of its TF relation from the distortions introduced by the peculiar velocity field. Clusters are thus well suited for the generation of a TF template relation. Second, the combination of independent distance estimates of several galaxies in a cluster provides a more accurate determination of the cluster distance: to the extent that N galaxies in a cluster can be considered to be at the same distance, the cluster distance can be found $\sim N^{1/2}$ times more accurately than as determined for a single galaxy. Well–sampled clusters and groups can thus provide "hard points" in the measurement of the large–scale peculiar velocity field.

The determination of a TF template relation on the base of a single cluster of galaxies is however flawed. Consider the case of a nearby cluster first. Assume that a cluster TF relation is a random realization of some universal relation, defined (in a logarithmic version of eqn. 1) by a zero point offset and a slope. Uncertainty is contributed by the scatter, intrinsic and experimental, of the data about the mean relation, and by the *a priori* unknown amplitude of the cluster motion with respect to the comoving reference frame. By measuring a large sample of galaxies, which is possible in a nearby cluster, we can reduce the uncertainty arising from the scatter while at the same time extending the dynamic range of the TF variables, thus obtaining a good estimate of the TF slope. However, if V_{pec} is the (a priori unknown) peculiar velocity of the cluster, the TF zero point offset remains indetermined to no less than $\Delta m \simeq -2.17 V_{pec}/cz$ mag, an amount that increases as the cluster distance cz diminishes. In other words, if the chosen cluster is nearby, the TF zero point uncertainty remains high, independent of the size of the sample. Consider now the case of a distant cluster. While the zero point offset obtained for such a system would be less uncertain, the dynamic range of the TF variables would be restricted, the slope more poorly determined and the sample size reduced, as flux, size or other selection criteria tend to deplete the faint end of the TF diagram.

A more attractive approach is that of building a template by combining data of many clusters. It is reasonable to expect that the mean peculiar motion of a cluster set, if well distributed over all the sky, approximates null velocity better than any single cluster. If for example the cluster peculiar velocity distribution function is characterized by a r.m.s. figure $< V_{pec}^2 >^{1/2}$, the average of the peculiar velocities of N randomly chosen clusters, with a mean redshift $< cz >$, will yield a TF zero point offset uncertainty of

$$|\Delta m| \sim 2.17 < V_{pec}^2 >^{1/2} < cz >^{-1} N^{-1/2} \qquad (5)$$

The TF relation parameters can be statistically better determined from the cluster set, as the larger number of galaxies sampled helps to fight the scatter and characterize the slope, while a large number of clusters improves the quality of the zero point. In order for the TF relation of each cluster to be adequately "spliced" to that of the others, it is necessary to measure its peculiar velocity with respect to the reference frame defined by the cluster set. Thus, the construction of a TF template relation and the determination of the peculiar motions of the cluster set need to be carried out in iterative fashion.

The best TF templates yield accuracies on the order of 3–5% for the slope and ~ 0.05 mag for the zero point; much of the latter is contributed by the quantity in eqn. (5). The impact of bias is also important; for details see G97a and refs. therein.

6   Applications

Here we review some recent applications of the TF relation. We use the results of two large samples of field and cluster spirals presented by G97a, for the discussions regarding the peculiar velocity field and the measurement of H_0 , in sections 6.1 through 6.3. Finally, in section 6.4 we briefly review recent results on estimates of M/L of spirals at high z.

6.1   Cluster Motions

While earlier studies reported relatively high values for cluster peculiar velocities, in some cases exceeding 3000 km s^{-1} (e.g Mould *et al.* 1991; Lucey *et al.* 1991), the results of G97a suggest a significantly more quiescent cluster peculiar velocity distribution. Of the 24 clusters in our sample, none exhibit 1-d peculiar velocities in excess of 600 km s^{-1} . Bahcall & Oh (1996) and Moscardini *et al.* (1996) compared our cluster velocities with expectations of numerical simulations in the framework of different cosmological scenarios, and obtained agreement only with models with $\Omega_0 \leq 0.4$. These data indicate that the 1-d r.m.s. peculiar velocity of clusters is $< v_{1d}^2 >^{1/2} = 293 \pm 28$ km s^{-1} .

The cluster set can also be used to estimate the LG reflex motion. When the subset of 14 clusters in G97a with cz between 4000 and 9000 km s^{-1} is used, the reflex motion of the LG mimics very well the CMB dipole:

$$V_{clu} = 577 \pm 101 \quad \text{vs.} \quad V_{cmb} = 627 \pm 22$$
$$l_{clu} = 272 \pm 20 \quad \text{vs.} \quad l_{cmb} = 276 \pm 2$$
$$b_{clu} = +31 \pm 17 \quad \text{vs.} \quad b_{cmb} = +30 \pm 2,$$

suggesting that most of our motion with respect to the CMB arises within the shell occupied by the clusters. This result is to be compared with that of Branchini *et al.* (1996) and Tini–Brunozzi *et al.* (1995), who estimated the LG reflex motion with respect to the Abell/ACO cluster set, from a redshift catalog alone, on the assumption that light traces mass. They found, as we did, that most of the LG motion arises within $cz \sim 5000$ km s^{-1} , but their results also suggested a sizable contribution to the motion arising at $cz > 10,000$, with convergence not being reached until $cz \sim 16,000$ km s^{-1} . The ongoing work of Dale *et al.* (1997 and these proceedings) will provide a direct comparison of the two methods, over volumes of comparable size.

6.2   The Peculiar Velocity Field

Similarly to what is observed for clusters, the peculiar velocities of field objects measured in the LG reference frame clearly reflects the motion of the LG with respect to the CMB, suggesting that the LG motion with respect to the CMB is largely a local phenomenon, not involving the majority of the volume subtended by the sample. Table 1 exhibits the dipole amplitudes and apices of the field spirals, segregated by redshift shells, and compared to the analogous quantities for the CMB dipole. It is clear that the outer shells of the sampled volume approach rest with respect to the CMB and that, according to these data, the coherence length of the LG motion would not appear to significantly exceed the radius of the sampled volume. The apex alignment of these data with the CMB is better than that obtained from the light distribution (see, e.g. Hudson 1993). It should however be kept in mind that these data do not sample distances $cz > 8000$ km s^{-1} , and are unable to adequately detect the convergence in $V_{pec,LG}(R)$.

Table 1. Dipoles of the Peculiar Velocity of Field Galaxies

V_{cmb} Window	$V_{pec}(LG)$	l_{apex}	b_{apex}	Nr.
All	417 ± 28	253 ± 09	37 ± 03	1585
$2000 - 3000$ km s^{-1}	333 ± 41	268 ± 16	41 ± 08	235
$3000 - 4000$ km s^{-1}	437 ± 57	242 ± 20	24 ± 05	303
$4000 - 5000$ km s^{-1}	551 ± 62	236 ± 24	37 ± 05	372
$5000 - 6000$ km s^{-1}	566 ± 81	281 ± 15	24 ± 15	270
CMB	627 ± 22	276 ± 03	30 ± 03	

The application of the TF relation to field galaxies for the measurement of peculiar velocities becomes difficult at $cz \sim 10000$ km s^{-1} or greater. For a typical dispersion of 0.3 mag, the peculiar velocity of a single galaxy cannot be measured to better than 1500 km s^{-1} at $cz = 10000$ km s^{-1} , restricting the application of the technique to volumes where averaging will not completely smear out the details of the velocity field. At large distances, the technique is the only suitable for application to clusters and large groups. For further analysis of these results, see Freudling (these proceedings), Giovanelli et al. 1996 and da Costa et al. (1996).

6.3   The Hubble Constant

Recent HST measurements of Cepheid distances to galaxies in the Virgo and Fornax clusters (see Livio et al. 1996) have made possible better estimates of the value of H_0. One of the thorny issues encountered is the uncertainty on V_{pec} of the clusters in which the galaxies with measured distances are assumed to reside.

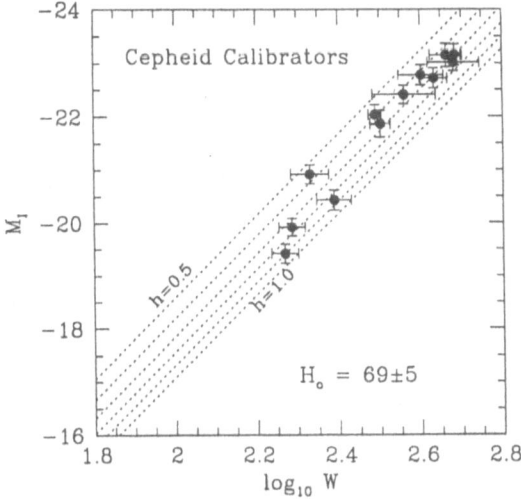

Fig. 2. TF Calibrators with available Cepheid distances, superimposed on a grid of TF template relations plotted for different values of h.

Rather than using a single cluster for this exercise, a better approach consists in using the TF template relation which, based as it is on averaging the motions of N distant clusters, yields a reference frame which is kinematically $N^{1/2}$ times more accurate than that based on any single cluster at the average distance of the cluster set. Now, the TF template relation is designed to yield peculiar velocities $cz - H_o d$, which do not require the specification of the value of H_o. Thus, calibrating the template relation by using galaxies endowed with a Cepheid distance is equivalent to specifying the value of H_o. For this purpose, any galaxy suitable for TF use and with a Cepheid distance can be used. Currently, 12 such objects exist with available I band photometry, including galaxies in the LG such as M31 and M33, and galaxies as far as the Virgo and Fornax clusters. In fig. 2, the template relation of G97a is plotted for different values of $h = H_o/100$, together with the location of suitable galaxies with Cepheid distances. The best match between template and Cepheid distances is for $H_o = 69 \pm 6$ km s^{-1} (see Giovanelli et al. 1997b). The completion of the Dale et al. (1997) survey will allow a significantly improved determination of the Tf template and, with it, a tighter definition of the kinematical reference frame on which both peculiar velocity and distance scale determinations fundamentally rely.

6.4   Evolution of M/L with Redshift

In the last few years, internal kinematics measurement have become available for spiral galaxies at increasingly higher redshift (Vogt et al. 1993; Bershady 1995; Simard & Pritchet 1996; Rix et al. 1996; Vogt et al. 1996). It is thus becoming possible to explore the potential of the TF relation as a probe of evolutionary changes in M/L, and to test the standard candle capacity of spiral galaxies at $z > 0.1$. Several of the surveys extend to $z \sim 0.3$, and that of Vogt et al. (1996) to $z \sim 1$. The currently available results exhibit indications of somewhat conflicting nature. While the measurements of Bershady (1995) and those of Vogt et al. (1996) suggest no significant changes in M/L to $z \sim 1$, Simard & Pritchet (1996) and Rix et al. (1996) find evidence for significant brightening, of up to ~ 2 mag, by $z \sim 0.3$. Bershady (1995) discusses these apparently conflicting results and suggests that they are partly to be ascribed to basic differences, in their rest frames, of the galaxies sampled by each study. In fig. 3 the magnitude deviations from the $z = 0$ TF relation in the B–band is shown for several samples, exhibiting the discrepancies discussed above. Progress along these lines will require the analysis of larger samples selected according to strict and uniform criteria, and the use of independent techniques to allow a separation of cosmological and evolutionary effects, such as the application of the Tolman test.

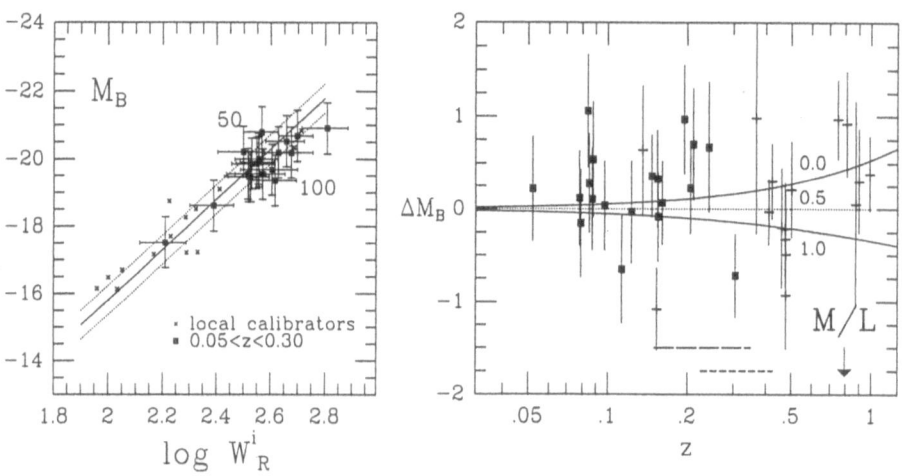

Fig. 3. Deviations from a $z = 0$ B–band TF relation for several samples: Bershady 1995 (filled squares), Vogt et al. 1996 (+), Rix et al. 1996 (long–dash) and Simard & Pritchet 1996 (short–dash). Smooth curves refer to changes expected for different values of q_0. After Bershady 1996, with permission.

This work was supported by the National Science Foundation grant AST94-20505.

References

Aaronson, M. & Mould, J. 1983, *ApJ* **265**, 1

Aaronson, M. *et al.* 1986, *ApJ* **302**, 536

Bahcall. N.A., & Oh, S.P. 1996, *ApJ* **462**, L49

Bershady, M. 1996, in *Dark and Visible Matter in Galaxies and Cosmological Implications*, ed. by M. Persic & P. Salucci, ASP Conference Series, San Francisco, in press.

Branchini, E. *et al.* 1996, *ApJ* **461**, L17

Broeils, A. 1992, Ph. D. Thesis, U. of Groningen

Casertano, S. 1983, *MNRAS* **203**, 735

Chiba, M. & Yoshii, Y. 1995, *ApJ* **442**, 82

da Costa, L. *et al.* 1996, *ApJ* **468**, L5

Dale, D.A. *et al.* 1997, *AJ*, submitted

Djorgovski, S. & Davis, M. 1987, *ApJ* **313**, 59

Dressler, A. *et al.* 1987, *ApJ* **313**, L37

Eisenstein, D. & Loeb, A. 1996, *ApJ* **459**, 432

Faber, S.M. & Jackson, R.E. 1976, *ApJ* **204**, 668

Franx, M. & de Zeeuw, T. 1992, *ApJ* **392**, L47

Freeman, K.C. 1970, *ApJ* **160**, 411

Giovanelli, R. *et al.* 1994, *AJ*, **107**, 2036

Giovanelli, R. *et al.* 1995, *AJ*, **110**, 1059

Giovanelli, R. *et al.* 1996, *ApJ*, **464**, L99

Giovanelli, R. *et al.* 1997a, *AJ*, in press [G97a]

Giovanelli, R. *et al.* 1997b, *ApJ (Letters)*, in press

Giovanelli, R., Avera, E. & Karachentsev, I.D. 1997c, preprint

Hudson, M. 1993, *MNRAS* **265**, 72

Karachentsev, I. 1989, *AJ* **97**, 1566

Karachentsev, I., Karachentseva, V. & Parnovskij, S. 1993, *Astr. Nachr.* **314**, 97

Knapen, J.H. & van der Kruit, P.C. 1992, *AA* **248**, 57

Kormendy, J. 1977, *ApJ* **218**, 333

Livio, M., Donahue, M. & Panagia, N. 1996, *Extragalactic Distance Scale*, in press

Lucey, J.R., Gray, P.M., Carter, D. & Terlevich, R.J. 1991, *MNRAS* **248**, 804

Moscardini, L. *et al.* 1996, preprint astro-ph/9511066

Mould, J.R. *et al.* 1991, *ApJ* **383**, 467

Navarro, J.F. 1996, preprint astro-ph/9611108

Navarro, J.F., Frenk, C.S. & White, S.D. 1996, preprint astro-ph/9611107

Pierce, M. & Tully, R.B, in *The Extragalactic Distance Scale*, ed. by M. Livio, M. Donahue & N. Panagia, in press

Rix, H.-W., Guhathakurta, P., Colless, M. & Ing, K. 1996, preprint astro-ph/9605204

Roberts, M.S. 1978, *AJ* **83**, 1026

Sandage, A. 1994, *ApJ* **430**, 1

Schechter, P. 1980, *AJ* **85**, 801

Schroeder, A. 1996, Ph. D. Thesis, Univ. Basel

Simard, L. & Pritchet, C.J. 1996, preprint astro-ph/9606006

Tini-Brunozzi, P. *et al.* 1995, *MNRAS* **277**, 1210

Tully, R.B. & Fisher, J.R. 1977, *AA* **54**, 661

Vogt, N.P. *et al.* 1996, *ApJ* **465**, L15

Willick, J.A. 1990, *ApJ* **351**, L5

A Comparison of Tully-Fisher and Fundamental Plane Distances for Nearby Clusters

Marco Scodeggio[1,2], Riccardo Giovanelli[1], and Martha P. Haynes[1]

[1] Dept. of Astronomy, Cornell University, Ithaca, NY 14853
[2] ESO, Karl-Schwarzschild-Str. 2, D-85748 Garching b. München, Germany

Abstract. With the aim at obtaining a quantitative measure of the degree of universality of the Tully-Fisher and Fundamental Plane relations, we are comparing accurate redshift-independent distance estimates of nearby clusters of galaxies derived using both relations. Systematic differences in the two distance scales, once cluster membership and sample bias related issues have been resolved, would indicate a lack of universality in the relations, and seriously question their use in mapping the large-scale peculiar velocity field, and in studying the cosmological evolution of early and late-type galaxies in high redshift clusters. New accurate Fundamental Plane distance estimates for 6 nearby clusters are presented here, and compared with comparable Tully-Fisher distance estimates. The agreement between the two sets of distances provides good support to the universality of the Tully-Fisher and Fundamental Plane relations.

1 Introduction

The use of the Tully-Fisher (TF) relation (Tully & Fisher 1977), and of the Fundamental Plane (FP) relation (Djorgovski & Davis 1987; Dressler et al. 1987) as distance-indication relations relies heavily on an assumption of universality: for all galaxies in all environments the correlation between magnitude and rotation velocity, or among effective radius, effective surface brightness, and velocity dispersion is described by the same linear relation (the TF or FP template), within observational uncertainties, and a small intrinsic scatter. However little effort has been made to precisely quantify the degree of universality of the TF and FP (see Aaronson et al. 1986; Djorgovski et al. 1988; Guzmán et al. 1992; Jørgensen et al. 1996).

An invalid or biased template relation will yield spurious peculiar velocities. It has even been argued, for example, that the very evidence suggesting the existence of a Great Attractor might result from bias (Silk 1989). A comparison by Mould et al. (1990) between cluster peculiar velocities obtained using the TF and FP (or $D_n - \sigma$) techniques shows disagreements which exceed those expected from the reported uncertainties associated with the two methods. The case of A2634 is even more notorious: Aaronson et al. (1986) applied the TF relation to find the cluster roughly at rest in the Cosmic Microwave Background (CMB) reference frame, while Lucey et al. (1991) used the $D_n - \sigma$ relation to obtain a cluster peculiar velocity of -3400 km s^{-1}.

The most direct way to test the reliability of a distance-indication relation is to compare its predictions with those of another independent, and at least equally reliable, method. The TF and FP relations provide the opportunity for such a comparison. They have similar accuracy, and they are totally independent methods, because they are applied to galaxies of different morphological type. Also, these galaxies populate different environments (E and S0 in the cluster cores, spirals preferentially at the cluster peripheries), and it seems likely that, even in the case both the TF and FP might not be universal relations, they would be biased by different amounts by any environmentally-related mechanisms. Therefore a comparison between the two methods would still provide an opportunity to quantify their universality. Because of the limited accuracy of the distance estimates obtained with the TF and FP relation, a direct comparison of the two methods can be done only using clusters of galaxies, where large samples can be used to reduce the statistical uncertainty in the estimate of the cluster distances. Also, the existing correlation between cluster richness and the properties of the intra-cluster medium (see for example Edge & Stewart 1991) offers the opportunity to perform the TF–FP comparison under different environmental conditions, using clusters of different richness.

Here we present the first results of a project aimed at obtaining a comparison of very accurate TF and FP distance estimates in nearby clusters, and therefore at providing a reliable quantification of the universality of the two relations.

2   The Data

A large data-set of TF distance-estimates in 24 nearby clusters and rich groups has been recently collected by Giovanelli et al. (1997a,b). We have selected 10 of the clusters and groups in that sample to build a comparable data-set of FP distance-estimates. These are: Coma, A1367, Hercules (A2147 and A2151), A2634, Pegasus, A262, Cancer, and the NGC 383 and NGC 507 groups. Here we present results for 6 of those clusters (Coma, A1367, A2634, A262, Pegasus, and NGC 507 group).

Photometric parameters for early-type galaxies in those clusters were derived from I band CCD observations obtained with the KPNO 0.9m telescope. The core of each cluster was mapped with a mosaic of frames on a regularly spaced grid, and photometric parameters were measured for all E and S0/S0a galaxies that are to be considered cluster members (see Giovanelli et al. 1997a for membership criteria), for a total of 624 galaxies. Exposures were all 10 minutes long. A galaxy's light distribution was fitted with elliptical isophotes to derive its radial surface brightness profile. This was fitted with a de Vaucouleurs $r^{1/4}$ profile, yielding an effective radius r_e and effective surface brightness μ_e (the mean surface brightness within r_e). The fit was performed from a radius equal to twice the seeing radius, out to the outermost isophotes for E galaxies; for S0 and S0a galaxies only the central core was fitted. The median uncertainty on the determination of r_e is 5%, and 0.06 mag on that of μ_e.

Stellar velocity dispersion measurements for a subset of the photometric sample, composed of 273 galaxies, were obtained from moderate dispersion optical spectra. All spectroscopic observations were obtained with the Hale 5m telescope of the Palomar Observatory. A 1200 lines mm^{-1} grating and a 2″ x 128″ slit were used to obtain spectra with a dispersion of 0.86 Å per pixel, and a resolution of 2.2 Å (corresponding to a velocity resolution of 129 km s^{-1} at 5300 Å). The spectral coverage was approximately from 5000 to 5600 Å, centered on the Mg Ib lines at ~5175 Å. Exposure times ranged from 15 to 90 minutes, depending on the brightness of the target galaxy, with a median value of 30 minutes. Late G and early K type giant stars were used as velocity standards and as templates for the velocity dispersion measurements. One-dimensional galaxy spectra, to be used for the velocity dispersion determinations, were extracted using a 6″ wide window (in the cross-dispersion dimension), centered on the peak of the galaxy continuum. All measurements of velocity and velocity dispersion were obtained using the cross-correlation technique of Tonry & Davis (1979), implemented in the IRAF task *fxcor*, by assuming that the line-of-sight velocity dispersion is Gaussian. The median uncertainty on the determination of the velocity dispersion is 5%, for $\sigma > 90$ km s^{-1}.

3   The Fundamental Plane

It has been argued that cluster samples for TF and FP work are almost bias-free, since all galaxies are roughly at the same distance. While it is certainly true that the classical Malmquist bias is avoided, a different and equally important bias affects cluster samples. This is generally referred-to as the "cluster population incompleteness bias" (Teerikorpi 1987). Sandage (1994) provided a detailed analysis of this bias and of the corrections it requires before unbiased distance estimates can be obtained with cluster TF samples. Here we have derived incompleteness bias corrections following the spirit of that treatment, but using Monte Carlo simulations to reproduce as closely as possible the completeness, measurement uncertainties, and scatter characteristic of our samples.

The FP template was obtained combining all individual bias-corrected cluster samples, under the assumption that, on average, these clusters are at rest with respect to the CMB reference frame. A chi-square fitting that minimizes the residuals in the orthogonal distance of each data-point from the plane was used. The resulting FP template is given by the relation

$$\log R_e = 1.55 \log \sigma + 0.326(\mu_e - 19.45) - 2.919 - \log h \qquad (1)$$

(or equivalently $\log R_e = 1.55 \log \sigma - 0.815 \log I_e - 9.26 - \log h$), where R_e is in kiloparsec. Each cluster sample mean deviation from the global template provides an estimate of that cluster peculiar velocity (see Tab. 1).

Figure 1 shows an edge-on view of the FP relation for the combined sample, composed of 257 galaxies. The solid line is the projection of the template relation (1). The dispersion around this relation, measured as the rms scatter of the residuals in $\log R_e$, is 0.088, equivalent to an uncertainty of 20% in the distance estimate of a single galaxy, or of 0.44 magnitudes in the estimate of its distance modulus. No correlation is present between the residuals from the FP relation and the most important galaxy parameters, like luminosity and ellipticity, or

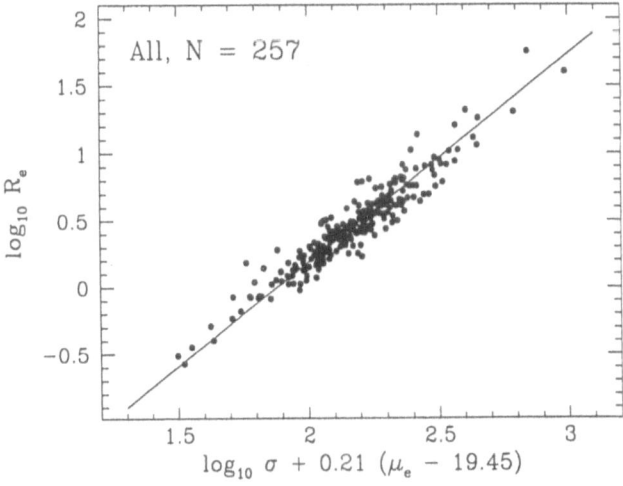

Fig. 1. Edge-on projection of the FP.

the FP parameters themselves. Also no significant difference is found in the FP template when only E or only S0 galaxies are considered.

A simple test for the presence of environmental effects can be performed analyzing the distribution of FP parameters as a function of cluster properties like the velocity dispersion, and the X-ray luminosity and temperature. All these quantities provide a measure of the richness of the cluster, and of its degree of dynamical evolution, and therefore should provide a measure of the strength of the environmental effects that might bias the FP relation. In agreement with the results of Jørgensen et al. (1996), we find no significant correlation between FP parameters and cluster properties, for a sample that spans two orders of magnitude in X-ray luminosity, and approximately a factor of three in velocity dispersion. No significant differences in the FP relation as a function of position within the cluster are found, either.

4   TF–FP Comparison

Giovanelli et al. (1997b) have obtained cluster peculiar velocities based on TF measurements, using the same approach described in the previous paragraph. The main difference between those peculiar velocities, and the ones derived here using the FP, is that their TF template was obtained using a larger, more evenly distributed in the sky, cluster sample. We list in Tab.1 their estimates of the peculiar velocity for the six clusters discussed here, together with our estimates based on the FP relation. A comparison of the six pairs of measurements is shown in Fig.2, where the solid line represents the one-to-one correspondence of the measurements. The good agreement between the two methods is evident.

Table 1. TF and FP peculiar velocities

Cluster	V_{CMB} (km s^{-1})	FP V_{pec} (km s^{-1})	TF V_{pec} (km s^{-1})
N507 gr.	4808	422±238	206±262
A262	4642	-156±203	0±187
A1367	6735	-117±352	-62±206
Coma	7185	43±196	228±214
Pegasus	3519	58±247	16±168
A2634	8895	-307±336	61±378

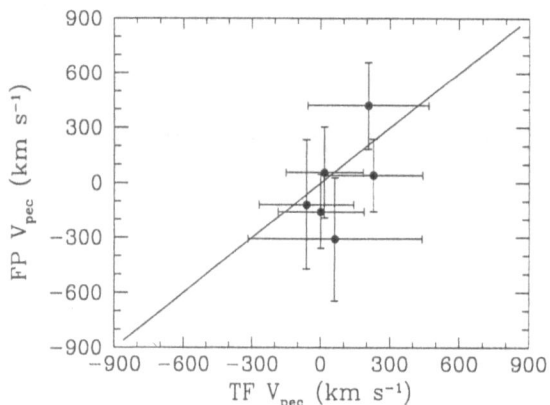

Fig. 2. Comparison of the TF and FP peculiar velocity estimates

The new FP measurements have contributed to a solution of the A2634 discrepancy, by showing that the large peculiar velocity estimate obtained by Lucey et al. (1991) was a spurious result (Scodeggio et al. 1997). There is also much better agreement than before for the remaining 5 clusters in the sample, although no discrepancy comparable to the one in A2634 was evident within the older, less accurate, distance estimates for those clusters. The agreement between the FP and TF sets of measurements could be even better if the different cluster samples used in deriving the two relation templates are taken into consideration. Within the larger Giovanelli et al. (1997b) TF sample, the six clusters discussed here have an average peculiar velocity of 75 km s^{-1}, while the FP peculiar velocities average to zero by definition. If we were to impose for this last sample a mean peculiar velocity equal to that obtained with the TF relation, this would slightly improve the agreement between the two methods.

We conclude that the absence of any significant difference in the form of the FP and TF relation as a function of cluster environment (this work, and Giovanelli et al. 1997b), and the good agreement between the distance-scales based on the two relations (this work) give strong support to the hypothesis of universality for both the TF and the FP relation. In particular no discrepancy of statistical significance has been observed in the comparison of cluster peculiar velocities, for clusters that have very different richness and therefore represent a rather diversified environment for both the early and late-type galaxy populations. It seems at this point undesirable to try to observe fainter galaxies in the attempt to improve on the accuracy achieved here for individual cluster measurements, because of the possible breakdown of the TF and FP relations for low mass galaxies, and because of many observational complications. Any further study of the universality of the two relations would therefore require a much larger sample than the present one, to obtain reliable statistics on the distribution of the differences in the two distance-scales. However, given the results discussed here, the existence of any environmental bias that could alter the TF or FP template by more than 5% seems extremely unlikely.

This research is supported by the NSF grants AST94–20505 to RG and AST92–18038 to MPH.

References

Aaronson, M., Bothun, G., Mould, J., Huchra, J., Schommer, R.A., & Cornell, M.E. 1986, ApJ, 302, 536

Djorgovski, S., & Davis, M. 1987, ApJ, 313, 59

Djorgovski, S., De Carvalho, R., & Han, M.S. 1988, in "The Extragalactic Distance Scale", ASP Conf. Ser., ed. S. van den Bergh & J. Pritchet, p. 329

Dressler, A., Lynden-Bell, D., Burstein, D., Davies, R.L., Faber, S.M., Terlevich, R.J., & Wegner, G. 1987, ApJ 313, 42

Edge, A.C., & Stewart, G.C.. 1991, MNRAS, 252, 428

Giovanelli, R., Haynes, M.P., Herter, T., Vogt, N.P., Wegner, G., Salzer, J.J., da Costa, L.N. and Freudling, W., 1997a, AJ, in press (Jan.97 issue)

Giovanelli, R., Haynes, M.P., Herter, T., Vogt, N.P., da Costa, L.N., Freudling, W., Salzer, J.J. and Wegner, G., 1997b, AJ, in press (Jan.97 issue)

Guzmán, R., Lucey, J.R., Carter, D., & Terlevich, R.J. 1992, MNRAS, 257, 187

Jørgensen, I., Franx, M., & Kjærgaard, P. 1996, MNRAS, 280, 167

Lucey, J.R., Gray, P.M., Carter, D., & Terlevich, R.J. 1991, MNRAS, 248, 804

Scodeggio, M., Giovanelli, R., & Haynes, M.P. 1997, AJ, in press (Jan.97 issue)

Mould, J.R., et al. 1990, ApJ, 383, 467

Sandage, A. 1994, ApJ, 430, 1

Silk, J. 1989, ApJ, 345, L1

Teerikorpi, P. 1987, A&A, 173, 39

Tonry, J.L., & Davis, M. 1979, AJ, 84, 1511

Tully, R.B., & Fisher, J.R. 1977, A&A, 54, 661

The Infrared Tully-Fisher Relation in Abell 1367 and Cancer

R.F. Peletier[1] and S.P. Willner[2]

[1] Kapteyn Laboratorium, Groningen, The Netherlands
[2] Harvard-Smithsonian Center for Astrophysics, Cambridge, MA, USA

Abstract. We analyze new H and I-band data of a complete sample of galaxies in two nearby galaxy clusters. We find that the Tully-Fisher scatter in H is slightly smaller than in I, but that the difference is so small that it is not worthwhile to redo large I-band surveys in the infrared. We find that the galaxies of both clusters form a tight sequence in the H vs. $I - H$ diagram, which shows that the galaxies display a large range (up to 2 decades) in metallicity.

1 Introduction

In 2 previous article (Peletier & Willner 1991, 1993) we have been studying the scatter in the Tully-Fisher relation by measuring the observed scatter in the nearby clusters Virgo and Ursa Major, using many different kinds of infrared H-band magnitudes and velocity widths, and various ways to determine the inclination. We found that the scatter in Virgo was larger than in Ursa Major, and in both cases it was larger than can be explained by errors in the observations. The scatter in Virgo is probably large because of subclustering, but it is not excluded that this is also the case for Ursa Major. Therefore, we decided to observe two other, nearby clusters, in which possibly regions could be found where the scatter is lower than in the Ursa Major cluster, which could imply that the intrinsic scatter in the Tully-Fisher relation itself is lower. Part of this study is presented here. We have tried to obtain samples of galaxies in both clusters that are as complete as possible, and have taken images of all of them in H. Because of the fact that our Ursa Major study has shown that our H-band images, which reach a 1σ noise limit of $H=21$ mag $(\text{arcsec})^{-2}$, are not deep enough to determine good inclinations, we also obtained images of the whole sample in the I-band. This immediately offers the possibility to compare the scatter in both bands. Although images in the I-band are much easier to obtain than in H, they can suffer substantially from extinction by dust, which increases the scatter. Here we investigate whether for large surveys (e.g. Matthewson & Ford 1996, Giovanelli et al. 1996, Han 1992) it pays off to also obtain H-band observations of all galaxies.

2 Observations and Sample

The samples were selected as follows. For Abell 1367 it consisted of all galaxies available in the NASA Extragalactic Database (NED) within 1.5° from the clus-

ter center of 11h42m +20d07m, and with heliocentric velocities between 4500
and 8300 km s^{-1}. This region is smaller than selected for this cluster by e.g.
Bothun *et al.* (1985). For Cancer we took all galaxies within 3 degrees of 8h18m
+21d14m (1950) and with velocities between 3300 and 7700 km^{-1}. From this list
were removed all galaxies with evidence for close interactions, and all ellipticals,
S0's and S0/a galaxies. All others were observed in I, to determine inclinations
from the I-band ellipticity profiles in the outer parts. Finally, only galaxies were
taken with inclinations larger than 45°.

Most of the optical observations were taken at the 1.2m telescope of the
Whipple Observatory at SAO, and the remaining part at the 1m JKT at La
Palma, all in 1994. The infrared H-band observations were obtained using the
Smithsonian Obs. Near-Infrared Camera SONIC on the 1.2m telescope at FLWO.
The data were reduced in the same way as in our previous papers, and calibrated
using standard stars. On the final mosaic, after removing some foreground stars,
magnitudes were determined in circular and elliptical apertures. To establish
their accuracy , we compared the data with aperture photometry by Bothun *et
al.* (1985) (Fig. 1). The comparison shows that the accuracy is better than 0.1
mag for magnitudes brighter than $H=12$ mag, and around 0.15 mag RMS for
magnitudes up to $H=14$. It has to be mentioned that, certainly for the fainter
magnitudes, a large fraction of the differences are due to errors in the aperture
photometry.

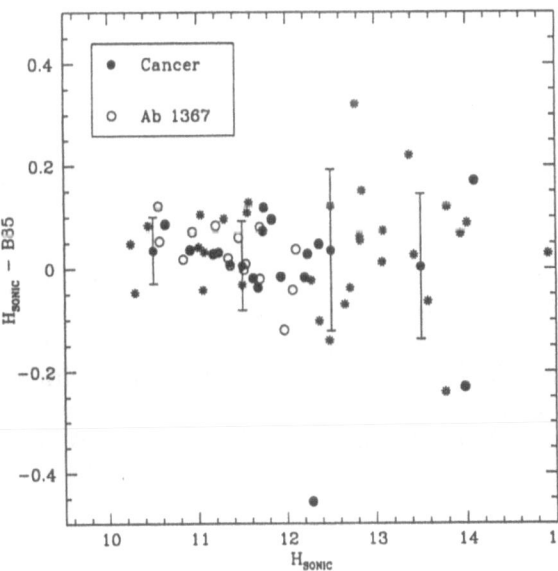

Fig. 1. Comparison with the H-band data of Bothun *et al.* (1985). Points with errorbars
are averages between resp. $H=10$ and 11, 11 and 12, 12 and 13, and 13 and 14 mag.

3   The Tully-Fisher Relation in Cancer and Abell 1367

To use the Tully-Fisher relation one needs, apart from magnitudes and inclinations, also velocity widths for each galaxy. We decided to take these from the HI measurements published in the literature.

3.1   Cancer

It has been known for a long time that the Cancer cluster consists of many subgroups (Bothun *et al.* 1983) in position and velocity. Since we are interested in finding a (sub)cluster with small scatter, we have taken all galaxies with heliocentric velocities smaller than 5000 km s^{-1}. They include the subgroups A, D and E of Bothun *et al.* (1983). Curiously, although groups D and E have a velocity about 800 km/s lower than A, their Tully-Fisher distance seems to be the same. For these galaxies we have determined various circular and elliptical magnitudes in both I and H, and determined their Tully-Fisher relation slope and scatter (see Table 1 [1]). The relations themselves are given in Fig. 2.

Table 1. Tully-Fisher solutions to $\alpha + \beta (\log W_{20}^c) = Mag$

Cluster	Band	Type	Aperture	α	β	N	scatter (mag)
Cancer	I	ell.	23.5	-10.88	8.73	10	0.31
	H	ell.	23.5	-13.18	10.10	13	0.26
	H	ell.	-0.5	-20.25	12.55	13	0.32
	H	circ.	-0.5	-15.59	10.89	13	0.28
Ab. 1367	I	circ.	23.5	-5.80	6.40	14	0.36
	I	circ.	-0.5	-6.75	6.56	14	0.35
	H	ell.	23.5	-4.25	6.30	11	0.32
	H	ell.	-0.5	-8.78	7.73	11	0.42
	H	circ.	-0.5	-6.49	6.99	11	0.36

We find for both clusters that the scatter is significantly larger than zero, in agreement with our results for Virgo and Ursa Major. The scatter however is not as large as 0.4 mag, which is quoted by Willick (1996) to be the typical value for the scatter in large CCD samples. In agreement with Pierce & Tully (1988) we find that the scatter in H is slightly smaller than in I. If one goes to smaller apertures, the scatter increases slightly as well. We also confirm our result for the Ursa Major cluster that if we use the small -0.5 apertures, the scatter is smaller for circular than for elliptical magnitudes.

---

[1] In Table 1 an aperture of 23.5 means that the aperture is defined within the isophote I=23.5 mag arcsec^{-2}, as in Han (1992). The value -0.5 indicates an aperture which is $10^{-0.5}$ times as large. The scatter in the last column is after correcting for observational uncertainties. α and β are based on fits taking into accounts uncertainties in both variables (see Peletier & Willner 1991).

3.2   Abell 1367

Here there have been no previous indications of subclustering. We have been trying to isolate groups here as well, but were unsuccessful, probably because of a lack of galaxies. For the Tully-Fisher slopes and scatter see also Table 1.

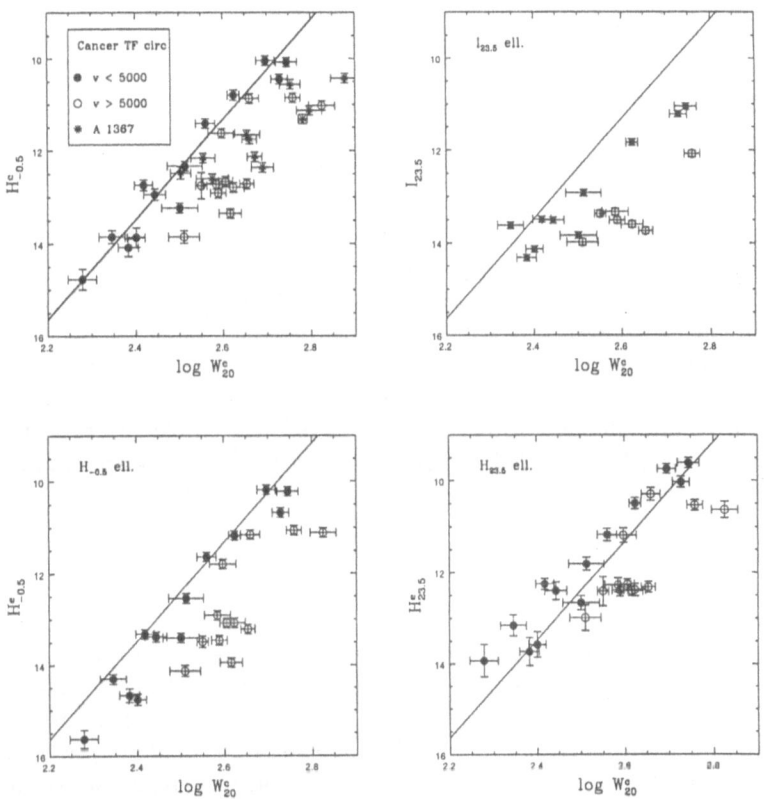

Fig. 2. Tully-Fisher relations for various types of magnitudes for Cancer. In Fig. (a) also the T-F relation for Abell 1367 is plotted.

It is curious that we find such a large difference between the Tully-Fisher slopes in the two clusters (see Fig. 2a). Comparison with for example the sample of Giovanelli et al. (1996), consisting of galaxies in a larger area on the sky, shows that the apparent difference in slope is presumably due to a lack of faint galaxies in our sample for this cluster. We should note that the selection criteria for Cancer and Abell 1367 have been the same, and that there is a clear lack of faint spirals in the latter. If one increases the area, there is an increased chance of subclustering in the sample.

4 The Tully-Fisher Relation: H vs. I

The fact that we have data available in both H and I allows us to address the question whether it is worthwhile to spend the extra effort to observe in H instead of I. First we look at the correction for inclination. In Fig. 3a we plot $(I - H)_{23.5}$ as a function of ellipticity, without and with extinction correction. The correction is the one used by Han (1992) in I. This correction is type-dependent, and more or less the same as the one used by Giovanelli et al. (1996). It can be as high as 0.7 mag for edge-on galaxies of type 4 and 5. For the H band we have applied the same correction, scaled according to the galactic extinction law. After correcting, there should not be any ellipticity-dependence any more, provided that the galaxies at low and high ellipticity are drawn from the same population. We see that after correcting the average $(I - H)_c$ decreases slightly with ellipticity. This decrease is mainly due to three blue galaxies at high inclination. These happen to be the faintest galaxies of the sample, and no galaxies of comparable colour are found at lower inclinations. The distribution of the rest of the galaxies gives us the indication that the required extinction correction might be somewhat larger than applied by Han (1992).

Next, we analyze the colour-magnitude diagram of the Cancer galaxies, after inclination correction (Fig. 3b). We find here a very narrow sequence, on which all faint galaxies are blue and all bright galaxies red. Our explanation for this effect is the following. Since both I and H are very red bands, they are both not very sensitive to light from young, generally blue, stars, and therefore the range in $I - H$ is small, similar to the range in $J - K$ (see e.g. Frogel et al. 1.978). It means that the $I - H$ colour of a galaxy is basically determined by the metallicity of the dominant old stellar population, or, in short, its metallicity. What we see in Fig. 3b is that fainter galaxies have lower metallicities than brighter galaxies. Also added to the Figure are lines indicating the $I - H$ color of old stellar populations of various metallicities from Vazdekis et al. (1996) and for low metallicities Bressan et al. (1994).. These lines are only drawn to give an indication of the range in metallicity that this interpretation would imply: probably about a factor 50. If we accept that part of the color difference can be due to differences in age, this factor is less, but in that case we also have to accept that all fainter galaxies are younger.

5 Conclusions

We find that:

- In two clusters the scatter in the Tully-Fisher relation in H is slightly smaller than in I. However, this difference is so small that one wins very little accuracy doing large Tully-Fisher surveys in the infrared. The scatter for both clusters is about 0.3 mag, and is the lowest for H-band magnitudes inside the 23.5^{th} I-band isophote.

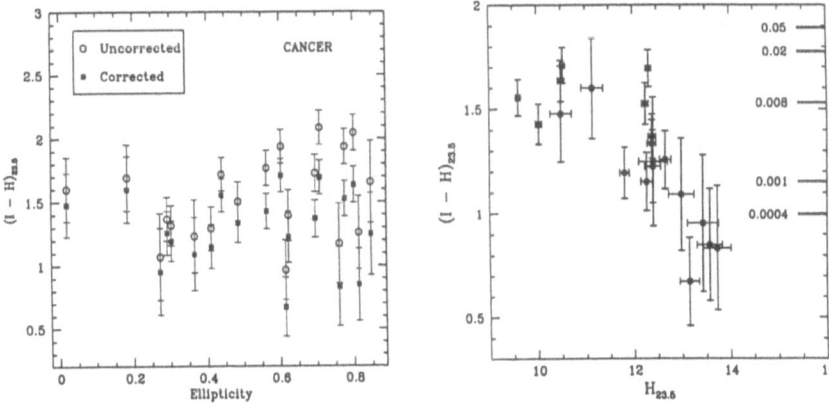

Fig. 3. a: Relation between $I - H$ and ellipticity for Cancer galaxies, before and after correcting for extinction. **b:** $(I-H)_{23.5}$, corrected for extinction, as a function of galaxy H-magnitude. The lines indicate colours of models of old stellar populations (17 Gyr) at various metallicities Z.

- From the dependence of $I - H$ as a function of inclination we find that the extinction correction for the I-band data applied by Han (1992) and Giovanelli *et al.* (1996) is probably slightly underestimated.
- The H vs. $I - H$ colour-magnitude relation shows very little scatter after correcting for extinction. The fact that our faint galaxies are about 0.8 mag bluer than the bright ones probably indicates that the galaxies in these clusters display a large range in metallicity, which in turn is responsible for the large difference in slope of the Tully-Fisher relation between H and I.

References

Bressan, A., Chiosi, C. & Fagotto, F., 1994, ApJS, 94, 63

Bothun, G.D., Geller, M., Beers, T.C. & Huchra, J.P., 1983, ApJ, 268, 47

Bothun, G.D., Aaronson, M., Schommer, B., Mould, J., Huchra, J. & Sullivan, W.T. III, 1985, ApJS, 57, 423

Frogel, J.A., Persson, S.E., Matthews, K. & Aaronson, M., 1978, ApJ, 220, 75

Giovanelli, R., Haynes, M., Herter, T., Vogt, N., da Costa, L., Freudling, W., Salzer, J. & Wegner, G., 1997, AJ, in press.

Han, M.S., 1992, ApJS, 81, 35

Matthewson, D.S. & Ford, V.L., 1996, ApJS, 107, 97

Peletier, R.F. & Willner, S.P., 1991, ApJ, 382, 382

Peletier, R.F. & Willner, S.P., 1993, ApJ, 418, 626

Pierce, M.J. & Tully, R.B., 1988, ApJ, 330, 579

Vazdekis, A., Casuso, E., Peletier, R.F. & Beckman, J.E., 1996, ApJ, 106, 307

Willick, J., 1996, preprint Astro-ph/9610200

Calibration of the Tully-Fisher Relation: An Interim Report on the HST Key Project

Jeremy Mould

Mount Stromlo & Siding Spring Observatories
Institute of Advanced Studies
Australian National University

Abstract. Analysis is now complete of approximately half the galaxies to be studied in the Hubble Space Telescope Key Project on the extragalactic distance scale. Galaxy scaling relations are among the secondary distance indicators calibrated by the project. An interim calibration of the Tully-Fisher relation yields $H_0 = 73 \pm 10$ km/sec/Mpc.

1 Introduction

This year the Tully Fisher relation is twenty years old. Early calibrations either offered a moderate number of galaxies with poorly determined distances or a small number of galaxies with well-determined distances (Aaronson *et al.* 1986). It has turned out that the earliest calibrations gave approximately correct distances (e.g. Tully & Fisher 1977) right from the beginning.

In 1996 Hubble Space Telescope has made it possible to measure an accurate Cepheid distance anywhere within the Local Supercluster (Kennicutt, Freedman, & Mould 1995). Therefore, a good Tully-Fisher calibration, possibly including any second parameter effects, such as Hubble type, hydrogen content and colour, and possibly taking the form envisaged by Ralf Bender this morning, is *achievable*.

2 Interim Calibration of Tully-Fisher

Mould *et al.* (1996) describe two calibrations, one for total magnitudes in the I band, another for the H band aperture magnitudes of Aaronson *et al.* (1986). Assuming $H_0 = 50$ km/sec/Mpc, Figure 1 compares the absolute magnitudes of galaxies from six Arecibo clusters and two Parkes clusters (only those with more than ten members in the sample) of Aaronson *et al.* (1986, 1989).

The poor fit of the new calibration with the cluster galaxies suggests one of three possibilities. Either $H_0 > 50$ km/sec/Mpc, or a sample selection effect has caused us to miss galaxies like the calibrators in the cluster sample, or some environmental effect has modified the Tully-Fisher relation in these clusters.

Mould *et al.* (1996) have examined the question of bias arising from the magnitude limit of the Arecibo and Parkes samples. They argue that the Cluster Population Incompleteness Bias corrections of Sandage, Tammann, & Federspeil (STF) (1996) tend to produce spurious results (see Table 1 and Figure 2).

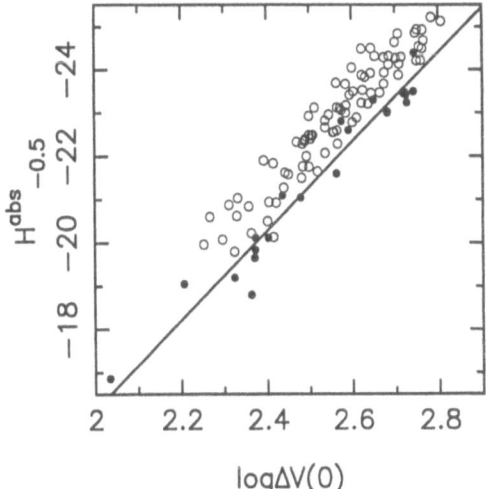

Fig. 1. Calibrators of the Tully Fisher relation (solid symbols). Cluster galaxies from the Arecibo and Parkes samples (open symbols). The absolute magnitudes of the cluster galaxies were obtained assuming $H_0 = 50$ km/sec/Mpc, and do not fit the interim calibration.

A better way to determine the effect of magnitude selection is to observe to a fainter magnitude limit. Mould *et al.* have also examined the question of bias arising from the 21 cm flux limits of the Arecibo and Parkes samples. Again, we can see the null dependence of intercept in the Tully-Fisher relation on flux limit (Figure 3).

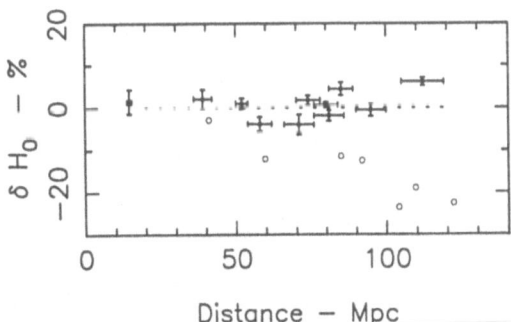

Fig. 2. CPIB corrections suggested by Sandage, Tammann & Federspeil (1996) tend to produce a spurious decrease in H_0 with increasing distance. Solid symbols are the original data of Aaronson *et al.* (1986); open symbols shown the data corrected following Table 1.

Table 1: CPIB corrections proposed by STF

Cluster	CPIB (mag)
Pisces	0.3
A400	0.55
A539	0.55
A1367	0.3
Coma	0.3
Pegasus	0.1
Hercules	0.55
A2634/66	0.55

MKW11

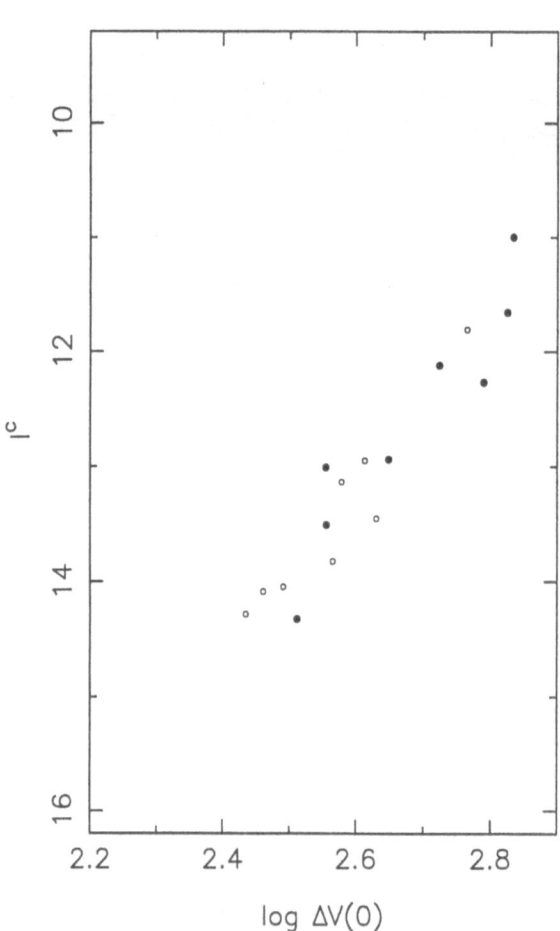

Fig. 3. Solid symbols: HI flux > 3 Jy-km/sec; open symbols: HI flux < 3 Jy-km/sec. MKW11 is one of the clusters studied by Mould *et al.* (1993).

A number of tests have been performed for environmental dependence in the Tully-Fisher relation (Han, Mould, & Bothun 1990). Figure 4 shows no correlation between Tully-Fisher *intercept* and cluster richness or cluster velocity dispersion. The reference just cited shows no correlation between Tully-Fisher *slope* and the same parameters.

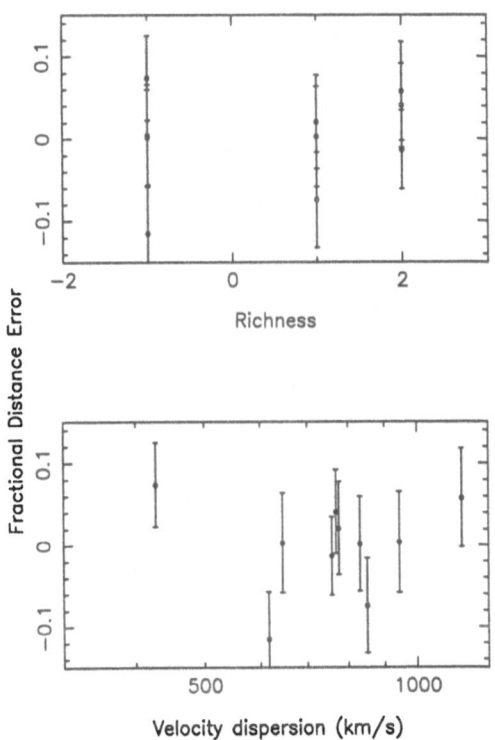

Fig. 4. There is no correlation of cluster environment parameters with distance deviation from the smooth Hubble flow in the sample of Aaronson *et al.* 1986.

Although we cannot present an exhaustive rejection of other possibilities, the interim conclusion by Mould *et al.* (1996) is that $H_0 = 73 \pm 10$ km/sec/Mpc. Freedman, Kennicutt, & Madore (1996) and Mould *et al.* (1996) show that consistent results are obtained from other scaling relations and from other galaxy standard candles. Further work is desirable to strengthen the present calibration, to sharpen the various bias tests, and to search for relevant hidden parameters in the Tully-Fisher relation.

Acknowledgements
I want to thank fellow team members Robert C. Kennicutt, Wendy L. Freedman, Barry F. Madore, Nancy Silberman, Robert Hill, Randy Phelps, Laura Ferrarese, Holland C. Ford, Garth D. Illingworth, Sandra M. Faber, Daniel Kelson,

James E. Gunn, John A. Graham, Shoko Sakai, Mingsheng Han, Shaun Hughes, John G. Hoessel, John P. Huchra, Lucas Macri, Anne Turner, Paul Harding, Fabio Bresolin, Abhijit Saha, Peter B. Stetson, Daya Rawson and Brad Gibson.

References

Aaronson, M. *et al.* 1986, ApJ, 302, 536

Aaronson, M. *et al.* 1989, ApJ, 338, 654

Freedman, Kennicutt, & Madore 1996, in The Extragalactic Distance Scale, eds. M. Livio & M. Donahue, Cambridge University Press.

Kennicutt, R., Freedman, W. & Mould, J. 1996, AJ, 110, 1476

Mould, J. , Han, M., & Bothun, G. 1990, ApJ, 347, 112

Mould, J., Sakai, S., Hughes, S., & Han, M. 1996, in The Extragalactic Distance Scale, eds. M. Livio & M. Donahue, Cambridge University Press.

Mould, J., Akeson, R., Bothun, G., Han, M., Huchra, J., & Schommer, R. 1993, ApJ, 409, 14

Sandage, A., Tammann, G., & Federspeil, M. (1996), ApJ, 452, 1

Tully, R. B. & Fisher, J.R. 1977, A& A, 54, 661

The Fundamental Plane at z = 0.18

Inger Jørgensen[1]* and Jens Hjorth[2]

[1] McDonald Observatory, The University of Texas, Austin, TX 78712, USA
[2] NORDITA, DK-2100 Copenhagen Ø, Denmark

Abstract. We present preliminary results regarding the Fundamental Plane (FP) for galaxies in the two rich clusters Abell 665 and Abell 2218. Both clusters have a redshift of 0.18. We have compared the FP for A665 and A2218, and for the cluster CL0024+16 at z=0.39, with the FP for the Coma cluster. The scatter around the FP is similar for all four clusters. There may be indications that the slope of the FP is more shallow for the intermediate redshift clusters than for the Coma cluster. More complete samples of galaxies in intermediate redshift clusters are needed to map in detail the possible change of the slope as function of redshift.

The mass-to-light (M/L) ratio as measured by the FP changes with redshift. At z=0.18 the M/L ratio (in Gunn r) is $16 \pm 9\%$ smaller than for the Coma cluster. Together with earlier results reported for CL0024+16 this implies that the M/L ratio changes with redshift as $\Delta \log M/L_r \sim -0.4\Delta z$.

The results presented here are in agreement with passive evolution of a stellar population, which formed at a redshift larger than one. However, the possible presence of more recent bursts of star formation complicates the interpretation of the data.

1 Introduction

Observational studies show that the formation and the evolution of galaxies are complex processes, which may involve interactions, starbursts and infall (e.g., Dressler et al. 1994ab; Lilly et al. 1995; Moore et al. 1996). Some nearby E and S0 galaxies may also have experienced recent star formation. Caldwell et al. (1993) found a fraction of the E and S0 galaxies in the Coma cluster to have post-starburst spectra, and Faber et al. (1995) suggest that nearby field E and S0 galaxies have a substantial variation in the mean age of their stellar populations.

In order to investigate the evolution of galaxies, studies are needed of both the morphological evolution and the evolution of the luminosities and the mass-to-light (M/L) ratios of the galaxies. High-resolution imaging with the *Hubble Space Telescope* (HST) and from the ground, combined with spectroscopy from the ground make it possible to carry out this kind of studies. Studies of the morphological evolution show that the fraction of spiral galaxies and irregular galaxies in clusters was higher at larger redshifts (e.g., Dressler et al. 1994ab; Oemler et al. 1997). The luminosity evolution of disk galaxies has recently been studied by Vogt et al. (1996) who established the Tully-Fisher (1977) relation for a sample of field galaxies with redshifts between 0.1 and 1.

The Fundamental Plane (FP) (Dressler et al. 1987; Djorgovski & Davis 1987) for elliptical galaxies makes it possible to study how the M/L ratios change with

---

* Hubble Fellow

redshift. Also, S0 galaxies in nearby clusters follow the FP (e.g., Jørgensen et al. 1996, hereafter JFK96). The FP relates the effective radius, r_e, the mean surface brightness within this radius, $<I>_e$, and the (central) velocity dispersion, σ, in a tight relation, which is linear in log-space. For 226 E and S0 galaxies in nearby clusters JFK96 found $\log r_e = 1.24 \log \sigma - 0.82 \log <I>_e$ +cst. This relation can be interpreted as $M/L \propto M^{0.24}$, see also Faber et al. (1987). The scatter of the FP is very low, equivalent to a scatter of 23% in the M/L ratio (e.g., JFK96). Thus, the FP offers the possibility of detecting even small differences in the M/L ratios by observing five to ten galaxies in a distant cluster.

The FP for intermediate redshift clusters has been established by van Dokkum & Franx (1996) and Kelson et al. (1997). Both studies find the M/L ratios of the galaxies to increase between redshifts of $z = 0.3$ to 0.6 and the present.

In this paper we establish the FP for the two rich clusters Abell 665 and Abell 2218. We use the data for these two clusters together with earlier published data for CL0024+16 (van Dokkum & Franx 1996) and data for the Coma cluster to study how the M/L ratios of the E and S0 galaxies change with redshift.

2   Observational Data for A665 and A2218

The central parts of the two rich clusters Abell 665 and Abell 2218 were both observed with the Nordic Optical Telescope (NOT), La Palma, in March 1994. Observations were done in the I-band and the V-band. The total size of the observed fields are $3\rlap{.}'4 \times 2\rlap{.}'7$ for A665 and $2\rlap{.}'8 \times 3\rlap{.}'5$ for A2218. The basic reductions and the standard calibration are described in detail in Jørgensen et al. (1997). Determination of the effective radius, r_e, and the mean surface brightness within this radius, $<\mu>_e$, was done following the technique described by van Dokkum & Franx (1996). This technique uses the full 2-dimensional information in the image, and takes the point-spread-function into account. The magnitudes were calibrated to Gunn r in the rest frame of the clusters. We are in the process of analysing HST images of the two clusters. The final discussion of the FP for these clusters will be based on photometry from both the NOT and the HST, cf. Jørgensen et al. (1997).

The spectra of 7 galaxies in A665 were obtained with the Multiple Mirror Telescope in January 1991. The total integration time was 9 hours. The observations include an E+A galaxy. The spectra of 10 galaxies in A2218 were obtained with the KPNO 4m telescope in June 1994; total integration time 12 hours. We have photometry of 8 of these galaxies. Detailed description of the reductions and the determination of the velocity dispersions can be found in Jørgensen et al. (1997). Figure 1 shows two of the spectra obtained of galaxies in A2218.

Determination of velocity dispersions of galaxies is usually done with spectra of template stars obtained with the same instrumentation as the galaxy spectra. Due to the large redshifts of A665 and A2218 this will not work for these clusters. A further complication is that the instrumental resolution of the spectra of A665 and A2218 varies with wavelength and from slit-let to slit-let. Therefore, the instrumental resolution as a function of wavelength and slit-let was mapped based on calibration lamp exposures. Spectra of template stars were then convolved to the exact same resolution. The velocity dispersions were determined by fitting the galaxy spectra with the template spectra using the Fourier Fitting Method (Franx et al. 1989). The velocity dispersions were aperture corrected following the technique described by Jørgensen et al. (1995b). We corrected the velocity dispersions to a circular aperture with diameter $1.19\,h^{-1}\,$kpc, equivalent to 3.4 arcsec at the distance of the Coma cluster.

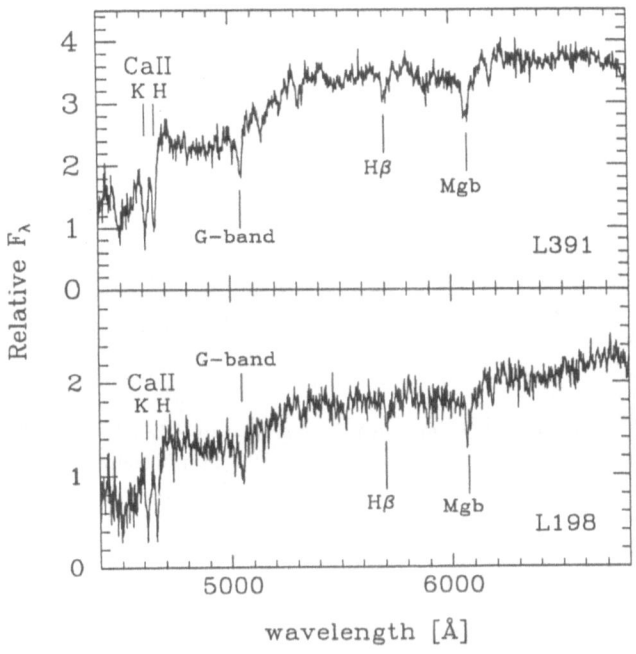

Fig. 1. Spectra for two of the galaxies in A2218. L391 is the brightest cluster galaxy. L198 is the faintest galaxy we observed in the cluster. Galaxy numbers are from Le Borgne et al. (1992). The main absorption features are labeled.

3   Data for the Coma Cluster

The Coma cluster is in the following used as a low redshift reference for the purpose of determination of changes in the FP as function of redshift. We use the sample of E and S0 galaxies, which has photometry in Gunn r from Jørgensen et al. (1995a), see also Jørgensen & Franx (1994). The sample covers the central $64' \times 70'$ of the cluster. The velocity dispersions from the literature (Davies et al. 1987; Dressler 1987; Lucey et al. 1991) were calibrated to a consistent system and aperture corrected. We use the values as listed by Jørgensen et al. (1995b).

Further, we use new velocity dispersions measurements from Jørgensen (1997). A total of 116 galaxies have available photometry and spectroscopy. The sample is 93% complete to a magnitude limit of $m_T = 15\overset{m}{.}05$ in Gunn r, equivalent to $M_{r_T} = -20\overset{m}{.}75$ ($H_0 = 50$ km s^{-1} Mpc^{-1} and $q_0 = 0.5$).

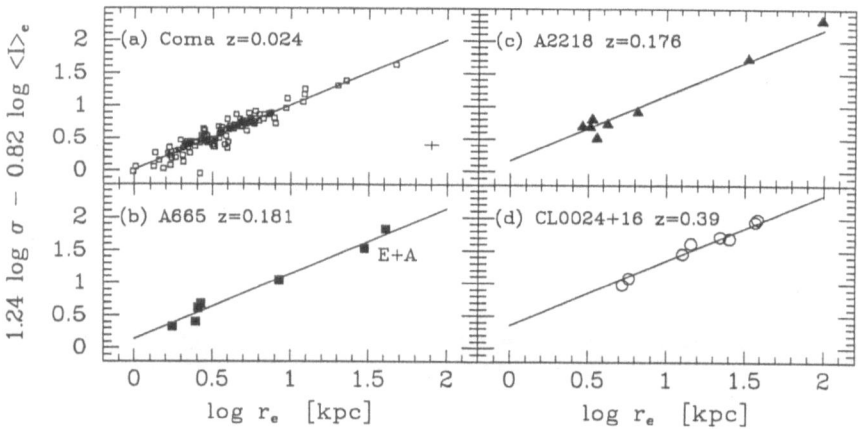

Fig. 2. The FP edge-on for Coma, A665, A2218 and CL0024+16. The data for CL0024+16 are from van Dokkum & Franx (1996). The photometry is calibrated to Gunn r in the rest frames of the clusters, and is not corrected for the dimming due to the expansion of the Universe. $\log <I>_e = -0.4(<\mu>_e - 26.4)$ is in units of $L_\odot/\text{pc}^2$. The solid lines are the FPs with coefficients adopted from JFK96 and zero points derived from the data presented in the figure. The E+A galaxy in A665 is labeled.

4   The Fundamental Plane

Figure 2 shows the FP for A665 and A2218. The figure also shows the FP for the Coma cluster and for the cluster CL0024+16 with a redshift of 0.39. The data for CL0024+16 are from van Dokkum & Franx (1996), and the photometry has been calibrated to rest frame Gunn r, see Jørgensen et al. (1997).

We adopt the coefficients for the FP from JFK96, $(\alpha,\beta)=(1.24\pm0.07,-0.82\pm0.02)$. The coefficients were derived for a sample of 226 galaxies in 10 nearby clusters. Photometry in Gunn r was used. The relations shown on Figure 2 are FPs with these coefficients. Figure 3 shows the FP face-on and in two edge-on views. This figure includes the Coma cluster galaxies as well as the galaxies in A665, A2218 and CL0024+16. The mean surface brightnesses have been corrected for the dimming due to the expansion of the Universe. The effective radii are in kpc ($H_0 = 50$ km s^{-1} Mpc^{-1} and $q_0 = 0.5$ were used).

The scatter around the FP is the same for A665 and A2218 (0.091 and 0.115 in $\log r_e$) as for the Coma cluster (0.095 in $\log r_e$). The scatter for CL0024+16

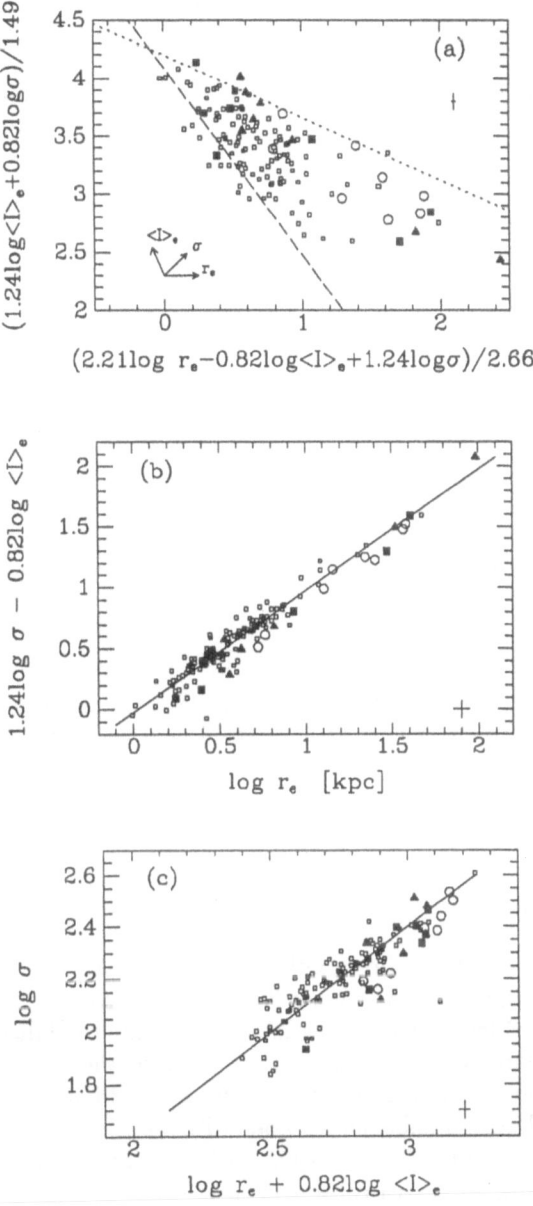

Fig. 3. The FP for Coma, A665, A2218 and CL0024+16 in Gunn r in the rest frames of the clusters. Open boxes – Coma; solid boxes – A665; solid triangles – A2218; circles – CL0024+16. (a) The FP face-on. The dashed line marks the luminosity limit for the completeness of the Coma cluster sample, $M_{r_T} = -20\overset{m}{.}75$. The dotted line is the so-called exclusion zone, $y \approx -0.54x + 4.2$, first noted by Bender et al. (1992). (b) & (c) The FP edge-on. The solid line is the FP for the Coma cluster with coefficients adopted from JFK96. Typical error bars are given on the panels.

is slightly lower, 0.064, though the difference is not statistically significant. The scatter for the Coma cluster reported here is somewhat larger than previous estimates based on smaller samples (e.g., Jørgensen et al. 1993, 1996; Lucey et al. 1991). The difference is due to inclusion of more galaxies, which have post-starburst spectra (cf., Caldwell et al. 1993), see also Jørgensen (1997).

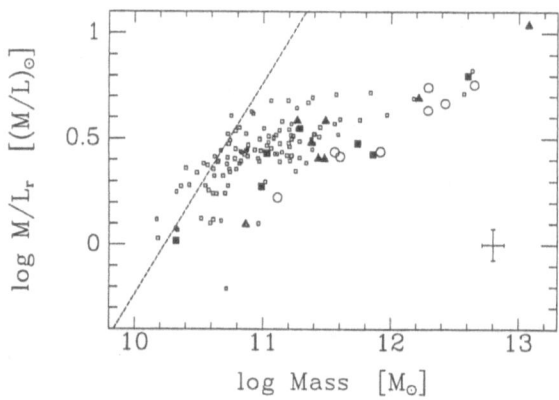

Fig. 4. The M/L ratio (in Gunn r) as function of the mass. Symbols as on Fig. 3. The dashed line marks the luminosity limit for the completeness of the Coma cluster sample, $M_{r_T} = -20^m\!.75$.

4.1   The Slope of the FP

The samples in the intermediate redshift clusters are heavily biased towards intrinsically bright galaxies. This makes it difficult to test for possible differences in the slope of the FP for Coma and for the intermediate redshift clusters. The selection bias is clearly visible in the face-on view of the FP, Figure 3a. In addition to the difference in selection effects the intermediate redshift clusters seem to contain a larger number of galaxies with low surface brightness and large effective radius than the Coma cluster. The samples are still too small and incomplete to perform any statistical test of this.

Figure 4 shows the FP as the relation between the M/L ratio and the mass. This figure and Figure 3c indicate that the slope of the FP may be slightly different for the intermediate redshift clusters than for the Coma cluster.

In order to test if the coefficients of the FP depend on the redshift we fit the FP to the three clusters A665, A2218, and CL0024+16 as parallel planes, under the assumption that the FPs for these clusters have the same coefficients. This gives $(\alpha,\beta)=(0.89\pm0.14,-0.78\pm0.04)$. Omitting the E+A galaxy in A665 does not change the fit. Formally the value of α is different from the coefficient for nearby cluster galaxies (JFK96). A fit to the Coma cluster galaxies alone gives $(\alpha,\beta)=(1.28 \pm 0.08, -0.83 \pm 0.03)$. In order to limit the effect of the different selection criteria we repeated the fit to the Coma cluster galaxies enforcing a magnitude limit of $M_{r_T} = -21^m65$. This does not give a result significantly different from the fit for the whole sample. The difference in α between the fit to the Coma cluster sample and the fit to the three intermediate redshift clusters is formally significant on the 2.5σ level. Still, the weight of the low luminosity galaxies is much larger in the Coma cluster sample than for the intermediate redshift clusters.

The coefficients we find for the FP for the intermediate redshift clusters imply (assuming structural homology of the galaxies) that

$$M/L \propto r_e^{0.28}\sigma^{0.86} \propto M^{0.43}r_e^{0.15} \qquad (1)$$

This should be compared with $M/L \propto M^{0.24}r_e^{-0.02}$ found for nearby cluster (JFK96). The difference may indicate that the low mass galaxies show a stronger luminosity evolution that the more massive galaxies. We emphasize that these results regarding the slope of the FP are preliminary. We discuss the issue in greater length in Jørgensen et al. (1997), where also data from Kelson et al. (1997) are included in the analysis.

4.2   The Evolution of the M/L Ratio

The zero point of the FP depends on the cosmological effects (surface brightness dimming and the value of q_0), and the evolution of the galaxies. The FP can in principle be used to test for the expansion of the Universe (Kjærgaard et al. 1993). In the following we correct the data for the expansion of the Universe, and we assume $q_0 = 0.5$. The FP shown in Figure 3 is corrected for the expansion of the Universe. The zero point of the FP can then be used to study the mean evolution of the galaxies in the clusters. Figure 5 shows the zero point differences between the intermediate redshift clusters and the Coma cluster as the offset in the M/L ratio.

The uncertainties of the zero point offsets have contributions from the photometric calibration, the calibration of the velocity dispersion, and uncertainties in the zero points for the FP for the Coma cluster and for the intermediate redshift clusters. The error bars shown on Figure 5 include these contributions, see also Jørgensen et al. (1997). The evolution of the M/L ratio with redshift is consistent with $\Delta \log M/L \approx (-0.4 \pm 0.1)\Delta z$.

The change of the M/L ratio with redshift is expected based on stellar populations models. Models of single burst populations show that for passive evolution

$$M/L \propto t_{age}^{\kappa} \qquad (2)$$

From models by Vazdekis et al. (1996) we find that $\kappa \approx 0.9 - 0.2x$ for photometry in Gunn r. Here x is the slope of the initial mass function (IMF) of the stars. A

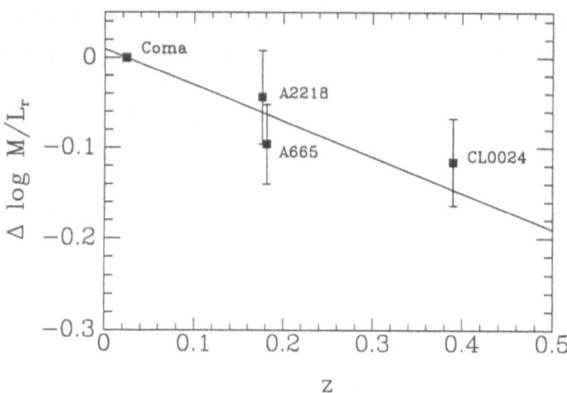

Fig. 5. The evolution of the M/L ratio (in Gunn r) as function of the redshift of the clusters. The result for CL0024+16 is based on data from van Dokkum & Franx (1996). The solid line has a slope of −0.4.

Salpeter (1956) IMF has $x = 1.35$. Eq. 2 can be used to relate the change in the M/L ratio with redshift to the formation redshift, z_{form}, and the value of q_0

$$\Delta \ln M/L = -\kappa(1 + q_0 + z_{form}^{-1})\,\Delta z \qquad (3)$$

(Franx 1995; van Dokkum & Franx 1996).

The slope we find for the change of the M/L ratio with redshift implies

$$\kappa(1 + q_0 + z_{form}^{-1}) = 0.9 \pm 0.2 \qquad (4)$$

The result is in agreement with the result found by van Dokkum & Franx (1996) based on CL0024+16, only. Eq. 4 is consistent with passive evolution of a single stellar population model with a Salpeter IMF, $q_0 = 0.5$ and $z_{form} = \infty$. A model with $z_{form} = 1.2$ deviates on the 3σ level.

The model constraints given here should only be taken as rough guidelines. The correct interpretation of the data is most likely rather more complicated than indicated here. The correct value of κ is not known. Further, the evolution of E and S0 galaxies cannot be viewed as a single burst event. The presence of younger stellar populations in the galaxies would imply a larger formation redshift for the old stellar populations in the galaxies. The most fundamental assumption for the interpretation of the data is that the observed E and S0 galaxies in the intermediate redshift clusters in fact evolve into galaxies similar to the present day E and S0 galaxies. It is possible that our selection of E and S0 galaxies in the intermediate redshift clusters is biased to select already aged galaxies, see the discussion by Franx (1995, and this volume). Larger and more complete samples of galaxies in intermediate redshift clusters are needed in order to address this problem in detail.

5   Conclusions

We have established the Fundamental Plane for the two rich clusters Abell 665 and Abell 2218, both at a redshift of 0.18. The photometric parameters were derived from ground based observations obtained with the Nordic Optical Telescope, La Palma. The photometry was calibrated to Gunn r in the rest frame of the clusters. Central velocity dispersions were measured for seven galaxies in A665 and ten galaxies in A2218. The results presented here are preliminary. The final analysis of the FP for the two clusters will include photometry based on HST data, see Jørgensen et al. (1997).

The FP for the two clusters were compared to the FP for nearby clusters derived by JFK96, and to the FP for the Coma cluster (Jørgensen 1997). The scatter around the FP for A665 and A2218 is similar to the scatter found for nearby clusters. We have used the data for A665 and A2218 together with data for CL0024+16 (van Dokkum & Franx 1996) to test if the slope of the FP changes with redshift. All photometry was calibrated to Gunn r in the rest frame of the clusters. There may be indications that the coefficient for $\log \sigma$ is significantly smaller for the intermediate redshift clusters than for the Coma cluster. However, severe selection effects and possibly also real differences in the distributions within the FP make it difficult to draw definite conclusions based on the current data. If the smaller value of the coefficient is confirmed in later studies it implies that $M/L \propto M^{0.43}$ for the intermediate redshift clusters, while $M/L \propto M^{0.24}$ for nearby clusters. This may indicate that the low mass galaxies in the clusters evolve faster than the more massive galaxies.

The zero point offsets in the FP for the intermediate redshift clusters were used to investigate the average evolution of the M/L ratio of the galaxies. The M/L ratios of the galaxies in A665 and A2218 are $16 \pm 9\%$ smaller than the M/L ratios of galaxies in the Coma cluster. From all four clusters we find that the evolution of the M/L ratio with redshift is consistent with $\Delta \log M/L \approx (-0.4 \pm 0.1)\Delta z$. This can be used to constrain the formation redshift of the galaxies. The interpretation depends on the crucial assumption that the E and S0 galaxies observed in the intermediate redshift clusters evolve into galaxies similar to present day E and S0 galaxies, and that the samples in the intermediate redshift clusters form a representative sample of all the galaxies that will end as present day E and S0 galaxies. For a single burst population, a Salpeter IMF and $q_0 = 0.5$ the evolution in the M/L ratio implies that the formation redshift is larger than one. A more complete analysis of our data for A665 and A2218, which also involves photometry based on HST images, is presented in Jørgensen et al. (1997).

Acknowledgements: We thank M. Franx for permission to use spectroscopic data for A665 before publication. The staff at NOT, KPNO and MMT are thanked for assistance during the observations. Financial support from the Danish Board for Astronomical Research is acknowledged. This research was supported through Hubble Fellowship grant number HF-01073.01.94A from the

Space Telescope Science Institute, which is operated by the Association of Universities for Research in Astronomy, Inc., under NASA contract NAS5-26555.

References

Bender, R., Burstein, D., Faber, S. M. 1992, ApJ, 399, 462

Caldwell, N., Rose, J. A., Sharples, R. M., Ellis, R. S., Bower, R. G. 1993, AJ, 106, 473

Davies, R. L., Burstein, D., Dressler, A., Faber, S. M., Lynden-Bell, D., Terlevich, R. J., Wegner, G. 1987, ApJS, 64, 581

Djorgovski, S., Davis, M. 1987, ApJ, 313, 59

Dressler, A. 1987, ApJ, 317, 1

Dressler, A., Lynden-Bell, D., Burstein, D., Davies, R. L., Faber, S. M., Terlevich, R. J., Wegner, G. 1987, ApJ, 313, 42

Dressler, A., Oemler, A. Jr., Butcher, H., Gunn, J. E. 1994a, ApJ, 430, 107

Dressler, A., Oemler, A. Jr., Sparks, W. B., Lucas, R. A. 1994b, ApJ, 435, L23

Faber, S. M., Dressler, A., Davies, R. L., Burstein, D., Lynden-Bell, D., Terlevich, R. J., Wegner, G. 1987, in Nearly Normal Galaxies, ed. S. M. Faber, Springer-Verlag, New York, p. 175

Faber, S. M., Trager, S. C., Gonzales, J. J., Worthey, G. 1995, in Stellar Populations, eds. van der Kruit, P. C., Gilmore, G., IAU coll. 164, Kluwer, Dordrecht, p. 249

Franx, M. 1995, in Stellar Populations, eds. van der Kruit, P. C., Gilmore, G., IAU coll. 164, Kluwer, Dordrecht, p. 269

Franx, M., Illingworth, G., Heckman, T. 1989, ApJ, 344, 613

Jørgensen, I. 1997, in preparation

Jørgensen, I., Franx, M. 1994, ApJ, 433, 553

Jørgensen, I., Franx, M., Hjorth, J., van Dokkum, P. G. 1997, in preparation

Jørgensen, I., Franx, M., Kjærgaard, P. 1993, ApJ, 411, 34

Jørgensen, I., Franx, M., Kjærgaard, P. 1995a, MNRAS, 273, 1097

Jørgensen, I., Franx, M., Kjærgaard, P. 1995b, MNRAS, 276, 1341

Jørgensen, I., Franx, M., Kjærgaard, P. 1996, MNRAS, 280, 167 (JFK96)

Kelson, D., van Dokkum, P. G., Franx, M., Illingworth, G. D., Fabricant, D. 1997, ApJ, in press

Kjærgaard, P., Jørgensen, I., Moles, M. 1993, ApJ, 418, 617

Le Borgne, J. F., Pelló, R., Sanahuja, B. 1992, A&AS, 95, 87

Lilly, S. J., Tresse, L., Hammer, F., Crampton, D., Le Fèvre, O. 1995, ApJ, 455, 108

Lucey, J. R., Guzmán, R., Carter, D., Terlevich, R. J. 1991, MNRAS, 253, 584

Moore, B., Matz, N., Lake, G., Dressler, A., Oemler, A. Jr. 1996, Nature, 379, 613

Oemler, A. Jr., Dressler, A., Butcher, H. 1997, ApJ, 474, 561

Salpeter, E. E. 1956, ApJ, 121, 161

Tully, R. B., Fisher, J. R. 1977, A&A, 54, 661

van Dokkum, P. G., Franx, M. 1996, MNRAS, 281, 985

Vazdekis, A., Casuso, E., Peletier, R. F., Beckman, J. E. 1996, ApJS, 106, 307

Vogt, N. P., Forbes, D. A., Phillips, A. C., Gronwall, C., Faber, S. M., Illingworth, G. D., Koo, D. C. 1996, ApJ, 465, L15

Measuring the Evolution of the Mass-to-Light Ratio from $z = 0$ to $z = 0.6$ from the Fundamental Plane

Marijn Franx[1], Dan Kelson[2], Pieter van Dokkum[1], Garth Illingworth[2], Dan Fabricant[3]

[1] University of Groningen, Netherlands
[2] University of California, Santa Cruz, USA
[3] Center for Astrophysics, Cambridge, USA

Abstract. Galaxy evolution is probably a complex process. Mergers, infall, and star bursts may change galaxy properties systematically with time. As a result, the interpretation of the luminosity function is ambiguous, and information on the mass evolution of galaxies is needed. Such information can be retrieved from the evolution of the Tully-Fisher relation, Faber-Jackson relation, or the Fundamental Plane with redshift.

Observations of this kind have recently become possible. We present the Fundamental Plane relation measured for galaxies in rich clusters out to $z = 0.58$. The galaxies satisfy a tight Fundamental Plane, with relatively low scatter (17 %). The M/L evolves slowly with redshift, $\ln M/L_V \propto 0.8z$. This result is consistent with simple evolutionary models if the bulk of the stellar population formed at high redshift.

It is not clear yet how these results can be made consistent with the rapid evolution of galaxies in intermediate redshift clusters as indicated by the Butcher-Oemler effect. Observations of post star burst galaxies ("E+A" galaxies) indicate that these systems are dominated by disks. Some have smooth spiral arms without signs of star formation. The E+A galaxies may evolve into galaxies which are underrepresented in most "normal" Fundamental Plane samples.

1 Measuring the Evolution of Mass $F(M_*, z)$, or $F(v_c, z)$

Galaxy evolution may be a complex process, with a large role for mergers, interactions, infall, and star bursts triggered by these events. Such processes complicate the interpretation of observations of high redshift galaxies, as galaxies can change rapidly in luminosity, due to star bursts, and can change morphology due to mergers, infall of gas, and enhanced star formation. The progenitors of certain types of galaxies at some redshift may be of different type at some other redshift, and their luminosities may be quite different.

In order to quantify these effects, more information is needed than the evolution of luminosity and color of galaxies, as measured by the evolution of the luminosity functions. Detailed information on the morphological evolution, and the evolution of the mass function of galaxies, is essential. The evolution of the mass function is possibly the most important, as it gives direct insight into the mass evolution of individual galaxies, and can directly determine when typical galaxies were assembled.

Unfortunately, the total masses of galaxies are notoriously difficult to measure. However, there exist good relations between circular velocity, and velocity dispersion, and photometric parameters: the Tully-Fisher relation for spirals (Tully & Fisher 1977), the Faber-Jackson relation (Faber & Jackson 1976), and the Fundamental Plane for early-types (Djorgovski & Davis 1987, Dressler et al. 1987). These relations are very suitable for evolutionary studies, because their intrinsic scatter is low at $z=0$. Furthermore, the kinematic parameters may be related to the dark matter halos of galaxies. As a result, the evolution of these parameters may also be related to the halo evolution.

These considerations provide the rationale for observational studies of the evolution of the Tully-Fisher relation, Faber-Jackson relation, and Fundamental Plane with redshift. The combination of the observations with the evolution of the luminosity function of galaxies, can be used to constrain the evolution of the distribution of circular velocities $F(v_c, z)$, which is less sensitive to star bursts than the equivalent $F(L, z)$. Similarly, the stellar mass locked up in early type galaxies can be measured in a similar way. Such results will provide direct constraints on theories of galaxy formation and evolution.

2   Evolution of the Fundamental Plane

Here we present new results on a program to measure the evolution of the Fundamental Plane relation with redshift. Early results can be found in Franx (1993a,b, 1995). The Fundamental Plane is a relation between effective radius r_e, effective surface brightness I_e, and central velocity dispersion σ of the form $r_e \propto \sigma^{1.24} I_e^{-0.82}$ (e.g., Bender et al. 1992, Jørgensen et al 1996, JFK). Its scatter is low, at 17% in r_e (Lucey et al. 1991, JFK). The implication of the Fundamental Plane is that the M/L ratio of galaxies is well behaved (e.g., Faber et al. 1987). Under the assumption that galaxies are a homologous family, the implied M/L scaling is $M/L \propto r_e^{0.22} \sigma^{0.49} \propto M^{0.24}$. Such scaling is sufficient for the existence of the Fundamental Plane, and vice versa. The cause of the variation in M/L with mass is not well understood, but it is thought to be mainly due to variations in metallicity (see also, e.g., Renzini & Ciotti 1993).

Observations at higher redshifts will yield the evolution of the M/L ratio as a function of redshift. Below we explore the expected variation of M/L.

2.1   Models for the Evolution of the M/L Ratio

The luminosity of a co-eval stellar population is expected to evolve with time. Tinsley (1980) showed that the luminosity evolves like

$$L \propto 1/(t - t_{form})^{\kappa}$$

where $\kappa = 1.3 - 0.3x$, and x is the slope of the IMF. The Miller–Scalo IMF implies $x=0.25$, and $\kappa \approx 1.2$. Recent studies indicate that the value of κ depend on passband and metallicity (Buzzoni 1989, Worthey 1994). These authors find $0.6 < \kappa < 0.95$ for the V band.

To first order, this evolution implies that the M/L ratio evolves like

$$\ln M/L(z) = \ln M/L(0) - \kappa(1 + q_0 + 1/z_{form})z,$$

where z_{form} is the formation redshift (Franx 1995). Hence the logarithm of the M/L ratio is expected to decrease linearly with redshift, and the coefficient depends on κ(IMF), q_0, and z_{form}. This equation is valid for $q_0 \approx 0$, and high z_{form}. The equation implies that the rate at which the M/L ratio decreases is a function of several unknown variables, and a direct interpretation of the observed decrease of the M/L ratio may not be very straightforward.

Fig. 1a shows the expected evolution of the L/M ratio if all galaxies form at the same redshift. As can be seen, the evolution depends strongly on the formation redshift. It is unlikely that galaxies formed in such a simple way. For Fig. 1b we explored models in which galaxies form at a range of redshifts. As a result, scatter is introduced in the L/M ratio, which increases with look back time. This is due to the fact that the relative age difference increases with look back time.

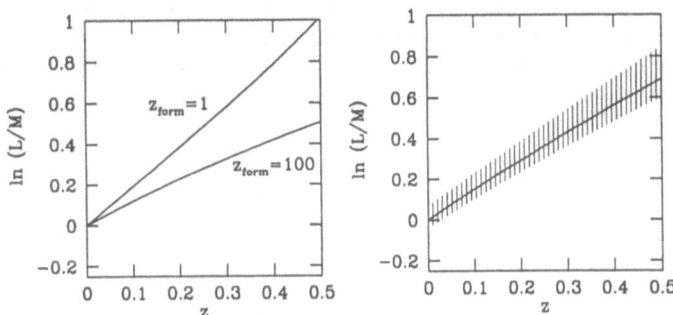

Fig. 1. The evolution of galaxies with a simple star formation history. a) shows the luminosity evolution for galaxies with co-eval populations. Galaxies which formed recently evolve faster. b) the evolution of the mean L/M ratio for a sample of early-types which formed at a random time between $z=1$ and $z=2$. The scatter in the relation increases with redshift, as the relative age difference increases with lookback time.

2.2 Complex Evolution

Even the last model is probably an over–simplification of the formation of early types. There is no good reason to assume that all stars in an early-type galaxy formed in a very short burst. A single galaxy may have had a complex formation history, with star formation extending over a long time. The evolution of the M/L ratio will be more complex if such age differences are taken into account.

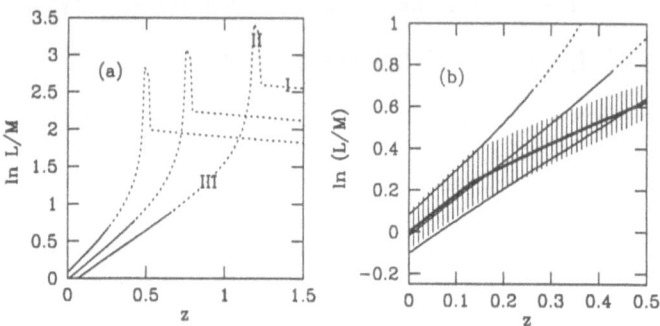

Fig. 2. The evolution of galaxies which undergo three distinct phases: I regular star formation in a disk, II star burst, III quiescent evolution. a) shows the L/M evolution for three such galaxies. The mass is taken to be the stellar mass after the burst. The galaxies are classified as regular early-types after 1.5 Gyr after the burst. This epoch is indicated by the solid curve. b) the evolution of the mean L/M ratio for a sample of early-types which formed in this complex way. The star burst is assumed to occur at a random time between $z=0.5$ and $z=2$. The thick line indicates the median L/M, the shaded area is bounded by the upper and lower quartile of the sample. The median L/M ratio bends at $z=0.2$, as more and more galaxies drop out from the sample. The sample becomes more and more biased to the oldest early-types at higher redshifts.

We have created models in which early type galaxies form by transformation of galaxies with continuous star formation. These models are rather ad hoc, but are qualitatively similar to the more physical models explored by Kauffmann (1996). It is assumed that the progenitors form stars in a continuous way, until a burst of star formation occurs, and the galaxies are transformed into non-star forming galaxies. These will appear as post star burst galaxies for another 1.5 Gyr, and then appear to be early types.

This type of evolution implies that the morphologies of galaxies evolve with time, from, perhaps, spiral, to post star burst galaxy, to early-type. This has important consequences, since the set of early-types at higher redshifts will be a special subset of the set of early-types at $z=0$. If we select early-types at higher and higher redshift, we are selecting a subsample that is more and more biased towards the oldest early-types. In short, we may be selecting the oldest galaxies, and find that they are old.

The effect is illustrated in Fig. 2. Fig. 2a shows the evolution of 3 galaxies. The solid curve is the phase in which they appear as early-types. Clearly, the oldest early-types appear as early-type for the longest time. Fig. 2b demonstrates the effect on the observed L/M ratios of a large sample. At low redshifts, all galaxies appear as early types, and the evolution of the median L/M ratio remains normal. The scatter around the mean increases rapidly with redshift. Around $z = 0.2$, some of the galaxies appear as post star burst galaxies, and they would be excluded. The median L/M ratio is biased towards low values.

This effects increases at higher redshifts. The bias is as strong as 30 % at z=0.5. As galaxies disappear from the sample, the scatter in the L/M ratio may even decrease at higher redshifts.

The relevance of this bias is best quantified for the field, where we can measure the evolution of the number density of early type systems with redshift directly. When studying clusters, we will have to search for all possible progenitors of current day ellipticals in order to constrain the relevance of this bias.

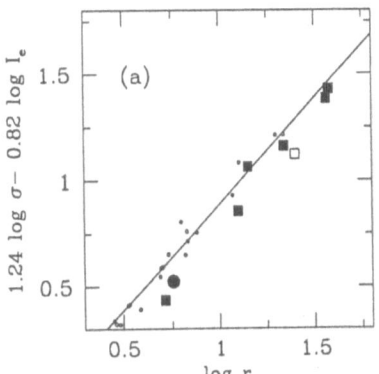 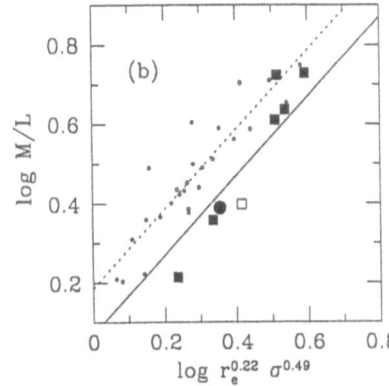

Fig. 3. a) The Fundamental Plane for galaxies in CL0024+16 at z=0.391 in the redshifted V band. The small symbols are galaxies in Coma. The Fundamental Plane in CL0024 is very similar to that in Coma, with similar low scatter (15%). b) The M/L ratio against $r_e^{0.22}\sigma^{0.49} \propto M^{0.24}$, for CL0024 and Coma. The lines are fits to the data points. The M/L ratio in CL0024 is lower by 31$\pm$ 12 %.

2.3 The Fundamental Plane in CL0024+16 at z=0.39

CL0024 is a rich cluster at z=0.39, and has been extensively observed (e.g., Dressler et al. 1985). We have obtained a deep, 19 hour integration at the MMT to measure the internal velocity dispersions of luminous galaxies in the cluster. HST images were used to measure the structural parameters of the galaxies. Full details of the observations and the analysis can be found in van Dokkum and Franx (1996).

Fig. 3a shows the resulting Fundamental Plane. There is a very clear relation, with relatively low scatter (15 %). The slope is very similar to that for nearby cluster galaxies (e.g., JFK). In short, *early-type galaxies exist at z=0.4 which are very similar to galaxies at z=0.*

Fig. 3b shows the observed M/L ratios for Coma and CL0024 against the parameter $r_e^{0.22}\sigma^{0.49}$. The Fundamental Plane implies that galaxies lie along a line in the plot. We see a clear offset between the two data sets. The lines indicate fits to both data sets. The mean difference in the M/L ratio is 31 %. The error is dominated by systematic effects, and is estimated at 12 %. It is clear that the sample for CL0024 is biased towards the most luminous galaxies, and this observational bias is partly the cause for the systematic uncertainty.

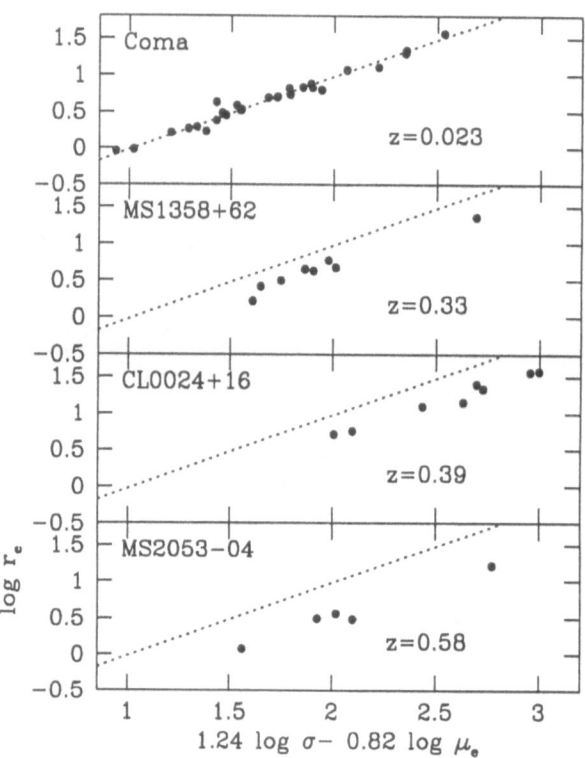

Fig. 4. The evolution of the Fundamental Plane from $z = 0$ to $z = 0.58$, based on data from Kelson et al (1997) and van Dokkum and Franx (1996). The Fundamental Plane relations for the clusters are shown. The drawn line is the relation for Coma. The high redshift clusters follow a relation with a similar slope, but with an offset. The offset is mostly due to cosmological surface brightness dimming.

3   Using Keck to Extend to $z = 0.58$

With modern telescopes and efficient instrumentation it is possible to extend the Fundamental Plane work out to higher redshifts. We have recently used Keck to measure the Fundamental Plane in two clusters at z=0.33 and z=0.58, CL 1358+62 and MS 2053-04 respectively. A full description can be found in Kelson et al 1997. Typical integration times were 2 hours on the Keck telescope. Fig. 4 shows the resulting Fundamental Planes from $z = 0$ to $z = 0.58$. The figure demonstrates how well the relation is defined at each redshift interval.

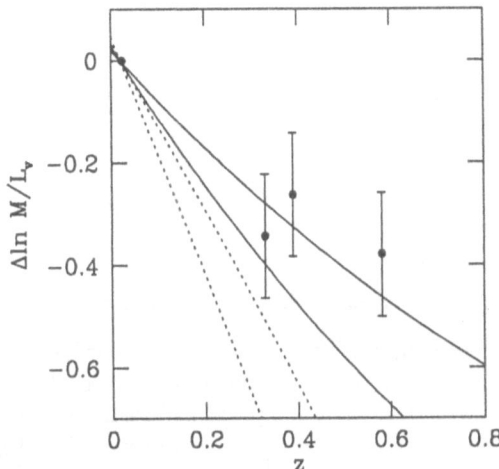

Fig. 5. The evolution of the mean M/L_V ratio, after correction for surface brightness dimming ($q_0 = 0.5$). The M/L_V ratio decreases slowly with redshift. The drawn lines indicate stellar population models with formation redshifts of ∞. The dashed lines indicates models with a formation redshift of 1. The data are consistent with high formation redshift. If lower values for q_0 are used, the datapoints move downward, and the model predictions move upward.

Surprisingly, the scatter in the relation remains low. We have now 22 galaxies with Fundamental Plane parameters, and we find a scatter of 17 %. This is quite comparable to the scatter in nearby rich clusters.

The evolution of the M/L_V ratio is shown in Fig. 5. The data are consistent with a slow evolution of $\ln M/L_V \propto 0.8z$. Both the low evolution, and the small scatter are consistent with high formation redshifts for cluster early types ($z_{form} > 2$).

4   How about the Butcher–Oemler Effect and E+A's ?

It has been well established that distant clusters have a high proportion of blue members (Butcher & Oemler, 1984). Furthermore, some galaxies have post star burst spectra (Dressler and Gunn, 1983, 1992). These galaxies have spectra which can be modeled as the superposition of a young component and an old component, and were named "E+A" by Dressler and Gunn. The "E" stands for early type, and "A" for A-star. These galaxies were defined to have no emission lines, i.e., little or no star formation. The population models invoked a peak in the star formation rate 1 Gyr earlier, and a subsequent drop in the star formation rates.

The relatively high fraction of such post star burst galaxies in clusters (Dressler and Gunn 1992) and the short lifetime of the phenomenon implies that a large fraction of galaxies in clusters underwent such a phase at intermediate redshifts. This would inevitably cause a rather large spread in the mean stellar ages of the galaxies. This appears inconsistent with the low scatter and slow evolution of the Fundamental Plane. How can these two observations be brought into agreement?

To answer this question, it is necessary to determine the morphologies of the "E+A" galaxies. This can be done on the basis of imaging, and kinematics. We discuss two samples below.

4.1   A Pair of E+A Galaxies in Abell 665 at $z = 0.18$?

Abell 665 contains a very bright, blue E+A galaxy. It was included by chance in the study of the kinematics of cluster members (Franx 1993a,b). The rotation curve of the galaxy proved that the galaxy was supported by rotation, and that it was likely a disk. Recent HST imaging has confirmed this. Fig. 6 shows the HST image, in combination with the rotation curve. The galaxy is strongly dominated by the disk, and has a very small bulge. It shows weak, smooth spiral arms, without any signs of star formation. The colors along the arms are also very smooth. The rotation curve is typical of a disk, and symmetric.

These data demonstrate that the galaxy is a disk. It will probably evolve into a disk dominated S0, given the lack of star formation. We notice that the optical morphology is rare in the nearby universe: a blue disk, and smooth spiral arms without star formation. These properties may be related to the recent cutoff in star formation.

There are two other blue galaxies with the same morphology very close to the E+A in Abell 665. One of them has signs of dust, the other is smooth. Furthermore, there is a fourth blue galaxy with strong star formation nearby. This association suggests that we may have a group of galaxies, undergoing the transformation from the star formation phase into a early type phase. Surprisingly, recent spectroscopy has demonstrated that most of these other blue galaxies have very different velocities from the large E+A galaxy. Only one of them has a very similar velocity, and this is the galaxy with the very same morphology (smooth arms, no features of dust). It has very strong Balmer lines, and it is a genuine E+A. We can draw two conclusions: i) the E+A spectroscopic classification corresponds to a "smooth-spiral arm" morphology; ii) we have a pair of E+A galaxies, which are apparently undergoing a transformation at the same time.

The mechanism of the transformation is not clear: it could be triggered by the small group itself, or it could be triggered by infall into the rich cluster. It

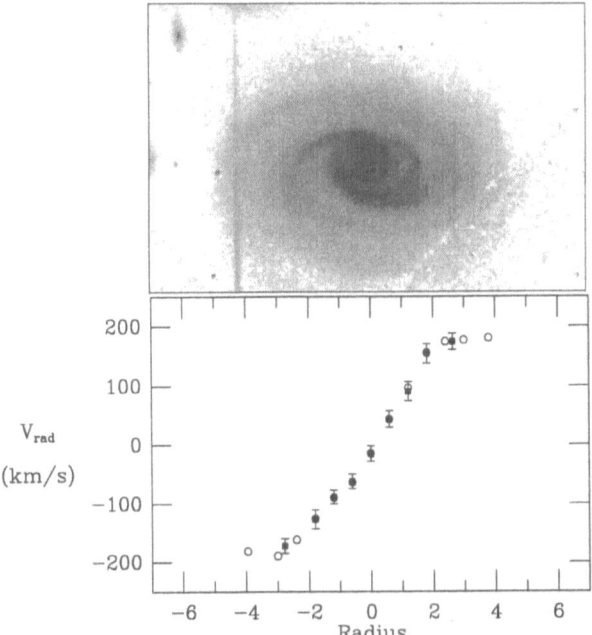

Fig. 6. A bright E+A galaxy in Abell 665 at $z = 0.18$. The upper panel shows the HST image. The galaxy has smooth spiral arms, without regions of star formation. This morphology is rare in the nearby universe, but does occur (e.g., Strom, Strom and Jensen 1976). The lower panel shows the rotation curve of the galaxy, at the same scale. The galaxy is dominated by rotation at large radii. This E+A galaxy has virtually no bulge.

is clearly not triggered by a major merger, as such a merger would not produce disk dominated galaxies. Minor mergers, or tidal interactions, could very well be the driving mechanism.

4.2   E+A Galaxies in CL 1358+62

We have obtained a large HST mosaic of 8x8 arcmin on the cluster CL 1358+62 at $z = 0.33$. We have spectra of 190 cluster members in the mosaic, and have selected E+A galaxies on the basis of this spectroscopy. Again, as in Abell 665, the E+A galaxies are disk dominated. Figure 7 shows a HST image of one of the E+A galaxies, and the rotation curve. This galaxy is similar to the E+A galaxy in Abell 665: it is disk dominated, it shows smooth spiral arms, but no signs of star formation, and the stellar kinematics are dominated by rotation.

This particular galaxy has strong Balmer lines, with a mean equivalent width of $< H_{\gamma,\delta} > \approx 8$ Å. In general, the colors are smooth across the galaxies, indicating that the young component is smoothly distributed. More detailed spectroscopy is needed to characterize the galaxies better. This will be presented in Franx et al, in preparation.

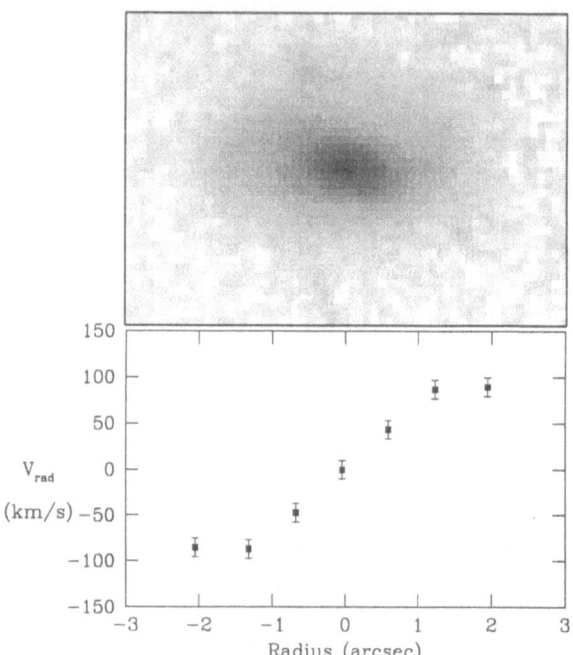

Fig. 7. An E+A galaxy in MS 1358 at $z = 0.33$. This galaxy is similar to that in Abell 665. The upper panel shows the HST image. The galaxy has smooth spiral arms, without regions of star formation. The lower panel shows the rotation curve of the galaxy, at the same scale. The galaxy is dominated by rotation at large radii. This E+A galaxy has a weak bulge.

4.3   Where do E+A's go ?

The above evidence suggests that E+A's evolve into disk dominated S0's. We have to note that we cannot be certain about this: the E+A's may also be in groups which merge to form Ellipticals, or bulge dominated S0's. We can think of several ways to explain the low evolution and low scatter in the Fundamental Plane and the evidence for recent bursts and star formation from E+A's:

1. The E+A's may evolve into disk dominated S0's which are possibly under-represented in the current sample for the Fundamental Plane.

2. The E+A's may have generally low central dispersions, and such galaxies are usually excluded from Fundamental Plane analysis.

3. The stellar population models may need to be modified.

4. The E+A's may merge with older systems so that the influence of the burst is "diluted".

More studies based on larger samples are needed to distinguish between these possibilities. Such studies are now in progress.

5   Discussion

We have shown that it is possible to determine the Fundamental Plane at intermediate redshifts, all the way up to $z = 0.58$. The relation is well defined at these redshifts, with a relatively low scatter of 17 %. The mean evolution of the M/L_V ratio is low, at about 45 % at $z = 0.58$. This evolution is consistent with an early star formation epoch for early type galaxies in clusters. We note, however, that the current measurement may be biased, mostly due to the fact that non-star forming galaxies were selected. This selection may exclude the star forming progenitors of current-day early type galaxies. All studies of the evolution of cluster early types may suffer from this bias.

We have analyzed the structure of E+A galaxies in our program clusters. We find that a large fraction of our E+A's are strongly disk dominated. They would likely evolve into strongly disk dominated S0's, unless they merge with other galaxies. It is still not quite clear how to explain the low scatter in the Fundamental Plane on the one hand, and the high fraction of E+A's on the other hand. It is possible, however, that most of the E+A's have a low central velocity dispersion, and are mostly omitted from Fundamental Plane samples in nearby clusters.

These observations demonstrate that information on galaxy masses can be obtained from deep, ground based spectroscopy. Parallel studies are underway which obtain very similar results (e.g., Bender et al 1997, Pahre 1996). The next step is to extend this work to higher redshift, and to the field. Furthermore, similar studies have demonstrated that the Tully-Fisher relation can be used for spiral galaxies at high redshift (Vogt et al 1996, Rix, Colless, & Guhathakurta 1996). The new generation optical telescopes will allow a rapid extension of this type of work to larger samples, lower galaxy masses, and higher redshifts, The fast progress that we are witnessing raises the prospect that we can apply these techniques in the future on the progenitors of current day galaxies.

Eventually, these observations can be used to determine the evolution of the distribution of circular velocities for galaxies $F(v_c, z)$, and the evolution of stellar masses locked up in early type galaxies $F(M_*, z)$. This requires accurate observations of the evolution of the luminosity function and the "mass relations" in the field. The final outcome of such a program can be expected to provide strong constraints on the models of galaxy formation and evolution.

It is a pleasure to thank ESO for a stimulating workshop. DK, GI and DF acknowledge support for this work from NASA through grants GO-05989.01-94A and GO-05991.01-94A from the Space Telescope Science Institute, which is operated by AURA, Inc., under NASA contract NAS5-26555.

References

Bender R., Burstein D., Faber S. M., 1992, ApJ, 399, 462
Bender R., Saglia R. P., Ziegler B., 1997, J. Bergeron ed., The Early Universe with the VLT. Springer, Berlin, p. 105.

Buzzoni A., 1989, ApJS, 71, 817

Djorgovski S., Davis M., 1987, ApJ, 313, 59

Dressler A., Gunn J. E., Schneider D. P., 1985, ApJ, 294, 70

Dressler A., Gunn J. E., 1992, ApJS, 78, 1

Dressler A., Lynden-Bell D., Burstein D., Davies R. L., Faber S. M., Terlevich R. J., Wegner G., 1987, ApJ, 313, 42

Faber S. M., Dressler A., Davies R. L., Burstein D., Lynden-Bell D., Terlevich R., Wegner G., 1987, Faber S. M., ed., Nearly Normal Galaxies. Springer, New York, p. 175

Faber S. M., Jackson R. E., 1976, ApJ, 204, 668

Franx M., 1993a, ApJ, 407, L5

Franx M., 1993b, PASP, 105, 1058

Franx M., 1995, van der Kruit P. C., Gilmore G., eds, Proc. IAU Symp. 164, Stellar Populations. Kluwer, Dordrecht, p. 269

Jørgensen I., Franx M., Kjærgaard P., 1996, MNRAS, 280, 167 [JFK]

Kauffmann, G., 1996, MNRAS, 281, 487

Kelson, D. D., van Dokkum, P. G., Franx, M., Illingworth, G. D., Fabricant, D., 1997, ApJL, in press, astro-ph 9701115

Lucey J. R., Guzmán R., Carter D., Terlevich R. J., 1991, MNRAS, 253, 584

Pahre M., 1996, M. Arnaboldi, G. S. Da Costa, P. Saha, eds., The Nature of Elliptical Galaxies, preprint, astro-ph 9611115

Renzini A., Ciotti L., 1993, ApJ, 416, L49

Rix, H. W., Colless, M. M., Guhathakurta, P., 1996, Bender, R., Davies, R. R., eds, Proc. IAU Symp. 171, New Light on Galaxy Evolution. Kluwer, Dordrecht, p. 241

Strom, S. E., Strom, K. M., Jensen, E. B., 1976, ApJL, 206, L11

Tinsley B. M., 1980, Fundamentals of Cosmic Physics, 5, 287

Tully R. B., Fisher J. R., 1977, AA, 54, 661

van Dokkum P. G., Franx M. 1996, MNRAS, 281, 985

Vogt, N. P., et al, 1996, ApJ, 465, L15

Worthey G., 1994, ApJS, 95, 107

The Near–Infrared Fundamental Plane of Elliptical Galaxies and Its Evolution

M.A. Pahre[1], S.G. Djorgovski[1], and R.R. de Carvalho[1,2]

[1] Palomar Observatory, Caltech, MS 105–24, Pasadena, CA 91125, USA
[2] Observatório Nacional–CNPq–DAF, Rio de Janeiro, Brazil

Abstract. We present the first Fundamental Plane (FP) that has been constructed in the K–band using near–infrared imaging source data. The slope of this IR FP differs systematically both from the optical relation and the virial expectation (assuming homology and constant M/L) in a way that cannot be explained by stellar populations effects alone. We also show observations of the optical and IR FP at higher redshift which demonstrates the same effect. A slight but systematic breakdown of homology along the elliptical galaxy sequence, in addition to stellar populations effects, could explain the FP at all wavelengths.

1 Introduction

The optical form of the Fundamental Plane (FP) for low–redshift cluster elliptical galaxies can be represented by the scaling relation

$$r_e \propto \sigma_0^{1.24\pm0.07} \langle \Sigma \rangle_e^{-0.82\pm0.02} \tag{1}$$

(Jørgensen, Franx, & Kjærgaard 1996, for the Gunn r–band). If we assume a constant M/L for all ellipticals, and that elliptical galaxies form a homologous family, then the virial theorem implies

$$r_e \propto \sigma_0^2 \langle \Sigma \rangle_e^{-1} \ . \tag{2}$$

It is clear that Equation (1) deviates significantly from the virial expectation in Equation (2). This could be due to a breakdown of either assumption—M/L could be varying systematically along the FP, or ellipticals could be deviating systematically from a homologous family along the FP.

We first address the possibility of variations in M/L. Systematic variations in the dark matter content could cause a variation of M/L along the sequence, but this effect would be in contradiction to the galactic wind model (Arimoto & Yoshii 1987) that has been successful in explaining qualitatively the Mg–σ correlation (see Ciotti, Lanzoni, & Renzini 1996 for a discussion). Stellar populations parameters (age and metallicity will be discussed here; for IMF variations see Chiosi, this volume) could be varying among ellipticals in the sense that the most luminous ellipticals are older or more metal–rich than the low luminosity ellipticals. If stellar populations are the cause, then the slope of the FP should become steeper in the near–infrared, i.e. the observed relation should lie closer to the virial expectation of Equation (2). Furthermore, if the stellar populations effect is age, then the slope of the FP would be predicted to evolve with redshift, while it would evolve more weakly if the effect is due to metallicity.

Elliptical galaxies could be deviating systematically from a homologous family along the FP in a way that could produce the observed correlation in Equation (1). This non–homology could be structural in that ellipticals are better described by a Sersic $r^{1/n}$ law profile than a de Vaucouleurs $r^{1/4}$ profile, in the sense that more luminous ellipticals have a larger n (Graham & Colless 1997, and references therein). The non–homology could also be dynamical in that the velocity structures of ellipticals are varying systematically along the FP, as found in N–body simulations of dissipationless merging (Capelato, de Carvalho, & Carlberg 1995). Velocity anisotropy could also contribute, but investigations of some analytical models (Ciotti, Lanzoni, & Renzini 1996) and of correlations involving velocity anisotropy (Djorgovski & Santiago 1993; Prugniel & Simien 1996) are inconclusive. In any case, such effects would be strictly independent of wavelength, and the IR FP should have the same slope as its optical counterpart.

Observations of the FP in the IR could thus help distinguish between the breakdown of either of the two assumptions—homology or constant M/L. Observations of the FP at different redshifts can provide an additional discriminant between the stellar populations effects of age and metallicity. For these reasons, we initiated two separate studies: one of the IR FP in the local universe (§2), and the other of the high–redshift FP (§3).

2   The Near–Infrared FP in the Local Universe

We have observed in the near–infrared K–band a large sample of early–type galaxies in 10 nearby clusters and groups of galaxies. We have utilized wide-field IR imaging cameras on the Palomar 60" and Las Campanas 40" telescopes. The data are of high quality, comparable to CCD imaging of 5–10 years ago. We fit elliptical isophotes to the images, and use the galaxy profiles to estimate the half–light effective radii r_e and mean surface brightness $\langle \mu_K \rangle_e$. Central velocity dispersions σ_0 have been drawn from the literature. The results from the first 1/3 of the survey can be found in Pahre, Djorgovski, & de Carvalho (1995); nearly the entire sample are shown in Figure 1.

We find that the FP in the K–band is well-represented by the scaling relation

$$r_e \propto \sigma_0^{1.66\pm0.09} \langle \Sigma \rangle_e^{-0.75\pm0.03} . \tag{3}$$

A comparison of Equations (2) and (3) demonstrates that the IR FP deviates from the virial expectation (under the assumptions of constant M/L and homology) by 0.34 ± 0.09 dex. A comparison of Equations (1) and (3) further demonstrates that the IR FP has a steeper slope (i.e. the σ_0 exponent) than the optical relation by 0.42 ± 0.11 dex. Finally, the observed scatter of the IR FP is $\sim 23\%$ in distance, which is similar to that of it optical counterpart; significantly smaller scatter is found for subsets of the data (i.e. the richest clusters or the highest–quality velocity dispersions). Most of the observed scatter can be explained by the observational errors, suggesting that the IR FP is intrinsically thin.

The FP in the K-Band

$\triangle$ A194 (14) $\bullet$ Perseus (15)

$\times$ N4373 Gr. (2) $\varhexagon$ A2199 (13)

$\blacksquare$ Coma (52) $\pentagon$ Hydra (14)

$\square$ A2634 (10) $\bowtie$ Centaurus Lo (13)

$\circ$ Virgo (32) $*$ Centaurus Hi (4)

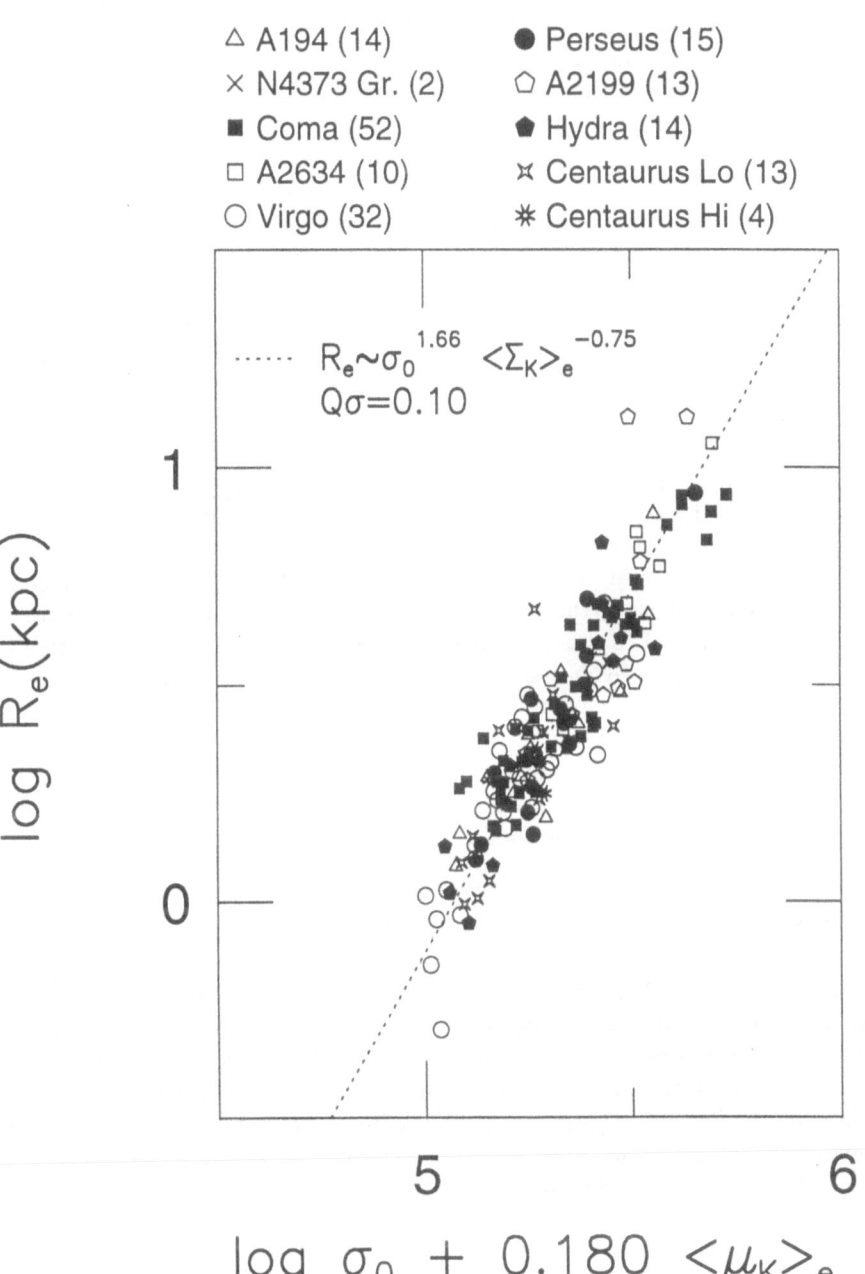

Fig. 1. The near–infrared K-band FP for galaxies in 10 nearby clusters and groups. The best–fitting bivariate relation, and its quartile–estimated scatter in $\log R_e$, are shown in the upper left of the panel.

3   The Near–Infrared FP at Higher Redshifts

The FP correlations are the optimal tool by which to study elliptical galaxy evolution. Variations in the surface brightness (SB) intercept at fixed metric size and velocity dispersion can provide a measure of the combined effect of cosmological SB dimming and luminosity evolution. The rate of evolution of the slope of the FP, if any, could also provide valuable insight as to the origins of the FP. For these reasons we have undertaken a large survey to observe the FP out to $z \sim 0.6$ using a combination of HST optical imaging (from the archive), wide–field multicolor CCD imaging, high–resolution IR imaging, and moderate dispersion spectroscopy at the Palomar and Keck Observatories. Our survey is distinguished from other investigations of the FP at higher redshifts (van Dokkum & Franx 1996; Bender, Ziegler, & Bruzual 1996; and Franx, this volume) due to our inclusion of multicolor information (both optical and IR), the large number of galaxies per cluster, and the large number of clusters surveyed.

An example of the FP at $z = 0.18$ in both the optical V–band (restframe B) and near–infrared K–band (restframe H) is presented in Figure 2 for only those galaxies within the HST/WFPC–2 field–of–view. As can be seen from this figure, the IR FP relation is steeper at $z = 0.18$ than its optical counterpart, as is seen locally (see §2). The scatter is also similar to, or less than, what is found locally. A preliminary analysis of the SB intercept (at $R_e = 3$ kpc and $\sigma_0 = 150$ km/s) for three clusters at $0.18 < z < 0.4$ suggests that there is ~ 0.2 mag evolution to $z \sim 0.2$ and ~ 0.5 mag evolution to $z \sim 0.4$, where there is slightly less evolution in the IR than found in the optical. This is fully consistent with the result found using the Kormendy relation between r_e and $\langle\mu\rangle_e$ in the same redshift range (Pahre, Djorgovski, & de Carvalho 1996). The SB dimming is modest and easily explained by simple passive evolution models for ellipticals that form at high redshift in an expanding world model.

4   Implications for the Origins of the FP

The slope of the FP changes significantly between the optical and IR both locally and at higher redshifts. This suggests that stellar populations are playing an important role in causing the slope of the FP to deviate from the virial expectation in the sense that the more luminous elliptical galaxies are older, more metal–rich, or both. Pahre & Djorgovski (1997), however, found no reasonable stellar populations models *alone* which can simultaneously explain the slope of the FP in both the optical and the IR. A wavelength–independent effect is required *in addition to the stellar populations effects.* Slight but systematic deviations of elliptical galaxies from a homologous family is one plausible explanation.

The result that the scatter of the IR FP is similar to that in the optical at all redshifts suggests that there is not a significant variation in both age and metallicity at any point along the FP. This argues against a model in which the scatter of the FP is kept thin at low redshifts by the correlated effects of metallicity and age, as the scatter of the FP would then increase significantly with

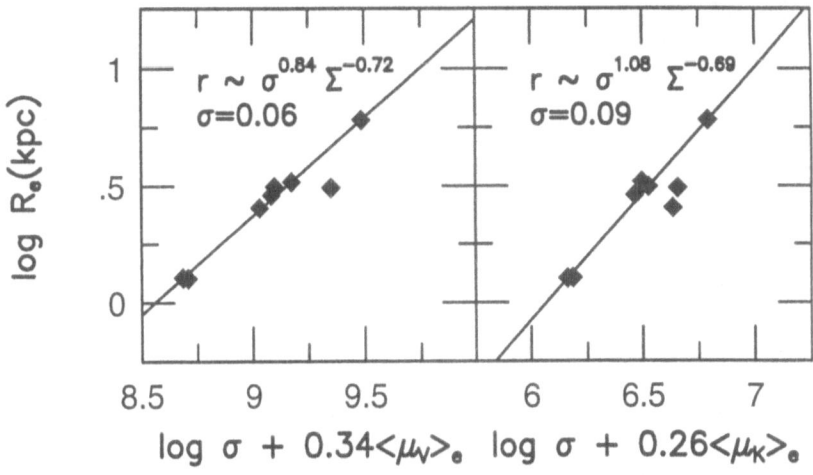

Fig. 2. The FP at $z = 0.18$ in the optical [left] and IR [right] for galaxies in cluster Abell 665. The scatter is similar to, or less than, that in the local universe. The slope of the FP (i.e. the index A in $r_e \propto \sigma^A \langle \Sigma \rangle_e^B$) is steeper in the IR than the optical, as is found locally.

redshift. While there appears to be evidence for such a model (Worthey, Trager, & Faber 1995) based on line–indices and model comparisons for nearby galaxies, those data are dominated by early–type galaxies in low–density environments. Future work on line indices for galaxies in dense cluster cores will be needed to allow a proper comparison with the FP.

Continued work on the FP at higher redshifts will probe whether or not the slope of the FP is evolving with redshift. This is crucial for breaking the age–metallicity degeneracy that confronts observations in the local universe, assuming that we can reliably trace the same population of early–type galaxies back in time to observe them when they were younger. A large number of galaxies (~ 20) must be observed in order to measure the slope of the FP reliably in a single cluster. Our survey has successfully obtained moderate dispersion, high S/N spectra of ~ 200 galaxies in 10 rich clusters at $0.08 < z < 0.55$, the results of which will make up a future contribution.

Observations of the FP at higher redshifts, however, will not necessarily provide an easy measurement of the relative importance of variations in M/L and deviations from homology. For example, if the stellar populations effects contributing to the slope of the FP at low redshifts is due at least in part to a systematic age effect along the FP, then the slope of the FP should become shallower with redshift (i.e. the exponent A in $r_e \propto \sigma_0^A \langle \Sigma \rangle_e^B$ becomes smaller). On the other hand, it was reported at this meeting (see Capelato et al., this volume) that the slope of the FP for second generation mergers of dissipationless systems is steeper than for first generation mergers. In this merging picture, as we trace the FP back to higher redshifts we are observing, on average, earlier generations of mergers which should be accompanied by a shallower slope for the FP. Both of these simple arguments—age and dynamical evolution—result in similar predictions for the slope of the FP, although the magnitude of the effect may differ. It will thus be extremely important to investigate all such scenarios through detailed models which simultaneously account for both stellar populations and dynamical effects as a function of redshift.

This work was supported in part by the NSF PYI Award AST–9157412, NASA HST Grant GO–06391, and the Bressler Foundation. We thank the staffs of Palomar, Keck, and Las Campanas Observatories for their expert assistance during our observing runs.

References

Arimoto, N., & Yoshii, Y. (1987): Chemical and Photometric Properties of a Galactic Wind Model for Elliptical Galaxies. A&A **173**, 23–38

Bender, R., Ziegler, B., & Bruzual, G. (1996): Redshift Evolution of the Stellar Populations in Elliptical Galaxies. ApJ **463**, L51–L54 (astro–ph/9604125)

Capelato, H. V., de Carvalho, R. R., & Carlberg, R. G. (1995): Mergers of Dissipationless Systems—Clues About the Fundamental Plane. ApJ **451**, 525–532

Ciotti, L., Lanzoni, B., & Renzini, A. (1996): The Tilt of the Fundamental Plane of Elliptical Galaxies: 1. Exploring Dynamical and Structural Effects. MNRAS **282**, 1–12

Djorgovski, S., & Santiago, B. (1993): The Meaning and Implications of the Fundamental Plane, in Proceedings of the ESO/EIPC Workshop on Structure, Dynamics, and Chemical Evolution of Early–Type Galaxies, eds. J. Danziger, et al., ESO publication No. 45, 59–70

Graham, A., & Colless, M. (1997): Some Effects of Galaxy Structure and Dynamics on the Fundamental Plane. MNRAS, in press (astro–ph/9701020)

Jørgensen, I., Franx, M., & Kjærgaard, P. (1996): The Fundamental Plane for cluster E and S0 galaxies. MNRAS **280**, 167–185 (astro–ph/9511139)

Pahre, M. A., Djorgovski, S. G., & de Carvalho, R. R. (1995): The Near–Infrared Fundamental Plane of Elliptical Galaxies. ApJ **453**, L17–L20 (astro–ph/9508127)

Pahre, M. A., Djorgovski, S. G., & de Carvalho, R. R. (1996): A Tolman Surface Brightness Test for Universal Expansion and the Evolution of Elliptical Galaxies in Distant Clusters. ApJ **456**, L79–L82 (astro–ph/9511061)

Pahre, M. A., & Djorgovski, S. G. (1997): Observational Constraints on the Origins of the Fundamental Plane, in *The Nature of Elliptical Galaxies*, eds. M. Arnaboldi, G. Da Costa, & P. Saha (ASP, San Francisco), in press (astro–ph/9611115)

Prugniel, P., & Simien, F. (1996): The Fundamental Plane of Early–Type Galaxies: Stellar Populations and Mass–to–Light Ratio. A&A **309**, 749–759

van Dokkum, P. G., & Franx, M. (1996): The Fundamental Plane in CL0024 at $z = 0.4$: Evolution of the Mass–to–Light Ratio. MNRAS **281**, 985–1000 (astro–ph/9603063)

Worthey, G., Trager, S. C., & Faber, S. M. (1995): The Galaxian Age-Metallicity Relation, in *Fresh Views of Elliptical Galaxies*, eds. A. Buzzoni, A. Renzini, & A. Serrano (ASP, San Francisco), 203–206

Surface Brightness Evolution of Cluster and Field Galaxies

David Schade[1]

National Research Council of Canada, Dominion Astrophysical Observatory,
5071 West Saanich Road, Victoria, Canada

Abstract. Photometric studies of faint galaxy morphology is an economical alternative to kinematic analysis in the pursuit of an understanding of the physical changes taking place in individual galaxies from $z \sim 1$ to the present day. Two-dimensional surface photometry extends the capabilities of classical morphological analysis because it provides not only a galaxy type but also a number of other physically meaningful parameters. Measurements of size, surface brightness, and color, for individual (bulge and disk) components in the field and in clusters at moderate and high redshift has yielded direct evidence of evolution of the galaxy populations. Elliptical galaxies of a given size are found to have surface brightnesses that increases with redshift as $\Delta M_B \sim -z$. Disk galaxies have surface brightnesses that evolve by similar amounts. Over the redshift range $0.2 < z < 0.6$ where comparisons have been made, no difference between the evolution of cluster and field populations has been detected (for either disk or elliptical galaxies). This similarity in surface brightness evolution is accompanied by other parallels. In both cluster and field environments, a population of blue galaxies emerges with increasing redshift and an increasing fraction of the population exhibits "peculiar" structure. Thus, the evolutionary processes in cluster and field galaxies are parallel in several important ways. Understanding environmental contributions to galaxy evolution requires the comparison of cluster galaxy properties with those of field galaxies at the same redshift.

1   Introduction

The central problem in pinpointing the physical processes that govern the evolution of galaxies is establishing the correspondence between a particular high-redshift galaxy and its local counterpart, or evolutionary endpoint. The observed evolution of the galaxy population is a sum of the physical transformations of individual galaxies. Many transformations are possible; it is challenging to determine which ones actually take place. Some invariant property is needed to establish the link between present-day and distant objects. Two (imperfect) candidates for this property are mass and morphology. Mass is imperfect because it can change either via merging or infall but one might hope (and try to demonstrate) that these effects are small over some relevant interval of time. Morphology is unlikely to be invariant with time. Changes in star-formation

[1] Guest User, Canadian Astronomy Data Centre, which is operated by the Dominion Astrophysical Observatory for the National Research Council of Canada's Herzberg Institute of Astrophysics.

rate, differential stellar evolution of bulge and disk components, or merging may result in changes of Hubble type with epoch.

Although morphological type is unlikely to be a good "tag" for galaxies, a promising candidate is physical size. Component size (disk scale length or bulge effective radius) has the same desirable property as galaxy mass in that it probably changes slowly (except for major merging events) and is static or monotonically increasing. Large, massive galaxies do not disintegrate. A very attractive feature of size is that it can be determined from photometry alone, in contrast to kinematically determined masses which are expensive in terms of telescope time because they require spectroscopy of faint objects.

Two-dimensional surface photometry using parametric models (bulge and disk) provides a classification (arguably as good as visual classification for faint galaxies), an objective estimate of the asymmetry/peculiarity, and several other physically significant photometric parameters: size, surface brightness, and luminosity for each of the bulge and disk components individually. The analysis of the evolution of these components yields conclusions that are beyond the reach of qualitative morphological analysis.

Here we present results of studies of the evolution of galaxies over the redshift range $0 < z < 1.2$ using photometry for hundreds of galaxies from the Canada-France Redshift Survey (CFRS; Lilly et al. 1995), the Canadian Network for Observational Cosmology survey (CNOC; Yee, Ellingson, & Carlberg 1996), and from HST archival imaging.

2   Elliptical Galaxies

A number of studies of elliptical galaxies (Aragon-Salamanca et al.; 1993 and Rakos and Schombert; 1995, Dressler and Gunn; 1990, Oke, Gunn, & Hoessel; 1996) have detected color evolution at moderate and high redshift, consistent with what is expected from an old stellar population evolving passively (Bruzual & Charlot 1993). Studies of the fundamental plane at moderate redshift (Bender, Ziegler, & Bruzual; 1996, van Dokkum & Franx; 1996, Jorgensen this volume), although based on small numbers of galaxies, are also consistent with passive evolution models. Quantitative morphological measurements of ellipticals in the cluster A851 at $z = 0.41$ (Pahre, Djorgovski, & de Carvalho; 1996; Barrientos, Schade, & López-Cruz; 1996) lead to the same conclusion.

We have measured the evolution in the size-luminosity relation $(M_B - \log R_e)$ for two samples of elliptical galaxies. The first consists of 166 field and cluster galaxies with redshifts $(0.2 < z < 0.6)$ from the Canadian Network for Observational Cosmology (CNOC) survey (Schade et al. 1996a) The second sample is 225 early-type galaxies in the fields of clusters with $0.2 < z < 1.2$ observed with HST (Schade et al. 1997). Cluster membership information is not available for these galaxies. The evolution in luminosity (or, equivalently, in surface brightness) is $\Delta M_B = -0.55 \pm 0.15$ mag at $z = 0.55$ and $\Delta M_B = -0.95 \pm 0.2$ at $z = 0.9$. Field and cluster ellipticals at $z < 0.6$ appear to evolve in very similar fashion from these studies. An extension of the field elliptical measurements to $z \sim 0.9$ from the Canada-France Redshift Survey is in progress.

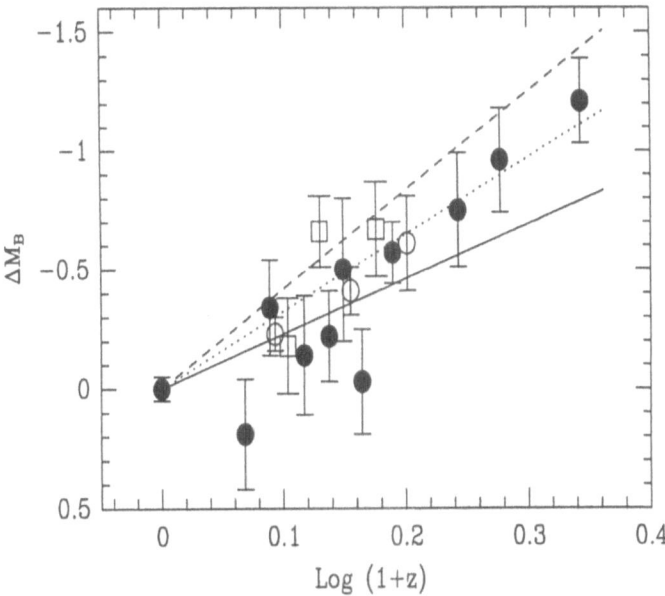

Fig. 1. ΔM_B is the shift in luminosity (or equivalently the shift in surface brightness) at a given size as measured from the average shift of the locus (in the $M_B - \log R_e$ plane) of luminous elliptical galaxies in clusters. Solid symbols are for cluster elliptical galaxies using HST imaging (but no membership information) (Schade et al. 1997) and open symbols are from ground-based imaging of CNOC fields (Schade et al. 1996a) where all of the galaxies have redshift information (open circles=cluster E's,open squares=field E's.)

3   Disk Galaxies

Surface brightness evolution of disks is been reported by Schade et al. (1995) from the analysis of 32 galaxies with HST imaging from the CFRS with $0.5 < z < 1.2$. The mean disk central surface brightness was found to be 1.2 ± 0.25 mag brighter than the Freeman (1970) value, a result confirmed by ground-based work (Schade et al. 1996c). Comparisons between cluster and field disk surface brightness in a sample of 351 CNOC galaxies with redshifts (Schade et al. 1996b) at $0.2 < z < 0.6$ reveals no difference between field galaxies and those in clusters, although any environmental effect might be very diluted in this widely-distributed (as opposed to cluster core-concentrated) sample of disks. The CNOC work (based on ground-based imaging of less-than-ideal quality) suggests that disks in clusters the field have surface brightness enhanced by roughly 1 mag at $z = 0.55$ relative to local objects.

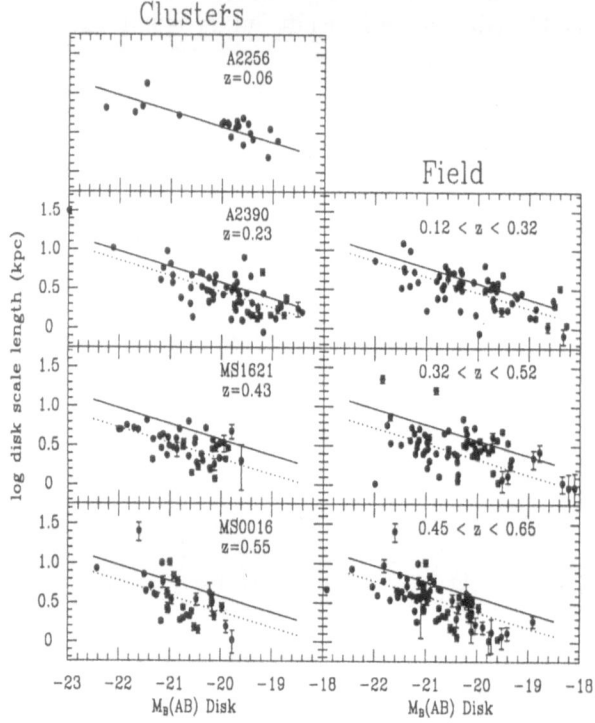

Fig. 2. The evolving size-luminosity relation for galactic disks in clusters and the field. Solid lines show the Freeman law and dotted lines indicate the mean evolved relation.

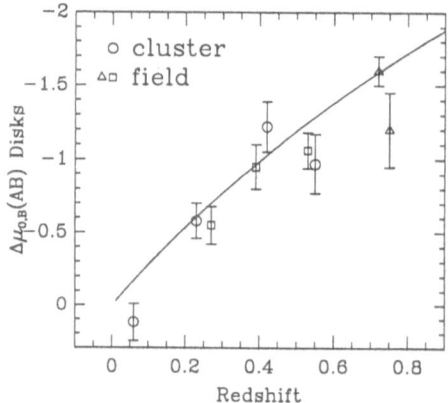

Fig. 3. $\Delta\mu$ is the shift in surface brightness relative to the Freeman law measured, e.g. from plots like Figure 3. Open circles are CNOC cluster galaxies and squares are CNOC field galaxies. Triangles are field galaxies from Schade et al. The line is the evolution of the luminosity density ($(1+z)^{2.7}$) in the B-band from Lilly et al. (1996).

Some kinematic work (Vogt et al. 1996) based on small numbers of galaxies shows evolution in the Tully-Fisher relation that is broadly consistent (perhaps with milder evolution) with our surface-brightness work. On the other hand, kinematic studies by Simard & Pritchet (1997) and by Rix et al. (1997) suggest stronger luminosity evolution than is seen in our photometric work, even at modest redshifts ($z < 0.5$).

4   Cluster versus Field Populations at High Redshift

The observed evolution of the surface brightness of disk and elliptical galaxies supports the view that galaxy populations in the field and in clusters evolve in a similar manner. Two reasons to be cautious about this conclusion are 1) the cluster population in our work is distributed over a wider region of space (particularly the disk work) compared to most other studies which are usually restricted to the cluster core region where environmental effects should be most pronounced, and 2) the similarity of cluster and field has only been tested over a limited range of redshift. It is important to extend this redshift range and to analyse the trends of galaxy properties with cluster-centric radius over the full redshift range.

There are other ways in which the evolutionary properties of cluster galaxy populations parallel those of the field. Both populations are dominated at high-redshift by a "blue-excess" population (the Butcher-Oemler 1984) effect in clusters) that appears to be absent from the local universe (Lilly 1993, Dressler et al. 1994). This population shows indications of elevated rates of star formation (Couch & Sharples 1987, Hammer et al. 1997). Furthermore, both populations have been claimed to show a greatly enhanced frequency of peculiar or asymmetric structure (Glazebrook et al. 1995, Couch et al. 1995) relative to the present-day galaxy population.

In summary, the galaxy population at high-redshift ($z > 0.5$) is dramatically different (in the respects enumerated above) from the local galaxy population but the evolutionary phenomena we see are remarkably independent of cluster/field environment. The contrast between galaxy populations in clusters and the field is much weaker at earlier epochs. An effective study of galaxy evolution needs to take account of these facts, and to make careful comparisons of cluster galaxies with those in the field *at the same redshift* so that the effects of environment can be disentangled from those due to evolution.

Acknowledgments

It is a pleasure to thank the many collaborators in the Canada-France Redshift Survey and the Canadian Network for Observational Cosmology groups, in particular Simon Lilly, Ray Carlberg, and Felipe Barrientos.

References

Aragon-Salamanca,A., Ellis,R., Couch,W., & Carter,D., 1993 MNRAS 262,764

Barrientos,F., Schade, D., & López-Cruz, O. 1996 ApJ 460,89

Bender,R. Ziegler, B., & Bruzual, G. 1996 ApJ 463,51

Butcher, H., & Oemler, A. 1984 ApJ 285,426

Carlberg, R. G., et al. 1994 JRASC 88,39

Couch, W., & Sharples, R. 1987 MNRAS 229,423

Couch, W., Ellis, R., Sharples, R., and Smail, I., 1994 ApJ 430,121

Dickinson, M. 1995 in *Fresh Views of Elliptical Galaxies*, A.S.P. Conference Series, ed. Buzzoni,A., Renzini, A., & Serrano, A. p. 283

Dressler, A., Oemler, A., Butcher, H., and Gunn, J. 1994 ApJ 430,1

Dressler, A., & Gunn, J. 1990, in *Evolution of the Universe of Galaxies*, San Francisco,Astronomical Society of the Pacific

Ellis,R., Smail,R., Dressler,A., Couch,W., Oemler,A., Butcher, H., & Sharples,R. 1996, preprint, astro-ph/9607154

Freeman, K. 1970 ApJ 160,811

Glazebrook, K., Ellis, R., Santiago, B., & Griffiths, R. 1995 MNRAS 275,L19

Lilly, S. 1993 ApJ 411,501

Lilly, S., Le Fevre, O., Hammer, F., Crampton, D. 1996 ApJ 460,L1

Lilly, S., Le Fevre, O., Crampton, D., Hammer, F., & Tresse, L. 1995 ApJ 455,50

Lavery, R., Henry, P. 1994 ApJ 426,524

Hammer, F., Flores, H., Lilly, S., Crampton, D., LeFevre, O., Rola, C., Mallen-Ornelas, G., Schade, D., Tresse, L. 1996 A & A, submitted

Oke, J., Gunn, J., & Hoessel, J. 1996 AJ 111,29

Pahre, M., Djorgovski, S., & de Carvalho, R. 1996 ApJ 456,79

Rakos,K., & Schombert,J. 1995 ApJ 439,47

Rix, H., Guhathakurta, P., Colless, M. & Ing, K. 1997 astro-ph/9605204 submitted to MNRAS

Schade, D., Lilly, S., Crampton, D., Le Fèvre, O., Hammer, F., & Tresse, L. 1995 ApJ 451,L1

Schade, D., Carlberg, R., Yee, H., López-Cruz, O., & Ellingson, E. 1996a ApJ 464,L63

Schade, D., Carlberg, R., Yee, H., López-Cruz, O., & Ellingson, E. 1996b ApJ 465,L103

Schade, D., Lilly, S., Le Fèvre, O., Hammer, F., & Crampton, D. 1996c ApJ 464,79

Schade, D., Barrientos,F., & López-Cruz, O. 1997 ApJ Letters, in press

Simard, L. & Pritchet 1997 astro-ph/9606006 submitted to ApJ

Thompson, L. 1988 ApJ 324,112

Vogt,N., et al. 1996 ApJ 465,L15

van Dokkum, P. & Franx, M. 1996 MNRAS 281,985

Yee, H.K.C., Ellingson,E., & Carlberg, R. G., 1996 ApJS, 102, 269

The Mg–σ Relation of Elliptical Galaxies at Various Redshifts

Bodo L. Ziegler

Universitätssternwarte, Scheinerstraße 1, D–81679 München, Germany

Abstract. The correlation between the Mg absorption index and the velocity dispersion (σ) of local elliptical galaxies is very tight. Because the Mg absorption depends on both metallicity and age of the underlying stellar population the observed Mg–σ relation constrains the possible variation in metallicity and age for a given velocity dispersion. For a time interval with no change in metallicity any variation of the Mg index is caused only by the aging of the stars.

We have measured the Mg absorption and velocity dispersion of ellipticals in three clusters at a redshift of $z = 0.37$ and established their Mg–σ relation. For any given σ, the measured Mg absorption is weaker than the mean value for local ellipticals. Since the evolution of bright cluster ellipticals between $z = 0.4$ and today is most probably only 'passive' this reduction in Mg can be attributed solely to the younger age of the stellar population. The small weakening of the Mg absorption of the distant galaxies compared to the local values implies that most of the stars in cluster ellipticals must have formed at high redshifts ($z_f > 2 \ldots 4$).

The Mg–σ test is a very robust method to investigate the evolution of elliptical galaxies and has several advantages over traditional methods using luminosities. A remaining problem is the aperture correction necessary to calibrate observations of galaxies at different distances. Here, we show that our general conclusions about the epoch of formation still hold when aperture corrections are calculated assuming a dependence of the radial gradient of σ on the galaxy's effective radius rather than assuming no dependence as was done in all previous studies.

1 The Local Mg–σ Relation

It is well known that all dynamically hot galaxies in the local universe follow the same linear relationship between the Mg absorption around $\lambda_0 \approx 5170$ Å and the internal velocity dispersion σ (Dressler et al. 1987, Bender et al. 1993). Although the galaxies span a wide range in Mg and σ, the Mg–σ relation is very tight. A sample of luminous Coma ellipticals ($\lg \sigma \geq 2.3$), e. g., with $Mg_2 \in [0.25, 0.36]$ mag has a standard deviation from the linear fit of only $\sigma_{int} = 0.011$ mag. Mg_2 as defined in the Lick system (Faber et al. 1985) comprises mainly the molecular absorption of MgH. Because the measurement of this index in redshifted galaxies is very noisy we use the atomic Mg_b index instead. A linear transformation from Mg_2 to Mg_b enables us to still use the 7 Samurai sample of Coma and Virgo ellipticals as the comparison at zero redshift. Both from observational (Gonzalez 1993) and theoretical data (stellar population synthesis of Worthey 1994) we derived consistently: $Mg_b/\text{Å} = 14.9 \pm 0.5 \cdot Mg_2/\text{mag}$. In figure 1 small circles present the Mg$_b$–σ relation of the local comparison sample. A principal component analysis yields as best fit:

$$Mg_b = 2.7 \lg \sigma_0 - 1.65 \qquad (1)$$

The dependence of Mg$_b$ on age and metallicity can be explored using stellar population synthesis models. From Worthey's 1994 models we derived the following formula that holds for ages $t > 3$ Gyrs and metallicities $-2 < \lg Z/Z_\odot < +0.25$:

$$\lg Mg_b = 0.20 \lg t + 0.31 \lg Z/Z_\odot + 0.37 \qquad (2)$$

This formula allows to determine the maximum variation of both relative age and relative metallicity as it is constrained by the very tightness of the Mg$_b$–σ relation. For a given velocity dispersion σ (with $\lg \sigma \geq 2.3$) and zero variation in metallicity or age, resp., we find:

$$\Delta t/t < 0.17 \quad \text{and} \quad \Delta Z/Z < 0.11 \qquad (3)$$

This narrow constraint on the age spread of cluster ellipticals implies that they did not form continuously at the same rate but that there was a rather short formation epoch of these galaxies. If, e.g., the majority of ellipticals were formed 12 Gyrs ago, then the scatter in age would be about 2 Gyrs. However the formation epoch itself can only be determined by comparing the relative ages of galaxies at zero and a significantly higher redshift.

2   The Mg$_b$–σ Relation at Redshift $z = 0.37$

As a first step we have established the Mg$_b$–σ relation of elliptical galaxies in three clusters at a redshift of $z = 0.37$ (*Abell 370, CL 0949+44* and *MS 1512+36*) (Bender et al. 1996). The spectra were taken with the 3.5m telescope at the Calar Alto observatory and the 3.6m and the NTT at ESO with total integration times per galaxy on the order of 8 hours. Very careful data reduction had been applied because in the relevant wavelength range ($\lambda(Mg_b, z = 0.37) \approx 7080$ Å) the spectra are heavily contaminated by strong night sky emission lines and telluric absorption bands (Ziegler and Bender 1997).

Figure 1 presents the datapoints for the distant galaxies together with the local Mg$_b$–σ relation. The first thing to note is that all distant ellipticals have lower Mg$_b$ line strengths than the local mean value at the same velocity dispersion. This is clear evidence for evolution of the stellar populations of elliptical galaxies between $z = 0.37$ and now. On the other hand the reduction in Mg$_b$ is very weak, on average $<\Delta Mg_b> = -0.37 \pm 0.08$ Å. This can be reconciled with current stellar population models only if there was virtually no new star formation in today's cluster ellipticals since $z = 0.37$ but only 'passive' evolution of the aging stars. Therefore, the metallicity did most probably not change at all and the reduction in Mg$_b$ can be fully attributed to the younger age of the distant galaxies. Setting $\Delta Z = 0$, equation 2 together with equation 1 can be

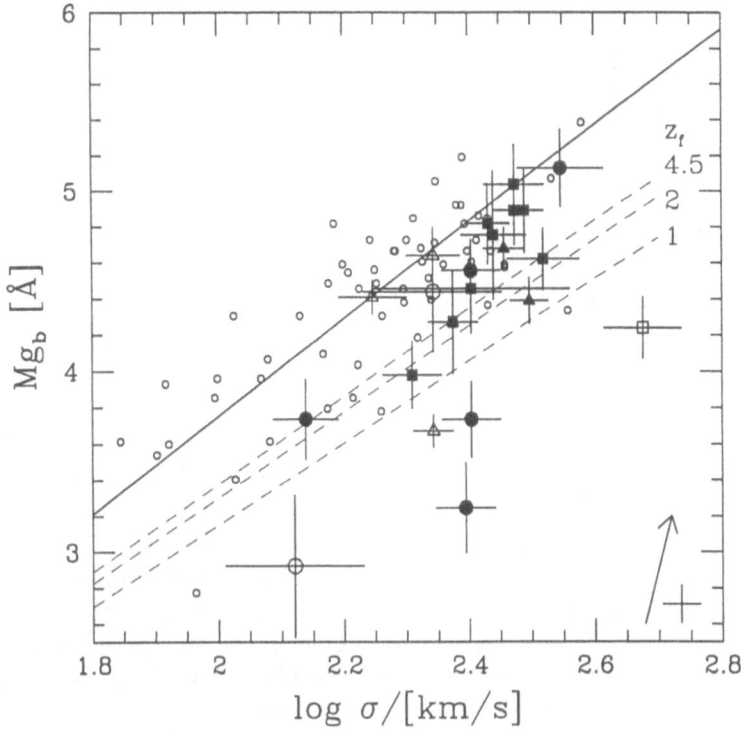

Fig. 1. Mg$_b$ – σ pairs at $z = 0.37$ (big symbols with errorbars) compared to the local Mg$_b$–σ relation (small circles: Coma and Virgo ellipticals, typical errorbar in lower right corner). Arrow: aperture correction applied. Dashed lines: expected Mg$_b$–σ relations at $z = 0.37$ for $z_f = 1, 2, 4.5$ and $H_0 = 50, q_0 = 0.5$.

transformed to deduce theoretical curves of the Mg$_b$ σ relation at the observed redshift $z = 0.37$:

$$\frac{\mathrm{Mg}_b\,(z{=}0)}{\mathrm{Mg}_b\,(z)} = \left(\frac{\mathrm{age}\,(z{=}0)}{\mathrm{age}\,(z)}\right)^{0.20} \tag{4}$$

The age of an object depends mainly on its redshift of formation z_f and less on the cosmology (H_0, q_0, Λ). In figure 1, the dashed lines correspond to the expected Mg$_b$–σ relation at $z = 0.37$, if z_f were 1, 2 or 4.5, resp.. The small weakening of Mg$_b$ to a look–back time of $\approx$ 5 Gyrs constrains the age of the stellar populations so that the majority of the stars in elliptical cluster galaxies were formed at high redshifts $z_f \geq 2$. For the most luminous ellipticals, where the reduction in Mg$_b$ is even lower, z_f could be as high as 4. This imposes great problems on current theories on the structure formation using Cold Dark Matter models, because contrary to these models smaller objects seem to be younger than big ones and formed at later times. But a large dynamically relaxed stellar system most probably did not form within only 1 Gyr after the Big Bang as is fixed by $z_f = 4$ in all reasonable cosmologies. A way out of this paradigm is a model in which the stars indeed existed already at $z \geq 4$ in a common gravitational potential but the galaxies were formed only at much later times through the collapse of these dark matter halos and/or through the dissipationless merging of smaller halos. The smaller elliptical galaxies could either have a much more extended formation epoch or experience events during their evolution which add new populations of stars and therefore lower their mean age.

The evolution of the stellar population as measured by the Mg_b–σ test can be transformed into a change in luminosity with the help of population synthesis models. From the Worthey and Bruzual & Charlot (1997) models we found consistently the following linear relation between the reduction in Mg_b and the increase in the B–band luminosity valid for ages greater than 1.5 Gyrs and metallicities between half and twice solar:

$$\Delta M_B \,[\text{mag}] \quad \approx \quad 1.35 \pm 0.1 \, \Delta Mg_b \,[\text{Å}] \tag{5}$$

Thus, the distant cluster ellipticals are on average brighter in the B–band by $<\Delta M_B> = -0.50 \pm 0.11 [\text{mag}]$ than local ones. This is quantitatively consistent with the brightening obtained via the Faber–Jackson relation.

The Mg_b–σ test is, therefore, a powerful and reliable tool to investigate the evolution of elliptical galaxies. Over all other methods using luminosities or colors it has the advantage to be free of problems like foreground and internal extinction and K–corrections. It also depends only very weak on the initial mass function. The accurate determination of the luminosity evolution by this method, therefore, makes it possible for the first time to calibrate elliptical cluster galaxies as standard candles. Together with the fundamental plane relations the cosmological parameter q_0 can be significantly constrained (see the contribution by Bender et al., this conference or 1997).

The weak part of the Mg_b–σ test is the aperture correction necessary to calibrate observations of galaxies at different distances. Even if our aperture size were the same as was used for the observations of the local comparison sample we would average our measured values over a much greater fraction of the galaxies. Indeed, the 7 Samurai determined central values whereas we integrate out to more than the effective radius (r_e). Because both Mg_b and σ have radial gradients an aperture correction has to be applied. From spectroscopic observations of ellipticals with sufficient S/N out to several r_e (Saglia et al., unpublished) we found the following gradients:

$$Mg_b = -0.87 \lg(r/r_e) + c \tag{6}$$

$$\lg \sigma(r) = -0.11 \, (r/r_e)^{3/4} + c \tag{7}$$

The gradient of $\lg \sigma$ depends on the effective radius in a non logarithmic manner unlike those used in previous studies but which were determined from data with a much lower radial extent (e.g. Jørgensen et al. 1994) and falls off steeply at r_e. Because we could not determine the effective radii of all the observed distant ellipticals from our ground–based photometry (Ziegler 1997) we chose a representative value and applied the same aperture correction for all galaxies in figure 1. Recently, we have been able to accurately measure r_e for 9 galaxies from our HST images of *Abell 370* and *MS 1512+36* and, thus, apply individual aperture corrections. Figure 2 presents the same diagram as figure 1 but with Mg_b and σ extrapolated to integrated mean values within r_e. Although individual galaxies changed their position in the new diagram our general conclusions about the age (z_f) and evolution are still valid with an average reduction in Mg_b of $<\Delta Mg_b> = -0.34 \pm 0.19$ Å.

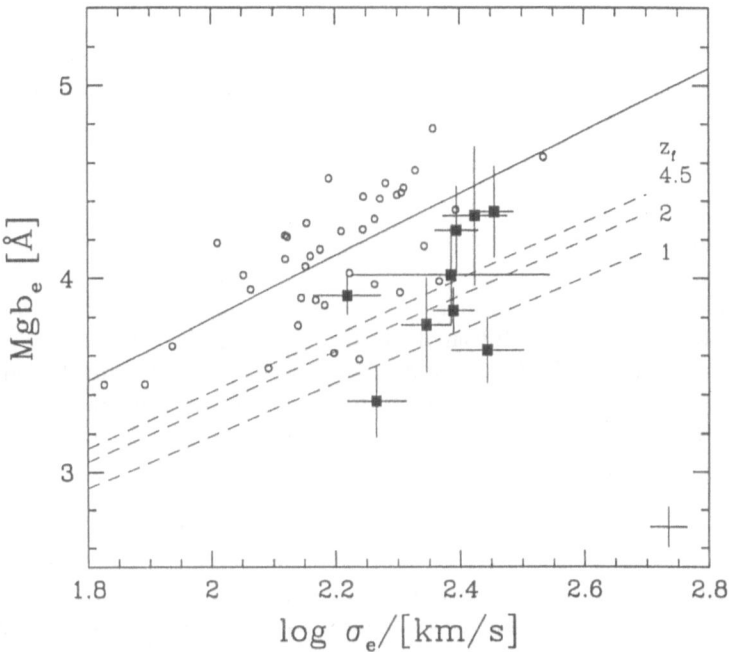

Fig. 2. Same diagram as figure 1 but with Mg_b and σ extrapolated to integrated mean values within r_e.

3 The Mg$_b$–σ Relation at Higher Redshifts

Going to higher redshifts we expect to find a more prominent evolution in Mg_b because the stellar population gets still younger and the formation epoch is approached. The higher amount of evolution and the wider range in look–back time should enable us to reduce further the errors in the calibration of elliptical cluster galaxies as standard candles and allow us to even better constrain the value of the cosmological parameter q_0. But if future data at high redshifts would show evolution allowing merging/accretion events in addition to passive evolution, then possible variations on the velocity dispersions must be taken into account.

Clearly, the spectroscopy of very faint objects will be a major challenge. Observing at higher redshifts means that the Mg_b absorption line moves more and more into the red wavelength range with a strongly increasing sky contamination. The next favourable redshift bins where there is a slight depression in the sky emission are: $0.575 < z < 0.583$ and $0.754 < z < 0.780$. The decrease in brightness of the galaxies due to their greater distance ($1...2$ mags) might just be compensated by their evolutionary brightening. In order to get spectra with S/N as good as we have obtained at $z = 0.37$ the suggested observations can only be done with a big enough telescope like the VLT together with a spectrograph of superb efficiency like FORS. Exposure times would be on the order of $1.5...2$ hours. Multi–object spectroscopy will be the appropriate method if the apertures are large enough to collect a sufficiently large area of the sky around each object in order to be able to cope with the sky subtraction.

4   Conclusions

Determining the Mg_b–σ relations at various redshifts is a powerful and robust method to measure the evolution of elliptical galaxies. From the first application at a redshift of $z = 0.37$ we could already make two firm conclusions: first, the stellar population of elliptical cluster galaxies evolves predominantly passive between this redshift and today and second, the epoch of formation of the stars that mainly make up the ellipticals today is at high redshifts. The evolution in Mg_b can be reliably transformed into an increase of the B–magnitude, thus, allowing the calibration of elliptical cluster galaxies as standard candles.

Acknowledgements: This work is a collaboration with Drs. Bender, Belloni, Bruzual and Saglia and was supported by the "Sonderforschungsbereich 375–95 für Astro–Teilchenphysik der Deutschen Forschungsgemeinschaft" and by DARA grant 50 OR 9608 5.

References

Bender, R., Burstein, D., Faber, S. (1993): ApJ 411, 153
Bruzual, G., Charlot, S. (1997): in preparation
R. Bender *et al.* (1997): Nature, subm.
Bender, R., Ziegler, B., Bruzual, G. (1996): ApJ 463, L51
Dressler, A. *et al.* (1987): ApJ 313, 42
Faber, S. M., Friel, E. D., Burstein, D., & Gaskell, C. M. (1985): ApJS 57, 711
Gonzalez, J.J. (1993): PhD Thesis, University of Santa Cruz
Jørgensen, I., Franx, M., Kjærgaard P. (1994): MNRAS 276, 1341
Worthey, G. (1994): ApJS 95, 107
Ziegler, B. L., Bender, R. (1997): MNRAS subm.
Ziegler, B. L. (1997): in preparation

Properties of High Redshift Cluster Ellipticals

Mark Dickinson

Space Telescope Science Institute, Baltimore MD 21218, USA

Abstract. Cluster ellipticals are often thought to be among the oldest galaxies in the universe, with the bulk of their stellar mass formed at early cosmic epochs. I review recent observations of color evolution in early–type cluster galaxies at high redshift, which show remarkably little evolution in the color–magnitude relation out to $z \approx 1$. Spectra of elliptical galaxies from $1.15 < z < 1.41$ demonstrate the presence of a dominant old stellar population even at these large lookback times, although there is some evidence for a tail of later star formation. The Kormendy relation, a photometric/morphological scaling law for elliptical galaxies, is used to extend fundamental plane investigations to $z = 1.2$ and to test for luminosity evolution. While the evidence, overall, is consistent with simple and mild passive evolution, I review some caveats and briefly consider the data in the light of hierarchical galaxy formation models.

1   Color Evolution at $z < 1$

Rich clusters remain the most "efficient" place to find and study high redshift elliptical galaxies – they are abundant in such environments, even at $z > 0.5$, and the richness of the cluster ensures that a statistical analysis of their colors or image properties should not be too strongly biased by field galaxy "interlopers" in the absence of extensive spectroscopy. Until recently, studies of distant cluster ellipticals were predicated on selecting them by color or spectral properties, e.g. by analyzing the colors of the "red envelope" of the galaxy distribution and assuming that any changes reflected the evolution of the early–type galaxy population (e.g. Aragón–Salamanca *et al.* 1993). While plausible, this requires a reliable statistical means of identifying and measuring properties of galaxies in a one sided distribution – e.g. the "reddest objects" in a given cluster – and risks missing important results if, for example, some *bona fide* ellipticals are substantially bluer than their reddest counterparts.

Now, however, a very extensive collection of WFPC2 images of high redshift clusters has amassed in the HST archive, enabling the armchair astronomer to ponder hundreds of high redshift ellipticals at will, selecting them on the basis of image morphology rather than color. The "MORPHS" collaboration (Dressler, Oemler, Ellis & many co–workers) have been responsible for collecting much of this data, and have recently analyzed the properties of early–type galaxies in three clusters at $z \approx 0.55$ (Ellis *et al.* 1997). In particular, they have examined the scatter in the elliptical galaxy color–magnitude (c–m) relation. Bower *et al.* (1992) used the very small scatter in the c–m relation for Coma and Virgo ellipticals to argue that these galaxies have had highly synchronous evolutionary histories: they either formed at very large redshifts, or with nearly identical

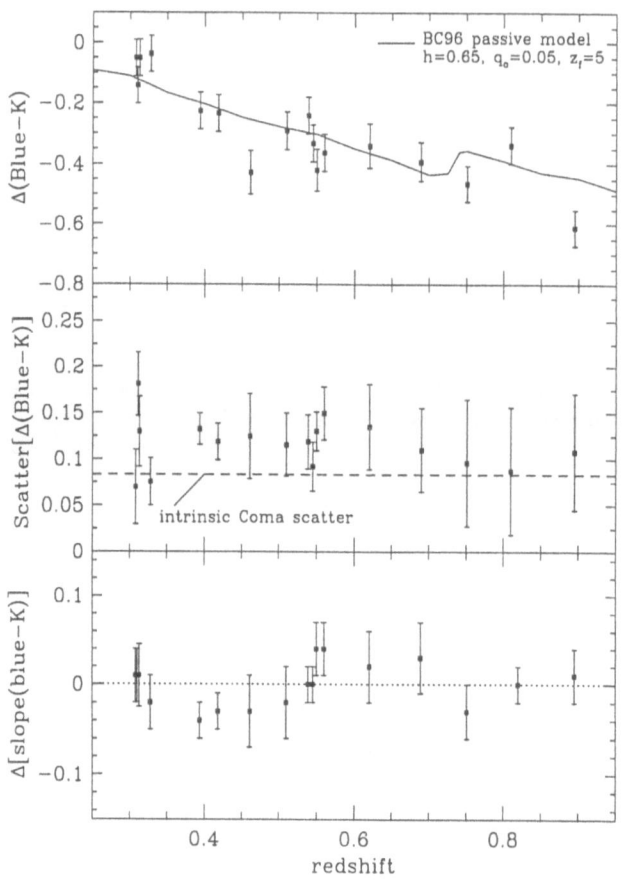

Fig. 1. Color evolution of early–type cluster galaxies out to $z = 0.9$, from Stanford, Eisenhardt & Dickinson 1997. The data consists of optical and infrared photometry of early–type galaxies selected from WFPC2 images of high redshift clusters. The "blue" band is chosen for each cluster to remain approximately fixed at the rest–frame U–band, while the K band remains fixed in the observer's frame. *Top:* differences in mean *blue* $-$ K colors relative to the same rest–frame colors in the Coma cluster. "No evolution" would be therefore be a horizontal line at $\Delta(blue - K) = 0$. *Middle:* intrinsic scatter in early–type galaxy colors, after removing the mean slope of the color–magnitude relation and the component of scatter due to photometric errors. *Bottom:* differences in the slope of the $(blue - K)$ vs. K color–magnitude relation, plotted relative to the slope measured for Coma.

subsequent star formation histories, or both. Ellis *et al.* find the same to be true in the three clusters they examined at $z \approx 0.55$, reinforcing the evidence for synchronized evolution.

Adam Stanford, Peter Eisenhardt and I have collected 5–band optical/IR imaging of a large sample of distant galaxy clusters (46 to date from $0 < z < 0.9$), taking care to match both the field of view (in Mpc) and the rest–frame limiting luminosity of the data for each cluster in order to provide a uniform data set. The optical/CCD observations were taken through two filters which are matched to the cluster redshift in order to approximately sample the rest–frame U and V regions of the spectrum, while the near–IR JHK data provide a long wavelength baseline for photometry and a very red rest–frame wavelength even for our most distant clusters in order to ensure that galaxies may be selected uniformly by the luminosities of their old stellar populations, even at $z = 1$.

Returning from the telescopes to our armchairs, we have used archival WFPC2 data to identify early–type galaxies, independently from their colors, in 19 clusters, and to study their photometric evolution (Stanford, Eisenhardt & Dickinson 1997). We have used new wide-field imaging of Coma cluster in the $UBVRIzJHK$ bands (Eisenhardt *et al.* 1997) to provide a present–day reference sample, ensuring that we can compare the properties of the distant cluster galaxies to data on nearby ellipticals at the same rest–frame wavelengths with only minimal differential k–corrections. Figure 1 summarizes this work, showing the evolution of the galaxy colors, and the scatter and slope of the color–magnitude relation, out to $z = 0.9$. The key results are:

- The mean color of early–type cluster galaxies becomes gradually and mono-tonically bluer toward higher redshifts, in a fashion broadly consistent with simple passive evolution of the stellar populations.
- The scatter of galaxy colors around the mean color–magnitude relation remains small and nearly constant with redshift out to $z = 0.9$.
- The slope of the color–magnitude relation remains unchanged from $z = 0.9$ to the present. This strongly supports the hypothesis that the c–m slope is primary due to differences in the mean metallicity of elliptical galaxies as a function of luminosity/mass, and not a consequence of differing ages (Kodama & Arimoto 1997; see also contribution of Arimoto to this volume).

The small and constant scatter in the galaxy colors is remarkable and some-what difficult to explain. In particular, if ellipticals in all clusters were a com-pletely coeval population, and if the small but non–zero scatter observed in their colors at $z = 0$ (Bower *et al.* 1992) were due to age variations, then one would ex-pect the scatter to increase at higher redshift. Although the mean scatter in the distant clusters is slightly higher than that measured for Coma, no other strong trend with redshift is observed. It may be that the intrinsic scatter at $z = 0$ is due, in part, to metallicity variations at a fixed mass rather than age differences. This would set a "floor" value to the scatter, suppressing the expected decrease with decreasing redshift. In this case, however, intrinsic age variations must be even smaller than the values inferred directly from the color scatter. Alterna-tively, small episodes of later star formation due to mergers, etc. may "reinflate" the scatter in a more–or–less continuous fashion, but again the amount of late star formation is strongly constrained by the small amplitude of the measured scatter. The 0.11 magnitude scatter observed for the most distant cluster in our sample, GHO 1603+4313 at $z = 0.895$, limits small (e.g. $\leq 10\%$ by mass) star formation episodes to have occurred no less than 2 Gyr prior to the epoch of observation.

2   Spectroscopic Characteristics of Ellipticals at $z > 1$

The lack of dramatic change in the color properties of early–type cluster galaxies out to $z = 0.9$ suggests that, by that redshift, we have not closely approached the era in which those galaxies formed the bulk of their stars. This therefore encourages us to extend the search for distant clusters and cluster ellipticals to higher redshifts. Peter Eisenhardt and I have been studying the environments of distant radio galaxies in order to search for distant galaxy clusters (cf. Dickinson 1995, 1997a, 1997b). Spinrad, Dey, Stern, LeFèvre and I have obtained extensive spectroscopy of one such cluster, that around 3C 324 at $z = 1.206$. Using deep WFPC2 imaging and infrared photometry, we have identified a number of elliptical galaxies in this cluster, as well as several in a "foreground" structure at $z = 1.15$ and one background elliptical at $z = 1.41$.

Figures 2 and 3 present coadded Keck/LRIS spectra of three of these galaxies, showing the rest–frame optical and near–UV regions and identifying some of the prominent spectral "breaks" which are useful as diagnostics of the stellar populations. The $z \approx 1.2$ galaxies exhibit a strong 4000Å break and CaII H+K lines, features prominent in the spectra of old stellar populations today. The near–UV spectra are strikingly similar to that of M32, although they are somewhat bluer (more flux at $\lambda_0 < 2600$Å relative to that at $\lambda_0 \approx 3100$Å) and the discontinuities at 2640Å and 2900Å have somewhat smaller amplitudes.

Dunlop et al. (1996) have recently presented spectra of a faint, red radio galaxy 53W091 at even larger redshift, $z = 1.552$, which exhibits very similar UV spectral breaks at 2640Å and 2900Å. Spinrad et al. (1997) have analyzed the 53W091 data, making extensive comparisons of the UV spectral features with the properties of stars in the IUE spectral atlas, with those in nearby elliptical galaxies, and with population synthesis models – the reader is referred to that paper for a much more thorough discussion, as well as to Charlot, Worthey & Bressan (1996) for an excellent review of potential pitfalls in the use of the evolutionary models.

Here, we restrict our comparison to a single set of evolutionary models, those of Bruzual & Charlot (1996), which utilize model stellar atmospheres and evolutionary tracks to investigate the spectrophotometric evolution of stellar populations with various metallicities. Figure 4 shows the dependence of four indices (the three spectral breaks plus the $R - K$ color) on age and metallicity for simple Salpeter IMF models (10^7 yr burst followed by purely passive evolution). The horizontal dashed lines mark the index values for the 3C 324 galaxies. All four indices exhibit the familiar age–metallicity degeneracy (e.g. Worthey 1994)

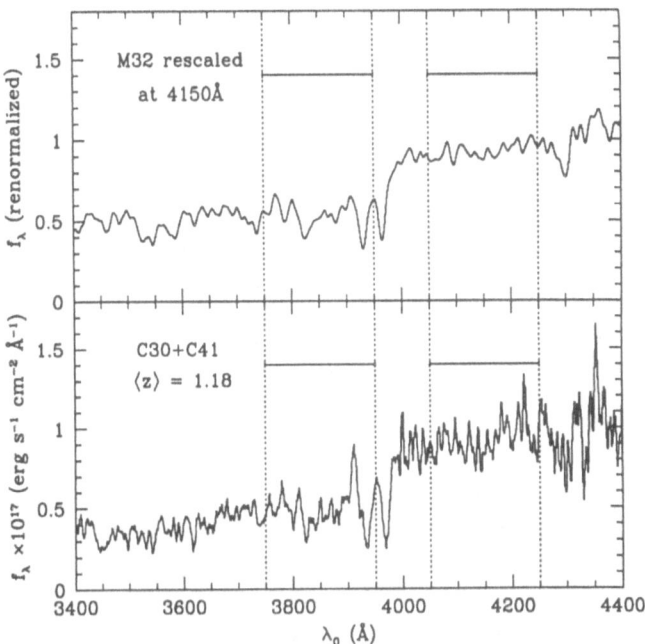

Fig. 2. Coadded spectra in the rest–frame 4000Å region of two elliptical galaxies at $z = 1.15$ and 1.21 in the field of 3C 324 (bottom) compared to that of M32 (top). A very prominent 4000Å break and CaII H+K lines are visible; the spectral regions used to define the D4000 break amplitude are marked.

– redder colors or larger break value may be matched either by older or more metal rich stellar populations. Spinrad *et al.* review evidence that an approximately solar metallicity is an appropriate average value for giant ellipticals; we leave this problem unresolved here, but note that by studying the spectral properties of galaxies over a large range of cosmic time (rather than at a single, large redshift) we may eventually hope to break this degeneracy, as different rates of evolution might be expected for different metallicities.

Inspecting the four panels, one finds that the "age" derived from the various indices becomes "younger" the farther into the UV we look. Moving from red to blue indices (D4000, $R - K$, B2900 and B2640), and taking the solar metallicity model for reference, the ages derived for the single–burst model are approximately 4.5, 3.5, 1.7 and 0.5 Gyr, respectively. This suggests that the single, short–burst star formation history is not strictly correct, and that younger starlight plays an increasing role as one moves farther into the rest–frame UV. Models with more protracted star formation histories (e.g. exponentially decaying) or with later bursts can be constructed which reasonably match the data. However, the large 4000Å break amplitude and $R - K$ color strongly indicate that the bulk of the stars in these galaxies must have formed *at least* 2 *Gyr* prior

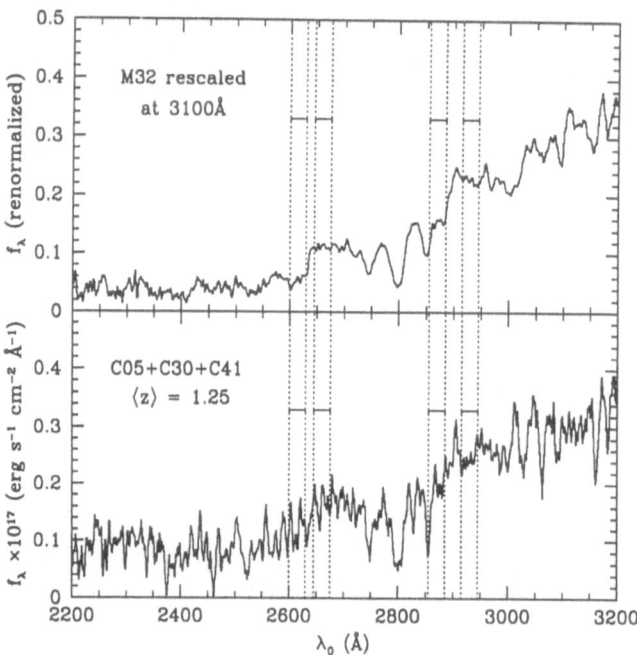

Fig. 3. Coadded near–UV rest–frame spectra of three elliptical galaxies at $z = 1.15$ to 1.41 in the field of 3C 324 (bottom) compared to IUE spectra of of M32 (top). The strong MgII 2800Å absorption feature is clearly seen in the high redshift objects, flanked by MgI 2852 and the absorption blend (primarily Fe+Cr) at 2746. The dotted lines mark the 2640Å and 2900Å break features defined by Spinrad *et al.* 1997.

to $z = 1.2$, even for populations with super–solar metallicities. For an open, 13 Gyr–old universe ($H_0 = 75$, $q_0 = 0$) this places the formation redshift for the bulk of the stellar mass very conservatively at $z > 2.3$, and most probably higher – for a closed universe, formation can be pushed back to almost arbitrarily large redshifts.

3   Scaling Relations to $z = 1.2$

Some of the most exciting data shown at this meeting has been the fundamental plane and Mg–σ observations at $0.1 < z < 0.6$, presented in the contributions of Bender, Franx, Jørgensen, Pahre, van Dokkum, and Ziegler to these proceedings (cf. also van Dokkum & Franx 1996 and Kelson *et al.* 1997). The conclusions of these various studies have been strikingly uniform: that the mass–to–light ratios of early–type galaxies in distant clusters have evolved very mildly, in a manner approximately consistent with expectations from simple, passive evolution of their stellar populations.

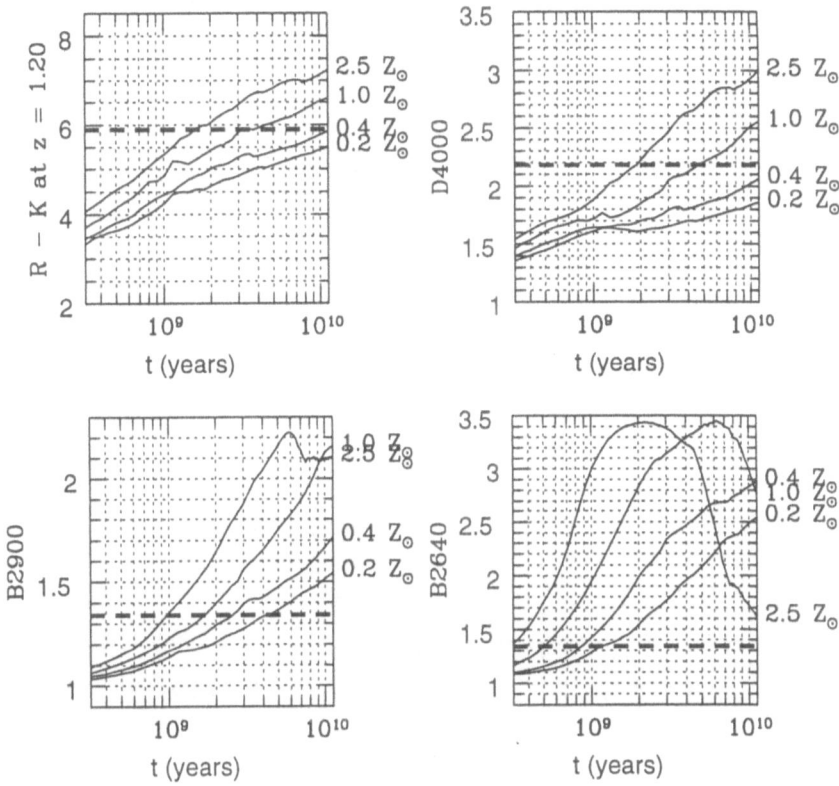

Fig. 4. Simple models of passive evolution of a stellar population formed in a 10^7 year burst with a Salpeter IMF, computed using the code of Bruzual & Charlot (1996) for various metallicities. The panels show the expected amplitudes of three spectral "breaks" and the $R - K$ broad band color – the R-band measures rest–frame light at $\approx 3150\text{Å}$ for $z = 1.2$, and thus is primarily sensitive to evolution at UV wavelengths intermediate between those measured by the 4000Å and 2900Å breaks. The horizontal dashed lines indicate the values measured for the $z > 1$ ellipticals discussed in the text.

At present it is still difficult to extend this work to much larger redshifts, even with 10m–class telescopes, because the galaxies become extremely faint and the spectral features of interest for measuring velocity dispersions or Mg indices shift into the near infrared where they are horrendously impacted by OH night sky emission. In the absence of high–S/N, high dispersion spectroscopy at $z > 0.8$, we may at least take advantage of purely photometric/morphological projections of the fundamental plane which allow us to push studies of galaxy scaling relations to higher redshifts. For example, the Kormendy relation is the projection of the fundamental plane onto the size vs. surface brightness axes, and is thus highly suitable for study using HST WFPC2 images alone. Dickinson (1995), Pahre *et al.* (1996), Barrientos *et al.* (1996), and Schade *et al.* (1996

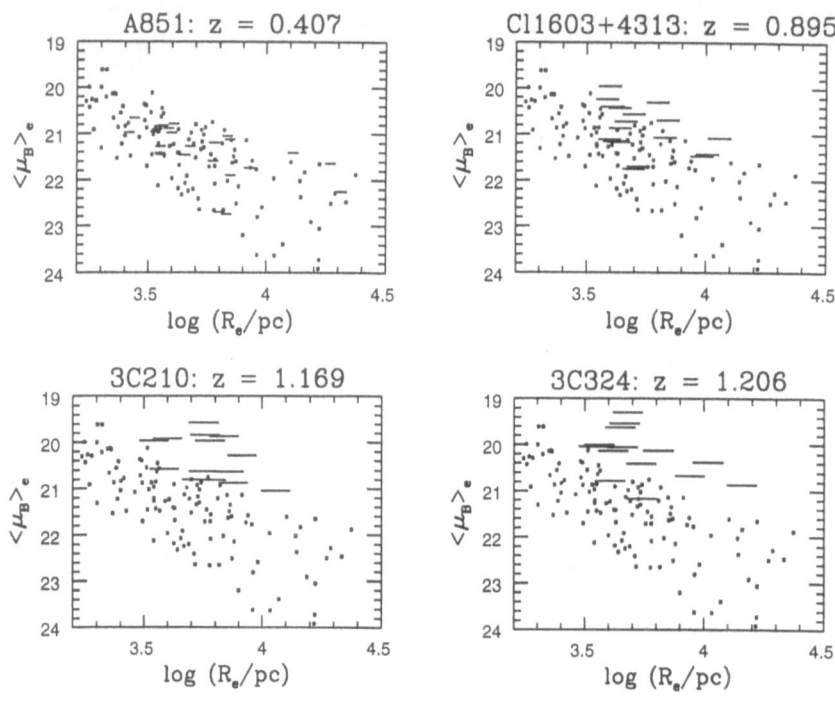

Fig. 5. The Kormendy relation for four high redshift galaxy clusters observed with HST. The small points are a sample of local cluster galaxies observed by Jørgensen *et al.* (1995). The horizontal lines represent individual high redshift ellipticals, and connect the R_e values for $q_0 = 0$ and $q_0 = 0.5$ ($H_0 = 50$ assumed throughout).

and this volume) have all employed variants on this technique. The method has the advantage of requiring only high–resolution imaging data which is readily obtained for a large number of cluster galaxies per WFPC2 field, but it lacks the precision of the direct fundamental plane measurements (due to the larger scatter in the Kormendy relation); moreover without the kinematic data offered by spectroscopy one cannot directly connect the observables to evolution in the mass–to–light ratio of distant galaxies.

Figure 5 presents an update on the data shown in Dickinson (1995), demonstrating the Kormendy relation in the rest–frame B–band for four high redshift clusters, including two at $z \approx 1.2$. As one moves to higher redshifts, the cluster galaxies lie increasingly "above" the locus of low–redshift ellipticals, indicating either higher rest–frame surface brightnesses, larger effective radii, or some combination thereof. It is simplest, but not definitive, to interpret this as a manifestation of passive luminosity evolution. Figure 6 plots the median offsets of the Kormendy relation for each cluster from the local values, and compares them

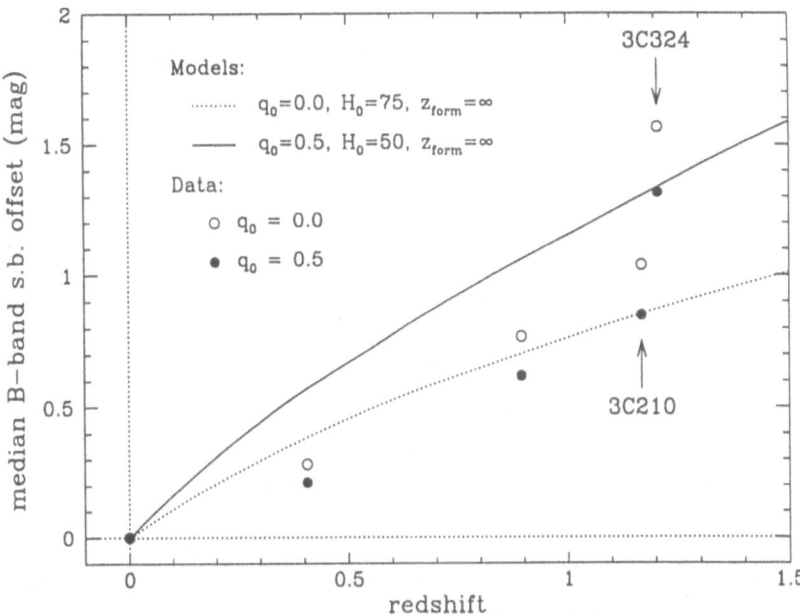

Fig. 6. Median rest–frame B–band surface brightness offsets of distant cluster ellipticals relative to the local Kormendy relation.

with expectations from passively evolving Bruzual–Charlot (1996) models for two different cosmologies, each assuming present–day ages for galaxies (and the universe as a whole) of 13 Gyr. The two $z \approx 1.2$ clusters seem to differ significantly, from one another, but it is important to note that the 3C324 data was obtained with the WFPC2 F702W filter, which samples the cluster galaxy light at a rest frame of $\lambda_0 \approx 3200$Å, requiring a large (and uncertain) k–correction to rest–frame B. 3C210 is at slightly lower redshift, and was observed with F814W (rest frame 3700Å), resulting in a much smaller k–correction. Overall, the implied luminosity evolution out to $z \approx 1$ is quite small compared to the stellar population models (see also van Dokkum & Franx 1996), and more consistent with the open universe models than the closed ones.

4   Discussion

Most of the effort directed toward studying distant elliptical galaxies has been aimed either at verifying the gospel of monolithic formation and passive evolution, or at falsifying this scenario as a heinous fiction which flies in the face of the cold logic of hierarchical merging. While most of the evidence presented above would seem consistent with the idea that early–type cluster galaxies have primarily undergone quiescent and passive evolution from $z \approx 1$ to the present, it is important to remember the distinction between the galaxies which we *observe* as ellipticals at high redshift and the exact progenitors of today's elliptical galaxies. While avoiding color–dependent selection effects, morphological selection of high redshift, early–type galaxies may itself introduce biases. By studying things that *look* like ellipticals, one is not necessarily tracing the past history of the actual ancestors to *all* of today's elliptical galaxies. In particular, if substantial merging has taken place to form the elliptical population, then the objects which are recognizably elliptical galaxies at any redshift may be only the oldest descendents (at that epoch) of that merging process, while more recent (or still–to–be) mergers may not enter into a morphologically–selected catalog. We may only be studying the most dynamically evolved galaxies in each cluster at each redshift, and thus those for which "recent" merger–induced star formation is pushed back to earlier and earlier cosmic times. Moreover, we have not attempted to separate ellipticals from S0 galaxies in our data. Recently, Dressler & Smail (1997) have suggested that there is a marked absence of S0 galaxies in distant clusters imaged by WFPC2. The implication is that the Butcher–Oemler effect may be, in part, a consequence of disk galaxies being transformed into S0s. Depending on the extent of star formation associated with this transformational process, this could provide a mechanism for "reinflating" the scatter in the combined E/S0 color–magnitude relation at late times, thus giving the impression of no evolution in that scatter.

Regardless, it seems to be clear that a substantial population of early–type galaxies was present in rich clusters out to $z \approx 1$ and beyond, and that the bulk of their stars must have formed at substantially larger redshifts. Whether *all* ellipticals formed at such early times is less certain, but there is little evidence from the cluster data to indicate that it is not so. It is important to emphasize that the data considered here has primarily been for elliptical galaxies in rich clusters, and that field galaxies may have had rather different evolutionary histories. Kauffmann *et al.* (1996) have used data from the Canada–France Redshift Survey to argue for strong number–density evolution in the population of early–type field galaxies since $z \approx 1$. The argument is basically that the apparent *lack* of strong evolution in the luminosity function of early–type (or at least red) galaxies in the CFRS (Lilly *et al.* 1995) contradicts expectations for passive stellar evolution (see figure 6), and thus must be counterbalanced by extensive number–density evolution. As for cluster ellipticals, the hierarchical merging models (e.g. as reviewed by White in this volume) predict that while many or most of their *stars* formed at $z > 1$, the galaxies themselves assembled late, with ∼70% forming after $z = 1$. This is difficult to test from the existing cluster data – the ellipticals already present in clusters at $z \approx 1$ cannot tell us how many more will join them by $z = 0$. Moreover, the richest clusters, which are generally those which we observe at high redshift, are quite probably the most unusual and overdense environments at those epochs, and thus are the places where the galaxy merging history is "pushed back" to the earliest cosmic times. The future of this work lies in the investigation of clusters at still larger redshifts, but equally importantly in the analysis of early–type galaxies across a broad range of environments at $z < 1$, from rich clusters through poorer systems and groups and into the field.

5   Acknowledgements

I would like to thank my collaborators, particularly Adam Stanford, Peter Eisenhardt, Hy Spinrad, Arjun Dey, Dan Stern, and Olivier LeFèvre, for allowing me to present material here in advance of publication. My thanks also to Stephane Charlot for assistance with the population synthesis models; to the conference organizers for their hospitality and financial support, and to the editor of these proceedings, for his patience.

References

Aragón–Salamanca, A., Ellis, R.S., Couch, W.J., and Carter, D., 1993, MNRAS, 248, 128.

Barrientos, F., Schade, D., and López–Cruz, O., 1996, Ap.J., 460, 89.

Bower, R.G., Lucey, J.R., and Ellis, R.S., 1992, MNRAS, 254, 601.

Bruzual, G., and Charlot, S. 1996, private communication.

Charlot, S., Worthey, G., and Bressan, A., 1996, Ap.J., 457, 625.

Dickinson, M. 1995, in *Fresh Views of Elliptical Galaxies,* eds. A. Buzzoni, A. Renzini, & A. Serrano, ASP, San Francisco, p. 283.

Dickinson, M., 1997a, in *The Early Universe with the VLT,* ed. J. Bergeron, Springer–Verlag, Berlin, p. 274.

Dickinson, M., 1997b, in *HST and the High Redshift Universe,* eds. N. Tanvir, A. Aragon-Salamanca, and J.V. Wall, World Scientific, in press.

Dressler, A., and Smail, I., 1997, in *HST and the High Redshift Universe,* eds. N. Tanvir, A. Aragon-Salamanca and J.V. Wall, World Scientific, in press.

Dunlop, J., Peacock, J., Spinrad, H., Dey, A., Jimenez, R., Stern, D., and Windhorst, R., 1996, Nature, 381, 581.

Eisenhardt, P.R.M., Stanford, S.A., Dickinson, M., and de Propris, R. 1997, in preparation.

Ellis, R.S., Smail, I., Dressler, A., Couch, W.J., Oemler, A., Butcher, H., and Sharples, R.M., 1997, Ap.J, in press.

Kauffmann, G., Charlot, S. and White, S.D.M., 1996, MNRAS, 283, 117.

Kelson, D.D., van Dokkum, P., Franx, M., Illingworth, G.D., and Fabricant, D. 1997, Ap.J., in press.

Lilly, S.J., Tresse, L., Hammer, F., Crampton, D., and LeFèvre, O., 1995, ApJ, 455, 108L.

Pahre, M.A., Djorgovski, S.G., and DeCarvalho, R.R. 1996, ApJ, 456, L79.

Schade, D., Barrientos, L.F., and López–Cruz, O., 1997, Ap.J., in press.

Spinrad, H., Dey, A., Stern, D., Dunlop, J., Peacock, J., Jimenez, R., and Windhorst, R. 1997, Ap.J., in press.

Stanford, S.A., Eisenhardt, P.R.M., and Dickinson, M. 1997, ApJ, submitted.

Jørgensen, I., Franx, M., and Kjaergaard, P. 1995, MNRAS, 273, 1097.

Kodama, T., and Arimoto, N. 1997, A&A, in press.

Van Dokkum, P.G., and Franx, M., 1996, MNRAS, 281, 985.

Worthey, G., 1994, ApJS, 95, 107.

The Nature of Compact Galaxies at $z \sim 0.2$–1.3: Implications for Galaxy Evolution and the Star Formation History of the Universe

R. Guzmán[1], A.C. Phillips[1], J. Gallego[1,2], D.C. Koo[1] and J.D. Lowenthal[1]

[1] UCO/Lick Observatory, University of California, Santa Cruz, CA 95064, USA
[2] Departamento de Astronomía, Universidad Complutense, 28040 Madrid, Spain

Abstract. We study the global scaling-laws of 51 compact field galaxies with redshifts $z \sim 0.2 - 1.3$ and apparent magnitudes $I_{814} < 23.74$ in the flanking fields of the Hubble Deep Field. Roughly 60% of the 45 compact emission-line galaxies have sizes, surface brightnesses, luminosities, velocity widths, excitations, star formation rates (SFR), and mass-to-light ratios characteristic of young star-forming HII galaxies. The remaining 40% form a more heterogeneous class of evolved starbursts, similar to local disk starburst galaxies. Without additional star formation, HII-like distant compacts will most likely fade to resemble today's spheroidal galaxies such as NGC 205. Our sample implies a *lower limit* for the global comoving SFR density of $\sim$0.004 $M_\odot$ yr^{-1} Mpc^{-3} at $z = 0.55$, and $\sim$0.008 $M_\odot$ yr^{-1} Mpc^{-3} at $z = 0.85$. These values, when compared to a *similar* sample of local galaxies, support a history of the universe in which the SFR density declines by a factor $\sim$10 from $z = 1$ to today. From the comparison with the SFR densities derived from previous data sets, we conclude that compact emission-line galaxies, though only $\sim$20% of the general field population, may contribute as much as $\sim$45% to the global SFR of the universe at $0.4 < z < 1$.

1 Introduction

Galaxies exhibit a wide variety of correlations among global parameters (such as luminosity, size, surface brightness, velocity dispersion, colors, and line strength indices). These empirical scaling-laws have been widely used to constrain current theories of galaxy formation, and to measure distances and peculiar velocities of galaxies to map the large-scale distribution of matter in the nearby universe. With the advent of the Hubble Space Telescope (HST) and the new generation of 10-m class telescopes, it is now possible to extend the study of scaling-laws to galaxies at high redshift (e.g., Koo et al. 1995; Van Dokkum & Franx 1996; Bender et al. 1996; Guzmán et al. 1996; Vogt et al. 1996). These new studies are proving key in our understanding of one of the major unresolved questions in modern cosmology: how galaxies evolve with look-back time.

One of the most controversial issues related to this question is the nature of the numerous faint blue galaxies observed in deep images of the sky (see reviews by Koo 1996, Ellis 1996, and references therein). The high surface density and weak clustering of these galaxies argue against their being either the progenitors or the merging components of present-day bright galaxies (Lilly et al. 1991; Efstathiou et al. 1991). Various theoretical scenarios have instead suggested that the faint blue galaxies are low-mass stellar systems experiencing their initial starburst at redshifts $z \leq 1$, some of which turn into the present population of spheroidal galaxies (Sph), such as NGC 205 (Babul & Ferguson 1996). Given their likely starburst nature, faint blue galaxies may also be major contributors to the global star formation rate (SFR) density already found to increase with lookback time to at least redshift $z \sim 1$ (Cowie et al. 1995, Lilly et al. 1996).

In this project we investigate the ideas above on the nature of the faint blue galaxies by comparing the scaling-laws of distant low-mass starbursts to those of nearby galaxies. The goals are: to identify their local counterparts, to assess their evolution with look-back time, and to study their role on the star formation history of the universe. A full description of the results summarized here can be found in Koo et al. (1995), Guzmán et al. (1996,1997), and Phillips et al. (1997).

2   The Data

The galaxy sample consists of 51 compact galaxies selected from I_{814} HST images of the flanking fields around the Hubble Deep Field (HDF; Williams et al. 1996). These objects are compact in the sense that they have small apparent half-light radii ($r_{1/2} \leq 0.5$ arcsec) and high surface brightnesses ($\mu_{I814} \leq 22.2$ mag arcsec^{-2}). With no color information, the "compactness" criterion optimizes the selection of dwarf stellar systems which are likely to be low-mass starbursts. Spectra for these objects were obtained using LRIS at the Keck telescope with a slitwidth of 1.1 arcsec and a 600 l/mm grating. The effective resolution is ~ 3.1 Å FWHM. Typical exposure times were 3000s. The total spectral range is ~ 4000-9000 Å. In addition, we obtained two 300s direct V-band exposures with LRIS in order to provide some color information. Our final data set includes: redshifts, $V_{606} - I_{814}$ colors, absolute blue magnitudes (M_B), half-light radii (R_e), surface brightnesses (SB_e), velocity widths (σ), masses (M), mass-to-light ratios (M/L), O[III]/Hβ line ratios, and SFRs.

Of the 51 galaxies, 6 (or 12%) show absorption-line spectra characteristic of elliptical and S0 galaxies, while the remaining 45 (88%) exhibit prominent oxygen and/or Balmer emission lines and blue continua characteristic of vigorous star-forming systems or narrow-line active galaxies. Most of the emission-line objects are very blue with nearly constant $V_{606} - I_{814} \sim 0.9$, while those with early-type spectra form a reasonably tight red sequence just blueward of the color track expected for non-evolving elliptical galaxies (Figure 1). Hereafter we focus our study on the emission-line compact galaxies. For convenience, we divide this sample into intermediate- ($z < 0.7$) and high-redshift ($z > 0.7$) samples.

3   Scaling-Laws

The global structural properties of galaxies can be adequately described using the $M_B - SB_e$ and $R_e - \sigma$ diagrams. In these diagrams, various galaxy types define distinct correlations, albeit with large scatter (Figures 2a and 2b). Most

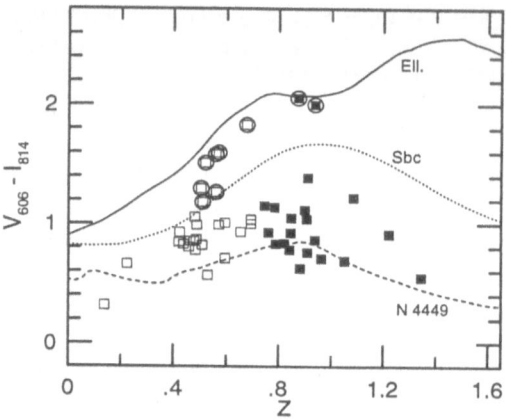

Fig. 1. $V_{606} - I_{814}$ color as a function of redshift. The tracks of local elliptical, Sbc and starburst (NGC 4449) galaxies are based on observed SEDs (no color evolution). Objects with absorption-line dominated spectra are marked with (o).

of our sample galaxies follow the sequence defined by young starbursts, such as nearby HII galaxies or distant compact narrow emission-line galaxies (CNELGs). To parametrize the starburst properties of the compact galaxy sample we use the $[OIII]/H\beta - M_B$ and the $M - SFR/M$ diagrams (Figures 3a and 3b). These diagrams discriminate among various types of starburst and active galaxies. Most of the compacts with $[OIII]/H\beta$ measurements lie in the moderate to high excitation regime populated by HII galaxies and CNELGs. These objects also have SFR/M characteristic of HII galaxies. A second group of distant compacts have $[OIII]/H\beta$ and SFR/M similar to those of more evolved disk starbursts such as local DANS and SBNs. Based on this, we have classified the sample into HII-like and disk starburst-like galaxies, depending on their SFR/M (see Figure 3b). With this simple criterion, we find that ~60% of distant compact galaxies have stellar *and* structural properties consistent with those of nearby HII galaxies, while the remaining 40% are similar to more evolved disk starbursts.

Two other major results can be drawn from the analysis of the $M - SFR/M$ diagram. First, the highest values of the SFR/M exhibited by compact galaxies are similar to those of local HII galaxies. Thus we do not find evidence for an increase in the peak of the SFR/M activity with redshift in our sample. Second, compact galaxies at $z > 0.7$ are, on average, ~10 times more massive than their counterparts with similar SFR/M at $z < 0.7$. Although selection effects may account for the lack of low-mass compact objects at high-z, they cannot explain why the massive star-forming systems are not present in the intermediate-z sample. This is not the result of a volume-richness effect either, since the volumes mapped in both samples differ by only a factor ~ 1.5. The apparent lack of massive starbursts in the intermediate-z sample suggests a steep evolution of the global SFR with redshift, although this result should be taken with caution given the small number of galaxies involved in this analysis.

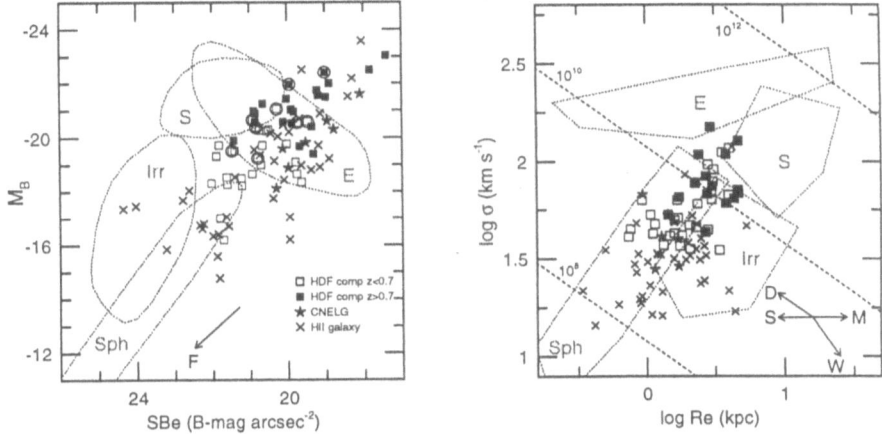

Fig. 2. (a): SB_e vs. M_B. Dotted lines indicate the general regions occupied by different classes of local galaxies; the arrow (F) represents the direction of fading. (b): R_e vs. σ. Dashed lines represent constant mass-lines in $M_{\odot}$; the arrows represent the effect of dissipation (D), mergers (M), stripping (S) and winds (W) on R_e and σ. We adopt $H_0 = 50$ km s^{-1} Mpc^{-1} and $q_0 = 0.05$.

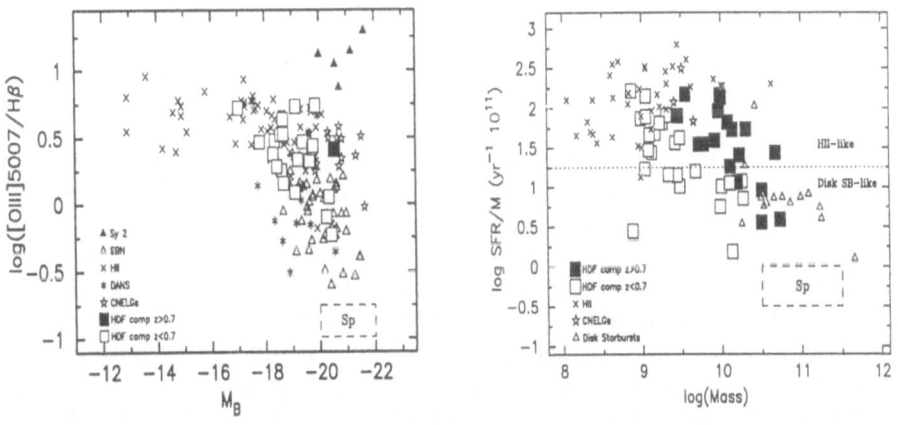

Fig. 3. (a): M_B vs. $[OIII]/H\beta$. Local galaxy sample is from Gallego et al. (1997); DANS: Dwarf Amorphous Nuclear Starbursts; SBN: Starburst Nuclei; Sy2: Seyfert 2 galaxies; HII: HII galaxies. Dashed lines represent the approximate location of spiral galaxies. (b): M vs. SFR/M. The dotted line represents the division between HII-like and disk starburst-like galaxies adopted in our classification.

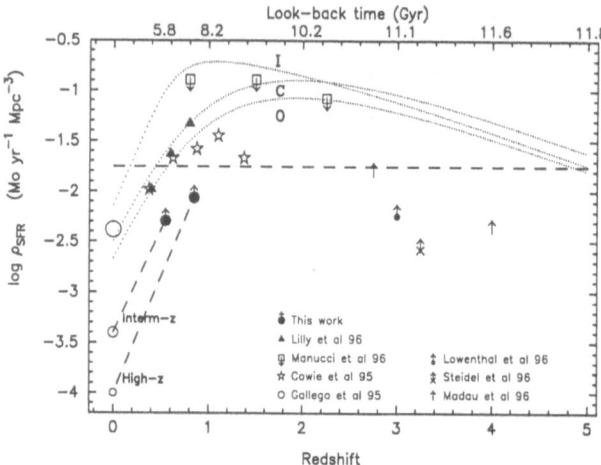

Fig. 4. SFR density vs. redshift. Filled circles are the estimates for compact galaxies. These values should be compared to the open circles labelled "Interm-z" and "High-z, which represent the values for *similar* samples of nearby compact galaxies. Dotted lines represent Pei & Fall's models (1995). The dashed line represents the fiducial value. We adopt $H_0 = 50$ km s^{-1} Mpc^{-1} and $q_0 = 0.5$.

4   Discussion

Distant compact emission-line galaxies are young, low-mass star-forming systems. Unless reignited by new star formation, they should fade within a few Gyrs. The issue of fading and transformation of one galaxy class to another is quite complex. Perhaps one of the most useful tools we have to study how distant young galaxies relate to nearby evolved stellar systems is the $R_e - \sigma$ diagram, since neither R_e nor σ depend strongly on the fading of the stellar population. Although there are several physical processes that may modify these parameters during galaxy evolution (see Figure 2b), we find no evidence against the idea that HII-like compact galaxies (most of those with $M < 10^{10} M_\odot$) are related structurally and kinematically to the nearby population of Sph and Irr galaxies. Their evolution into one galaxy class or another may depend critically on their ability to retain part of their interstellar medium in the likely event of starburst-driven galactic winds. The extremely low mass-to-light ratios of HII-like compacts (i.e., $M/L \sim 0.3$ solar) suggest that the kinetic energy supplied by the current starburst is large enough, compared to their binding energy, to blow out most of the gas, thus preventing future star formation. Without additional star formation, galaxy evolution models predict that these low-mass starbursts will fade enough to match the low luminosities and surface brightnesses of Sph galaxies (see Figure 2a). We thus conclude that a class of HII-like, faint blue galaxies may actually be among the progenitors of today's spheroidals.

The compact galaxy sample is also useful to investigate the role of low-mass starbursts on the evolution of the SFR density at redshifts $z < 1$. In Figure 4, we show a current overall picture of the evolution of the SFR density with redshift. The interpretation of this figure should be approached with caution, given the likely differences in the calibrations for the various SFR tracers, incompleteness of the data sets, and uncertainties in the models. Despite these caveats, most of the results summarized in this figure imply that the total SFR density of the universe decreased by a factor of $\sim$10 from $z \sim$1 to the present-day. Assuming our sample is representative of the general population of compact galaxies, we estimate that the total SFR densities associated to this class are: 0.004 $M_\odot$ yr^{-1} Mpc^{-3} at z=0.55, and 0.008 $M_\odot$ yr^{-1} Mpc^{-3} at z=0.85. These values, when compared to a *similar* sample of local galaxies, support a similar decline in the SFR density in the last $\sim$8 Gyrs. From the comparison with the SFR densities derived by Cowie et al. (1995), we conclude that compact emission-line galaxies, though only $\sim$20% of the general field population, may contribute as much as $\sim$45% to the global SFR of the universe at $0.4 < z < 1$.

Acknowledgements. This project is a collaborative effort of the DEEP team at UC Santa Cruz (http://www.ucolick.org/~deep/home.html). R. Guzmán would like to thank the organizing and scientific committees for their kind invitation and financial support to participate in this excellent meeting. Funding for this project is credited to NASA grants AR-06337.08-94A, AR-06337.21-94A, GO-05994.01-94A, AR-5801.01-94A, and AR-6402.01-95A from the Space Telescope Institute, and NSF grants AST 91-20005 and AST 95-29098.

References

Babul A. & Ferguson H.C., 1996, ApJ, 458, 100

Bender R., Ziegler B. & Bruzual G., 1996, MNRAS, 281, 985

Cowie L.L., Songaila A. & Hu E.M., 1991, Nature, 354, 460

Efstathiou G., Berstein G., Katz, N. et al. , 1991, ApJ, 380, L47

Ellis R.S., 1996, in ESO Workshop on *The Early Universe with the VLT*, in press

Gallego J., Zamorano J., Aragón-Salamanca A. & Rego M., 1995, ApJ, 455, L1

Gallego J., Zamorano J., Rego M. & Vitores A.G., 1997, ApJ, in press

Guzmán R., Koo D.C., Faber S.M. et al. , 1996, ApJ, 460, L5

Guzmán R., Gallego J., Phillips A.C. et al. , 1997, ApJ, submitted

Koo D.C., Guzmán R., Faber S.M. et al. , 1995, ApJ, 440, L49

Koo D.C., 1996, in IAU Symp. 168, ed. M. Kafatos (Dordrecht: Kluwer), p 201

Lilly S.J., Cowie L.L. & Gardner J.P., 1991, ApJ, 369, 79

Lilly S.J., Le Fevre O., Hammer F. & Crampton D., 1996, ApJ, 460, L1

Lowenthal J.D., Koo D.C., Guzmán R. et al. 1997, ApJ, in press

Madau P., Ferguson H.C., Dickinson M.E. et al. , 1996, MNRAS, 283, 1388

Mannucci F., Thompson D. & Beckwith S.V.W., 1996, in *HST and the High Redshift Universe*, 37th Herstmonceux Conf., Cambridge, in press

Pei Y.C. & Fall S.M., 1995, ApJ, 454, 69

Phillips A.C., Guzmán R., Gallego J. et al. , 1997, ApJ, submitted

Steidel C.C., Giavalisco M., Pettini M. et al. , 1996, ApJ, 462, L17

Van Dokkum P. & Franx M., 1996, ApJ, 463, L51.

Vogt N.P., Forbes D.A., Phillips A.C. et al. , 1996, 465, L15

Williams R.E. et al. 1996, AJ, 112, 1335

The 3CR Radio Galaxies at $z \sim 1$: Old Stellar Populations in Central Cluster Galaxies

Philip Best[1,2], Malcolm Longair[2], and Huub Röttgering[1]

[1] Sterrewacht Leiden, Huygens Lab, Postbus 9513, 2300 RA, Leiden, The Netherlands
[2] Cavendish Laboratory, Madingley Road, Cambridge CB3 0HE, England

Abstract. We investigate the old stellar populations of the 3CR radio galaxies at redshift $z \sim 1$ using observations made with the Hubble Space Telescope and the United Kingdom InfraRed Telescope. At radii $r \lesssim 35\,\mathrm{kpc}$, the infrared radial intensity profiles of the galaxies follow de Vaucouleurs' law, whilst at larger radii the galaxies show an excess of emission similar to that of low redshift cD galaxies. The locus of the high redshift 3CR galaxies on the Kormendy relation is investigated: passive evolution of the stellar populations is required to account for their offset from the relation defined by low redshift giant ellipticals and brightest cluster galaxies. The 3CR galaxies, on average, possess larger characteristic radii than the low redshift brightest cluster galaxies. Coupled with existing evidence, these results are strongly suggestive that distant 3CR galaxies must be highly evolved systems, even at a redshift of one, and lie at the centre of moderate to rich (proto–)clusters.

1 Introduction

The revised 3CR sample of radio sources defined by Laing et al. (1983) consists of the brightest radio sources in the northern sky, selected at $178\,\mathrm{MHz}$. It contains radio galaxies and quasars out to redshifts $z \sim 2$. The low redshift radio galaxies in the sample have long been known to be associated with giant elliptical galaxies containing old stellar populations; if the high redshift sources are similarly associated with giant ellipticals then these sources provide an ideal opportunity to study the evolution of stellar populations at early cosmic epoch, and thus to constrain models of galaxy formation and evolution.

Following the advent of infrared bolometers, Lilly and Longair (1982,1984) obtained infrared K–magnitudes for an almost complete sample of 83 3CR galaxies with redshifts $0 < z < 1.6$. Plotting K magnitude against redshift for these objects, they showed that the resulting relation has a remarkably small scatter ($\lesssim 0.6$ magnitudes). The tightness of this correlation was interpreted as indicating that the high redshift 3CR host galaxies are also giant elliptical galaxies. Lilly and Longair (1984) showed that, unless the deceleration parameter is as large as $q_0 \sim 3.5$, the shape of the $K-z$ relationship is not consistent with non–evolving stellar populations, but that at least passive evolution is required.

The optical–ultraviolet morphologies of the distant galaxies are, however, far more complicated. In 1987, McCarthy et al. and Chambers et al. discovered that the optical emission of powerful high redshift radio galaxies tends to be elongated and aligned along the direction of the radio emission. Our HST images have allowed us to study the morphologies of these galaxies on kpc scales, and demonstrate that the form of the alignment differs greatly from source to source (Longair et al. 1995, Best et al. 1996a,b); in some cases it arises from the elongation of a single central emission region, whilst in others strings of bright knots are seen to stretch along the radio axis. The most promising models for this aligned emission are star formation induced by the radio jet, scattering of quasar light by dust or electrons, or nebular emission from the warm ionised gas (eg. Röttgering and Miley 1996).

2   Spectral Energy Distributions

The critical point about all of the mechanisms for producing the aligned emission is that they possess relatively flat spectra, and so at the longer wavelengths of the infrared emission they are dominated by the emission of the underlying old stellar population.

To estimate the fraction of the K–band flux density that would arise from the aligned emission, we produce a relatively simple model for the spectral energy distributions (SED's) of the galaxies, using a combination of two components: (i) a passively evolving old stellar population, and (ii) a flat spectrum ($f_\nu \propto \nu^0$) emission component, representing the aligned emission. The SED of the first component was derived using the stellar synthesis codes of Bruzual and Charlot (1993,1996), assuming that the stars formed in a 1 Gyr burst beginning at a redshift $z = 10$. A Scalo IMF with upper and lower mass cut–offs of 0.1 and $65 M_\odot$ was adopted. The precise spectral shape of the second component of the fit depends upon the nature of the alignment effect; the adoption of a flat spectral index provides a good compromise and will not be too far wrong in any case.

For each galaxy, we calculate the sum of these two components which best matches the broad band flux densities of the galaxies measured at four different wavelengths from our UKIRT and HST images, taking account of any emission line contributions (Best et al. 1996b). 3C437 and 3C470, for which only two broad band flux densities were available, and 3C22 and 3C41 which both possess strong nuclear components in the infrared images (see Section 3) were omitted from this analysis. The fits obtained are generally good. To illustrate this, in Figure 1 we show the results for the first half of the sample. These indicate that the simple two component model provides a good representation of the SED's of these galaxies. The percentage of K–band light associated with the flat spectrum component ranges from only $\sim 1\%$ in the very passive sources 3C65 and 3C337, up to $\sim 22\%$ in the case of 3C368, with an mean value of $\sim 8\%$. Although these percentages would be higher if a redder spectral shape had been adopted for the aligned component, the assumption that the K–band light is dominated by the old stellar population seems reasonably secure.

3   Radial Intensity Profiles

Our observations can be used to compare the radial intensity profiles of these galaxies with de Vaucouleurs' law: $I(r) \propto \exp\left[-7.67 \left(r/r_e\right)^{-1/4}\right]$. For 8 of the

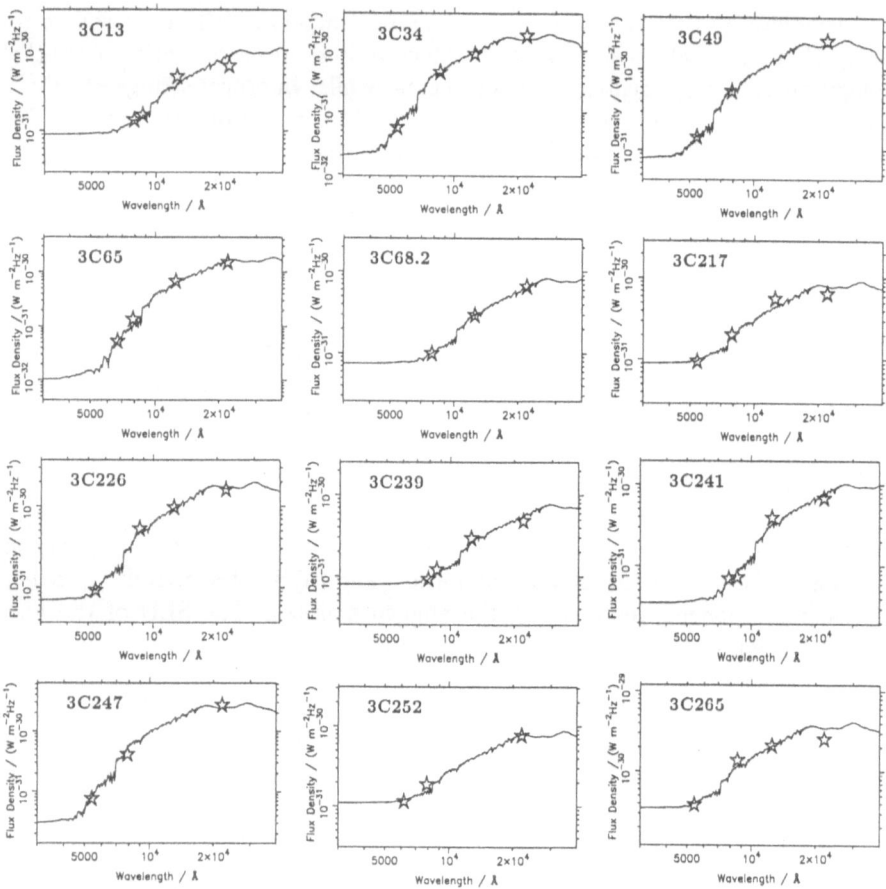

Fig. 1. SED fits to the broad band flux densities of the 3CR galaxies using an old stellar population and a flat spectrum component (see text for details).

galaxies in the sample there is little evidence for a significant ultraviolet emission component, in the sense that only a small ($\leq 5\mu$Jy) flat spectrum component is required in the fit to their SED, and the morphologies of the HST images are almost symmetrical. For these galaxies, the radial profiles of the optical emission were measured, with nearby companions objects being removed and replaced by the average of the background pixels at that distance from the centre of the galaxy. The radial profiles are shown in Figure 2: in 6 of the 8 cases a de Vaucouleurs profile provides an excellent match to the observed data. The cases of 3C22 and 3C41 are discussed below. The values of the characteristic radius can typically be measured to an accuracy of order 15%.

Using the characteristic radii determined from the HST images, the infrared profiles of these galaxies can also be investigated. For each of the galaxies, a de Vaucouleurs profile with the characteristic radius derived from the HST fit,

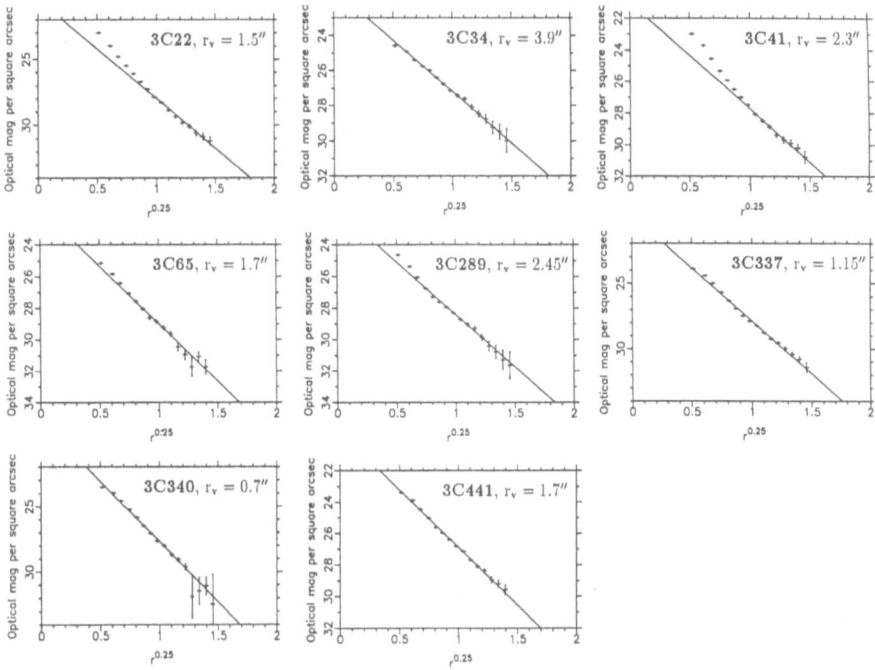

Fig. 2. De Vaucouleurs fits to the radial intensity profiles of the HST images of eight 3CR radio galaxies which do not show a significant active ultraviolet component. The characteristic radius of each, determined from the gradient of the best fitting straight line, is given.

and an unresolved emission source were each convolved with effects of the seeing (which was typically between 1 and 1.2 arcsec). The combination of these two components which provided the best match to the observed data was then determined in each case, the results being shown in Figure 3: the dashed line shows the radial profile of the de Vaucouleurs component, the dotted line shows that of the point source component, and the solid line shows the total intensity profile (note that in many cases, the best fit was produced using no point source component, and so this does not appear on the plot). In each case a good match is obtained.

Such fits could also be made for the remaining galaxies in the sample, for which no de Vaucouleurs fit to the HST data was possible due to the aligned emission. For these galaxies, the characteristic radius was allowed to be a further free parameter in the fit. For the five galaxies with redshifts $z > 1.4$, the low signal–to–noise ratio of the infrared images meant that the χ^2 of the fit varied only slowly with characteristic radius and so the best–fitting characteristic radius was not well–defined; these galaxies have therefore been omitted from the analysis. For the remaining galaxies the characteristic radius could typically be

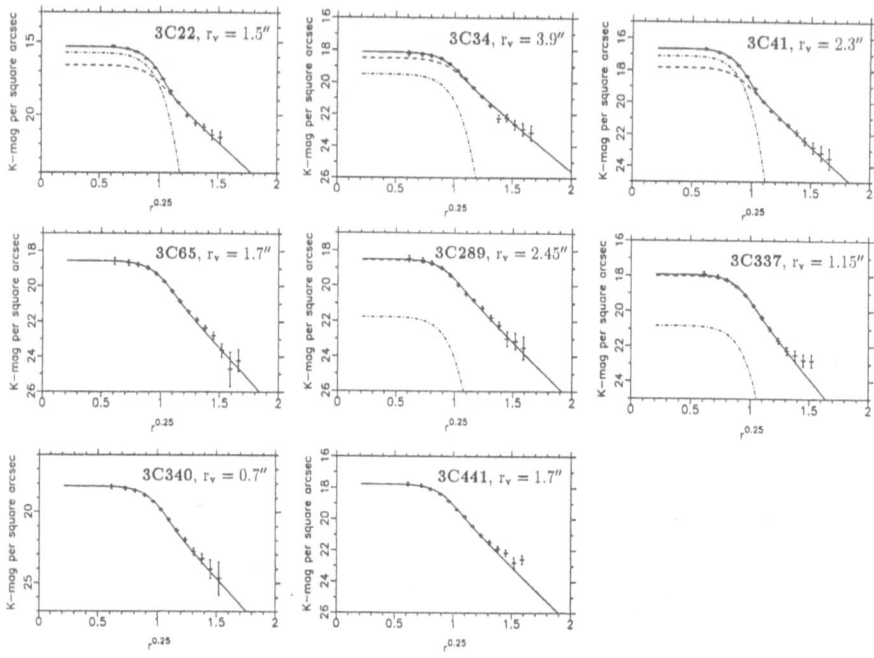

Fig. 3. Fits to the radial intensity profiles of the K–band images of the eight 3CR galaxies shown in Figure 2, using the sum of an unresolved point source (dash–dot line) and a de Vaucouleurs profile with the characteristic radius determined from the HST images (dashed line). For each of the profiles, the effect of seeing has been taken into account. The sum of the two components is indicated by the solid line. (Note that in many cases the best fit does not involve a point source component.)

determined to an accuracy of $\sim 35\%$, and the best fitting models are shown in Figure 4.

Figures 3 and 4 demonstrate that, except for the two cases of 3C22 and 3C41, there is only a small ($< 10\%$) point source component contributing to the total K–band flux density of the 3CR radio galaxies. Indeed, the point source contribution is consistent, within the 90% confidence limits, with being zero in all but these two cases. For 3C22 and 3C41, approximately 37% and 24% (respectively) of the K–band emission is associated with an unresolved emission source. It is interesting to note that these two galaxies are the brightest in our sample in the K–band, lying furthest from the mean $K-z$ relationship (Best et al. 1996c). Rawlings et al. (1995) have previously proposed that 3C22 possesses a significant unresolved K–band contribution and, together the detection of broad Hα emission (Economou et al. 1995) from this galaxy, this suggests that this source may be a reddened quasar.

It is noticeable that for a proportion of the galaxies the observed emission

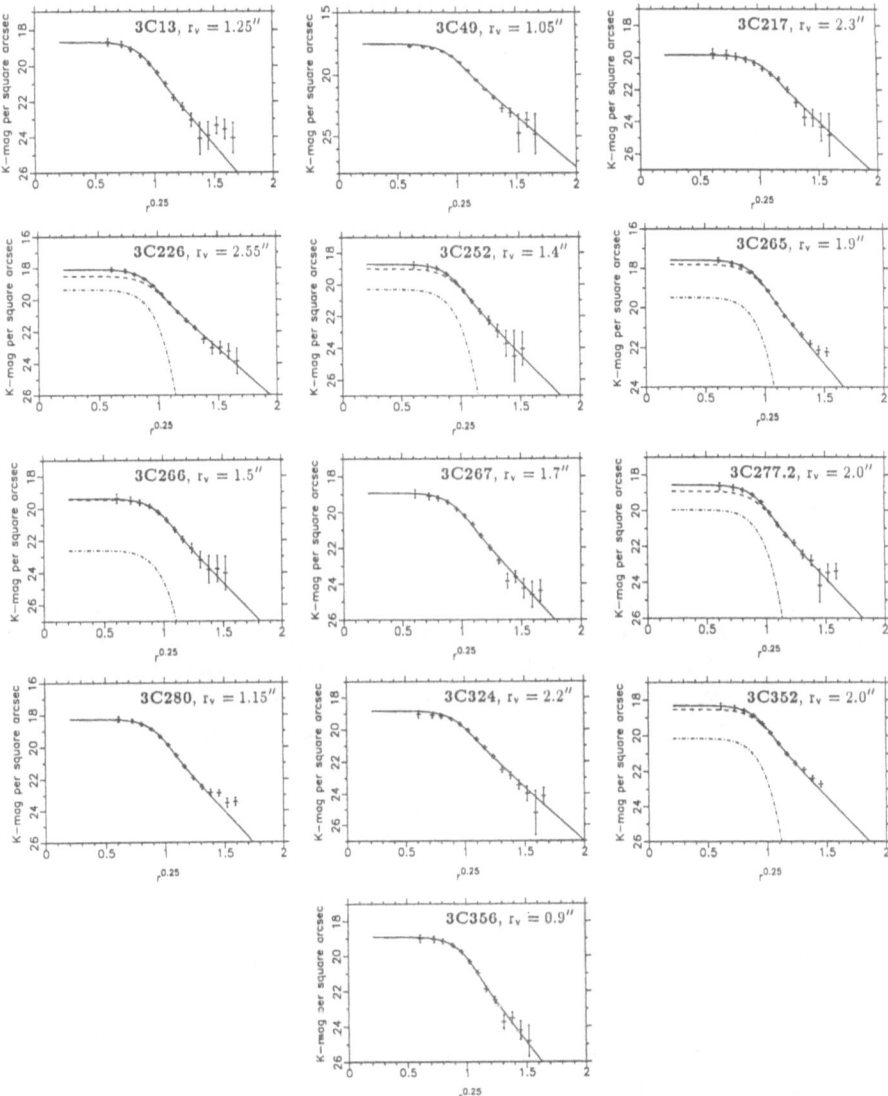

Fig. 4. Fits to the radial intensity profiles of the K–band images of the 13 3CR galaxies for which enhanced optical emission prevented such a fit to the optical intensity profile. The characteristic radii have been determined from the best fitting profiles. The same notation is used as in Figure 3.

profile becomes brighter than the predicted de Vaucouleurs profile at large radii, suggesting that these galaxies possess diffuse extended envelopes. To improve the signal–to–noise of this feature, we have scaled the radial profile of each galaxy with respect to the characteristic radius derived for it, and summed the

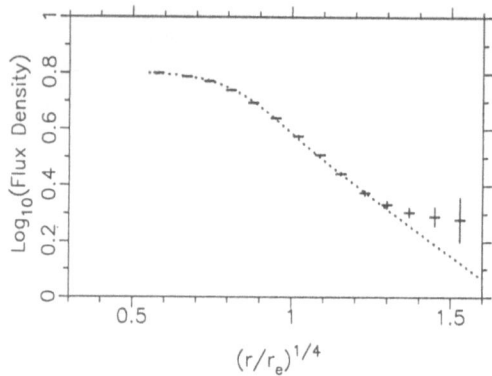

$(r/r_e)^{1/4}$

Fig. 5. A combined radial intensity profile in the K–band for the 12 galaxies with $1.0 \leq r_e \leq 2.0$. Units on the y–axis are arbitrary. Errors for each point are marked. The dotted line shows a combined de Vaucouleurs profile: halo emission is clearly visible at $(r/r_e)^{1/4} \gtrsim 1.25$.

intensity profiles. Only those galaxies for which $1.0 \leq r_e \leq 2.0$ were included: for galaxies with a smaller r_e there will still be a significant effect due to the seeing at radii of 2 to 3 r_e, whilst for galaxies with a larger r_e there is insufficient signal at the largest radii to accurately test for the presence of the halo. For the 12 galaxies which meet this criteria, the scaled radial profiles were summed, weighting each galaxy equally, and the results are presented in Figure 5. At radii $(r/r_e)^{1/4} \gtrsim 1.25$, corresponding to $r \gtrsim 35\,\mathrm{kpc}$, there is a clearly significant halo component.

This halo component is very similar to that of low redshift cD galaxies, which lie towards the centre of galaxy clusters. Together with the large (a few times $10^{11} M_\odot$) masses of these galaxies, this suggests that high redshift 3CR galaxies may live in moderately rich environments. Existing evidence supports this hypothesis, eg. the detection of cooling flows around these galaxies (Crawford and Fabian 1995), evidence for companion galaxies in narrow–band [OII] images (McCarthy 1988), and the fact that individual well–studied 3CR galaxies at high redshift appear to live in clusters (eg. 3C324, Dickinson et al. 1996).

4 The Kormendy Relation for the Redshift One 3CR Radio Galaxies

The Kormendy r_e *vs* μ_e projection of the fundamental plane alleviates the need for detailed spectroscopy, which is a difficult process for high redshift galaxies. In the previous section we obtained good estimates for the characteristic radii of the 3CR galaxies; by measuring also their surface brightnesses, it will be possible to compare their location on the Kormendy projection with those of low redshift brightest cluster galaxies and giant ellipticals. Thus, we can investigate how much evolution of the stellar populations must be occurring with cosmic epoch.

There are a number of details which need to be dealt with before the location of the 3CR galaxies on the Kormendy relation can be determined. The foremost of these is that we can not use the HST images to measure the surface brightnesses, because of the large component of light associated with the aligned emission at these wavelengths. Instead, the K–band images must be used, with the adoption of an appropriate k–correction. We must also take account of the effects of seeing in the K–band observations, of any point source or flat spectrum contributions to the flux density (as calculated in the earlier sections of this paper), and compensate for cosmological surface brightness dimming.

As a first, null hypothesis, we assume that the stellar populations of the 3CR radio galaxies are not evolving, and so we use the SED of low redshift giant elliptical galaxies to calculate the required k–correction. The rest–frame B–band surface brightnesses thus calculated for the 3CR radio galaxies are plotted against the measured de Vaucouleurs radii in Figure 6a, together with data from low redshift samples of giant ellipticals and brightest cluster galaxies (Oegerle and Hoessel 1991, Schombert 1987) and a sample of low redshift radio galaxies (Lilly and Prestage 1987). The 3CR galaxies have greater surface brightnesses than the low redshift galaxies in the fundamental plane, implying that some stellar evolution must have occurred between a redshift of one and the current epoch.

We can repeat the process, but instead make the more physical assumption that the stellar populations of the 3CR radio galaxies form at high redshift ($z = 10$) and then evolve passively. If evolution is occurring, the stellar populations observed in the high redshift radio galaxies will be younger, and hence brighter, than those seen in the nearby galaxies. The Bruzual and Charlot (1993, 1996) stellar synthesis codes were used to construct such passively evolving galaxies with the age that each 3CR galaxy would possess, and thus to calculate the required k–correction. The models were then used to determine the evolution in the rest–frame B–magnitude of the stellar populations that would occur for each galaxy between its observed redshift and a redshift of zero. This procedure therefore derived surface brightnesses of the 3CR galaxies that could be compared directly with low redshift giant ellipticals in the fundamental plane.

The derived surface brightnesses are plotted against the de Vaucouleurs radius in Figure 6b. It can be seen that they lie along the fundamental plane defined by the low redshift giant ellipticals, providing further evidence that the stellar populations evolve passively. There is an indication that the slope of the fundamental plane defined by the 3CR galaxies may be slightly steeper than that of the low redshift giant ellipticals, but this is likely to be just a selection effect: the dashed line on Figure 6b shows a line of constant total luminosity for the galaxies, that is, a line along which the product of the surface brightness and the square of the characteristic radius is constant; the 3CR galaxies lie closely along this line. That is, they lie within the fundamental plane, and also along the line of constant luminosity.

According to standard cannibalism models (eg. Hausman and Ostriker 1978) the position of a galaxy along the fundamental plane is interpreted as being

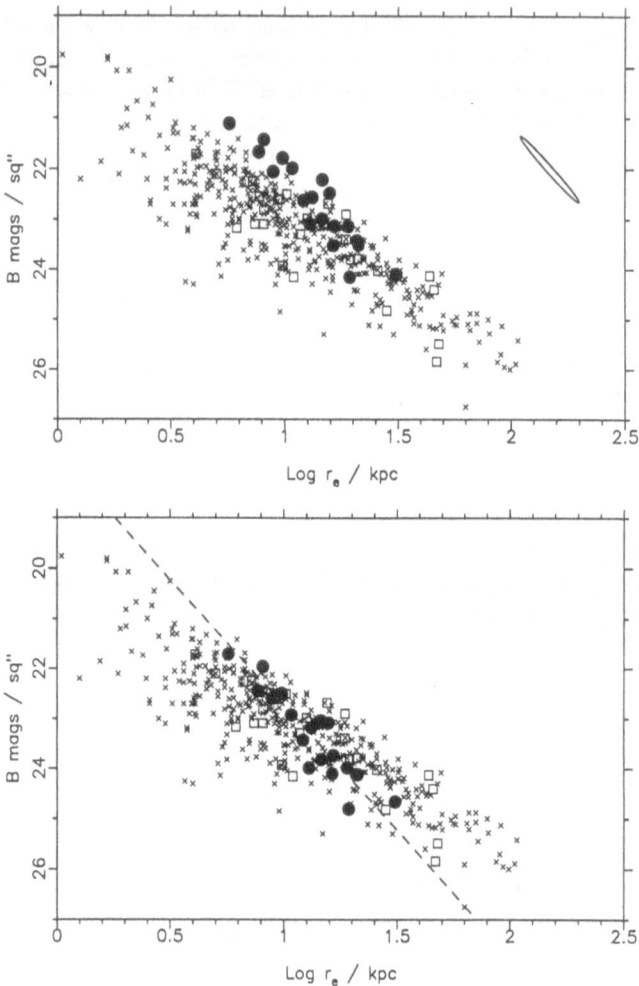

Fig. 6. Plots of B–band surface brightness vs characteristic radius for the 3CR galaxies (solid circles) compared with low redshift giant ellipticals and brightest cluster galaxies (crosses), and low redshift radio galaxies (open squares). The low redshift data are taken from Rigler and Lilly (1994) and references therein. The ellipse indicates the error ellipse for the 3CR galaxies whose characteristic radius was measured from the UKIRT data; those for which the measurement was from the HST data have much an error ellipse less than half of this size. (a) Assuming no evolution of the stellar populations of the 3CR galaxies. (b) Assuming that the stellar populations evolve passively. The dashed line shows a line of constant total luminosity.

related to its merger history. Galaxies which have undergone more mergers and are more highly evolved lie further to the right along the fundamental plane. In this respect it is interesting to note that: (i) on Figure 6, the high redshift 3CR galaxies possess a smaller spread of characteristic radii than those of the low redshift samples, indicating perhaps that these galaxies are all seen at a similar point in their evolutionary history, and (ii) the mean characteristic radius of the 3CR galaxies ($14.6 \pm 1.4 \, \mathrm{kpc}$) is larger than that of the low redshift samples ($11.0 \pm 0.5 \, \mathrm{kpc}$). The indicates that the 3CR galaxies must be highly evolved systems, even by a redshift of one.

5 Conclusions

From our study of the high redshift 3CR radio galaxies, we conclude the following points:

- The radial intensity profiles of the host galaxies of the 3CR radio sources are well matched by de Vaucouleurs' law.
- The galaxies possess extended cD type halos.
- Passive evolution of the stellar populations is required if the 3CR galaxies are to lie along the fundamental plane defined by low redshift giant ellipticals.
- Their characteristic radii are large, with relatively little scatter.

We conclude that the 3CR radio galaxies at redshift $z \sim 1$ are highly evolved galaxies, containing old stellar populations, which lie at the centre of moderately rich (proto–)clusters.

References

Best P. N., Longair M. S., Röttgering H. J. A., 1996a, MNRAS, **280**, L9

Best P. N., Longair M. S., Röttgering H. J. A., 1996b, MNRAS: *submitted.*

Best P. N., Longair M. S., Röttgering H. J. A., 1996c, MNRAS: *submitted.*

Bruzual G., Charlot S., 1993, ApJ, **405**, 538

Bruzual G., Charlot S., 1996, *submitted*

Chambers K. C., Miley G. K., van Breugel W. J. M., 1987, Nature, **329**, 604

Crawford C. S., Fabian A. C., 1995, MNRAS, **273**, 827

Dickinson M., Dey A., Spinrad H., 1996, *Galaxies in the Young Universe*, Hippelein H. ed, Springer Verlag, *in press*

Economou F., Lawrence A., Ward M. J., Blanco P. R., 1995, MNRAS, **272**, L5

Hausman M. A., Ostriker J. P., 1978, ApJ, **224**, 320

Laing R. A., Riley J. M., Longair M. S., 1983, MNRAS, **204**, 151

Lilly S. J., Longair M. S., 1982, MNRAS, **199**, 1053

Lilly S. J., Longair M. S., 1984, MNRAS, **211**, 833

Lilly S. J., Prestage R. M., 1987, MNRAS, **225**, 531

Longair M. S., Best P. N., Röttgering H. J. A., 1995, MNRAS, **275**, L47

McCarthy P. J., van Breugel W. J. M., Spinrad H., Djorgovski S., 1987, ApJ, **321**, L29

McCarthy P. J., 1988, PhD Thesis, University of California, Berkeley

Oegerle W. R., Hoessel J. G., 1991, ApJ, **375**, 15

Rawlings S., Lacy M., Sivia D. S., Eales S. A., 1995, MNRAS, **274**, 428

Rigler M. A., Lilly S. J., 1994, ApJ, **427**, L79

Röttgering H. J. A., Miley G. K., 1996, *The Early Universe with the VLT*, Bergeron J. ed, Springer Verlag, *in press*

Schombert J. M., 1987, ApJ Supp., **64**, 643

Part 3

APPLICATIONS

Cosmological Implications of Large-Scale Flows

Avishai Dekel[1,2]

[1]Racah Institute of Physics, The Hebrew University, Jerusalem 91904, Israel
[2]University of California, Berkeley & Santa Cruz

Abstract. Cosmological implications of the observed large-scale peculiar velocities are reviewed, alone or combined with redshift surveys and CMB data. The latest version of the POTENT method for reconstructing the underlying three-dimensional velocity and mass-density fields is described. The initial fluctuations and the nature of the dark matter are addressed via statistics such as bulk flow and mass power spectrum. The focus is on constraining the mass density parameter Ω, directly or via the parameter β which involves the unknown relation between galaxies and mass. The acceptable range for Ω is found to be $0.4 - 1.0$. The range of β estimates is likely to reflect non-trivial features in the galaxy biasing scheme, such as scale dependence. Similar constraints on Ω and Λ from global measures are summarized.

1 Introduction

A major goal of the analysis of cosmic flows is measuring the cosmological parameters, in particular the mass density parameter, Ω (or Ω_m). As illustrated in Figure 1, the data can be used to constrain Ω in several different ways. Methods

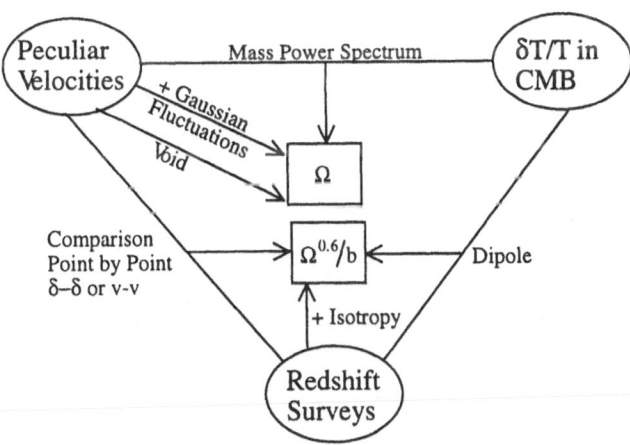

Fig. 1. Methods for measuring Ω and β from large-scale flows.

that are based on the peculiar velocity data alone (§ 4) are independent of the "biasing" relation between galaxy and mass density. They refer to the present-day large-scale structure (LSS) which is insensitive to the cosmological constant

Λ. They can thus serve to measure Ω directly, but they involve relatively large errors. These methods largely rely on the assumption (supported by observations, *e.g.*, Nusser, Dekel & Yahil 1995) that the initial fluctuations were drawn from a *Gaussian* random field.

All the methods that use the spatial distribution of galaxies must depend on the biasing relation between the densities of galaxies and mass. In the linear approximation to gravitational instability theory (GI), which is roughly valid for the fields when they are smoothed on very large scales, what is actually being measured by these methods is the degenerate parameter $\beta \equiv \Omega^{0.6}/b$ rather than Ω itself, where the density fluctuations of galaxies and mass are assumed to be related via a linear biasing relation, $\delta_G = b\delta$. These methods include, for example, measurements of redshift-space distortions from redshift surveys under the assumption of global isotropy, comparisons of the galaxy distribution with the CMB dipole, and comparisons of the observed peculiar velocities with the galaxy distribution (or the predicted velocities) deduced from redshift surveys (§ 5). The most recent best estimates of β for IRAS galaxies lie in the range $0.5 \leq \beta_I \leq 0.9$. As argued below, the contamination by non-trivial biasing introduces a significant uncertainty in the translation of the various β estimates to a reliable measurement of Ω.

Another way of determining Ω is by comparing the mass power spectrum from observed peculiar velocities with the spectrum of fluctuations in the CMB (§ 3.3, § 3.4). This comparison involves more detailed modeling of the formation of structure, and the current studies are limited to the CDM family of models, allowing for a non-zero cosmological constant, a possible tilt in the spectrum with or without tensor fluctuations, and a mixture of dark-matter species.

The outline is as follows: § 2 briefly describes reconstruction methods from peculiar velocities. § 3 discusses the statistics of mass-density fluctuations, including bulk flow and power spectrum, as determined by the velocity data alone or combined with CMB data. § 4 focuses on direct estimates of Ω from peculiar velocities alone. § 5 presents estimates of β by comparing velocity and galaxy density data, and addresses the issue of biasing. § 6 puts the results in perspective by summarizing other measures of the cosmological parameters. § 6.1 provides a summary of the results.

2   POTENT Reconstruction from Peculiar Velocities

2.1   Data for Velocity Analysis

The key for measuring peculiar velocities are the distance indicators (for review: Willick 1997). The current typical intrinsic scattering the Tully-Fisher (TF) distance indicator is at best $\sigma_m \sim 0.33$ mag, corresponding to a relative distance error of $\Delta = (\ln 10/5)\sigma_m \approx 0.15$.

The most comprehensive catalog of peculiar velocity data available today is the **Mark III** catalog (Willick *et al.* 1995; 1996; 1997a), which is a careful compilation of several data sets under the assumption that all galaxies trace the same underlying velocity field. The merger was non-trivial because the observers differ in their selection procedure, the quantities they measure, the method of measurement and the TF calibration techniques.

The original Mark II catalog, which was used in the first application of PO-TENT (Dekel *et al.* 1990; Bertschinger *et al.* 1990), consisted of about 1000 galaxies (mostly Lynden-Bell *et al.* 1988 and Aaronson *et al.* 1982). The extended Mark III catalog consists of ~ 3400 galaxies (dominated by Mathewson *et al.* 1992). This sample enables a reasonable recovery of the dynamical fields with $12\,h^{-1}$Mpc smoothing in a sphere of radius $\sim 60\,h^{-1}$Mpc about the Local Group (LG), extending to $\sim 80\,h^{-1}$Mpc in certain regions (§ 2.7). More uniformly sampled data of more than 1000 spiral galaxies in the north is in preparation, and was already subject to preliminary analysis (SFI, Giovanelli *et al.* 1997).

2.2   Methods of Velocity Analysis

One way of classifying the methods of velocity analysis is as follows:

	Inferred Distance Space	Redshift-Space + V Model
Forward TF	POTENT *selection + distance bias*	VELMOD *selection bias*
Inverse TF	POTINV *distance bias*	MFPOT *z-space smoothing*

The "forward" and "inverse" methods refer to whether the TF relation is interpreted as $M(\eta)$ or $\eta(M)$ (M being the absolute magnitude and η the rotation velocity). The difference is crucial because the apparent magnitude depends on distance while η is not, and because the selection depends on magnitude and is independent of η. On the other hand, the velocity field can be computed either from peculiar velocities that were evaluated at each galaxy's TF-inferred position, d, or by fitting a parametric model for the potential (and thus the velocity) field in redshift space, z. Each method is affected by different systematic errors, and techniques have been developed for statistically correcting them. The success of each technique has been tested using mock catalogs, and the goal is to have the different methods recover consistent results.

The original POTENT described below is a *forward* TF method in d-space. It has to deal with *selection* bias of the TF parameters because of the magnitude limit, and with *Malmquist* bias in the inferred distances and velocities arising from the random distance errors convolved with the geometry of space and the clumpiness of the galaxy distribution. Methods have been developed for statistically correcting these biases. In particular, the correction of the Malmquist bias requires external information about the underlying number density of galaxies in the samples from which galaxies were selected for the peculiar velocity catalogs (see Willick 1997).

One can alternatively infer distances using the *inverse* TF relation. This eliminates the selection bias, but there is still an inferred-distance Malmquist bias. In the inverse-TF case, the distance bias can in principle be corrected using information that is fully contained in the catalog itself (Landy & Szalay 1992). In practice, the quality of the correction is limited by the sparseness of the sampling. The POTENT analysis of the inverse data corrected this way is termed POTINV (Eldar, Dekel & Willick 1997).

If the selection does not explicitly depend on η, Malmquist bias can be eliminated by minimizing η residuals in redshift space without ever inferring actual distances to individual galaxies. The distance is replaced by $r = z - u_\alpha(z)$, where u_α is a parametric model for the radial peculiar velocity field. If the forward TF relation is used, as in VELMOD (Willick *et al.* 1997b; § 5.4), the method still has to correct for selection bias. The use of the *inverse* TF relation (Schechter 1980) guarantees in this case that both the selection bias and the distance bias are eliminated, at the expense of over-smoothing due to the representation of the fields in redshift space. Recent implementations of such methods were termed MFPOT (Blumenthal, Dekel & Yahil 1997) and ITF (Davis, Nusser, & Willick 1996).

Another way of distinguishing between the methods is by their goals. The methods working in d-space can serve for reconstruction of 3D maps of the velocity and mass-density fields, unbiased and uniformly smoothed with equal-volume weighting throughout the volume. These fields can then be straightforwardly compared to other data and to theory in order to obtain cosmological implications (*e.g.*, § 5.2). Alternatively, one may direct the method to estimating certain parameters of the model (*e.g.*, β) without ever reconstructing uniform maps. The redshift-space methods serve this purpose well (*e.g.*, § 5.4).

Yet another characteristic of some of the methods is the usage of a whole-sky redshift survey (such as IRAS 1.2 Jy) as an intrinsic part of the reconstruction from peculiar velocities. This is the case in the SIMPOT, VELMOD and ITF methods (§ 5). These methods are geared towards determining β, with SIMPOT also providing uniform reconstruction maps.

Finally, one can focus on optimal formal treatment of the random errors, which are in fact the main obstacle. A method based on Wiener Filtering has been developed for recovering the most probable mean field from the noisy peculiar-velocity data in d-space (Zaroubi, Dekel & Hoffman, in preparation). This can serve as a basis for constrained realizations of uniform smoothing, each of which being an equally good guess for the structure in our real cosmological neighborhood.

2.3   Correcting Malmquist Bias

The selection bias in the calibration of the forward TF relation can be corrected once the selection function is known (see Willick 1994). But then, the TF inferred distance, d, and the mean peculiar velocity at a given d, suffer from a *Malmquist* or *inferred-distance* bias. The distances, either forward or inverse, are corrected for Malmquist bias in a statistical way before being fed as input to POTENT-like procedures.

If M is distributed normally for a given η, with standard deviation σ_m, then the forward inferred distance d of a galaxy at a true distance r is distributed log-normally about r, with relative error $\Delta \approx 0.46\sigma_m$. Given d, the expectation value of r is

$$E(r|d) = \frac{\int_0^\infty r P(r|d)dr}{\int_0^\infty P(r|d)dr} = \frac{\int_0^\infty r^3 n(r) \exp\left(-\frac{[\ln(r/d)]^2}{2\Delta^2}\right) dr}{\int_0^\infty r^2 n(r) \exp\left(-\frac{[\ln(r/d)]^2}{2\Delta^2}\right) dr}, \qquad (1)$$

where $n(r)$ is the number density in the underlying distribution from which galaxies were selected. The deviation of $E(r|d)$ from d reflects the bias. The homogeneous part arises from the geometry of space — the inferred distance d underestimates r because it is more likely to have been scattered by errors from $r > d$ than from $r < d$, the volume being $\propto r^2$. If $n = const$, equation 1 reduces to $E(r|d) = de^{3.5\Delta^2}$, in which the distances should simply be multiplied by a factor, 8% for $\Delta = 0.15$, equivalent to changing the zero-point of the TF relation. Fluctuations in $n(r)$ are responsible for the inhomogeneous bias (IM), which systematically enhances the inferred density perturbations and thus the value of Ω inferred from them.

In one version of the Mark III data for POTENT analysis, the forward IM bias is corrected in two steps. First, the galaxies are grouped in z-space (Willick et al. 1995), reducing the distance error of each group of N members to $\Delta/\sqrt{N}$ and thus significantly weakening the bias. With or without grouping, the noisy inferred distance of each object, d, is replaced by $E(r|d)$, with an assumed $n(r)$ properly corrected for grouping if necessary. This procedure has been tested using realistic mock data from N-body simulations, showing that IM bias can be reduced to the level of a few percent. The practical uncertainty is in $n(r)$, which can be approximated for example by the high-resolution density field of IRAS or optical galaxies, or by the recovered mass-density itself in an iterative procedure under certain assumptions about how galaxies trace mass. The resultant correction to the density recovered by POTENT is $< 20\%$ even at the highest peaks.

Distances are alternatively inferred via the *inverse* TF relation between internal velocity parameter η and magnitude m, $\eta = \eta^0(m - 5\log d)$. Under the assumption that the selection was independent of η and was not an explicit function of distance, the expectation value of the true distance r given d is

$$E(r|d) = d\,e^{3\Delta^2/2}\,f(de^{\Delta^2})/f(d), \qquad (2)$$

where $\Delta \equiv (\ln 10/5)\sigma_\eta/\eta^0$. In this case, the required density function, $f(d)$, is in d-space, and is derivable from the sample itself (Landy & Szalay 1992). Eldar, Dekel & Willick (1997) have applied this correction to the inverse distances in the Mark III catalog, to serve as input for a POTENT analysis (POTINV). The agreement between the forward POTENT and POTINV results are well within the level of the random errors.

2.4   Smoothing the Radial Velocities

The goal of the POTENT analysis is to recover from the collection of Malmquist-corrected, radial peculiar velocities u_i at inferred positions d_i the underlying

3D velocity field $v(x)$ and the associated mass-density fluctuation field $\delta(x)$, smoothed with a Gaussian of radius R_s (we denote hereafter a 3D Gaussian window of radius $12\,h^{-1}$Mpc by G12, etc.). The first, most difficult step is the smoothing, or interpolation, into a radial velocity field with minimum bias, $u(x)$. The desire is to reproduce the $u(x)$ that would have been obtained had the true $v(x)$ been sampled densely and uniformly and smoothed with a spherical Gaussian window of radius R_s. With the data as available, $u(x_c)$ is taken to be the value at $x = x_c$ of an appropriate *local* velocity model $v(\alpha_k, x - x_c)$. The model parameters α_k are obtained by minimizing the weighted sum of residuals,

$$S = \sum_i W_i \left[u_i - \hat{x}_i \cdot v(\alpha_k, x_i) \right]^2 , \qquad (3)$$

within an appropriate local window $W_i = W(x_i, x_c)$. The window is a Gaussian, modified such that it minimizes the combined effect of the following three types of errors.

Tensor Window Bias. Unless $R_s \ll r$, the u_is cannot be averaged as scalars because the directions $\hat{x}_i$ differ from $\hat{x}_c$, so $u(x_c)$ requires a fit of a local 3D model as in Eq. 3. The original POTENT used the simplest local model, $v(x) = B$ of 3 parameters, for which the solution can be expressed explicitly in terms of a tensor window function (Dekel *et al.* 1990). However, a bias occurs because the tensorial correction to the spherical window has conical symmetry, weighting more heavily objects of large $\hat{x}_i\hat{x}_c$. A way to reduce this bias is by generalizing the zeroth-order B into a 9-parameter first-order velocity model, $v(x) = B + \bar{\bar{L}} \cdot (x - x_c)$, with $\bar{\bar{L}}$ a symmetric tensor that automatically ensures local irrotationality. The linear terms tend to "absorb" most of the bias, leaving $v(x_c) = B$ less biased. Unfortunately, a high-order model tends to pick undesired small-scale noise. The optimal compromise for the Mark III data was found to be a 9-parameter model fit out to $r = 40\,h^{-1}$Mpc, smoothly changing to a 3-parameter fit beyond $60\,h^{-1}$Mpc (Dekel *et al.* 1997).

Sampling-Gradient Bias. If the true velocity field is varying within the effective window, the non-uniform sampling introduces a bias because the smoothing is galaxy-weighted whereas the aim is equal-volume weighting. The simplest way to correct this bias is by weighting each object with the local volume that it "occupies", or the inverse of the local density. A crude estimate of this volume is $V_i \propto R_n^3$, where R_n is the distance to the n-th neighboring object (*e.g.*, $n = 4$). This procedure is found via simulations to reduce the sampling-gradient bias in Mark III to negligible levels typically out to $60\,h^{-1}$Mpc as long as one keeps out of the Galactic zone of avoidance. The $R_n(x)$ field can serve later as a flag for poorly sampled regions, to be excluded from any quantitative analysis.

Reducing Random Errors. The ideal weighting for reducing the effect of Gaussian noise has weights $W_i \propto \sigma_i^{-2}$, where σ_i are the distance errors. Unfortunately, this weighting spoils the carefully designed volume weighting, biasing

u towards its values at smaller r_i and at nearby clusters where the errors are small. A successful compromise is to weight by both, *i.e.*

$$W(\boldsymbol{x}_i, \boldsymbol{x}_c) \propto V_i\, \sigma_i^{-2}\, \exp[-(\boldsymbol{x}_i - \boldsymbol{x}_c)^2/2R_s^2]\ . \tag{4}$$

The resultant errors in the recovered fields are assessed by Monte-Carlo simulations. We generate noisy data via full, realistic Monte-Carlo mock catalogs, where the noise is added as scatter in the TF quantities (Kolatt *et al.* 1996). The error in the final δ at a grid point is estimated by the standard deviation of the recovered δ over the Monte-Carlo simulations, σ_δ (and similarly σ_v). In the well-sampled regions, which extend in Mark III out to 40–60 h^{-1}Mpc, the errors are $\sigma_\delta \approx 0.1$–$0.3$, but they may blow up in certain regions at large distances. To exclude noisy regions, any quantitative analysis could be limited to points where σ_v and σ_δ are within certain bounds.

2.5   From Radial Velocity to Density Fields

If the LSS evolved according to GI, then the large-scale velocity field is expected to be *irrotational*, $\nabla \times \boldsymbol{v} = 0$. Any vorticity mode would have decayed during the linear regime as the universe expanded, and, based on Kelvin's circulation theorem, the flow remains vorticity-free in the mildly-nonlinear regime as long as it is laminar. Bertschinger & Dekel (1989) have demonstrated that irrotationality is valid to a good approximation when a nonlinear velocity field is properly smoothed over. Irrotationality implies that the velocity field can be derived from a scalar potential, $\boldsymbol{v}(\boldsymbol{x}) = -\nabla\Phi(\boldsymbol{x})$, so the radial velocity field $u(\boldsymbol{x})$ should contain in principle enough information for a full 3D reconstruction. In the POTENT procedure, the potential is computed by integration along radial rays from the observer,

$$\Phi(\boldsymbol{x}) = -\int_0^r u(r', \theta, \phi)dr'\ . \tag{5}$$

The two missing transverse velocity components are then recovered by differentiation.

The final step of the POTENT procedure is the derivation of the mass-density fluctuation field associated with the peculiar velocity field. This requires a solution to the equations of GI in the mildly-nonlinear regime with mixed boundary conditions.

Let $\boldsymbol{x}, \boldsymbol{v}$ be the position and peculiar velocity in comoving units (corresponding to $a\boldsymbol{x}$ and $a\boldsymbol{v}$ in physical units, with $a(t)$ the universal expansion factor). Let $\delta \equiv (\rho - \bar{\rho})/\rho$ be the mass-density fluctuation. The equations governing the evolution of fluctuations of a pressureless gravitating fluid in a standard cosmological background during the matter era are the *Continuity* equation, the *Euler* equation of motion, and the *Poisson* field equation.

In the *linear* approximation, the growing mode of the solution, $\delta \propto D(t)$, is irrotational and can be expressed in terms of $f(\Omega) \equiv H^{-1}\dot{D}/D \approx \Omega^{0.6}$. The corresponding linear relation between density and velocity is $\delta_1 = -f^{-1}\nabla\cdot\boldsymbol{v}$. The use of δ_1 is limited to the small dynamical range between a few tens of megaparsecs and the $\sim 100\, h^{-1}\mathrm{Mpc}$ extent of the current samples. However, the sampling of galaxies enables reliable dynamical analysis with a smoothing radius as small as $\sim 10\, h^{-1}\mathrm{Mpc}$, where $|\nabla \cdot v|$ obtains values larger than unity and therefore nonlinear effects play a role. Even reconstruction with $\sim 5\, h^{-1}\mathrm{Mpc}$ smoothing may be feasible in well-sampled regions nearby. Figure 2 shows that δ_1 becomes a severe underestimate at large $|\delta|$. Mild nonlinear effects carry crucial information about the formation of LSS, and should therefore be treated properly.

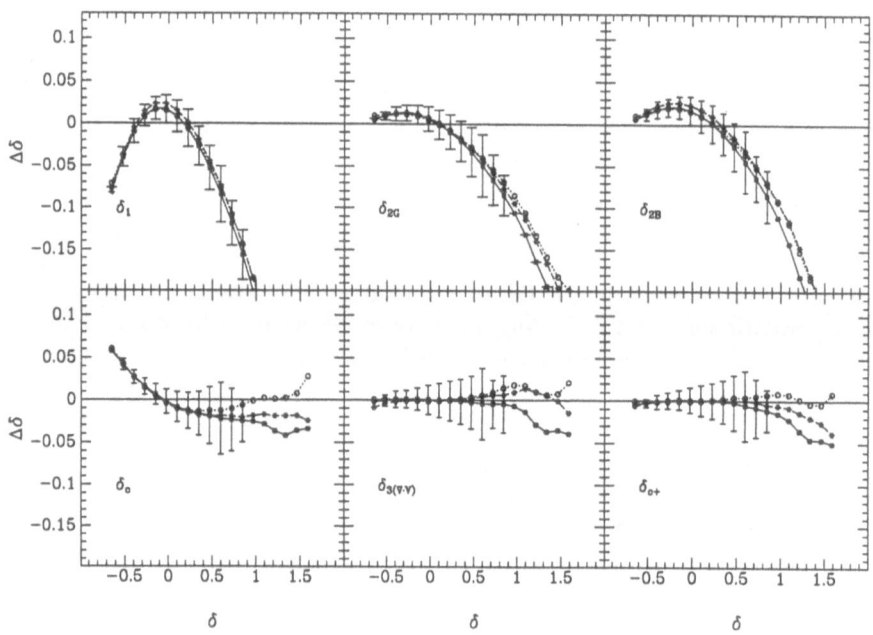

Fig. 2. Quasi-linear velocity-to-density approximations. $\Delta\delta \equiv \delta_{approx}(v) - \delta_{true}$. The mean and standard deviation are from large CDM N-body simulations normalized to $\sigma_8 = 1$, smoothed G12. The three curves correspond to different models: standard CDM (squares, solid), tilted CDM ($n = 0.6$, stars, dashed), and open CDM ($\Omega = 0.2$, open circles, dotted) (Ganon *et al.* 1997).

A basis for useful *mildly-nonlinear* relations is provided by the *Zel'dovich* (1970) approximation. The displacements of particles from their initial, Lagrangian positions q to their Eulerian positions x at time t are assumed to have a universal time dependence, and thus, $x(q) - q = f(\Omega)^{-1} v(q)$. For the purpose of approximating GI, the Lagrangian Zel'dovich approximation can be interpreted in *Eulerian* space, $q(x) = x - f^{-1} v(x)$, provided that the flow is *laminar* with no orbit mixing, or when multi-streams are appropriately smoothed over. The solution of the continuity equation then yields (Nusser *et al.* 1991)

$$\delta_c(x) = \|I - f^{-1}\partial v/\partial x\| - 1 \,, \tag{6}$$

where the bars denote the Jacobian determinant and I is the unit matrix. The

Zel'dovich displacement is first order in f^{-1} and $\boldsymbol{v}$ and therefore the determinant in δ_c includes second- and third-order terms as well, involving sums of double and triple products of partial derivatives which we term $\Delta_2(\boldsymbol{x})$ and $\Delta_3(\boldsymbol{x})$ respectively.

The approximation δ_c can be improved by slight adjustments of the coefficients of the nth-order terms,

$$\delta_{c+} = -(1 + \epsilon_1)f^{-1}\nabla \cdot \boldsymbol{v} + (1 + \epsilon_2)f^{-2}\Delta_2 + (1 + \epsilon_3)f^{-3}\Delta_3 . \tag{7}$$

The coefficients were empirically tuned to best fit the CDM simulation of 12 h^{-1} Mpc smoothing over the whole range of δ values, with $\epsilon_1 = 0.06$, $\epsilon_2 = -0.13$ and $\epsilon_3 = -0.3$. This approximation is found to be robust to uncertain features such as the value of Ω, the shape of the power spectrum, and the degree of non-linearity as determined by the fluctuation amplitude and the smoothing scale. Such robustness is crucial when a quasilinear approximation is used for determining Ω (§ 4). This is the approximation currently used in POTENT.

Fig. 2 compares the accuracy of the various approximations using the N-body simulations. δ_c is the best among the physically motivated approximations, which also include two second-order approximations (Bernardeau 1992; Gramman 1993). The latter do somewhat better at the negative tail, but they provide severe underestimates in the positive tail. δ_{c+} is an excellent robust fits over the whole mildly-nonlinear regime.

We note in passing that the relation 6 is not easily invertible to solve for $\boldsymbol{v}$ when δ is given, e.g., from redshift surveys, but a useful approximation derived from simulations is $\nabla \cdot \boldsymbol{v} = -f\delta/(1 + 0.18\delta)$.

2.6 Testing with Mock Catalogs

The way to optimize POTENT and other reconstruction methods is by minimizing the systematic errors when applied to mock catalogs. It is important that these mock catalogs mimic the real data as closely as possible. Such mock catalogs have been produced, for example, to mimic the Mark III and the IRAS 1.2Jy catalogs (Kolatt et al. 1996). They are publicly available and serve as standard "benchmarks" for the competing methods.

The procedure for making these mock catalogs involves two main steps: a dynamical N-body simulation that mimics our actual cosmological neighborhood, and the generation of galaxy catalogs from it.

Figure 3 demonstrate the quality of the POTENT reconstruction from the Mark III catalog by comparing the recovered density field to the true G12-smoothed field. This comparison is done at the points of a uniform grid inside a volume of effective radius 40 h^{-1}Mpc. The field shown is the average of the fields recovered from ten Monte Carlo mock catalogs of noisy galaxy velocities sampled sparsely and nonuniformly. One can see that the remaining systematic errors are small. The final systematic error is not correlated with the signal (slope ~unity in the scatter diagram) and is on the order of $\Delta\delta \sim 0.13$. The random errors are not a major obstacle in certain well-sampled regions (such as

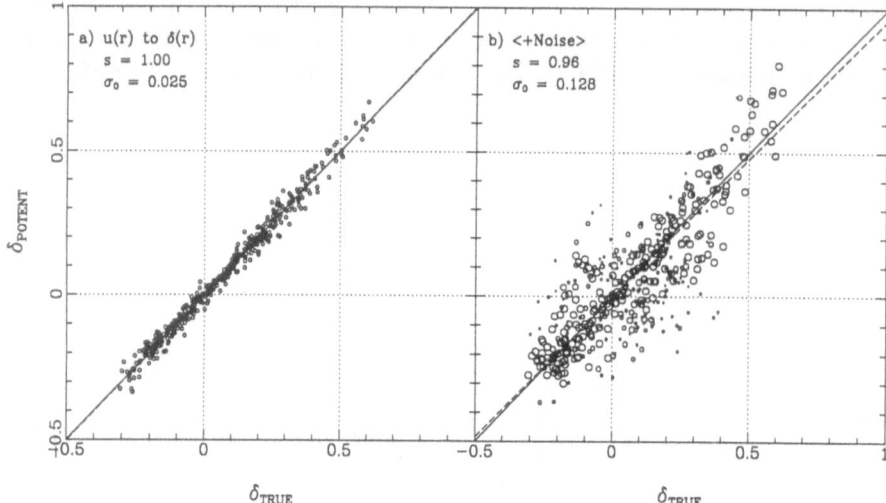

Fig. 3. Systematic errors in the POTENT analysis. The density field recovered by POTENT from the mock data is compared with the "true" G12 density. The comparison is at uniform grid points within a volume of radius $40\,h^{-1}$Mpc. Left: The input to POTENT is the true, G12-smoothed radial velocity. The small scatter of 2.5% reflects small deviations from potential flow, the scatter in the non-linear approximation Eq. 7, and numerical errors. Right: The input is noisy and sparsely-sampled mock data. Shown is the average of 10 random realizations. The global bias is only -4% (Dekel *et al.* 1997).

the Great Attractor), but they become severe in poorly-sampled regions (such as parts of the Perseus-Pisces region near the Galactic plane). The errors derived from the noisy mock catalogs are used to eliminate poorly-recovered regions from quantitative analyses.

2.7 Maps of Velocity and Density Fields

Figure 4 shows Supergalactic-plane maps of the velocity field in the CMB frame and the associated δ_{c+} field (for $\Omega = 1$) as recovered by POTENT from the Mark III catalog. The recovery is reliable out to $\sim 60\,h^{-1}$Mpc in most directions outside the Galactic plane ($Y = 0$). Both large-scale ($\sim 100\,h^{-1}$Mpc) and small-scale ($\sim 10\,h^{-1}$Mpc) features are important; *e.g.*, the bulk velocity reflects properties of the initial fluctuation power spectrum (§ 3), while the small-scale variations indicate the value of Ω (§ 4).

The velocity map shows a clear tendency for motion from right to left, in the general direction of the LG motion in the CMB frame ($L, B = 139°, -31°$ in Supergalactic coordinates). The bulk velocity within $60\,h^{-1}$Mpc is $300-350\,\mathrm{km\,s^{-1}}$ towards ($L, B \approx 166°, -20°$) (§ 3.1) but the flow is not coherent over the whole volume sampled, *e.g.*, there are regions in front of PP (bottom right) and behind the GA (far left) where the XY velocity components vanish, *i.e.*, the streaming relative to the LG is opposite to the bulk flow direction. The velocity field shows local convergences and divergences which indicate strong density variations on scales about twice as large as the smoothing scale.

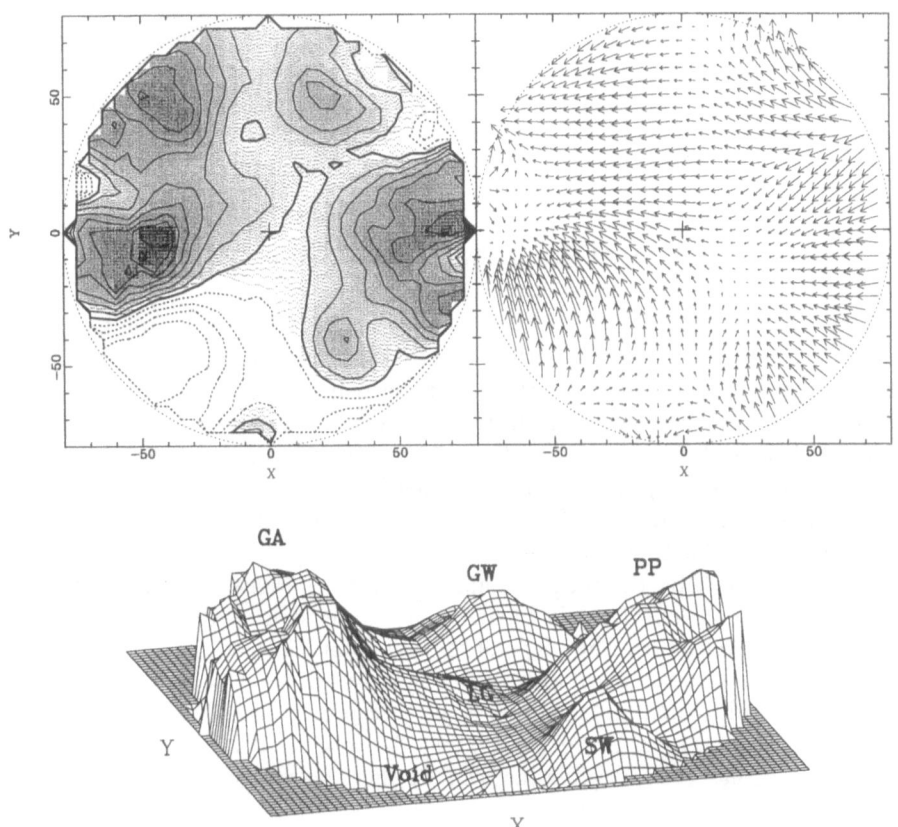

Fig. 4. POTENT dark-matter maps. The G12 fluctuation fields of peculiar velocity and *mass*-density in the Supergalactic plane as recovered by POTENT from the Mark III peculiar velocities. The vectors are projections of the 3D velocity field in the CMB frame, dominated by large derivatives embedded in a coherent bulk flow. Distances and velocities are in $100\,\mathrm{km\,s^{-1}}$. Contour spacing is 0.2 in δ, with the heavy contour marking $\delta = 0$ and dashed contours $\delta < 0$. The height in the surface plot is proportional to the total mass density contrast δ. The LG is at the center, GA on the left, PP and the Southern Wall on the right, Coma Great Wall at the top, and the Sculptor void in between (Dekel *et al.* 1997).

The bottom panel of Fig. 4 shows the POTENT mass density field in the Supergalactic plane as a landscape plot. The Great Attractor (with $12\,h^{-1}\mathrm{Mpc}$ smoothing and $\Omega = 1$) is a broad density ramp of maximum height $\delta = 1.4 \pm 0.3$ located near the Galactic plane $Y = 0$ at $X \approx -40\,h^{-1}\mathrm{Mpc}$. The GA extends towards Virgo near $Y \approx 10$ (the "Local Supercluster"), towards Pavo-Indus-Telescopium (PIT) across the Galactic plane to the south ($Y < 0$), and towards the Shapley concentration behind the GA ($Y > 0, X < 0$). The structure at the top is related to the "Great Wall" of Coma, with $\delta \approx 0.6$. The Perseus Pisces peak which dominates the right-bottom is peaked near Perseus with $\delta = 1.0 \pm 0.4$. PP extends towards the southern galactic hemisphere (Aquarius, Cetus), coinciding with the "Southern Wall" as seen in redshift surveys. Underdense regions separate the GA and PP, extending from bottom-left to top-right. The deepest region in the Supergalactic plane, with $\delta = -0.8 \pm 0.2$, roughly coincides with the galaxy-void of Sculptor and is useful in bounding Ω (§ 4).

3   Statistics of Mass-Density Fluctuations

Having assumed evolution by GI, the structure can be traced backward in time in order to recover the initial fluctuations and to measure statistics which character-ize them as a random field, e.g., the power spectrum, $P(k)$, and the probability distribution functions (PDF). "Initial" here may refer either to the *linear* regime at $z \sim 10^3$ after the onset of the self-gravitating matter era, or to the origin of fluctuations in the early universe before being filtered on sub-horizon scales dur-ing the plasma-radiation era. The spectrum is filtered on scales $\leq 100\, h^{-1}$Mpc by dark-matter dominated processes, but its shape on scales $\geq 10\, h^{-1}$Mpc is not affected much by the mildly-nonlinear effects (because the faster density growth in superclusters roughly balances the slower density depletion in voids at the same wavelength). The shape of the one-point PDF, on the other hand, is expected to survive the plasma era unchanged but it develops strong skewness even in the mildly-nonlinear regime. Thus, the present day $P(k)$ can be used as is to constrain the origin of fluctuations (on large scales) and the nature of the DM (on small scales), while the PDF needs to be traced back to the linear regime first.

The competing scenarios of LSS formation are reviewed for example by Pri-mack (1997). In summary, if the dark matter (DM) is all baryonic, then by nucleosynthesis constraints the universe must be of low density, $\Omega \lesssim 0.1$, and a viable model for LSS is the Primordial Isocurvature Baryonic model (PIB) with several free parameters. With $\Omega \sim 1$, the non-baryonic DM constituents are either "cold" or "hot", and the main competing models are CDM, HDM, and CHDM — a 7:3 mixture of the two. The main difference in the DM effect on $P(k)$ arises from free-streaming damping of the "hot" component of fluctuations on galactic scales. Currently popular variants of the standard CDM model ($\Omega = 1$, $n = 1$) include a tilted power spectrum on large scales ($n \lesssim 1$) and a flat, low-Ω universe with a non-zero cosmological constant such that $\Omega_{tot} + \Omega_\Lambda = 1$.

The peculiar velocities of the Mark III catalog enable direct derivations of the mass power spectrum itself, independent of galaxy biasing, roughly in the range $10-100\, h^{-1}$Mpc. The bulk velocity in spheres of radii up to $60\, h^{-1}$Mpc is sensitive to even larger wavelengths. In all standard theories, the power spectrum on large scales is expected to be a power law, $P_k \propto k^n$, with n of order unity. It is expected to turn around at $k_{peak} \sim 0.065(\Omega h)^{-1}(h^{-1}\text{Mpc})^{-1}$, corresponding to the horizon scale at the epoch of equal energy densities in matter and radiation. The dark matter type mostly affects the shape of the filtered spectrum in the "blue" side of the peak ($k > k_{peak}$). Once the fluctuation amplitude on very large-scales is fixed by COBE's measurements of CMB fluctuations, the bulk velocity is sensitive to n and is insensitive to Ω or the DM type. The steep slope of the CDM-like spectra at $k > k_{peak}$, where it is best constrained by the data, makes it more sensitive than the bulk velocity to Ωh.

We first describe the bulk velocity (§ 3.1). Then two ways of evaluating $P(k)$: a model-independent evaluation from the velocity field recovered by POTENT (§ 3.2), and a likelihood estimation from raw radial peculiar velocities under a prior model (§ 3.3). The $P(k)$ from the local velocities is then compared to sub-degree angular power spectrum of CMB fluctuations (§ 3.4).

3.1   Bulk Velocity

A simple and robust statistic related to the power spectrum is the bulk velocity — the amplitude of the vector average V of the R_s-smoothed velocity field $v(x)$ over a volume defined by a normalized window function $W_R(r)$ (*e.g.*, top-hat) of a characteristic scale R,

$$V \equiv \int d^3x \, W_R(x) \, v(x) \, , \quad \langle V^2 \rangle = \frac{f(\Omega)^2}{2\pi^2} \int_0^\infty dk \, P(k) \, \widetilde{W}_R^2(k) \, . \qquad (8)$$

We denote by V_r the bulk velocity in a top-hat sphere of radius $R = r\,h^{-1}\text{Mpc}$. The ensemble variance $\langle V^2 \rangle$ for a model that is characterized by $P(k)$ is an integral of $P(k)$ in which the wavelengths $\geq R$ are emphasized by $\widetilde{W}_R^2(k)$, the Fourier transform of $W_R(r)$. The bulk velocity can be obtained from the observed radial velocities by minimizing residuals as in Eq. 3. The first report by Dressler *et al.* (1987) of $V = 599 \pm 104$ for ellipticals within $\sim 60\,h^{-1}\text{Mpc}$ was interpreted prematurely as being in severe excess of common predictions, but it quickly became clear that the effective window was much smaller due to the nonuniform sampling and weighting (Kaiser 1988). The sampling-gradient bias can be crudely corrected by volume weighting as in POTENT (§ 2.4), at the expense of larger noise. Courteau *et al.* (1993) reported based on an early version of the Mark III data $V_{60} = 360 \pm 40$ towards $(L, B) = (162°, -36°)$. Alternatively, V_r can be computed from the POTENT v field by simple vector averaging from the grid.

The bulk velocity as a function of R, from several recent sources, is shown in Figure 5. The Mark III POTENT result at $R = 50\,h^{-1}\text{Mpc}$ is $V_{50} = 374 \pm 85\,\text{km s}^{-1}$ towards $(158°, -9°) \pm 10°$. The $\sim 20\%$ error bars are due to distance errors, and one should consider an additional uncertainty of similar magnitude due to the non-uniform sampling. The SFI sample of Sc galaxies yields at the same R a very similar result (contrary to premature rumors), $V_{50} \approx 364\,\text{km s}^{-1}$ towards $(172°, -14°)$ (da Costa *et al.* 1996). These samples are not large enough for a reliable estimate of V at larger radii.

Supernovae Type Ia provide more accurate distances, with only $\sim 8\%$ error, and they can be measured at larger distances. The current sample of 44 such

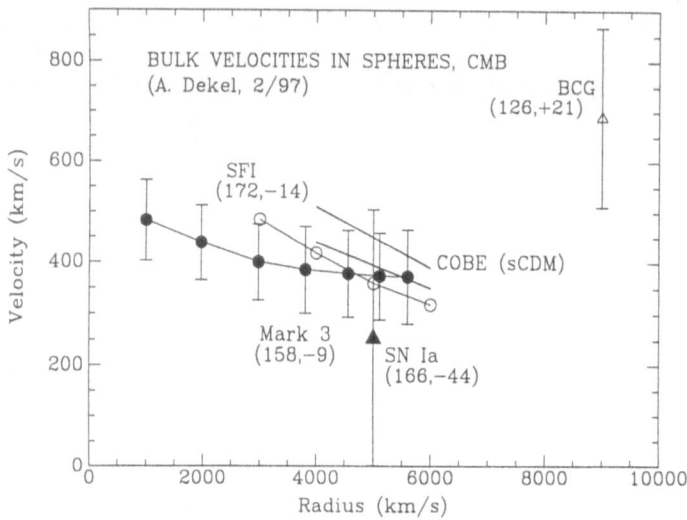

Fig. 5. Bulk velocity. The amplitude of the bulk velocity relative to the CMB frame in top-hat spheres about the LG, as derived from several data sets. The directions are indicated in supergalactic coordinates (L, B). The Mark III data and the SFI data yield consistent results. The new result from Supernovae type Ia on larger scales is a natural extrapolation. The result from brightest galaxies in clusters (LP) is discrepant at more than the $2 - \sigma$ level.

SNe by Riess & Kirshner (1997, following Riess, Press & Kirshner 1995), which extends out to $\sim 300\,h^{-1}$Mpc, shows a bulk flow of $V = 253 \pm 252\,\mathrm{km\,s^{-1}}$ towards $(166°, -44°)$. The effective radius of this data set for a bulk flow fit is in fact less than $50\,h^{-1}$Mpc because the data is weighted inversely by the errors. The SNe bulk flow is consistent with the results from the Mark III and SFI galaxy data. They all make a bulk of sense within the framework of standard isotropic and homogeneous cosmology.

The only apparently discrepant result comes from the velocities measured on a larger scale using brightest cluster galaxies (BCG) as distance indicators (Lauer & Postman 1994, LP). They indicate a large bulk velocity of $V = 689 \pm 178$ towards a very different direction $\sim (126, 21)$. An ongoing effort to measure BCG's in a larger sample of clusters and distances to clusters based on other distance indicators will soon tell whether this early result is a $\gtrsim 2\sigma$ statistical fluke (Watkins & Feldman 1995; Strauss *et al.* 1995), whether the errors were underestimated, or whether something is systematically different between the BCG distances and the distances measured by other indicators.

Shown in comparison are the expected *rms* bulk velocity in a standard CDM model ($\Omega = 1$, $n = 1$) normalized to COBE, for $h = 0.5$ (bottom) and 0.8 (top) (Sugiyama 1995). These theoretical curves would not change much if a 20% hot component is mixed with the cold dark matter, or if Ω is lower but still in the range $0.2 - 1.0$, as long as $n \approx 1$. The main effect of Ω and H_0 on $P(k)$ is via k_{peak}. The predicted bulk velocity over $\sim 100\,h^{-1}$Mpc is effectively an integral of

$P(k)$ over $k < k_{peak}$, and is therefore relatively insensitive to Ω while it is quite sensitive to n. When compared to a theoretical prediction, the error should also include cosmic scatter due to the fact that only one sphere has been sampled from a random field. These errors are typically on the order of the measurement errors.

The measurements of CMB fluctuations on scales $\leq 90°$ are independent of the local streaming motions, but GI predicts an intimate relation between their amplitudes. The CMB fluctuations are associated with fluctuations in gravitational potential, velocity and density in the surface of last scattering at $z \sim 10^3$, while similar fluctuations in our neighborhood have grown by gravity to produce the dynamical structure observed locally. The comparison between the two is therefore a crucial test for GI.

Before COBE, the local streaming velocities served to predict the expected level of CMB fluctuations. The local surveyed region of $\sim 100\, h^{-1}$Mpc corresponds to a $\sim 1°$ patch on the last-scattering surface. An important effect on scales $\geq 1°$ is the Sachs-Wolfe effect (1967), where potential fluctuations $\Delta\Phi_g$ induce temperature fluctuations via gravitational redshift, $\delta T/T = \Delta\Phi_g/(3c^2)$. Since the velocity potential is proportional to Φ_g in the linear and mildly-nonlinear regimes, $\Delta\Phi_g \sim V x$, where x is the scale over which the bulk velocity is V. Thus $\delta T/T \geq V x/(3c^2)$. A typical bulk velocity of $\sim 300\,\mathrm{km\,s^{-1}}$ across $\sim 100\, h^{-1}$Mpc (§ 3.1) corresponds to $\delta T/T \geq 10^{-5}$ at $\sim 1°$. If the fluctuations are roughly scale-invariant ($n = 1$), then $\delta T/T \geq 10^{-5}$ is expected on all scales $> 1°$. Bertschinger, Gorski & Dekel (1990) produced a crude $\delta T/T$ map of the local region as seen by a hypothetical distant observer, and predicted $\delta T/T \geq 10^{-5}$ from the local potential well associated with the GA. An up-to-date version of the $\delta T/T$ maps is provided by Zaroubi et al. (1997b), who added a proper treatment of the acoustic effects on sub-degree scales for various cosmological models.

Now that CMB fluctuations of $\sim 10^{-5}$ have been detected practically on all the relevant angular scales, the argument can be reversed: if one assumes GI, then the *expected* bulk velocity in the surveyed volume is $\sim 300\,\mathrm{km\,s^{-1}}$, i.e., the inferred motions of § 2 are most likely real. If, alternatively, one accepts the peculiar velocities as real for other reasons, then their consistency with the CMB fluctuations is a relatively sensitive and robust test of the validity of GI. This test is unique in the sense that it addresses the specific fluctuation growth rate as predicted by GI theory (§ 5.1). It is robust in the sense that it is quite insensitive to the values of the cosmological parameters and is independent of the complex issues involved in the process of galaxy formation.

3.2 Power Spectrum from the Velocity Field via POTENT

One way to compute the power spectrum is via the smoothed mass density field as recovered by POTENT (Kolatt & Dekel 1997). The key is to correct the result for systematic deviations from the true $P(k)$. The data suffers from distance errors and sparse, nonuniform sampling, and they were heavily smoothed. The $P(k)$ is computed from within a window of effective radius $\sim 50\, h^{-1}$Mpc, say, where the densities are weighted inversely by the squares of the local errors. The density field is zero-padded in a larger periodic box in order to enable an FFT procedure. The $P(k)$ is computed by averaging the amplitudes of the Fourier transforms in bins of k. This procedure yields an "observed" $P(k)$, which we term $O(k)$.

The systematic errors in the above procedure are then modeled by $O(k) = M(k)[S(k) + N(k)]$, where $S(k)$ is the true signal $P(k)$, $N(k)$ is the noise, and $M(k)$ represents the effects of sampling, smoothing, applying a window etc. The correction functions $M(k)$ and $N(k)$ can be derived from Monte Carlo mock catalogs (§ 2.6). The factor $M(k)$ is derived first by $M^{-1} = S/\langle O \rangle_{no-noise}$, where S here is the known power spectrum built into the simulations, and the averaging is over mock catalogs not perturbed by noise. Then $N(k)$ is computed by $N = M^{-1}\langle O \rangle_{noise} - S$, where the averaging is over noisy mock catalogs.

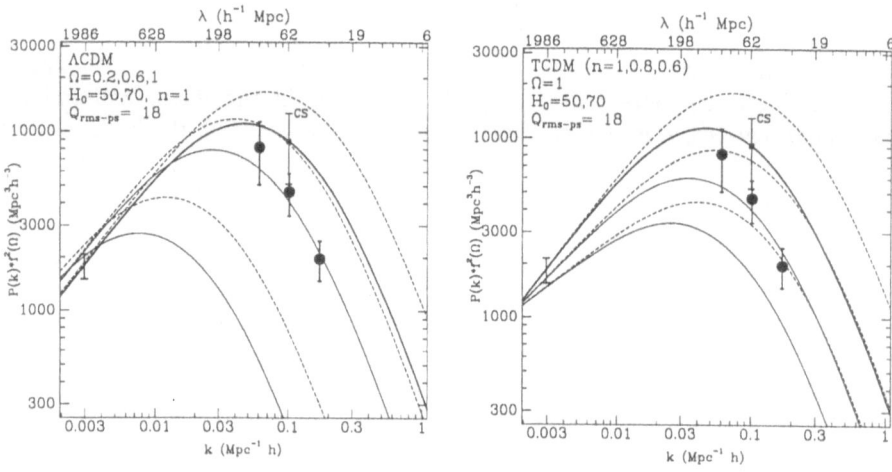

Fig. 6. The mass power spectrum $[\times f(\Omega)^2]$ from POTENT Mark III velocities (filled symbols), with 1σ random errors. The curves are COBE-normalized theoretical predictions for flat CDM with $h = 0.5$ (solid) and $h = 0.7$ (dashed). Left: Λ-CDM with $n = 1$ and Ω growing upwards. Right: tilted-CDM with $\Omega = 1$ and n growing upwards. Typical cosmic scatter (CS) for $\Omega = 1$ and $n = 1$ is indicated (Kolatt & Dekel 1997).

Equipped with the correction functions $M(k)$ and $N(k)$, the $P(k)$ observed from the real universe, $O(k)$, is corrected to yield the true $P(k)$ by $S(k) = M(k)^{-1}O(k) - N(k)$. The recovered mass-density $P(k)$ is shown in Figure 6 in three thick logarithmic bins covering the range $0.04 \leq k \leq 0.2\,(h^{-1}\mathrm{Mpc})^{-1}$, within which the results are reliable. The robust result is $P(k)f(\Omega)^2 = (4.6 \pm 1.4) \times 10^3 (h^{-1}\mathrm{Mpc})^3$ at $k = 0.1\,(h^{-1}\mathrm{Mpc})^{-1}$ (using the convention where the Fourier transform is defined with no 2π factors in it's coefficient). The logarithmic slope at $k = 0.1$ is -1.45 ± 0.5. This translates to $\sigma_8\,\Omega^{0.6} \simeq 0.7 - 0.8$, depending on where the peak in $P(k)$ is (see § 3.3).

The observed $P(k)$ is compared in the figure to the linear predictions of a family of Inflation-motivated flat CDM models ($\Omega + \Omega_A = 1$), normalized by the 4-year COBE data, with the Hubble constant fixed at $h = 0.5$ or $h = 0.7$. For $n = 1$, maximum likelihood is obtained at $\Omega \simeq 0.7 h_{50}^{-1.3} \pm 0.1$. For $\Omega = 1$, assuming no tensor fluctuations, the linear power index is $n \simeq 0.75 h_{50}^{-0.8} \pm 0.1$.

3.3 Power Spectrum from Velocities for COBE-CDM Models

The power spectrum, in a parametric form including Ω, h and n among the parameters, has alternatively been determined from the velocity data via a Baysian likelihood analysis (Zaroubi et al. 1997a; see also Kaiser & Jaffe 1995). According to Bayse, the probability of the model parameters (m) given the data (d), which is the function one wishes to maximize, can be expressed as $P(m|d) = P(d|m) P(m)/P(d)$. The probability $P(d)$ serves here as a normalization constant. Without any external constraints on the model parameters, one assumes that $P(m)$ is a constant in a given range. The remaining task is to maximize the likelihood $\mathcal{L} = P(d|m)$ as a function of the model parameters. This function can be written down explicitly.

Under the assumption that the velocities and the errors are both Gaussian random fields with no mutual correlations, the likelihood can be written as $\mathcal{L} = (2\pi|D|)^{-1/2} \exp(-d_i D_{ij}^{-1} d_j/2)$, where d_i are the data at points $i = 1, ..., N$, and D_{ij} is the covariance matrix, which can be split into covariance of signal (s) and covariance of noise (n), $D_{ij} \equiv \langle d_i d_j \rangle = \langle s_i s_j \rangle + \langle n_i n_j \rangle$. If the errors are uncorrelated, the noise matrix is diagonal. The signal matrix is computed from the model $P(k)$ as a function of the model parameters.

Zaroubi et al. (1997a) used a parametric model for the PS of the general form $P_k = A k^n T(\Gamma_i; k)$, where $T(k)$ is a small-scale filter of an assumed shape characterized by free parameters Γ_i, k^n is the initial $P(k)$ which is still valid on large scales today, and A is a normalization factor. The normalization can either be determined by COBE's data (for given Ω, A, h, n and tensor/scalar fluctuations), or be left as a free parameter to be fixed by the velocity data alone. The filter $T(k)$ can either be taken from a specific physical model (e.g., CDM, where $\Gamma = \Omega h$), or be an arbitrary function with enough flexibility to fit the data.

The robust result for all the models is a relatively high amplitude, with $P(k)f(\Omega)^2 = (4.8 \pm 1.5) \times 10^3 (h^{-1}\text{Mpc})^3$ at $k = 0.1 (h^{-1}\text{Mpc})^{-1}$. An extrapolation to smaller scales using the different CDM models yields $\sigma_8 \Omega^{0.6} = 0.88 \pm 0.15$ (for the dispersion in top-hat spheres of radius $8 h^{-1}\text{Mpc}$).

Within the general family of CDM models, allowing for a cosmological constant in a flat universe and a tilt in the spectrum, the parameters are confined by a 90% likelihood contour of the sort $\Omega h_{50}^{\mu} n^{\nu} = 0.8 \pm 0.2$, where $\mu = 1.3$ and $\nu = 3.4, 2.0$ for models with and without tensor fluctuations respectively. Figure 7 displays the likelihood map in the $\Omega - n$ plane for these models. For open CDM the powers are $\mu = 0.95$ and $\nu = 1.4$ (no tensor fluctuations). A Γ-shape model free of COBE normalization yields only a weak constraint: $\Gamma = 0.4 \pm 0.2$ (where Γ is not necessarily Ωh).

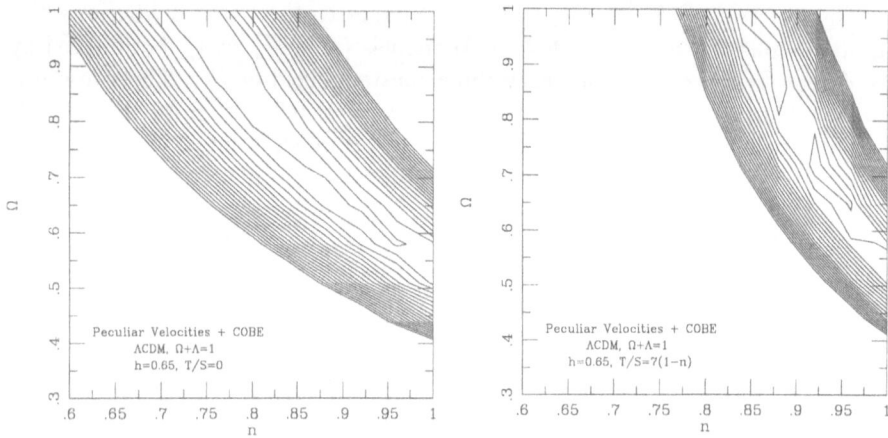

Fig. 7. Likelihood contour maps in the $\Omega - n$ plane for flat CDM with and without tensor fluctuations and $h = 0.65$. Contour spacing is unity in log-likelihood. Under the assumption of a χ^2 distribution, the two-dimensional 90 percentile corresponds to 2.3 contours, and the one-dimensional 90 percentile corresponds to 1.35 contours (Zaroubi *et al.* 1997a).

Both Ω and n obtained by the likelihood analysis from the raw peculiar velocities tend to be slightly higher ($\sim 20\%$) than their estimates based on the $P(k)$ recovered from the POTENT output. This difference may arise from the different relative weighting assigned to the different wavelengths in the two analyses. The difference between the results obtained in the two different ways is on the order of the errors in each analysis and the cosmic scatter. Very similar estimates of $P(k)$ are obtained from a preliminary analysis of the SFI sample of Sc galaxies (in preparation).

In summary: The "standard" CDM model is marginally rejected at the $\sim 2\sigma$ level, while each of the following modifications lead to a good fit to the peculiar velocities and large-scale CMB data: $n \lesssim 1$, $\Omega_\nu \sim 0.3$, or $\Omega \lesssim 1$. The strong implication on the dark matter issue is that values of Ω as low as ~ 0.2 are ruled out with high confidence (independent of Λ), leaving, in particular, no room for the baryonic PIB model.

3.4   Peculiar Velocities vs Small-Scale CMB Fluctuations

Sub-degree angular scales at the last scattering surface correspond to the $\leq 100\, h^{-1}\text{Mpc}$ comoving scales explored by peculiar velocities today. Thus, under the assumption that the local neighborhood is typical, the power spectrum on these scales is simultaneously constrained by the mass-density fluctuations in our cosmological neighborhood and by the CMB fluctuations.

The sub-degree CMB fluctuations are now being explored by many balloon-born experiments, and in less than a decade we expect accurate results from the

CMB satellites MAP and Planck. These measurements will eventually allow a simultaneous likelihood analysis of the two kinds of data. At this point, however, although there are already preliminary detections of the first acoustic peak in the angular power spectrum, the uncertainties are still large. Any current comparison is therefore limited to the semi-quantitative level. The range of parameters permitted by the peculiar velocity data for power spectra of the CDM family (Zaroubi *et al.* 1997a) can be translated to a range of angular power spectra, C_l. This range is plotted against current observations in Figure 8.

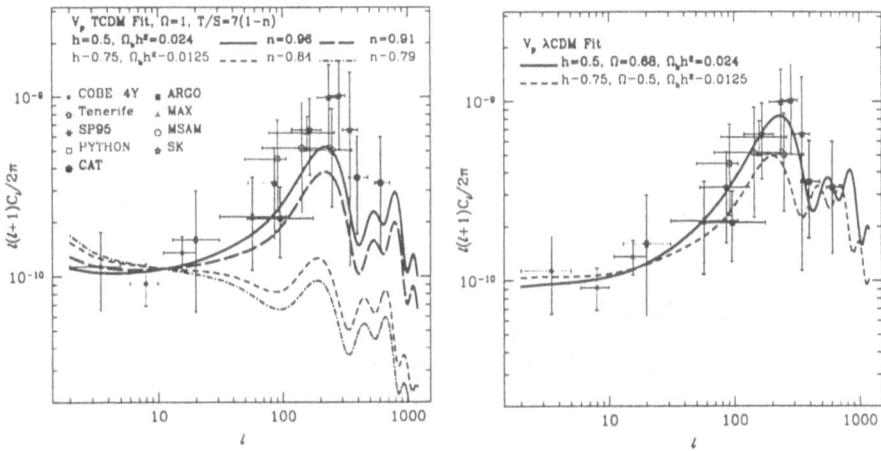

Fig. 8. Angular power spectrum in the CMB. The current data (symbols) are compared to certain CDM models that fit the peculiar velocity data at 90% likelihood. The extreme models shown are either with low h and high $\Omega_b h^2$ or vice versa. Left: tilted CDM with $\Omega = 1$, including tensor fluctuations. Right: Flat CDM with $n = 1$.

The immediate conclusions from a visual inspection of the figure are that a wide range of CDM models can simultaneously obey the two data sets, but there is a subset of models that fits the velocities well but seems to fail to produce a high enough acoustic peak in the CMB spectrum.

The acoustic peak in the CMB is sensitive to Ω_b and the observations prefer a high baryon content, $\Omega_b h^2 \sim 0.025$ (similar to the value measured by Tytler *et al.* 1996; Burles & Tytler 1996), while the peculiar velocities have little to add because $P(k)$ is hardly affected by Ω_b.

The power index n is important in both cases; the peculiar velocities allow values of n significantly lower than unity, but the current CMB data seem not to tolerate values of n below 0.9 or so.

The peculiar velocity data prefers $\Omega \geq 0.4$, and the location of the first acoustic peak in the sub-degree CMB data indicates in agreement a high value of $\Omega + \Omega_\Lambda$.

4   Direct Measurements of Ω from Peculiar Velocities

Assuming that the inferred motions are real and generated by GI, they can be used to estimate Ω in several different ways. Most of the evidence from virialized systems on scales $\leq 10\,h^{-1}$Mpc suggest a low mean density of $\Omega \sim 0.2$ (see Dekel, Burstein & White 1997). The spatial *variations* of the large-scale velocity field provide ways to measure the mass density in a larger volume that may be closer to a "fair" sample. One family of such methods is based on comparing the dynamical fields derived from velocities to the fields derived from galaxy redshifts (§ 5). These methods can be applied in the linear regime but they always rely on the assumed biasing relation between galaxies and mass often parameterized by b, so they actually provide an estimate of $\beta \equiv f(\Omega)/b$. Another family of methods measures β from redshift surveys alone, based on z-space deviations from isotropy (see Strauss 1997). In the present section, we focus first on methods that rely on non-linear effects in the peculiar velocity data *alone*, and they thus provide estimates of Ω independent of galaxy density biasing. These methods are based on the assumption that the initial fluctuations were Gaussian.

4.1   Divergence in Voids

A diverging flow in an extended low-density region can provide a robust dynamical lower bound on Ω, based on the fact that large gravitating outflows are *not* expected in a low-Ω universe (Dekel & Rees 1994). In practice, for any assumed value of Ω, the partial derivatives of the smoothed observed velocity field are used to infer a non-linear approximation for the mass density via the approximation δ_c (6). A key point is that this approximation is typically an overestimate, $\delta_c > \delta$ (when the true value of Ω is used). For fluctuations that started Gaussian, the probability that δ_c is an overestimate, in the range $\delta < -0.5$, is well over 99%. Analogously to $\delta_0 \approx -\Omega^{-0.6}\nabla \cdot v$, the δ_c inferred from a given diverging velocity field is more negative when a smaller Ω is assumed, so for a small enough Ω one may obtain $\delta < -1$ in certain void regions. Such values of δ are forbidden because mass is never negative, so this provides a lower bound on Ω.

The inferred δ_c field, smoothed at $12\,h^{-1}$Mpc, and the associated error field σ_δ, were derived by POTENT at grid points from the observed radial velocities of Mark III. Focusing on the deepest density wells, the input Ω was lowered until δ_c became significantly smaller than -1. The most promising "test case" provided by the Mark III data is a broad diverging region centered near the supergalactic plane at the vicinity of $(X, Y) = (-25, -40)$ in h^{-1}Mpc — the "Sculptor void" of galaxies between the GA and the "Southern Wall" extension of PP (Figure 9, compare to Fig. 4).

Values of $\Omega \approx 1$ are perfectly consistent with the data, but δ_c becomes smaller than -1 already for $\Omega = 0.6$. The values $\Omega = 0.3$ and 0.2 are ruled out at the 2.4σ and 2.9σ levels in terms of the random error σ_δ.

This result is still to be improved. The systematic errors have been partially corrected in POTENT, but a more specific investigation of the biases affecting

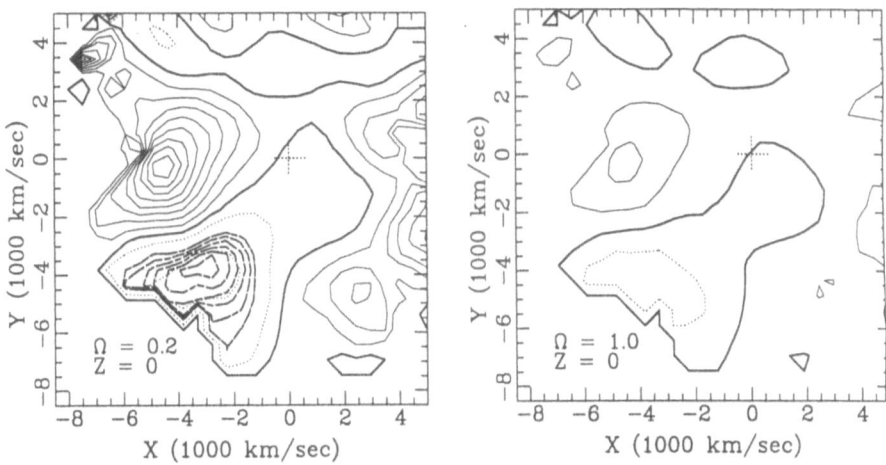

Fig. 9. Maps of δ_c inferred from the observed velocities near the Sculptor void in the Supergalactic plane, for two values of Ω. The LG is marked by '+' and the void is confined by the Pavo part of the GA (left) and the Aquarius extension of PP (right). Contour spacing is 0.5, with $\delta_c = 0$ heavy, $\delta_c > 0$ solid, and $\delta_c < 0$ dotted. The heavy-dashed contours mark the illegitimate downward deviation of δ_c below -1 in units of σ_δ, starting from zero (i.e., $\delta_c = -1$) and decreasing with spacing $-0.5\sigma_\delta$. The value $\Omega = 0.2$ is ruled out at the 2.9σ level (Dekel & Rees 1994).

the smoothed velocity field in density wells is still in progress. For the method to be effective one needs to find a void that is (a) bigger than the correlation length for its vicinity to represent the universal Ω, (b) deep enough for the lower bound to be tight, (c) nearby enough for the distance errors to be small, and (d) properly sampled to trace the velocity field in its vicinity.

Note that this method does not require that the void be spherical or of any other particular shape, and is independent of galaxy density biasing. Another pro is that there is no much cosmic scatter — one deep and properly sampled void is enough for a meaningful constraint. The main limitation is the poor (and perhaps biased) sampling of the velocity field in the vicinity of a void.

4.2 Deviations from Gaussian PDF

Assuming that the initial fluctuations are a Gaussian random field, the one-point probability distribution function (PDF) of smoothed density develops a characteristic skewness due to non-linear effects early in the non-linear regime (e.g., Kofman et al. 1994). The skewness of δ is given according to second-order perturbation theory by

$$\langle \delta^3 \rangle / \langle \delta^2 \rangle^2 \approx (34/7 - 3 - n) , \tag{9}$$

with n the effective power index of the power spectrum near the (top-hat) smoothing scale (Bouchet et al. 1992). Since this ratio of moments for δ is practically independent of Ω, and since $\nabla \cdot v \sim -f\delta$, the corresponding ratio for $\nabla \cdot v$

must strongly depend on Ω, and indeed in second-order it is (Bernardeau *et al.* 1995)

$$T_3 \equiv \langle (\nabla \cdot \boldsymbol{v})^3 \rangle / \langle (\nabla \cdot \boldsymbol{v})^2 \rangle^2 \approx -f(\Omega)^{-1}(26/7 - 3 - n) \ . \tag{10}$$

Using N-body simulations and $12\,h^{-1}$Mpc smoothing one indeed finds $T_3 = -1.8 \pm 0.7$ for $\Omega = 1$ and $T_3 = -4.1 \pm 1.3$ for $\Omega = 0.3$, where the error is the cosmic scatter for a sphere of radius $40\,h^{-1}$Mpc in a CDM universe ($H_0 = 75$, $b = 1$). An estimate of T_3 in the current POTENT velocity field within $40\,h^{-1}$Mpc is -1.1 ± 0.8, where the errors this time represent distance errors. With the two errors added in quadrature, $\Omega = 0.3$ is rejected at the $\sim 2\sigma$ level (somewhat sensitive to the assumed $P(\boldsymbol{k})$).

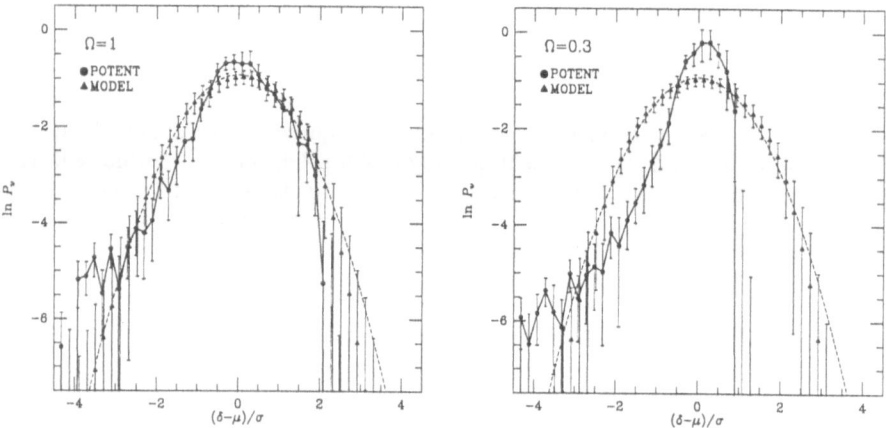

Fig. 10. Ω from IPDF. The density IPDF recovered from the G2 POTENT peculiar velocity field (solid), compared to a normal distribution (short dash) and with the IPDF recovered from the velocity field of Gaussian CDM simulations (triangles). The assumed Ω is 1.0 (left) or 0.3 (right). The simulations are of $\Omega_0 = \Omega$ accordingly (Nusser & Dekel 1993).

Since the present PDF contains only part of the information stored in the data and is in some cases not that sensitive to the initial PDF (IPDF), a more powerful bound can be obtained by using the detailed present velocity field $\boldsymbol{v}(\boldsymbol{x})$ to recover the IPDF, and using the latter to constrain Ω by measuring it's Ω-dependent deviation from the assumed normal distribution (Nusser & Dekel 1993). The necessary "time machine" is provided by the Eulerian interpretation of the Zel'dovich approximation (Nusser & Dekel 1992).

The velocity out of POTENT Mark II, within a conservatively selected volume, was fed into the IPDF recovery procedure with Ω either 1 or 0.3, and the errors due to distance errors and cosmic scatter were estimated. Figure 10 shows the recovered IPDF's. The IPDF recovered for $\Omega = 1$ is marginally consistent with Gaussian, while the one recovered for $\Omega = 0.3$ shows significant deviations. The largest deviation, bin by bin in the IPDF, is $\lesssim 2\sigma$ for $\Omega = 1$ and $> 4\sigma$ for $\Omega = 0.3$, and a similar rejection of $\Omega = 0.3$ is obtained with a χ^2-type statistic. The skewness and kurtosis are poorly determined because of noisy tails but the replacements $\langle x|x|\rangle$ and $\langle|x|\rangle$ allow a rejection of $\Omega = 0.3$ at the $(5-6)\sigma$ levels.

The main advantage of the methods based on the PDF is their insensitivity to galaxy density biasing. The main weakness is the need for a "fair" sample; the cosmic scatter is large due to the large smoothing scale within the limited volume.

5   Galaxy Density vs Velocities: Ω and Biasing

5.1   Galaxies vs Mass: Fit of GI and Linear Biasing Model

The theory of GI combined with the assumption of linear biasing for galaxies predict a correlation between the dynamical density field and the galaxy density field, which can be addressed quantitatively based on the mock catalogs and the estimated errors in the two data sets. Figure 11 compares density maps in the Supergalactic plane for IRAS 1.2 Jy galaxies (δ_G) and POTENT Mark III mass (δ), both G12 smoothed. The general correlation is evident — the GA, PP, Coma and the voids all exist both as dynamical entities and as structures of galaxies. To evaluate goodness of fit, Figure 12 shows the statistic $\chi^2 = N^{-1} \sum^N (\delta_G - b\delta)^2/\sigma^2$ as computed from the data in comparison with its distribution over pairs of Mark III and IRAS 1.2 Jy mock catalogs. The fact that the data lies near the center of this distribution indicates that the two data sets are consistent with being noisy versions of an underlying fluctuation field and that the data are in agreement with the hypotheses of GI plus linear biasing (Dekel *et al.* 1993; Sigad *et al.* 1997; more in § 5.2).

What is it exactly that one can learn from the observed $v - \delta_G$ correlation (Babul *et al.* 1994)? First, it argues that the velocities are real because it is hard to invoke any other reasonable way to make the galaxy distribution and the TF measurements agree so well. On the other hand, although it is true that gravity is the only long-range force that could attract galaxies to stream toward density concentrations, the fact that a $v - \delta_G$ correlation is predicted by GI plus linear biasing does not necessarily mean that the observation can serve as a sensitive test for either. Recall that converging (or diverging) flows tend to generate overdensities (or underdensities) simply as a result of mass conservation, independent of the source of the motions.

Let us assume for a moment that galaxies trace mass, *i.e.*, that the linearized continuity equation, $\dot{\delta} = -\nabla \cdot v$, is valid for the galaxies as well. The observed correlation (in the linear approximation) is then $\delta \propto -\nabla \cdot v$, and together they imply that $\delta \propto \delta$, or equivalently that $\nabla \cdot v$ is proportional to its time average. This property is not exclusive to GI; one can construct counterexamples where the velocities are produced by a non-GI impulse.

Even irrotationality does not follow from $\delta \propto -\nabla \cdot v$; it has to be adopted based on theoretical arguments in order to enable reconstruction from radial

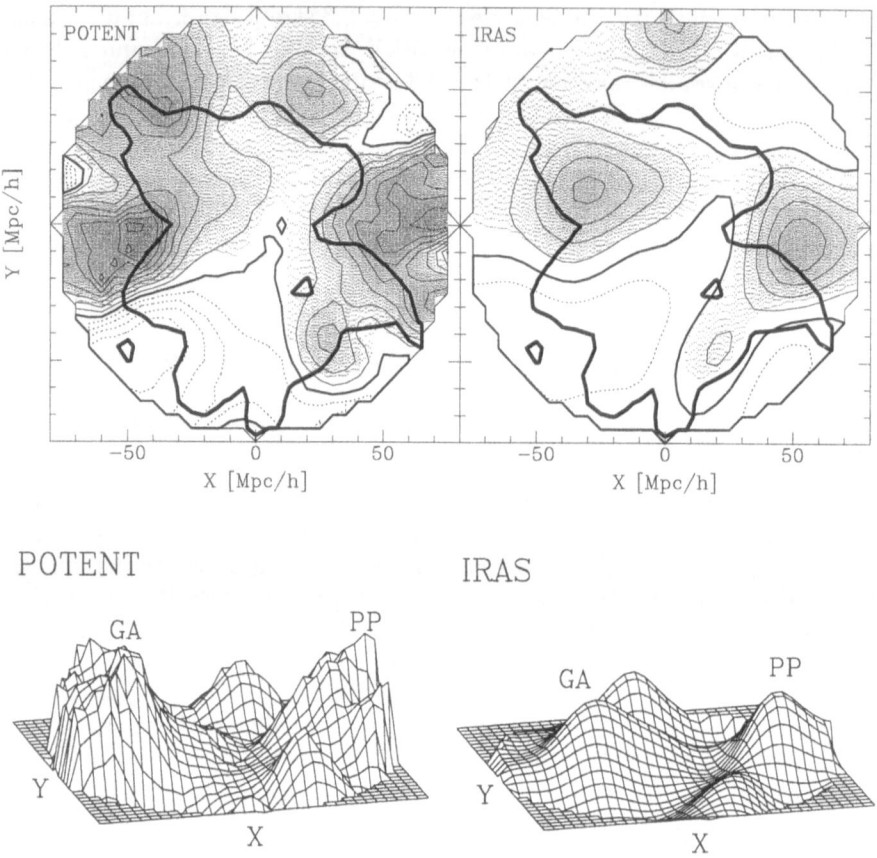

Fig. 11. Mass versus galaxies. POTENT mass ($\Omega = 1$) versus IRAS galaxy density fields in the Supergalactic plane, both smoothed G12. Contour spacing is 0.2. The heavy contour marks the boundary of the comparison volume of effective radius $46\,h^{-1}$Mpc. The height in the surface plot is proportional to δ. The LG is at the center, GA on the left, PP on the right, and the Sculptor void in between (Sigad *et al.* 1997).

velocities or from observed densities. Once continuity and irrotationality are assumed, the observed $\delta \propto -\nabla \cdot v$ implies a system of equations which is identical in all its *spatial* properties to the equations of GI, but can differ in the constants of proportionality and their temporal behavior. It is therefore impossible to distinguish between GI and a non-GI model which obeys continuity plus irrotationality based only on snapshots of *present-day* linear fluctuation fields. This makes the relation between CMB fluctuations and velocities an especially important test for GI.

On the other hand, the fact that the constant of proportionality in $\delta \propto -\nabla \cdot v$ is indeed the same everywhere is a non-trivial requirement from a non-GI model. For example, a version of the explosion scenario (Ostriker & Cowie 1981; Ikeuchi

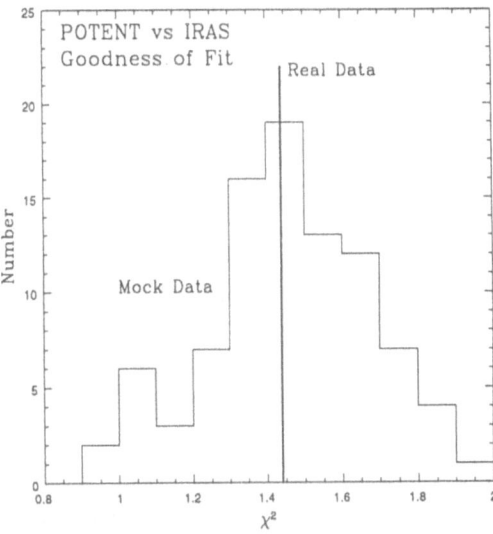

Fig. 12. Goodness of fit in the comparison of G12 density fields of POTENT mass and IRAS galaxies, as tested by a χ^2 statistic that is an error weighted sum of differences between the two fields (Sigad *et al.* 1997).

1981), which tested successfully both for irrotationality and $v - \delta$ correlation, requires special synchronization among the explosions (Babul *et al.* 1994).

So, what is the $v - \delta_G$ relation good for? While it's sensitivity to GI is only partial, this relation turns out to be quite sensitive to the validity of a *continuity-like* relation for the *galaxies*. When the latter is strongly violated all bets are off for the $v - \delta_G$ relation. A non-linear biasing scheme would make continuity invalid for the galaxies, which would ruin the $v - \delta_G$ relation even if GI is valid. The observed correlation is thus a sensitive test for density *biasing*. It implies, subject to the errors, that the $\sim 12\, h^{-1}$Mpc-smoothed density fields of galaxies and mass are related via a biasing relation that could be crudely approximated by a linear relation with b of order unity (but see a refinement of this in § 5.5).

Now that the data of peculiar velocities and the data of galaxy density are found to be compatible with the model of GI and linear biasing, they can be combined to constrain the degenerate parameter $\beta \equiv \Omega^{0.6}/b$. The comparison between the two data sets can be done in several different ways. In particular, it could be done by comparing density fields derived locally from the two data sets (§ 5.2), or by comparing velocities derived from the two data sets (§ 5.4). It can be done successively by first recovering fields from each data set and then combining them to obtain β, or by a simultaneous recovery of fields and beta determination from the two data sets (§ 5.3). It can be done by direct comparison of fields in r-space, or by comparing coefficients mode by mode in a model expansion in z-space (Davis, Nusser & Willick 1996).

5.2 Density-Density Comparison on Large Scales: POTIRAS

The main advantage of comparing densities is that they are *local*. The densities are independent of long-range effects due to the unknown mass distribution outside the sampled volume, which could affect the velocities. The densities are also independent of reference frame, and can be reasonably corrected for non-linear effects.

The POTENT analysis extracts from the peculiar velocity data a mildly-nonlinear mass density fluctuation field in a spatial grid, smoothed G12 (§ 2.5). The associated real-space density field of galaxies can be extracted with similar smoothing from a whole-sky redshift survey such as the IRAS 1.2 Jy survey (see Strauss & Willick 1995; Sigad *et al.* 1997).

A brief summary of the recovery of the IRAS density field is as follows. The solution to the linearized GI equation $\nabla \cdot v = -f\delta$ for an irrotational field is

$$v(x) = \frac{f}{4\pi} \int_{\text{all space}} d^3x' \, \delta(x') \frac{x' - x}{|x' - x|^3} \, . \tag{11}$$

The velocity is proportional to the gravitational acceleration, which ideally requires full knowledge of the distribution of mass in space. In practice, one is provided with a flux-limited, discrete redshift survey, obeying some radial selection function $\phi(r)$. The galaxy density is estimated by $1 + \delta_G(x) = \sum n^{-1}\phi(r_i)^{-1} \delta^3_{dirac}(x - x_i)$, where $n \equiv V^{-1}\sum \phi(r_i)^{-1}$ is the mean galaxy density, and the inverse weighting by ϕ restores the equal-volume weighting. Eq. 11 is then replaced by

$$v(x) = \frac{\beta}{4\pi} \int_{r < R_{\max}} d^3x' \, \delta_G(x') \, S(|x' - x|) \frac{x' - x}{|x' - x|^3} \, . \tag{12}$$

Under the assumption of linear biasing, the cosmological dependence enters through β. The integration is limited to $r < R_{max}$ where the signal dominates over shot-noise. $S(y)$ is a small-scale smoothing window ($\geq 500 \, \text{km s}^{-1}$) essential for reducing the effects of non-linear gravity, shot-noise, distance uncertainty, and triple-value zones.

The distances are estimated from the redshifts in the LG frame by

$$r_i = z_i - \hat{x}_i \cdot [v(x_i) - v(0)] \, . \tag{13}$$

Equations 12-13 can be solved iteratively: make a first guess for the x_i, compute the v_i by Eq. 12, correct the x_i by Eq. 13, and so on until convergence. The convergence can be improved by increasing β gradually during the iterations from zero to its desired value.

Even under $12 \, h^{-1}$Mpc smoothing, δ_G is of order unity in places, necessitating a mildly-nonlinear treatment. Local approximations from v to δ were discussed in § 2.5, but the non-local nature of the inverse problem makes it less straightforward. A possible solution is to find an inverse relation of the sort $\nabla \cdot v = F(\Omega, \delta_G)$, including non-linear gravity and non-linear biasing. This is a Poisson-like equation in which $-\beta\delta_G(x)$ is replaced by $F(x)$, and since the smoothed velocity field is still irrotational for mildly-nonlinear perturbations, it can be integrated analogously to Eq. 11. With smoothing of $10\,h^{-1}\mathrm{Mpc}$ and $\beta = 1$, the approximation based on an empirical inverse to δ_c (6) has an *rms* error $< 50\,\mathrm{km\,s^{-1}}$.

In recent applications, the galaxy density field is recovered from the noisy IRAS data via a Power-Preserving Filter (PPF, by A. Yahil, described in Sigad *et al.* 1997) — a modification of the Wiener Filter. The PPF returns a field that is not far from the Wiener, most probable field, but it makes the result more realistic by forcing the variance to be constant in space despite the fact that the errors vary.

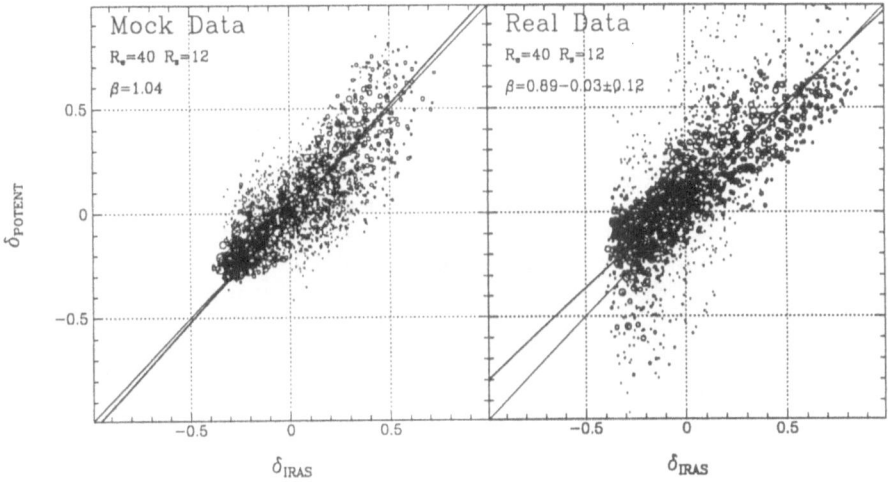

Fig. 13. β from POTENT vs IRAS density comparison. The smoothing is G12 and the comparison volume is of effective radius $40\,h^{-1}\mathrm{Mpc}$. Two-dimensional regression lines are marked. Left: Regression between the averages of 20 mock catalogs of each type, showing a bias as small as 4%. Right: Real data (Sigad *et al.* 1997).

The simplest way of comparing the POTENT and IRAS density fields is via a two-dimensional linear regression using the values of the fields at grid points within a local comparison volume. The errors of both fields enter the regression. The comparison volume is determined by equal-error contours.

The latest comparison of POTENT Mark III and IRAS 1.2 Jy data at $12\,h^{-1}\mathrm{Mpc}$ smoothing within a volume of $(65\,h^{-1}\mathrm{Mpc})^3$ yields $\beta_I = 0.86 \pm 0.12$ (Sigad *et al.* 1997). The corresponding scatter diagram is shown in Figure 13. The systematic error in this derivation, of only 4%, is deduced from the analogous scatter diagram for the averages of 20 random mock catalogs of each type. This is an update of the higher estimate $\beta_I = 1.3 \pm 0.3$ (Dekel *et al.* 1993) obtained based on an earlier version of POTENT with the Mark II velocities and the IRAS 1.9 Jy redshifts.

Similar comparisons of the mass density field with the density of optical galaxies indicate a similar correlation and a $\sim 30\%$ lower estimate for β_o (Hudson et al. 1995), in agreement with the ratio of biasing factors, $b_o/b_I \approx 1.3$, obtained by direct comparison of optical and IRAS galaxy densities.

A comparison of similar nature of the POTENT Mark III data with the density distribution of Abell/ACO $R \geq 0$ clusters and the corresponding predicted velocities at G15 smoothing yields similar consistency out to distances of $\sim 60\,h^{-1}$Mpc, and an estimate of $\beta_c = 0.26 \pm 0.11$ (Plionis et al. 1997). This is consistent with a linear biasing factor for the clusters that is about 4 times larger than that of galaxies, in accordance with the observed ratio of about 4^2 for the corresponding correlation functions (see Bahcall 1997).

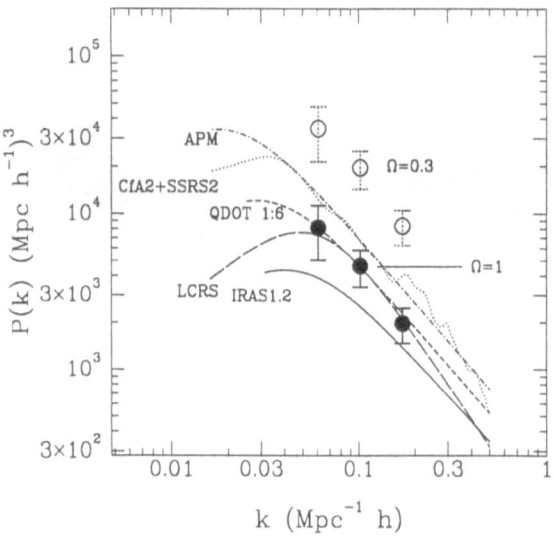

Fig. 14. β from power spectra of galaxies versus mass. The estimates from various galaxy density samples were all translated from redshift to real space using Kaiser's approximation and the best-fit value of β. The $P(k)$ from peculiar velocities (Fig. 6) is marked by solid symbols for $\Omega = 1$ (open symbols for $\Omega = 0.3$). The values of β (for any Ω) can be read directly from the vertical offset of the solid symbols and the corresponding curves (Kolatt & Dekel 1997).

A direct comparison of the mass power spectrum as derived from peculiar velocities (§ 3.2) with the galaxy power spectra as derived from different redshift and angular surveys is shown in Figure 14 (Kolatt & Dekel 1997). It demonstrates a similarity in shape and yields for the various galaxy types β values in the range $0.77 - 1.21$, with a typical error of ± 0.1. For IRAS galaxies typically $\beta \gtrsim 1$, and for optical galaxies $\beta \lesssim 1$. These estimates do not directly address the value of Ω, but it is clear from the figure that if Ω is as small as ~ 0.3, then all galaxy types must be severely antibiased.

In principle, the degeneracy of Ω and b is broken in the mildly-nonlinear regime, where $\delta(v)$ is no longer $\propto f^{-1}$. Compatible mildly-nonlinear corrections in POTENT and in the IRAS analysis allowed a preliminary attempt to separate these parameters using the Mark II and IRAS 1.9 Jy data (Dekel et al. 1993). Unfortunately, nonlinear biasing effects are hard to distinguish from non-linear gravitational effects, so a specific nonlinear biasing scheme is a prerequisite for such an analysis.

5.3   Simultaneous Fit of Velocity and Density: SIMPOT

The dynamical fields and β can be recovered simultaneously by a fit of a parametric model for the potential field to the combined data of the observed radial peculiar velocities and the distribution of galaxies in redshift space. This procedure takes advantage of the complementary features of the data in the recovery of the fields, it enforces the same effective smoothing on the data without preliminary reconstruction procedures such as POTENT, and it obtains a more reliable best fit by simultaneous rather than successive minimization. It has been implemented so far to the forward TF data, but it can be generalized in principle to minimize inverse TF residuals.

In the SIMPOT procedure by Nusser & Dekel (1997), the model for the potential field is taken to be an expansion in spherical harmonics Y_{lm} and Bessel functions j_l where the coefficients Φlmn are the free parameters,

$$\Phi(r) = \sum_{l=1}^{l_{max}} \sum_{m=-l}^{l} \sum_{n=1}^{n_{max}} \Phi_{lmn} \, j_l(k_n r) Y_{lm}(\hat{r}). \qquad (14)$$

The model radial velocity is derived from this potential by $u = -\partial\Phi/\partial r$, and the model density in redshift space is derived using linear theory via $\delta = f^{-1}\nabla^2\Phi + s^{-2}(\partial/\partial s)(s^2 \partial\Phi/\partial s)$. The second term reflects redshift distortions, where s is the radial variable in redshift space. The resulting models for $u(x)$ and $\delta(s)$ are expansions in certain functions A_{lmn} and B_{lmn} that are appropriate combinations of the original base functions.

The combined χ^2 to minimize as a function of the parameters Φ_{lmn} and β is the sum of

$$\chi_u^2 = \sum_i \sigma_{ui}^{-2} \left[u_i^{obs} - \sum_{lmn} A_{lmn}(r_i)\Phi_{lmn} \right]^2 \qquad (15)$$

and

$$\chi_\delta^2 = \int d^3 s \, \sigma_\delta^{-2}(s) \left[\delta^{obs}(s) - \sum_{lmn} B_{lnm}(s)\Phi_{lmn} \right]^2. \qquad (16)$$

The observations are the peculiar velocities u_i^{obs} and a continuous density field in redshift space $\delta^{obs}(s)$, that is somewhat more tricky to obtain. The β dependence enters only in χ_δ^2, both via the velocity-density relation and the redshift distortions.

A SIMPOT fit to the Mark III peculiar velocity data and the IRAS 1.2 Jy redshift survey provides first hints for scale-dependent biasing: $\beta_I \approx 0.6$ and 1.0 (± 0.1) for smoothings that are roughly equivalent to Gaussian with radii 6 and $12\,h^{-1}$Mpc respectively.

5.4 Velocity-Velocity Comparison on Small Scales: VELMOD

Earlier comparisons of the peculiar velocities from the Mark II catalog and the velocities predicted from the IRAS redshift surveys (QDOT, 1.9 Jy, and 1.2 Jy) yielded estimates of β_I in the range $0.4 - 1.0$ (Kaiser *et al.* 1991; Nusser & Davis 1994).

The more sophisticated recent VELMOD method of comparison (Willick *et al.* 1997b) compares the raw peculiar velocity data with a "model" velocity field that is predicted from the IRAS 1.2 Jy redshift survey. It is done without attempting to reconstruct a velocity field from the data. The key feature of VELMOD is that it explicitly allows for a non-unique mapping between real space and redshift space. Triple valuedness in the redshift field as well as non-negligible small-scale velocity "temperature" are treated in a unified way. This is done by expressing the probability that an object at distance r has a redshift z by

$$P(z|r) = \frac{1}{\sqrt{2\pi}\sigma_u} \exp\left[-\frac{1}{2}\frac{[z - r - u(r)]^2}{2\sigma_u^2}\right] \tag{17}$$

where $u(r)$ is the radial component of the model velocity field and σ_u is the small-scale velocity noise (which can in principle be a function of position). The above probability is then multiplied by the TF probability factor, $P(m, \eta, r)$, and integrated over the entire line-of-sight to obtain the probability of the *observable* quantities (m, η, z). One then maximizes that probability over the entire data set.

The method is computer-intensive because numerical integrals are required for each galaxy, and for each fit parameter (TF parameters, σ_u, velocity field model parameters, etc.). This effort is worthwhile to the degree that the velocity field is triple-valued or the small-scale noise σ_u is comparable to the TF error. In particular, VELMOD is more rigorous in an analysis of the very local ($z \leq 3000\,\mathrm{km\,s^{-1}}$) region.

VELMOD has been applied with a Gaussian smoothing of $3\,h^{-1}\mathrm{Mpc}$ to the IRAS 1.2 Jy redshift survey and a subset of 838 spiral galaxies from the Mark III catalog within $z \leq 3000\,\mathrm{km\,s^{-1}}$ of the Local Group. The method was tested successfully using mock catalogs drawn from the N-body simulation of Kolatt *et al.* (1996). When applied to the real data it yielded consistency with the model of linear GI and linear biasing once an artificial quadrupole was allowed, with $\beta_I = 0.5 \pm 0.1$ at $3\,h^{-1}\mathrm{Mpc}$. The catch is that it is not at all clear why linear GI and the simplified deterministic biasing should be valid for the densities and velocities at such high resolution. The estimated value of β should therefore be interpreted with caution.

It is interesting to note that Shaya *et al.* (1995) obtained a similarly low value for β_O from the same local neighborhood. They applied the least-action reconstruction method to a redshift survey of several hundred spirals in comparison with TF data. Their method is likely to underestimate β because it assumes that the mass is all concentrated in the centers of galaxies and groups and thus tends to overestimate the gravitational forces between them. This systematic effect is yet to be quantified.

Table 1 summarizes the estimates of β and Ω from cosmic flows.

Table 1. Ω and β from Cosmic Flows

Peculiar	Gaussian IPDF	Nusser & Dekel 93	$\Omega > 0.3 \ (> 4\sigma)$
Velocities	Skewness($\nabla \cdot \boldsymbol{v}$)	Bernardeau *et al.* 94	$\Omega > 0.3 \ (2\sigma)$
Alone	Void	Dekel & Rees 94	$\Omega > 0.3 \ (2.4\sigma)$
	Power spectrum	Kolatt & Dekel 97	$\sigma_8 \Omega^{0.6} = 0.7 \pm 0.15$
	+COBE	Zaroubi *et al.* 97	$\sigma_8 \Omega^{0.6} = 0.8 \pm 0.15$
Galaxy	M2-QDOT v	Kaiser *et al.* 91	$\beta_I = 0.9^{+0.2}_{-0.15}$
Density	M3-I1.2 v-dipole	Nusser & Davis 94	$\beta_I = 0.6 \pm 0.2$
vs.	M3-I1.2 v-inverse	Davis *et al.* 96	$\beta_I = 0.6 \pm 0.2(?)$
Velocities	M3-I1.2 v G3	Willick *et al.* 96	$\beta_I = 0.5 \pm 0.1$
	M3-I1.2 δ G12	Sigad *et al.* 97	$\beta_I = 0.86 \pm 0.15$
	M3-I1.2 δ/v G6-12	Nusser & Dekel 96	$\beta_I = 0.6 - 1.0$ scale
	M2-Optical v	Hudson 94	$\beta_O = 0.5 \pm 0.1$
	TF-Optical	Shaya *et al.* 94	$\beta_O = 0.35 \pm 0.1$
	M3-Optical δ G12	Hudson *et al.* 95	$\beta_O = 0.75 \pm 0.2$
	M3-clusters G15	Plionis *et al.* 97	$\beta_C = 0.26 \pm 0.11$
Redshift	ξ I1.2	Peacock & Dodds 94	$\beta_I = 1.0 \pm 0.2$
Distortions	ξ I1.2	Fisher *et al.* 94a	$\beta_I = 0.45^{+0.3}_{-0.2}$
	Y_{lm} I1.2	Fisher *et al.* 94b	$\beta_I = 1.0 \pm 0.3$
	P_k I1.2, QDOT	Cole *et al.* 95	$\beta_I = 0.5 \pm 0.15$
	ξ I1.2, QDOT	Hamilton 95	$\beta_I = 0.7 \pm 0.2$
	Y_{lm} I1.2	Heavens & Taylor 95	$\beta_I = 1.1 \pm 0.3$
	P_k I1.2	Fisher & Nusser 96	$\beta_I = 0.6 \pm 0.2$
CMB	vs galaxies angular	Yahil *et al.* 86	$\beta_I = 0.9 \pm 0.2$
Dipole	vs galaxies redshift	Strauss *et al.* 92	$\beta_I = 0.4 - 0.85$
		Rowan-Rob. *et al.* 91	$\beta_I = 0.8^{+0.2}_{-0.15}$
	vs galaxies angular	Lynden-Bell *et al.* 89	$\beta_O = 0.3 - 0.5$
	vs galaxies redshift	Hudson 93	$\beta_O = 0.7^{+0.4}_{-0.2}$
	clusters	Scaramella *et al.* 91	$\beta_C \sim 0.13$
		Plionis *et al.* 91	$\beta_C \sim 0.17 - 0.22$

$\beta \equiv \Omega^{0.6}/b$, $b_C : b_O : b_I \approx 4.5 : 1.3 : 1.0$, $\sigma_8 \Omega^{0.6} = (0.69 \pm 0.05)\beta_I$,
M3= Mark III,   I1.2 = IRAS 1.2 Jy,   G12 = Gaussian smoothing $12\,h^{-1}$Mpc

5.5   Galaxy Biasing as a Stochastic Process

In all the methods described in § 5, the cosmological parameter of interest Ω is contaminated by the uncertain relation between galaxy and mass density, the so called "galaxy biasing". Nontrivial galaxy biasing clearly exists. The fact that galaxies of different types cluster differently (Dressler 1980) implies that at least some do not trace the underlying mass. This is hardly surprising because any reasonable physical theory would predict non-trivial biasing (Kaiser 1984; Davis *et al.* 1985; Bardeen *et al.* 1986; Dekel & Silk 1986; Dekel & Rees 1987; Braun, Dekel & Shapiro 1988; Weinberg 1995). In particular, simulations of galaxy formation in a cosmological context (*e.g.*, Cen & Ostriker 1992; 1993; Lemson *et al.* 1997) indicate a biasing relation that is non-linear in density, is varying with scale, and has a statistical scatter reflecting dependencies on factors other than density.

One should therefore not be surprised by the fact that the various estimates of β span a large range, from less than one half to more than unity. Some of this scatter is due to the different types of galaxies involved, and some may be due to remaining effects of non-linear gravity or other systematic errors, but a significant fraction of the scatter in β is likely to reflect non-trivial properties of the biasing scheme. This means that translating a measured β into Ω is non-trivial; it requires a detailed knowledge of the relevant biasing scheme.

In order to strengthen this point, we demonstrate below that an obvious source of systematic variations in β is the inevitable *statistical* scatter in the biasing process (Dekel & Lahav 1997). This scatter in the relation between densities can be interpreted as reflecting the dependence of galaxy formation efficiency, or galaxy density, on physical properties of the protogalaxy environment other than density. These could be local properties such as the potential field, the deformation tensor, tidal effects, and angular momentum, or long-range effects carried by radiation or particles from neighboring sources. In the simple example below, we assume that this scatter in the biasing is local and neglect possible spatial correlations.

Let $\delta(\boldsymbol{x})$ be the field of mass-density fluctuations smoothed with a given window, and let $\delta_G(\boldsymbol{x})$ be the corresponding field for galaxies of a given type. We treat them as random fields, both with probability densities of zero mean by definition. Denote $\langle \delta^2 \rangle \equiv \sigma^2$ and $\langle \delta^3 \rangle \equiv S$. Consider the "biasing" relation between galaxies and mass to be a *random* process, specified by the *conditional probability* function $B(\delta_G | \delta)$. The common deterministic biasing relation, $\delta_G = b(\delta)\delta$, is replaced by the conditional mean,

$$\langle \delta_G | \delta \rangle \equiv b(\delta)\delta. \tag{18}$$

The statistical character of the relation is expressed by the conditional moments of higher order about the mean, such as

$$\langle (\delta_G - b\delta)^2 | \delta \rangle \equiv \sigma_b^2, \quad \text{and} \quad \langle (\delta_G - b\delta)^3 | \delta \rangle \equiv S_b. \tag{19}$$

This statistical nature of biasing leads to a different "biasing parameter" for each specific application.

Take for example the ratio of variances, $b_2^2 \equiv \langle \delta_G^2 \rangle / \langle \delta^2 \rangle$, such as being obtained by a ratio of power spectra or two-point correlation functions, or by comparing the mass function of clusters to the variance of δ_G at $8\,h^{-1}\text{Mpc}$ (White, Efstathiou & Frenk 1993). One can prove in general that $\langle \delta_G^m \rangle = \langle\, \langle \delta_G^m | \delta \rangle_{\delta_G} \rangle_\delta$, and therefore, $\langle \delta_G^2 \rangle = \langle b^2(\delta)\,\delta^2 \rangle + \langle \sigma_b^2(\delta) \rangle$. Thus, in the simple case where $b(\delta)$ is constant, b_2 is an overestimate of b by

$$b_2 = b\,(1 + \Delta_2)^{1/2}, \quad \Delta_2 \equiv \langle \sigma_b^2 \rangle / (\sigma^2 b^2). \tag{20}$$

Another common way of estimating β is via linear regression of the noisy field $-\nabla \cdot \boldsymbol{v}(\boldsymbol{x})\ [\approx f(\Omega)\delta(\boldsymbol{x})]$ on $\delta_G(\boldsymbol{x})$ (§ 5.2), or via a regression of the corresponding velocities. The slope of the forward regression of δ_G on δ is $b_f = \langle \delta_G \delta \rangle / \langle \delta^2 \rangle$, and the slope of the inverse regression of δ on δ_G is $b_i^{-1} = \langle \delta \delta_G \rangle / \langle \delta_G^2 \rangle$. In the case where b is constant, $b_f = b$, and b_i is an overestimate,

$$b_i = b\,(1 + \Delta_2). \tag{21}$$

The promising method of estimating β from large-scale redshift distortions measures yet a different quantity. It turns out that most methods for determining β lead similarly to an underestimate.

The level of the effect depends on the actual values of Δ_2 and similar parameters. One way to estimate the natural biasing scatter at a given smoothing scale is by investigating goodness of fit of the density fields of mass and light and the model of deterministic biasing. By requiring that $\chi^2 = 1$ per degree of freedom one can estimate the scatter needed in addition to the known errors. For example, Hudson et al. (1995) estimated for optical galaxies versus POTENT Mark III mass at $12\,h^{-1}\text{Mpc}$ smoothing $\sigma_b \sim 0.15$, which corresponds to $\Delta_2 \sim 0.25$. Alternatively, one can estimate σ_b from theoretical simulations. For example, preliminary hydro simulations (Cen & Ostriker 1993) yield $\sigma_b = 0.25$ under $10\,h^{-1}\text{Mpc}$ Gaussian smoothing, i.e. $\Delta_2 \sim 0.4$. If $b = \Omega = 1$, then the β values derived by the various methods are expected to span the range $0.7 \le \beta \le 1$, and this is solely due to the dispersion in the biasing relation. A large skewness in $B(\delta_G | \delta)$ may stretch this range even further.

A relevant moral from the biasing uncertainty is that methods for measuring Ω independent of density biasing (§ 4) are desirable. However, it has to be born in mind that the galaxies may also be biased tracers of the *velocity* field of the matter. Such a "velocity biasing" would affect any attempt to extract dynamical information from large-scale velocities. The expected magnitude of the velocity biasing in the standard scenarios of structure formation is a matter of debate, and even it's sign is unclear (*e.g.*, Summers et al. 1995). Based on recent simulations it seems likely to be limited to a $\sim 10 - 20\%$ effect.

6   Cosmological Parameters

The previous sections discussed measurements of Ω (or β) from large-scale structure on scales $10 - 100\,h^{-1}\text{Mpc}$. In this section we try to put these estimates in a wider perspective (see Dekel, Burstein & White 1997 for a review).

One very interesting large-scale constraint that has not been discussed here is based on cluster abundance, that can be predicted for a Gaussian field via the Press-Schechter formalism, and is quite insensitive to the shape of the power spectrum. The current estimates are $\sigma_8 \Omega^{0.6} \simeq 0.5 - 0.6$ (White, Efstathiou & Frenk 1993; Eke et al. 1996; Mo et al. 1996). This is only slightly lower than the estimates of $\sigma_8 \Omega^{0.6} \simeq 0.7 - 0.8$ from the power spectrum of the peculiar velocity data (§ 3.2, § 3.3). Note that this quantity is related to β_I via $\sigma_8 \Omega^{0.6} = \sigma_{8I} \beta_I$, where σ_{8I} is the rms fluctuation of IRAS galaxies in a top hat window of $8 \, h^{-1}$Mpc. With the estimate from the IRAS 1.2 Jy survey of $\sigma_{8I} = 0.69 \pm 0.05$ (Fisher et al. 1994a), the results from cluster abundance and from the $\delta - \delta$ POTENT-IRAS comparison (§ 5.2) are in pleasant agreement.

Constraints from virialized systems such as galaxies and clusters on smaller comoving scales of $1 - 10 \, h^{-1}$Mpc (e.g., Primack 1997; Bahcall 1997; Peebles 1997) typically yield low values of $\Omega \sim 0.2 - 0.3$, but with several loopholes. Most interesting among these is the constraint involving the baryonic fraction in clusters from X-ray data and the estimates of Ω_b from the observed deuterium abundance and the theory of big-bang nucleosynthesis. With baryonic fraction in the middle of the observed range $f_b = (0.03 - 0.08) h^{-3/2}$ (White et al. 1993; White & Fabian 1995), and with the recent estimates of $\Omega_b h^2 \simeq 0.025$ (Tytler et al. 1996; Burles & Tytler 1996), the current estimate is $\Omega \simeq 0.5 h_{65}^{-1/2}$. This result favors a low value of Ω, but $\Omega = 1$ cannot be definitively excluded.

The global cosmological measures commonly involve combinations of the cosmological parameters. Constraints in the $\Omega - \Omega_\Lambda$ plane are displayed in Figure 15, and briefly discussed below.

Occam's Razor. The above working hypotheses, and the order by which more specific models should be considered against observations, are guided by the principle of Occam's Razor, i.e., by simplicity and robustness to initial conditions. It is commonly assumed that the simplest model is the Einstein-deSitter model, $\Omega = 1$ and $\Omega_\Lambda = 0$. One property that makes it robust is the fact that Ω remains constant at all times with no need for fine tuning at the initial conditions. The most natural extension according to the generic model of inflation is a flat universe, $\Omega_{tot} = 1$, where Ω can be smaller than unity but only at the expense of a nonzero cosmological constant.

Classical Tests of Geometry. The parameter-dependent large-scale geometry of space-time is reflected in the volume-redshift relation. There are two classical versions of the tests that utilize this dependence: magnitude versus redshift (or "Hubble diagram") and number density versus redshift. The luminosity distance and the angular-diameter distance to a redshift z, which enter these tests, depend on Ω and Ω_Λ. At $z \sim 0.4$, these distances happen to be (to a good approximation) a function of the combination $\Omega - \Omega_\Lambda$ (not q_0) (Perlmutter et al. 1996).

The main advantage of such tests is that they are direct measures of global geometry. Supernovae type Ia are the popular current candidate for a standard candle, based on the assumption that stellar processes are not likely to vary much in time.

The first 7 supernovae analyzed by Perlmutter *et al.* (1996) at $z \sim 0.4$ yield $-0.3 < \Omega - \Omega_\Lambda < 2.5$ as the 90% two-parameter likelihood contour (Fig. 15). For a *flat* universe they find for each parameter $\Omega = 0.94^{+0.34}_{-0.28}$, and $\Omega_\Lambda < 0.51$ (or equivalently $\Omega > 0.49$) at 95% confidence.

Number Count of Quasar Lensing. This is a promising new version of the classical number density test. When Ω_Λ is positive and comparable to Ω, the universe should have gone through a phase of slower expansion in the recent cosmological past, which should be observed as an accumulation of objects at a specific redshift of order unity. In particular, it should be reflected in the observed rate of lensing of high-redshift quasars by foreground galaxies (Fukugita *et al.* 1990). The contours of constant lensing probability in the $\Omega - \Omega_\Lambda$ plane for $z_s \sim 2$ happen to almost coincide with the lines $\Omega - \Omega_\Lambda = const$. The limits from lensing are thus similar in nature to the limits from SNe Ia.

This test shares all the advantages of direct geometrical measures. The high redshifts involved bring about a unique sensitivity to Ω_Λ, compared to the negligible effect that Ω_Λ has on the structure observed at $z \ll 1$.

From the failure to detect the accumulation of lenses, the current limit for a *flat* model is $\Omega_\Lambda < 0.66$ (or $\Omega > 0.36$) at 95% confidence (Kochanek 1996) (Fig. 15).

Microwave Background Acoustic Peaks. This test is expected to provide the most stringent constraints on the cosmological parameters within a decade. The next generation of CMB satellites (MAP, to be launched by NASA in 2001, and in particular Plank, scheduled by ESA for 2004) are planned to obtain a precision at ~ 10 arc-minute resolution that will either rule out the current framework of GIfor structure formation or will measure the cosmological parameters to high precision. Detailed evaluation of Plank shows that nominal performance and expected foreground subtraction noise will allow parameter estimation with the following accuracy (ignoring systematics): $H_0 \pm 1\%$, $\Omega_{tot} \pm 0.005$, $\Omega_\Lambda \pm 0.02$, $\Omega_b \pm 2\%$.

Current ground-based and balloon-born experiments provide preliminary constraints on the *location* of the first acoustic peak on sub-degree scales in the angular power spectrum of CMB temperature fluctuations, $l(l+1)C_l$. In the vicinity of a flat model, the first peak is predicted at approximately the multipole $l_{peak} \simeq 220(\Omega + \Omega_\Lambda)^{-1/2}$. The results of COBE's DMR ($l \sim 10$) provide an upper bound of $\Omega + \Omega_\Lambda < 1.5$ at the 95% confidence level for a scale-invariant initial spectrum (and the constraint becomes tighter for any "redder" spectrum, $n < 1$) (White & Scott 1996). Several balloon experiments ($l \sim 50-200$) strengthen this upper bound (*e.g.*, the Saskatoon experiment, Scott *et al.* 1996). The Saskatoon experiment and the CAT experiment ($l \sim 350-700$) yield a preliminary lower bound of $\Omega + \Omega_\Lambda > 0.3$ (Hancock *et al.* 1996) (Fig. 15).

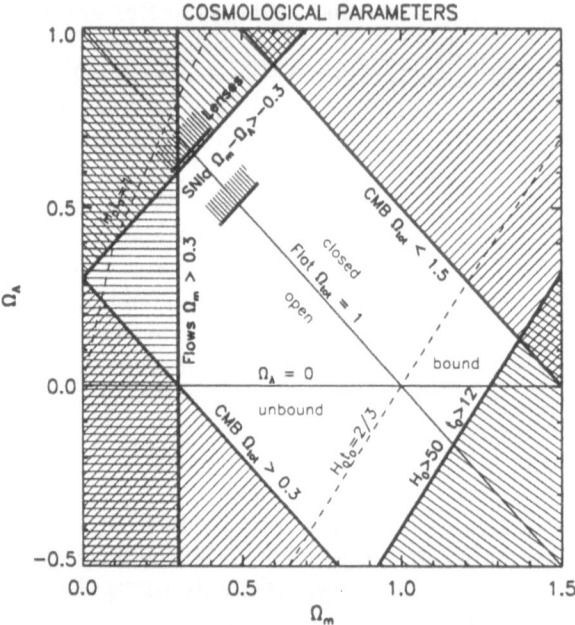

Fig. 15. Current limits ($\sim 2\sigma$) on the cosmological parameters Ω and Ω_Λ from global measures: luminosity distance of SNIa, lens count, the location of the CMB peak, and the age versus Hubble constant. The short marks are the one-parameter 95% limits from SNIa and lenses for a flat universe. Also shown (vertical line) is the 95% lower bound on Ω from cosmic flows. The most likely value of Ω lies in the range 0.5 to 1. The Einstein-deSitter model is permitted. An open model with $\Omega \simeq 0.2$ and $\Omega_\Lambda = 0$, or a flat model with $\Omega \simeq 0.3$ and $\Omega_\Lambda \simeq 0.7$, are ruled out (Dekel, Burstein & White 1997).

The Age of the Universe. Measured independent lower bounds on the Hubble constant and on the age of the oldest globular clusters provide a lower bound on $H_0 t_0$ ($= 1.05 ht$, where $H_0 \equiv 100h$ km s^{-1}Mpc^{-1} and $t_0 \equiv 10t$Gyr), and thus an interesting constraint in the $\Omega - \Omega_\Lambda$ plane. The exact expressions are computable in the various regions of parameter space. A useful crude approximation near $H_0 t_0 \sim 2/3$ is $\Omega - 0.7\Omega_\Lambda \simeq 5.8(1 - 1.3ht)$.

Progress has been made in measuring H_0 via the HST key project detecting Cepheids in nearby clusters for calibration of TF distances, and via accurate distances to SNe Type Ia. The new calibration of local Cepheids by the Hipparcos astrometric satellite (Feast & Catchpole 1997) seem to have reduced the estimates of H_0 by $\sim 10\%$. The indications from SNe velocities for a local void of radius $\sim 75\,h^{-1}$Mpc out to the Great Wall (Zehavi et al. 1997), takes another $\sim 5\%$ from H_0 as measured by TF distances from within the void, and brings the various estimates into agreement at $h \simeq 0.6 \pm 0.1$.

The Hipparcos calibration of the distances to local subdwarf stars (Reid 1997) had an even more dramatic effect on the estimates of the ages of the oldest globular clusters (*e.g.*, Van den Berg *et al.* 1996). The current estimates seem to be $t \simeq 1.2 \pm 0.2$ (*e.g.*, M. Bolte, private communication). Thus, the most likely value of $H_0 t_0$ is not far from $2/3$, consistent with the standard Einstein deSitter model.

6.1   Conclusion

We conclude with a summary of the main implications of the observed cosmic flows.

Gravitational Instability. The strongest evidence for gravitational origin of structure comes from the growth rate of fluctuations as indicated by the comparison of the $\delta T/T \sim 10^{-5}$ fluctuations at the last scattering surface and the $\sim 300 \, \mathrm{km \, s}^{-1}$ motions over $\sim 100 \, h^{-1} \mathrm{Mpc}$ scales in our local neighborhood.

Initial Fluctuations and Dark Matter. The COBE measurements of CMB fluctuations at large angular scales and the comparison to the observed flows indicate a power spectrum near scale invariance, $n \sim 1$.

The bulk velocity in a sphere of radius $50 \, h^{-1} \mathrm{Mpc}$ about the LG is $V_{50} = 375 \pm 85 \, \mathrm{km \, s}^{-1}$. The mass power spectrum deduced from the peculiar velocities has an amplitude of $P_{0.1} \Omega^{1.2} = (5 \pm 2) \times 10^3 (h^{-1} \mathrm{Mpc})^3$ at $k = 0.1 (h^{-1} \mathrm{Mpc})^{-1}$. This extrapolates to $\sigma_8 \Omega^{0.6} = 0.8 \pm 0.2$ on smaller scales.

For COBE-normalized CDM models, a likelihood analysis of the mass power spectrum yields $\Omega n^2 h_{65} = 0.7 \pm 0.2$. A comparison to preliminary detections of the first acoustic peak in the CMB angular power spectrum requires that $n \gtrsim 0.9$, and that $\Omega_b \sim 0.1$. Thus, within the family of CDM models, most successful in matching the current LSS data are either of the following variants: (a) $\Omega = 1$ with a tilted spectrum $n \sim 0.9$, (b) $\Omega \sim 0.5$, with or without a cosmological constant, and $n = 1$, and (c) $\Omega = 1$ with 20% hot dark matter. A high baryonic content (and relatively low Ω) may be required to explain a peak in the galaxy density power spectrum at $\sim 125 \, h^{-1} \mathrm{Mpc}$, if confirmed (Broadhurst *et al.* 1990; S. Landy based on LCRS, private communication; Cohen *et al.* 1996; Einasto 1997).

In view of the nucleosynthesis constraints on the baryonic density, the high Ω indicated by the motions requires non-baryonic dark matter. The observed mass-density power spectrum on scales $10 - 100 \, h^{-1} \mathrm{Mpc}$ does not yet allow a clear distinction between the competing models involving baryonic, cold and hot dark matter and possibly a cosmological constant. I do not think that any of the front-runner models can be significantly ruled out based on current tests, contrary to occasional premature statements in the literature about the "death" of a certain model. I predict that were the dark matter constituent(s) to be securely detected in the lab, the corresponding scenario of LSS would find a way to overcome the $\sim 2\sigma$ obstacles it may be facing now.

Galaxy Biasing and β. Generally speaking, galaxies trace mass. For each of the different smoothing scales, the data of velocities and redshift surveys are consistent with GI and linear biasing (properly modified in the tails). However, the best estimates of β_I span the range $0.5 - 1.0$. This can be explained by the fact that, when inspected in detail, the biasing scheme involves scale dependence, non-linear features, and intrinsic scatter. It is difficult to distinguish non-linear biasing from non-linear gravitational effects.

The Value of Ω. Methods based on virialized objects tend to favor low values of $\Omega \sim 0.2$, but with plausible loopholes.

The current peculiar-velocity data provide in several different ways a significant ($> 2\sigma$) lower bound of $\Omega > 0.3$. This bound is independent of Λ, H_0, and the biasing relation between galaxies and mass. The range of β values obtained on different scales by different methods may be partly due to underestimated errors and partly due to non-trivial biasing.

The global measures of geometry provide a lower bound of similar nature, $\Omega - \Omega_\Lambda > 0.3$. The age constraints, which used to favor low values of Ω until recently, seem to agree with $\Omega \sim 1$ according to the new calibration of the distance scale by Hipparcos.

The data is thus consistent with $\Omega = 1$. Based on the whole range of constraints, and ignoring the Occam's razor desire for simplicity, the most likely value may be argued to be $\Omega \sim 0.5$. Values of $\Omega = 0.3$ and below are significantly ruled out. The data are consistent with the general predictions of Inflation: flat geometry and Gaussian, almost scale-invariant initial fluctuations.

Acknowledgments:

This review is based on work with several close collaborators (see references), supported by grants from the US-Israel Binational Science Foundation, the Israel Science Foundation, the NSF and NASA.

References

Aaronson, M., Huchra, J., Mould, J. R., Tully, R. B., & Fisher, J. R., *et al.* 1982a, ApJS, 50, 241

Babul, A., Weinberg, D., Dekel, A., & Ostriker, J. P. 1994, ApJ, 427, 1

Bahcall, N. 1997, in Formation of Structure in the Universe, eds. A. Dekel & J.P. Ostriker (Cambridge Univ. Press) (astro-ph/9611148) in press

Bardeen, J., Bond, J. R., Kaiser, N., & Szalay, A. 1986, ApJ, 304, 1

Bernardeau, F. 1992, ApJ, 390, L61

Bernardeau, F., Juszkiewicz, R., Dekel, A., & Bouchet, F., 1995, MNRAS, 274, 20

Bertschinger, E., Dekel, A. 1989, ApJ, 336, L5

Bertschinger, E., Dekel, A., Faber, S.M., Dressler, A., & Burstein, D. 1990, ApJ, 364, 370

Bertschinger, E., Gorski, K., & Dekel, A. 1990, Nature, 345, 507

Blumenthal, G. R., Dekel, A., & Yahil, A. 1997, in preparation

Bouchet, F., Juszkiewicz, R., Colombi, S., & Pellat, R. 1992, ApJ, 394, L5

Braun, E., Dekel, A., & Shapiro, P. 1988, ApJ, 328, 34

Broadhurst, T. J., Ellis, R. S., Koo, D. C., & Szalay, A. S. 1990, Nature, 343, 726

Burles, S., & Tytler, D. 1996, ApJ, 460, 584

Cen, R., & Ostriker, J. P. 1992, ApJ, 399, L113

Cen, R., & Ostriker, J. P. 1993, ApJ, 417, 415

Cohen, J. G., Cowie, L. L., Hogg, D. W., Songaila, A., Blandford, R., Hu, E. M., & Shopbell, P. 1996, ApJ, 471, 5

Cole, S., Fisher, K. B., & Weinberg, D. 1995, MNRAS, 275, 515

Courteau, S., Faber, S. M., Dressler, A., & Willick J. A. 1993, ApJ, 412, L51

da Costa, L.N., Freudling, W., Wegner, G., Giovanelli, R., Haynes, M., Salzer, J.J. 1996, ApJ, 468, L5

Davis, M., Efstathieou, G., Frenk, C. S., & White, S. D. M. 1985, ApJ, 292, 371

Davis, M., Nusser, A., & Willick, J. A. 1996, ApJ, 473, 22

Dekel, A. 1981, A&A, 101, 79

Dekel, A. 1994, ARA&A, 32, 371

Dekel, A. 1997, in Formation of Structure in the Universe, eds. A. Dekel & J.P. Ostriker (Cambridge Univ. Press) in press

Dekel, A., Bertschinger, E., & Faber, S.M. 1990, ApJ, 364, 349

Dekel, A., Bertschinger, E., Yahil, A., Strauss, M., Davis, M., & Huchra, J. 1993, ApJ, 412, 1

Dekel, A., Burstein, D., & White, S. D. M. 1997, in Crtical Dialogs in Cosmology, ed. N. Turok (Princeton: Princeton University Press) in press (astro-ph/9611108)

Dekel, A., Eldar, A., Kolatt, T., Yahil, A., Willick, J.A., Faber, S.M., Corteau, S., & Burstein, D. 1997, ApJ, , in preparation

Dekel, A., & Lahav, O. 1997, in preparation

Dekel, A., & Rees, M. J. 1987, Nature, 326, 455

Dekel, A., & Rees, M. J. 1994, ApJ, 422, L1

Dekel, A., & Silk, J. 1986, ApJ, 303, 39

Dressler, A. 1980, ApJ, 236, 351

Dressler, A., Lynden-Bell, D., Burstein, D., Davies, R. L., Faber, S. M., Terlevich, R., & Wegner, G. 1987, ApJ, 313, 42

Einasto, J. et al. 1997, Nature, 385, 139

Eke, V. R., Cole, S., & Frenk, C. S. 1996, astro-ph 9601088

Eldar, A., Dekel, A., & Willick, J. A. 1997, in preparation

Feast, M. W., & Catchpole, R. M. 1997, MNRAS, , submitted

Fisher, K. B., Davis, M., Strauss, M. A., Yahil, A., & Huchra, J. P. 1994a, MNRAS, 266, 50

Fisher, K. B., Davis, M., Strauss, M. A., Yahil, A., & Huchra, J. P. 1994b, MNRAS, 267, 92

Fisher, K. B., & Nusser, A. 1996, MNRAS, 279, L1

Fukugita, M., Yamashita, K., Takahara, F., & Yoshii, Y. 1990, ApJ, 361, L1

Ganon, G., Dekel, A., Mancinelli, P., & Yahil, A. 1997, in preparation

Giovanelli, R. et al. 1997, in preparation

Gramman, M. 1993, ApJ, 405, L47

Hamilton, A. J. S. 1995, in Clustering in the Universe, eds. C. Balkowski & S. Morgatadeau (Editions Frontieres)

Hancock, S., Rocha, G., Lasenby, A. N., & Cutierrez, C. M. 1996, MNRAS, , submitted

Heavens, A.F., & Taylor, A.N. 1995, MNRAS, 275, 483

Hudson, M.J. 1993, MNRAS, 265, 72

Hudson, M. J. 1994, MNRAS, 266, 475

Hudson, M. J., Dekel, A., Courteau, S., Faber, S. M., & Willick, J. A. 1995, MNRAS, 274, 305

Ikeuchi, S. 1981, PASJ, 33, 211

Jaffe, A. H., & Kaiser, N. 1995, ApJ, 455, 26

Kaiser, N. 1988, MNRAS, 231, 149

Kaiser, N. 1984, ApJ, 284, L9

Kaiser,N., Efstathiou, G., Ellis, R., Frenk, C., Lawrence, A. , Rowan-Robinson, M., & Saunders, W. 1991, MNRAS, 252, 1

Kochanek, C. S. 1996, ApJ, 466, 638

Kofman, L., Bertschinger, E., Gelb, J., Nusser, A., & Dekel, A. 1994, ApJ, 420, 44

Kolatt, T. & Dekel, A. 1994, ApJ, 428, 35

Kolatt, T. & Dekel, A. 1997, ApJ, in press (astro-ph/9512132)

Kolatt, T., Dekel, A., Ganon, G., & Willick, J. 1996, ApJ, 458, 419

Kolatt, T., Dekel, A., & Primack, J. R. 1997, in preparation

Landy, S., & Szalay, A. 1992, ApJ, 391, 494

Lauer, T.R., & Postman, M. 1993, ApJ, 425, 41

Lemson, G., Dekel, A., Kauffmann, G., & White, S. D. M. 1997, in preparation

Lynden-Bell, D., Faber, S. M., Burstein, D., Davies, R. L., Dressler, A., Terlevich, R. J., & Wegner, G. 1988, ApJ, 326, 19

Lynden-Bell, D., Lahav, O., & Burstein, D. 1989, MNRAS, 241, 325

Mathewson, D. S., Ford, V. L., & Buchhorn, M. 1992, ApJS, 81, 41

Mo, H. J., Jing, Y. P., & White, S. D. M. 1996, MNRAS, 282, 1096

Nusser, A., & Dekel, A. 1992, ApJ, 391, 443

Nusser, A., & Dekel, A. 1993, ApJ, 405, 43

Nusser, A., & Dekel, A. 1997, in preparation

Nusser, A., Dekel, A., Bertschinger, E., & Blumenthal, G.R. 1991, ApJ, 379, 6

Nusser, A., Dekel, A., & Yahil, A. 1995, ApJ, 449, 439

Ostriker, J. P., & Cowie, L. L. 1981, ApJ, 243, L127

Peacock, J. A., & Dodds, S.J. 1994, MNRAS, 267, 102

Peebles, P. J. E. 1997, in Formation of Structure in the Universe, eds. A. Dekel & J.P. Ostriker (Cambridge Univ. Press) in press

Perlmutter, S., Gabi, S., Goldhaber, G. et al. 1996, ApJ, , submitted (astro-ph/9602122)

Plionis, M., Branchini, E., Zehavi, I., & Dekel, A. 1997, in preparation

Primack, J. R. 1997, in Formation of Structure in the Universe, eds. A. Dekel & J.P. Ostriker (Cambridge Univ. Press) in press

Reid, I. N. 1997, ApJ, , submitted

Riess, A. G., Press, W. H., & Kirshner, R. P. 1995, ApJ, 438, L17

Rowan-Robinson, M., Lawrence, A., Saunders, W., & Leech, K. 1991, MNRAS, 253, 485

Sachs, R. K., & Wolfe, A. M. 1967, ApJ, 147, 73

Scaramella, R., Vettolani, G., & Zamorani, G. 1991, ApJ, 376, L1

Schechter, P. 1980, AJ, 85, 801

Scott P. F., Saunders, R., Pooley, G. et al. 1996, ApJ, 461, L1

Shaya, E. J., Peebles, P. J. E., & Tully, R. B. 1995, ApJ, 454, 15

Sigad, Y., Dekel, A., Strauss, M. S., & Yahil, A. 1997, in preparation

Strauss, M. A. 1997, in Formation of Structure in the Universe, eds. A. Dekel & J.P. Ostriker (Cambridge Univ. Press) in press (astro-ph/9610033)

Strauss, M. A., Cen, R., Ostriker, J. P., Lauer, T. R., & Postman, M. 1995, ApJ, 444, 507

Strauss, M. A., Yahil, A., Davis, M., Huchra, J. P., & Fisher, K. B. 1992. ApJ, 397, 395

Strauss, M. A., & Willick, J. A. 1995, PhR, 261, 271

Sugiyama, N. 1995, ApJS, 100, 281

Summers, F. J., Davis, M., & Evrard, A. E. 1995, ApJ, 454, 1

Tytler, D., Fan, X. M., & Burles, S. 1996, Nature, 381, 207

Van den Berg, D., Stetson, P., & Bolte, M. 1996, ARA&A, 34, 461

Watkins, R., & Feldman, H. A. 1995, ApJ, 453, 73

Weinberg, D. 1995, in Wide-Field Spectroscopy and the Distant Universe, eds. S.J. Maddox & A. Aragon-Salamanca (World Scientific: Singapore)

White, M., & Scott, D. 1996, ApJ, 459, 415

White, S. D. M., Efstathiou, G., & Frenk, C. S. 1993, MNRAS, 262, 1023

White, S. D. M., Navaro, J., Evrard, A., & Frenk, C. S. 1993, Nature, 366, 429

White, S. D. M., & Fabian, A. 1995, MNRAS, 273, 72

Willick, J. 1994, ApJS, 92, 1

Willick, J.A. 1997, in Formation of Structure in the Universe, eds. A. Dekel & J.P. Ostriker (Cambridge Univ. Press) in press (astro-ph/9610200)

Willick, J.A., Courteau, S., Faber, S.M., Burstein, D., & Dekel, A. 1995, ApJ, 446, 1

Willick, J.A., Courteau, S., Faber, S.M., Burstein, D., Dekel, A., & Kolatt, T. 1996, ApJ, 457, 460

Willick, J.A., Courteau, S., Faber, S.M., Burstein, D., Dekel, A., & Strauss, M.A. 1997a, ApJS, in press (astro-ph/9610202)

Willick, J.A., Strauss, M.S., Dekel, A., & Kolatt, T. 1997, ApJ, in press (astro-ph/9612240)

Yahil, A., Walker, X., & Rowan-Robinson, M. 1986, ApJ, 301, L1

Zaroubi, S., Hoffman, Y., & Dekel, A. 1997, in preparation

Zaroubi, S., Sugiyama, N., Silk, J., Hoffman, Y., & Dekel, A. 1997b, ApJ, submitted (astro-ph/9610132)

Zaroubi, S., Zehavi, I., Dekel, A., Hoffman, Y., & Kolatt, T. 1997a, ApJ, in press (astro-ph/9610226)

Zehavi, I., Riess, A., Kirshner, R., & Dekel, A. 1997, in preparation

Zel'dovich, Ya.B. 1970, A&A, 5, 20

Large Scale Flows from the Mark III Tully-Fisher Data

Adi Nusser[1], Marc Davis[2] and Jeffrey A. Willick[3]

[1] Max-Planck-Institut für Astrophysik, Karl-Schwarzschild-Str. 1, 85740 Garching, Germany
[2] Departments of Astronomy and Physics, University of California, Berkeley, CA. 94720
[3] Department of Physics, Stanford University, Stanford, CA 94305-40560

Abstract. We use the Mark III Tully-Fisher data of 2900 spiral galaxies to derive an estimate of the large scale peculiar velocity field. A comparison of this field with the 1.2Jy IRAS gravity field yields the parameter $\beta = \Omega^{0.6}/b$, where b is the linear bias factor relating the distribution of the IRAS galaxies to the dark matter density fluctuations. By working in redshift space and the inverse Tully-Fisher relation, this comparison is essentially free of Malmquist biases. The velocity and gravity fields are expanded in the same set of smooth functions and therefore our estimate of β and the assessment of the agreement between the fields are free of ambiguities resulting from possible differences in the smoothing of the fields. We find a general good agreement, especially for data within 3000 km s^{-1}, between the velocity and gravity for $\beta \sim 0.6$. The statistical 1σ error in our estimate of β is 0.1. However, the fields do not agree in detail, precluding a firm determination of β from these data sets at present.

1 Introduction

Peculiar velocities of galaxies are a direct probe of the nature of the underlying dark matter in the Universe. Under the assumption that the large scale structure has formed via gravitational amplification of small initial density fluctuations, the observed velocity field yields a prediction for the present matter-density fluctuations given an assumed value for Ω. Furthermore, quasi-linear gravitational instability theory can be used to reconstruct the primordial density fluctuations from the observed flows. This enables us to address fundamental questions about the nature of the initial fluctuations and their relation to Ω. For example, Nusser & Dekel (1993) have demonstrated that unless Ω is around unity the initial density fluctuations were non-gaussian. Here we concentrate on comparing the peculiar velocity field from Tully-Fisher data with the gravity field computed from the distribution of galaxies in redshift space. Unlike quasi-linear methods for estimating Ω based on the observed velocity field alone combined with some assumption about the initial fluctuations such as Gaussianity (Nusser & Dekel 1993, Berbardeau *et al.* 1995), linear methods involving peculiar velocities and redshift surveys do not yield estimates of Ω independent of the bias factor b. The basic idea behind velocity-gravity comparisons is the following: assuming a linear biasing relation between the large scale fluctuations in the number density of galaxies and the mass fluctuations, we obtain from linear theory

$$\mathbf{v_L}(\mathbf{r}) = \frac{H_0\beta}{4\pi\bar{n}} \sum_i \frac{1}{\phi(r_i)} \frac{\mathbf{r_i} - \mathbf{r}}{|\mathbf{r_i} - \mathbf{r}|^3} + \frac{H_0\beta}{3}\mathbf{r} , \qquad (1)$$

where $\bar{n}$ is the true mean galaxy density in the sample, $\beta \equiv \Omega^{0.6}/b$, and where $\phi(r)$ is the radial selection function (Yahil et al. 1991). Note the sum in (1) is to be computed in real space, whereas the galaxy catalog exists in redshift space. Note also that the result is insensitive to the value of H_0, as the right hand side has units of velocity. We shall henceforth quote all distances in units of km s^{-1}. The density $\bar{n}$ and selection function $\phi(r)$ can be estimated from the redshift catalog used in the analysis, so that comparison of the measured velocity field to the predicted velocity field $\mathbf{v_L}(\mathbf{r})$ gives us a measure of β. The value of β is not the only ultimate outcome of such comparisons. With the advent of large peculiar velocity data sets, such comparisons can be indispensable in constraining the biasing relation (galaxy formation) and assessing the universality of intrinsic properties of galaxies, e.g. the Tully-Fisher relation.

Several comparisons of velocity and gravity fields have been made with older datasets (e.g. Hudson 1994; Kaiser et al. 1991; Yahil 1988; Strauss and Davis 1988) but these analyses were all meant to be preliminary and none of their conclusions were compelling. These studies did not included a proper treatment of the correlated noise in the analysis. Nusser & Davis (1994) computed the dipole component of peculiar velocity field on distant shells as predicted from the IRAS 1.2 Jy and compared it with the observed velocity dipole obtained from the POTENT compilation (Dekel et al. 1990) of the Mark III data (Willick et al. 1996). This comparison is especially useful since the IRAS dipole relative to the Local Group motion is entirely determined by the structure internal to that shell. This comparison showed a good alignment between the two dipole fields and yielded $\beta = 0.55 \pm 0.05$. This analysis however is not suitable for inspecting details of the flow patterns. Shaya et al. (1995) have compared the peculiar velocities of a set of 298 spiral galaxies within a redshift of 3000 km s^{-1} to the gravity field derived by least action analysis of the 1138 mass tracers derived from the Nearby Galaxies Catalog (Tully 1988). Their preliminary analysis gives a value $\Omega_0 = 0.2 \pm 0.2$, but they have not included covariance in the errors of the predicted mass model, nor have they realistically included possible influence from the mass distribution at redshifts $z > 3000$ km s^{-1}, nor is it clear how well their analysis compares with the linear theory techniques used by others. More recently, Willick et al. (1996) made a non-linear maximum likelihood analysis of the Mark III data and the IRAS 1.2Jy gravity. They restricted their consideration to data within $cz = 3000$ km s^{-1} and concluded $\beta = 0.49 \pm 0.07$. consistent with the Nusser & Davis (1994) result from the dipole analysis. Their approach does not naturally provide visual point by point comparison of the the fields.

A somewhat more direct way to estimate β can be done at the density level. Because of the covariance of the IRAS gravity field a comparison of the mass fluctuations as inferred from the flows versus the IRAS density field has considerable merits. Assuming the flow is irrotational, the POTENT algorithm reconstructs a three dimensional flow from the observed radial flow field, and then computes derivatives of this flow to infer the underlying mass field (Nusser *et al.* 1991). This then can be compared directly to the *IRAS* density field (or to any other survey). Such a comparison was done by Dekel *et al.* (1993), using an earlier version of the *IRAS* catalog, and the Mark II catalog of peculiar velocities for 493 objects (Faber and Burstein 1988). On the basis of this density-density comparison, Dekel *et al.* (1990) derived $\beta = 1.3 \pm 0.3$. More recently, Hudson *et al.* (1995) have compared POTENT mass densities reconstructed from the Mark III data (Willick *et al.* 1996) with the optical density field of Hudson (1994), finding $\beta = 0.74 \pm 0.13$. One difference between the velocity-gravity and density-density comparisons is that, in the presence of observational errors, the velocity field probes scales larger than those which the density field is sensitive to. Another difference is that the mass-fluctuations which are derived by POTENT increase with decreasing β. According to the quasi-linear approximation (Nusser *et al.* 1991) employed by POTENT, the mass-fluctuation field inferred from the velocities diverges for low β, thus invalidating the density-density comparison. Therefore the density-density comparison can not formally rule out very low values of β. However, the fluctuations in the velocity field predicted from the observed distribution of galaxies are smaller for lower β and the applicability of the linear approximation is guaranteed. Therefore, the velocity-velocity comparison provides a better tool for assessing the agreement between the fields for low β. The velocity-gravity comparison is therefore somewhat "orthogonal" to the density-density comparison.

In this paper we shall compare the observed radial velocity field derived from the Mark III Tully-Fisher data with the radial gravity field inferred from the 1.2 Jy IRAS redshift survey. We shall use the method of orthogonal mode expansion as described by Nusser and Davis (1995), which we shall henceforth label as the Inverse Tully-Fisher (ITF) method. We follow the the methods and analysis described in Nusser and Davis (1995) and Davis, Nusser and Willick (1996) (henceafter ND and DNW). In Section 2 we briefly review the ITF method for deriving peculiar velocities and discuss how to fit the velocity and gravity fields with the same set of functions to ensure identical smoothing for both fields. Section 3 gives details of the comparison of the Mark III velocity field with the *IRAS* inferred gravity field. Section 4 presents a summary and conclusions.

2   Reconstruction of Peculiar Velocities Using the Inverse Tully-Fisher Relation

Any of the available methods (e.g. Yahil *et al.* 1991, Fisher *et al.* 1995b) for generating *IRAS* predicted peculiar velocities from the redshift space distribution of galaxies can be used. However, the method of Nusser & Davis (1994) is particularly convenient, as it is easy to implement, fast, and requires no iterations. Moreover, this method closely parallels the derivation of peculiar velocities from Tully-Fisher data (ND, DNW) briefly outlined bellow.

Given a sample of galaxies with measured circular velocity parameters, η_i, apparent magnitudes, m_i, and redshifts, z_i, the goal is to derive an estimate for the smooth underlying peculiar velocity field. We assume that the circular velocity parameter, η, of a galaxy is, up to a random scatter, related to its absolute magnitude, M, by means of a linear inverse Tully-Fisher (ITF) relation, i.e.,

$$\eta = \gamma M + \eta_0. \tag{2}$$

We use the inverse relation because samples selected by magnitude, as most are, will not be plagued by selection Malmquist bias effects when analyzed in the inverse direction (Schechter 1980, Aaronson et al. 1982, Tully 1988, Lynden-bell 1991). We also work with a velocity model that is parameterized in terms of the observable redshifts and therefore our estimation of the underlying velocity field is free of the spatial Malmquist biases which arise when working in measured distance space. ND consider a very general model and write the absolute magnitude of a galaxy, $M_i = M_{0i} + P_i$, where $M_{0i} = m_i + 5\log(z_i) - 15$ and $P_i = 5\log(1 - u_i/z_i)$, where m_i is the apparent magnitude of the galaxy, z_i is its redshift in units of km s^{-1}, and u_i its radial peculiar velocity in the LG frame. In general, one can write the function P_i in terms of an expansion over orthogonal functions,

$$P_i = \sum_{j=0}^{j_{max}} \alpha^j \tilde{F}_i^j, \tag{3}$$

with orthonormality conditions, $\sum_{i=1}^{N_g} \tilde{F}_i^j \tilde{F}_i^{j'} = \delta_K^{j,j'}$, with the zeroth mode defined by $\tilde{F}_i^0 = 1/\sqrt{N_g}$, where N_g is the number of galaxies in the sample. The zeroth mode describes a Hubble-like flow in the space of the data set which is clearly degenerate with the zero point of the ITF relation. Here we arbitrarily set $\tilde{\alpha}^0 = 0$. The best fit parameters, α^j, the slope, γ, and the zero point η_0, are found by minimizing the χ^2 statistic

$$\chi^2 = \sum_i \frac{(\gamma M_{0i} + \gamma P_i + \eta_0 - \eta_i)^2}{\sigma_\eta^2}. \tag{4}$$

where σ_η is the rms scatter in η about the ITF relation.

The choice of the basis functions, $\tilde{F}_j$, the expansion of the modes can be made with considerable latitude. The functions should obviously be linearly independent, smooth, and and close to a complete set of functions up to a given resolution limit. ND chose spherical harmonics Y_l^m for the angular wavefunctions (Fisher et al. 1995b) and derivatives of spherical Bessel functions, $j_l[y(z)]$ for the radial basis functions, where the transformation from z to y in the argument of these Bessel functions is designed to make them oscillate nonuniformly with depth in order to match the spatial distribution of the TF data. The use of the coordinate y instead of z significantly reduces the number of the fit parameters necessary to describe the underlying velocity field in terms of our velocity model (DNW). As a demonstration of the use of the variable y, we tested the Mark III catalog which lists 2237 galaxies within $cz = 6000$ km s^{-1}, details of which will be described below. We took the linewidth scatter to be $\sigma_\eta = 0.05$ (Willick *et al.* 1996). For no velocity model, (i.e. setting all coefficients $\alpha^j = 0$), the scatter diagram of the Tully-Fisher regression of observed versus predicted linewidths η yields $\chi^2 = 3516$. Fitting a flow model of 69 degrees of freedom ($l_{max} = 3$, $n_{max} = 4$) in terms of of the coordinate z leads to a much reduced $\chi^2 = 2667$, while doing the same fit of 69 modes in terms of $y = [ln(1 + z/z_*)]^{1/2}$ with $z_* = 1000$ km s^{-1}, gives $\chi^2 = 2644$, a modest but significant improvement. Further tests on simulations with smaller σ_η demonstrate that the effect of the coordinate transformation is considerably more pronounced. There is clearly room for refinement here, but this choice of radial variable is sufficient for our purposes, especially since we shall use identical basis functions for the expansion of the velocity and gravity fields.

Using the machinery for computing a gravity field described in Nusser & Davis (1994), one can generate a linear theory predicted peculiar velocity v_L for any point in space as a function of its redshift for any value of β. We must ensure that the smoothing scales of the ITF and *IRAS* predicted peculiar velocities are matched to the same resolution. Therefore we expand the *IRAS* predicted peculiar velocity in terms of the modes used in the ITF velocity model (3). This will filter out higher frequency modes in the *IRAS* field, v_L, that are not described by the resolution of our basis functions. However, it is important to recall that the modes are chosen to be orthogonal to the mode $P_i = $ constant, which would describe a smooth Hubble flow. In the fitting for the ITF modes, pure Hubble flow is absorbed into a shift of the zero point η_0 and the orthogonality is ensured. Within a given set of test points occupying a volume smaller than that used to define the gravity field, it is possible for v_L to have a non-zero component of a Hubble flow like, which must be removed before we tabulate the mode coefficients. This component of the Hubble flow is not trivial in amplitude, and can be a 10% correction on the effective Hubble constant within simulated Mock catalogs.

3   Comparison of Mark III versus IRAS

In the comparison of the Mark III velocity and *IRAS* gravity fields discussed below, we shall use the Mark III sample only within a redshift limit of 6000 km s^{-1}. The basis functions, $\tilde{F}_j$, for our velocity model where chosen such that the radial and the transverse resolutions are approximately the same (cf. DNW for details). The functions were constructed to have varying resolution with depth. In order to match the sparseness of the Mark III data the resolution was 680 km s^{-1} at $z = 1000$ km s^{-1}, and 3400 km s^{-1} at $z = 4000$ km s^{-1} and varied as dz/dy. The total number of the resulting fit parameters, α^j, was 56. Figure 1 shows the correlation function of the η residuals, before and after the modal expansion. This plot demonstrates the residuals to have zero coherence on scales larger than the resolution scale of the mode expansion. It appears as though the mode expansion has done its job. The randomness of the $\delta\eta$ scatter suggests that the 56 mode expansion is sufficient for this dataset.

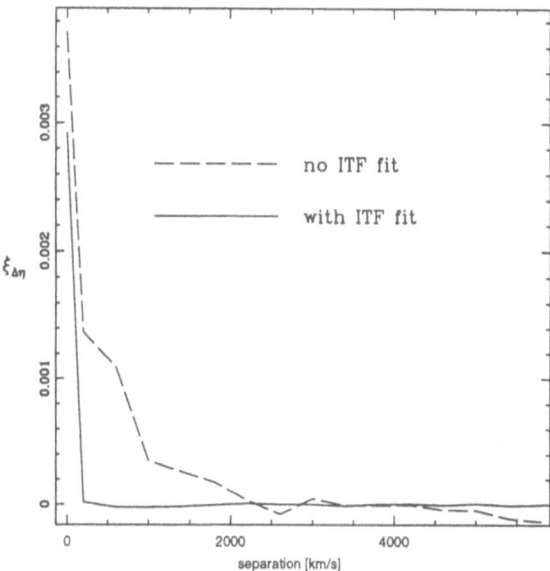

Fig. 1. The auto-correlation function of the η residuals of the Mark III galaxies versus redshift space separation. The dashed curve results when the 56 mode coefficients are set to 0, while the solid curve is after the 56 modes are fit to the Mark III data.

The resulting ITF velocity field of the Mark III galaxies is shown in slices of redshift space in Figure 2. This velocity maps is in the LG frame. In the nearby zone, the flow is dominated by infall to Virgo ($l = 284°$, $b = 74°$) and Ursa Major ($145°$, $65°$) in the North, and modest outflow from Fornax ($237°$, $-54°$) in the south. In the middle redshift zone, the dipole pattern persists, and is more pronounced. This is composed partially by reflex dipole of the motion of the LG, plus backside infall from behind the Virgo supercluster. The Hydra cluster seen at ($270°$, $26°$) is observed to be falling toward us, while the foreground of the Centaurus region ($310°$, $20°$) is moving away from us. In the most distant slice of redshift space, the dipole pattern is further enhanced, with the entire Southern galactic sky, including the Perseus-Pisces region ($150°$, $-15°$) and the Pavo-Indus-Telescopium region ($320°$, $-15°$) flowing away from us, the Centaurus cluster flowing away rather substantially, but the background of Hydra and the Northern galactic cap flowing toward us. All of this is a combination of the reflex of the motion of the LG, plus the motion of the individual regions. Note the very strong shear in the observed ITF field along the boundary $l = 300$, $b > 0$, which divides the Hydra and Centaurus regions, and amounts to a change of radial peculiar velocity of approximately 900 $\mathrm{km\,s^{-1}}$ in a transverse interval of 4100 $\mathrm{km\,s^{-1}}$, or 20% of a Hubble flow differential motion over a substantial volume! This is quite a major gradient, one that appears to have been confirmed by distance estimates of elliptical galaxies in Centaurus (Lynden-Bell *et al.* 1988,

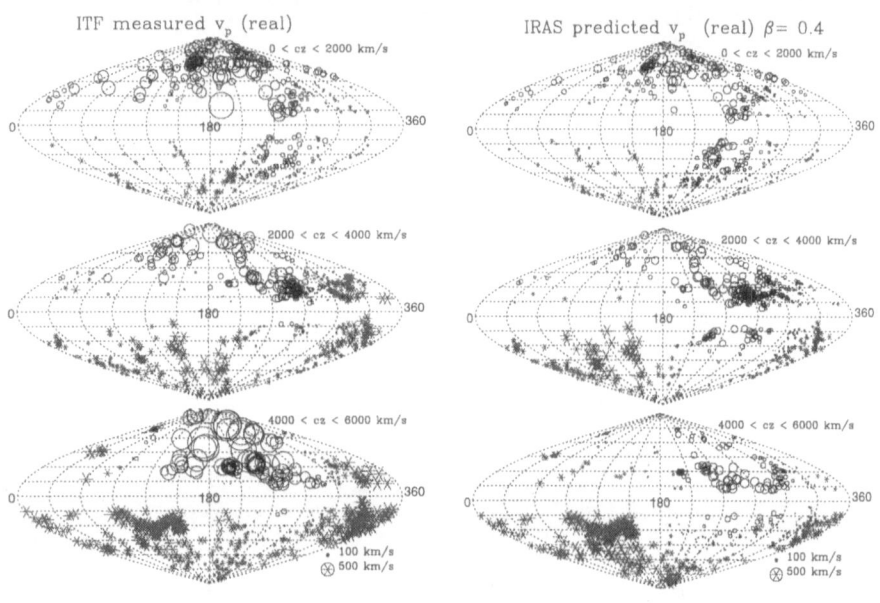

Fig. 2. The u_{itf} (left) and u_{iras} (right) sky projection as seen in the LG frame for the Mark III galaxies, in galactic coordinates. The *IRAS* predicted flow u_{iras} is for $\beta = 0.4$. The open circles are points that are flowing inward and the stars are points flowing outward. The size of the symbols is proportional to the flow velocity, with a key showing 500 km s^{-1} flow. Note the three separate panels of redshift interval. Note also the dominance of the dipole mode, especially for the outer shell. This is the signature of the reflex motion of the LG.

Dressler 1994). In fact the true gradient is limited by the angular resolution of our modal expansion, since the reported outflow of the Centaurus clusters is 500–2000 km s^{-1} in the LG frame, whereas the ITF measured flows for the entire Centaurus cluster are in the range 350–450 km s^{-1} in the LG frame. For comparison, Figure 3 shows the *IRAS* predicted field for $\beta = 0.4$. Figure 4 shows the difference of the fields in Figure 2 and 3. It is important to remember that the ITF velocity and *IRAS* gravity fields have been processed completely independently of each other. To a first approximation, the agreement between these Figures is a remarkable confirmation of the consistency between the gravity and velocity fields of our local Universe. The *IRAS* galaxies must have some close relationship to the underlying mass distribution. The qualitative alignment of the fields is consistent with the large scale structure having been formed by

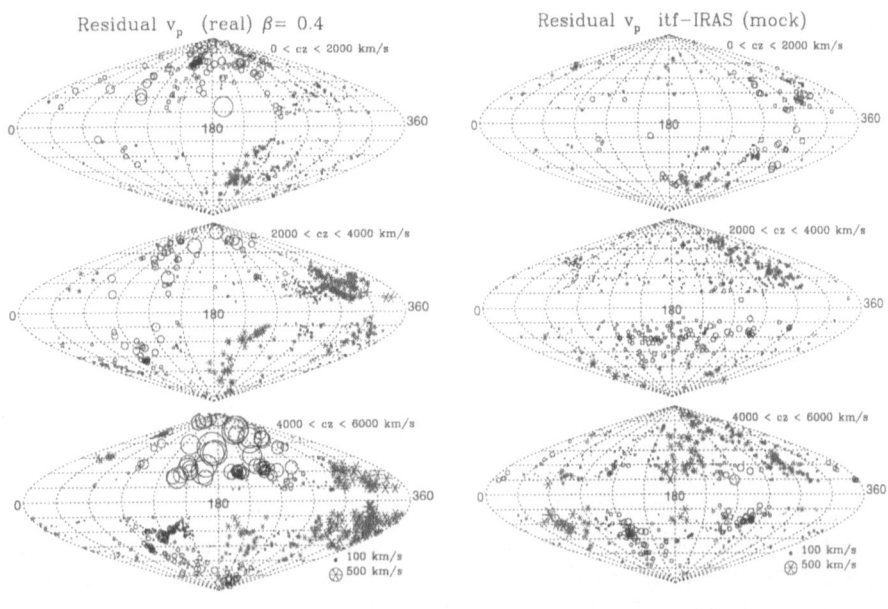

Fig. 3. The sky projection of the residuals $u_{itf} - u_{iras}$: *Left* for real Mark III and *IRAS* for $\beta = 0.4$. This is simply the difference between the fields in Figure 2. *Right* for a mock catalog. Note the small amplitude of the mock catalog residual on the right and the absence of significant dipole contribution to it.

the process of gravitational instability and the universality of the Tully-Fisher relation.

On the other hand, the residual field is not nearly as clean as seen in the mock catalogs in Figure 5, and is dominated by strong dipole patterns that increase linearly with redshift. Inspection of the *IRAS* maps for various values of β show that they never reproduce the strong shear seen in the ITF maps between the Hydra and Centaurus regions. They never reproduce the strong infall measured by ITF for the $l = 180°$, $b = 45°$, $cz = 5000$ km s^{-1} region. For $\beta \leq 0.3$, the local supercluster flow is much too small, and the flow toward Perseus-Pisces is too small. For $\beta = 0.4$, the residual field in the local zone is still completely coherent, but the flow toward Perseus-Pisces and the average of Hydra and Centaurus is approximately correct. For $\beta = 0.7$ the relative motion toward Virgo matches the ITF measurement (-340 km s^{-1} in the LG frame) but the flow away from Perseus-Pisces and toward the Great Attractor complex is now too large.

4 Summary and Conclusions

We have presented a method for comparing the velocity field derived from Tully-Fisher type data with the gravity field derived from full-sky redshift surveys of galaxies, such as the *IRAS* surveys. The chief strength of the method is that the two fields are filtered through the same set of low resolution modes, so that one is assured they indeed have the same resolution. In spite of working in the 'inverse' direction, we are able to produce pictures of the measured fields.

Although the method works extremely well in mock catalogs, with modal coefficients fully consistent with the expected noise, we find a much poorer agreement for the real data. This discrepancy is not consistent with random errors, and may be indicative of a systematic error in one or both of the datasets. We derive a "best" value of $\beta = 0.4 - 0.6$, in reasonable agreement with the value derived from the analysis of Willick *et al.* (1996), though considerably smaller than the values ($\beta \simeq 1.0$) typically obtained from density-density comparisons using the POTENT algorithm. Because the χ^2 for the fit of the ITF velocity field to the *IRAS* gravity field is 100 for 55 degrees of freedom, the fields cannot be said to agree in detail. We therefore urge extreme caution in concluding that β has been measured in large scale flows. It is worth emphasizing that there is qualitative agreement between the velocity and gravity fields, particularly at distances $\lesssim 3000$ km s^{-1}. However, until the discrepancies on larger scales can be resolved, and all methods give consistent answers, we must prudently consider the value of β to be an open question.

Future work should greatly improve this situation, as more data, and more precise data, become available. The Mark III catalog is not optimal for the analysis presented here, since the sky coverage is relatively nonuniform. More encouraging results are already achieved in work in progress (da Costa *et al.* 1997) where the analysis of ND and DNW is applied to a newly compiled homogeneous sample of Tully-Fisher data. Details of this project will be reported elsewhere.

This work was supported in part by NSF grant AST 92-21540 and NASA grant NAG 51360.

References

Aaronson, M. Huchra, J., Mould, J., Schechter, P., & Tully, R. B. 1982, *Astrophys. J.* 258, 64

Bernardeau, F., Juszkiewicz, R., Dekel, A. & Bouchet, F.R. 1995, *M.N.R.A.S.* 274, 20

da Costa, L., Nusser, A., Freudling, W., Giovanelli R., Haynes, M.P., Salzer, J.J., & Wegner, G. 1997, in preparation

Davis, M., Nusser, A. and Willick, J.A. 1996, *Astrophys. J.* 473 22

Dekel, A., Bertschinger, E., & Faber, S. M. 1990, *Astrophys. J.* 364, 349

Dekel, A., Bertschinger, E., Yahil, A., Strauss, M., Davis, M., & Huchra, J. 1993, *Astrophys. J.* 412, 1

Fisher, K.B., Huchra, J., Strauss, M.A., Davis, M. , Yahil, A., & Schlegel, D. 1995a, *Astrophys. J. (suppl.)* 100, 69

Fisher, K.B., Lahav, O., Hoffman, Y., Lynden-Bell, D., & Zaroubi, S. 1995b, *M.N.R.A.S.* 272, 885

Hudson, M. 1994, *M.N.R.A.S.* 266, 475

Hudson, M.J., Dekel, A., Courteau, S., Faber, S.M., & Willick, J.A. 1995, *M.N.R.A.S.* 274, 305

Kaiser, N., Efstathiou, G. Saunders, W., Ellis, R., Frenk, C., Lawrence, A. & Rowan-Robinson, M. 1991, *M.N.R.A.S.* 252, 1

Lynden-Bell, D. 1991, in *Statistical Challenges in Modern Cosmology*, eds. Babu, G.B & Feigelson, E.D.

Nusser, A., & Dekel, A. 1993, *Astrophys. J. (Lett.)* 405, 437

Nusser, A., & Davis, M. 1994, *Astrophys. J. (Lett.)* 421, L1

Nusser, A., & Davis, M. 1995, *M.N.R.A.S.* 276, 1391

Nusser, A., Dekel, A., Bertschinger, E., & Blumenthal, G. R. 1991, *Astrophys. J.* 379, 6

Schechter, P. 1980 *Astrophys. J.* 85, 801

Tully, R. B. 1988, *Nature* 334, 209

Willick, J.A., Courteau, S, Faber, S., Burstein, D., Dekel, A., & Kolatt, T. 1996a, *Astrophys. J.* 457, 460

Willick, J.A., Strauss, M.A., Dekel, A. & Kolatt, T. 1996, astro-ph/9610202

Peculiar Motions of Clusters
in the Perseus–Pisces Region

R.J. Smith[1], M.J. Hudson[1,2], J.R. Lucey[1] and J. Steel[1]

[1] Department of Physics, University of Durham, South Road, Durham DH1 3LE, U.K.
[2] Department of Physics and Astronomy, University of Victoria, P.O. Box 3055,
Victoria BC V8W 3PN, Canada

Abstract. We present results of a new study of peculiar motions of 7 clusters in the Perseus–Pisces (PP) region, using the Fundamental Plane as a distance indicator. The sample is calibrated by reference to 9 additional clusters with data from the literature. Careful attention is paid to the matching of spectroscopic and photometric data from several sources.

For six clusters in the PP supercluster no significant peculiar motions are detected. For these clusters we derive a bulk motion of 60 ± 220 km s^{-1}, in the CMB frame, directed towards the Local Group. This non-detection is in marginal conflict with previous Tully-Fisher studies. Two clusters in the background of the supercluster exhibit significant negative peculiar velocities, characteristic of backside infall into PP.

A bulk-flow fit to all 16 clusters reveals a statistically insignificant motion of 430 ± 280 km s^{-1} towards $l = 265°, b = 26°$ (CMB frame). Comparison with the velocity field predicted from the IRAS 1.2Jy survey yields $\beta = 1.0 \pm 0.5$. We find no evidence for residual bulk motions generated by mass concentrations beyond the limiting depth of the IRAS density field.

1   Introduction

The Perseus–Pisces (PP) supercluster, at $cz \sim 5000$ km s^{-1}, lies directly opposite the apex of the large-scale streaming detected by Lynden-Bell et al. (1988). If, as in the flow model of Faber & Burstein (1988), the local dynamics are dominated by a single 'Great Attractor' (GA) beyond Hydra–Centaurus, then the peculiar velocity at PP is predicted to be $\lesssim 150$ km s^{-1}. If however, very distant sources are responsible for large-scale motions, then the PP region would be expected to share in a bulk streaming velocity of ~ 500 km s^{-1}.

Measurements of bulk motion in PP have been made by Willick (1991) and by Han & Mould (1992). These studies, employing the Tully–Fisher (TF) relation for spiral galaxies, both argue for a large bulk motion (~ 400 km s^{-1}). Courteau et al. (1993) support this result, and invoke very large-scale density fluctuations to account for the large coherence length of the flow.

Motions in the PP region have not been well-studied using elliptical galaxy distance indicators. Here, we describe a programme to measure peculiar motions for 7 clusters in PP, using the Fundamental Plane (FP) relation for early-type galaxies. Spectroscopic and photometric data and methods are presented in full by Smith et al. (1997a). Hudson et al. (1997) present a detailed account of the FP fits and velocity field analyses.

2   A New Survey

The current survey is based on a sample of early-type galaxies in clusters. Clusters are chosen since *a priori* grouping in redshift-space reduces the magnitude of Malmquist bias effects. This is especially desirable in regions of high density contrast, such as PP, where corrections for inhomogeneous Malmquist bias would otherwise be large and uncertain. Early-type galaxies provide the best means of probing cluster cores with minimal contamination from field galaxies.

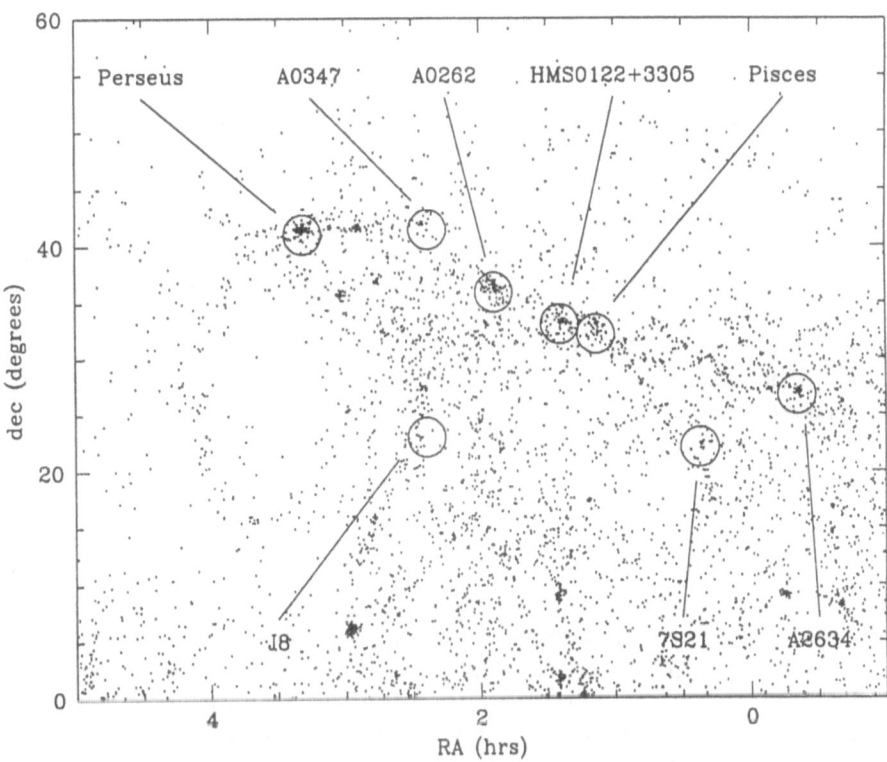

Fig. 1. Projected distribution of $cz < 12000\,\mathrm{km\,s^{-1}}$ galaxies in the PP region, selected from the ZCAT compilation of Huchra et al. (1993). Open circles mark the location of clusters observed in this programme.

Figure 1 shows the projected distribution of the cluster sample, along with the projected galaxy distribution. The supercluster 'ridge' at $cz \sim 5000\,\mathrm{km\,s^{-1}}$ is evident in the plot, and is traced by six observed clusters (Perseus, A0347, A0262, HMS0122+3305, Pisces and 7S21). The sample includes the cluster J8, located in the background of the ridge, at $cz \sim 9000\,\mathrm{km\,s^{-1}}$. The cluster A2634 is formally part of our calibration sample, but can also be considered as a PP background cluster.

2.1 Photometry

The fundamental plane parameters R_e and $\langle\mu\rangle_e$ are measured from R-band photometric observations obtained at the 1m Jacobus Kapteyn Telescope on La Palma. Standard reduction techniques are applied, with parameters finally being derived from an $R^{1/4}$-law fit to the aperture magnitude growth-curve. Internal and external comparisons indicate that measurement uncertainties in the photometric parameters contribute distance errors smaller than 1.5% per galaxy.

2.2 Spectroscopy

Velocity dispersions, σ, are derived from spectra obtained at the 2.5m Isaac Newton Telescope on La Palma. Internal comparisons indicate an average uncertainty of 7.6% per measurement. The aperture correction of Jørgensen, Franx & Kjærgaard (1995b) is adopted, with all σ measurements referred to a physical aperture of diameter $1.19\,h^{-1}$ kpc.

In order to construct large samples of peculiar velocity data, it is necessary to merge measurements from different observing runs, instrumental configurations, observers, etc. Despite careful attempts to correct for aperture effects, systematic offsets often persist at the 3-4% level between velocity dispersions measured on different systems. Whilst the precise cause of these offsets is not fully understood, it is vital that they are corrected for and (most importantly) that the uncertainties on these corrections are included in the overall error budget.

The removal of systematic offsets can be achieved by intercomparison of measurements for galaxies common to two or more systems. The set of overlap observations used here includes 1300 measurements for 350 different galaxies, on 19 systems. We adopt the (aperture-corrected) Lick system of Davies et al. (1987) as our 'standard system'. For all other systems, we determine corrections required to match onto this standard. Since many galaxies have data on more than two systems, the fit has been performed simultaneously for all the corrections.

From bootstrap realisations of this process, we estimate the error on each system correction and also the correlation between the errors for different systems. The mean systematic error in σ can then be computed for each cluster. For the PP ridge clusters, this 'system matching' error is $\sim 100\,\mathrm{km\,s^{-1}}$.

2.3 Fundamental Plane Fits

Since the PP cluster sample studied here has very limited sky coverage, it would be impossible, given the present data alone, to differentiate between a bulk motion of PP and a velocity zero-point error. In order to resolve this degeneracy, we utilise data for a well-distributed sample of 'calibration clusters'. Specifically, we include six rich clusters (A0194, A0539, A3381, A3574, DC2345-28, Hydra) from the work of Jørgensen, Franx & Kjærgaard (1996), and three well-studied clusters (Coma, A2199, A2634) from Lucey et al. (1997). Photometric parameters from the three sources can be brought onto a consistent system by the application of well defined offsets (Smith et al. 1997a).

Using the resulting sample of 16 clusters, we fit simultaneously for the global FP slope parameters, and for the individual cluster zero-points. The fit is performed by minimizing residuals in $\log \sigma$. This 'inverse fit' is unbiased by selection on photometric parameters (radius, surface brightness or any combination thereof, such as total magnitude). The inverse fit would, however, be biased by any explicit selection on $\log \sigma$. In the present work, no such selection occurs, since galaxies are not thrown out of the sample, *a posteriori*, based on their measured velocity dispersions.

The best fitting fundamental plane for the sample is given by

$$\log R_{\mathrm{e}} = 1.383(\pm 0.040)\log\sigma + 0.326(\pm 0.011)\langle\mu\rangle_{\mathrm{e}} + \gamma_{\mathrm{cl}}$$

with scatter equivalent to a distance error of 20% per galaxy. The cluster zero-points, γ_{cl}, are used to derive relative distances, as discussed below. In practice we find that an initial calibration, based on the assumption of zero peculiar motion for Coma, leads to a mean radial peculiar velocity for the sample which is indistinguishable from zero. Hereafter, we adopt this simply-defined zero-point for the present analysis.

In Figures 2 and 3 we show the FP fits for the 16 clusters separately. The slopes derived for individual clusters are consistent in every case with the slope of the globally-defined FP.

3   Results

3.1   Peculiar Velocities

Distances and peculiar velocities for the 16 cluster sample are presented in Table 1. Here, the distance estimates have been corrected for homogeneous Malmquist bias. The use of a cluster sample leads to corrections of only 0.5–2% for this effect. Inhomogeneous Malmquist bias corrections are expected to be smaller than this, and are neglected here. The velocity vectors for clusters in PP are shown in graphical form in Figure 4.

The present data do not indicate a significant peculiar motion for any of the six clusters in the ridge of PP. In the background of the supercluster, J8 and A2634 display large, significant peculiar velocities in the sense expected if they are involved in 'backside infall' into PP.

In Table 2, we present comparisons between the current results and peculiar velocities measured by other authors. The agreement is good for most clusters.

3.2   Bulk Flow Fits

As a simple description of the local velocity field, we consider a bulk flow model with three free parameters. Fitting for all 16 clusters, we find a bulk velocity of $430 \pm 198 \,\mathrm{km\,s^{-1}}$, in the CMB frame, directed towards $l = 264.6°, b = -25.6°$. Uncertainties in the matching of spectroscopic systems raise the total error on

Table 1. Cluster distances and peculiar velocities. Column 3 indicates the source of photometric and spectroscopic data, in that order. For each cluster, N_{gal} indicates the number of galaxies used in the fit. Distances and peculiar velocities (in the CMB frame) are given in $km\,s^{-1}$. The velocity error is a 1σ random error including a contribution from uncertainty in the cluster redshift.

	Cluster	Data Source	N_{gal}	Distance	Peculiar Velocity	Velocity error
PP ridge	7S21	1,1	7	5448	69	450
	Pisces	1,1	25	4583	131	208
	HMS0122	1,1	9	4680	−44	352
	A0262	1,1	10	4787	−259	340
	A0347	1,1	8	5590	−277	430
	Perseus	1,1	31	5176	−136	307
PP background	J8	1,1	13	10449	−1032	602
	A2634	2,2	35	10111	−960	364
Calibrators	A2199	2,2	36	9285	−342	325
	Coma	2,2	71	7200	0	204
	A0194	3,4+5	19	4379	743	230
	A0539	3,4	22	8380	234	402
	A3381	3,4	14	10883	578	593
	A3574	3,4	7	4217	655	369
	DC2345-28	3,5	27	8645	−174	362
	Hydra	3,4+5	18	3946	30	239

References for data sources: 1 – Smith et al. (1997a); 2 – Lucey et al. (1997); 3 – Jørgensen, Franx & Kjærgaard (1995a); 4 – Jørgensen, Franx & Kjærgaard (1995b); 5 – Lucey & Carter (1988).

Table 2. Comparisons of present results with peculiar velocities measured from the TF relation by Han & Mould (1992), and from the FP by Baggley (1996) and Scodeggio et al. (1997). Baggley's results are based on a subset of the EFAR data. Note that our 'HMS0122' is the 'N507 group' of Scodeggio et al.

Cluster	This study	Han & Mould (TF)	Baggley (FP)	Scoddegio (FP)
Pisces	131±208	76±296		
HMS0122	−44±352	311±374		420±238
A0262	−259±340	−576±391	782±429	−157±203
A2634	−960±364	−906±639	−978±967	−308±336
A2199	−342±325		697±497	
Coma	(0±204)	80±428	(0±356)	42±196
A0539	234±402	11±621		
DC2345-28	−174±362		591±403	
Hydra	30±239	184±296		

FP

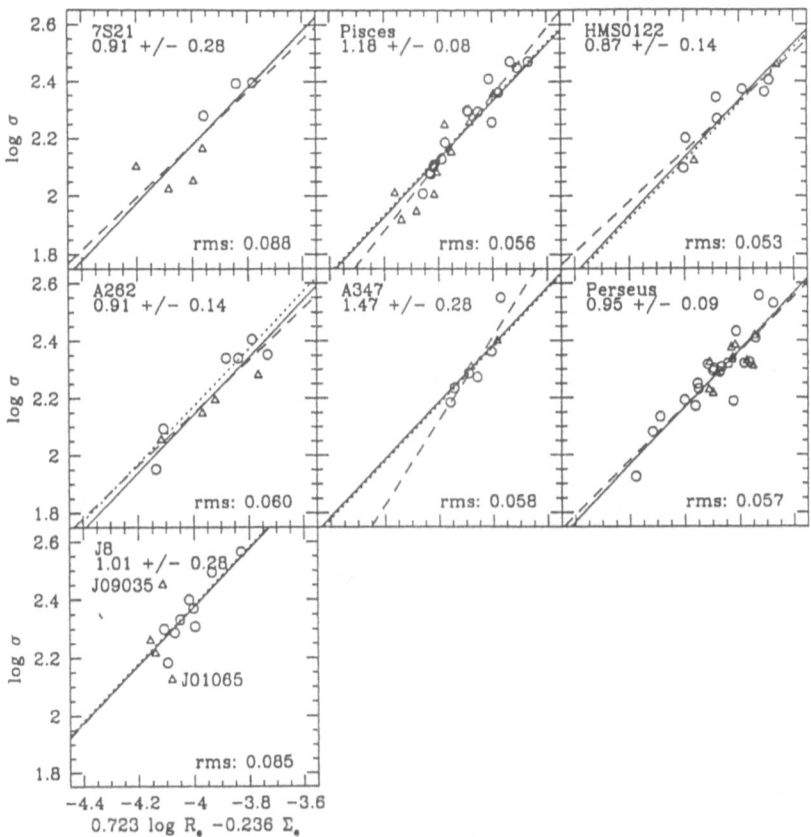

Fig. 2. FP data and fits for the clusters 7S21, Pisces, HMS0122, A0262, A0347, Perseus and J8. Early type galaxies (E, E/S0, D or cD) are indicated by circles, later types by triangles. The solid line shows the global inverse FP, found by minimising σ residuals simultaneously over the whole cluster sample with the same slope but varying zero-points for each cluster. Each cluster's measured scatter in σ around this global fit is given in the lower right-hand corner. Galaxies which deviate from the global fit by more than 2.5 times the global scatter are labelled. The dotted line shows the median of the residuals with the slope fixed from the whole cluster sample. The dashed line shows the best slope and zero-point fit to the individual cluster. The individual cluster slope relative to the global slope is given in the upper left-hand corner.

FP

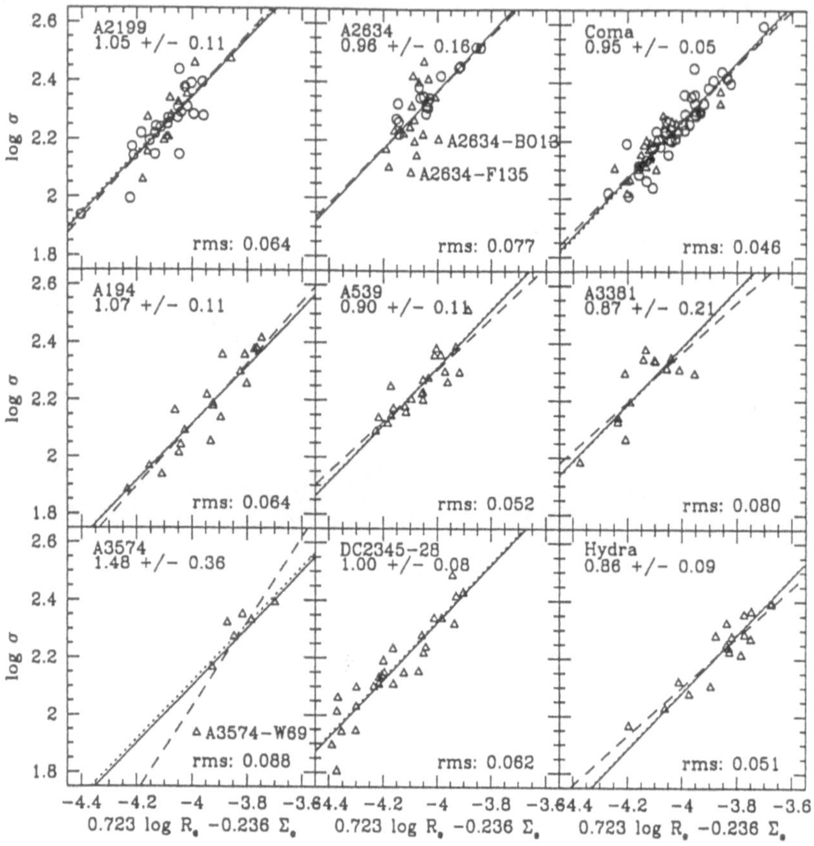

Fig. 3. FP data and fits for the calibration clusters A2199, A2634, Coma, A0194, A0539, A3381, A3574, DC2345-28 and Hydra. Symbols are as in Figure 2 for Coma, A2199 and A2634. For other clusters, all galaxies are indicated by triangles, irrespective of type. Lines are as in Figure 2.

the bulk-flow amplitude to $280 \, \mathrm{km \, s^{-1}}$. The bulk motion of this sample of clusters is therefore not statistically significant.

Fitting a bulk-flow model for only the six PP ridge clusters, no streaming motion of the supercluster is detected. The fit yields a PP bulk-flow of $59 \pm 221 \, \mathrm{km \, s^{-1}}$ (toward the Local Group). The quoted error, includes system matching uncertainties and a contribution from the uncertainty in setting a velocity zero-point for the sample.

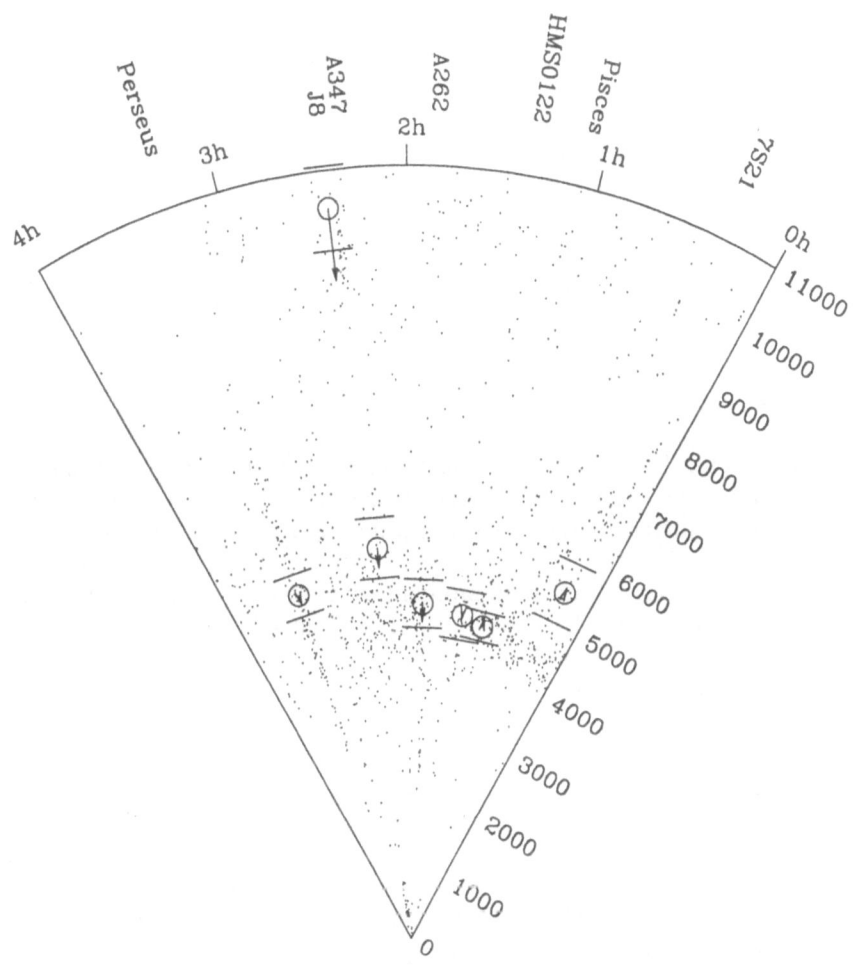

Fig. 4. Cone diagram showing peculiar velocity vectors deduced from this study. Small points indicate the CMB-frame redshift-space positions of galaxies from the ZCAT compilation (Huchra et al. 1993). PP clusters are shown at their inferred distances by open circles, with vectors extending to their CMB-frame velocities to indicate peculiar motions. 1σ position errors are indicated by the bracketing lines. The six ridge clusters have peculiar motions indistinguishable from zero. In the background of the supercluster, J8 has a significant peculiar velocity, as does A2634 which lies just outside the range of this plot.

3.3   Comparison with the Predicted Velocity Field

A more realistic model of the flow field can be obtained by predicting the peculiar velocity of each cluster from a suitable redshift survey, with $\beta = \Omega^{0.6}/b$ as a single free parameter. Here we employ the IRAS 1.2 Jy survey, within $12000\,\mathrm{km\,s^{-1}}$, the IRAS density field being kindly provided to us by M. Strauss.

The best fit to the observed cluster velocities is obtained for $\beta = 0.94 \pm 0.48$. While this represents a marginal detection of β, the errors are too large for useful constraints to be derived from the present sample.

After subtraction of the IRAS-predicted peculiar velocities from the observed motions, the residual (CMB frame) bulk motion of the sample is $383 \pm 294\,\mathrm{km\,s^{-1}}$, directed towards $l = 313.2°, b = -26.4°$. We conclude from this non-detection, that mass concentrations beyond $12000\,\mathrm{km\,s^{-1}}$ are not required to explain the observed velocities.

4   Conclusions

We have completed a new survey of cluster motions in the PP region, using the FP as a distance indicator. Careful attention has been paid to the construction of a homogeneous, merged dataset, and to accounting for systematic uncertainties in this procedure.

We derive an insignificant bulk motion (-60 ± 220 $\mathrm{km\,s^{-1}}$ towards the Local Group) for 6 clusters in the main ridge of the supercluster. This result is in marginal conflict with Tully–Fisher surveys by Willick (1991), Han & Mould (1992) and Courteau et al. (1993). Two clusters in the background of PP show evidence for 'backside infall' into the supercluster.

Comparing the observed cluster motions with the velocity field predicted from the IRAS 1.2Jy survey, we find a best fit for $\beta = 1.0 \pm 0.5$. There is no evidence for residual bulk flows generated by sources beyond the IRAS density field limit.

A new, all-sky, survey of ~ 50 Abell clusters within $12000\,\mathrm{km\,s^{-1}}$ is currently in progress (see Smith et al. 1997b, this volume). This improved sample, will yield useful constraints on β, in addition to a reliable measurement of the bulk motion.

References

Aaronson, M., Bothun, G., Mould, J.R., Huchra, J.P., Schommer, R., Cornell, M. 1986, ApJ, 302, 536

Baggley, G., 1996, PhD thesis, University of Oxford

Courteau, S., Faber, S.M., Dressler, A., Willick, J.A. 1993, ApJ, 412, L51

Davies, R.L., Burstein, D., Dressler, A., Faber, S.M., Lynden-Bell, D., Terlevich, R.J., Wegner, G. 1987, ApJS, 64, 581

Faber, S.M., Burstein, D. 1988, in Coyne, G., Rubin, V.C., eds, Proceedings of the Vatican Study Week, Large Scale Motions in the Universe. Princeton Univ. Press, Princeton, p. 135

Han, M.S., Mould, J.R. 1992, ApJ, 396, 453

Huchra, J.P., Geller, M.J., Clemens, C.M., Tokarz, S.P., Michel, A. 1993, Astronomical Data Center archives

Hudson, M.J., Lucey, J.R., Smith, R.J., Steel, J. 1997, MNRAS, submitted

Jørgensen, I, Franx, M., Kjærgaard, P. 1995a MNRAS, 273, 1097

Jørgensen, I, Franx, M., Kjærgaard, P. 1995b MNRAS, 276, 1341

Jørgensen, I, Franx, M., Kjærgaard, P. 1996 MNRAS, 280, 167

Lucey, J.R., Carter, D. 1988, MNRAS, 235, 1177

Lucey, J.R., Guzmán, R, Steel, J., Carter, D. 1997, MNRAS, submitted

Lynden-Bell, D., Faber, S.M., Burstein, D., Davies, R.L., Dressler, A., Terlevich, R.J., Wegner, G. 1988, ApJ, 326, 19

Scodeggio, M. et al. 1996, this volume

Smith, R.J., Lucey, J.R., Hudson, M.J., Steel, J. 1997a, MNRAS, submitted

Smith, R.J., Hudson, M.J., Lucey, J.R., Schlegel, D.J., Davies, R.L., Baggley, G. 1997b, this volume

Willick, J.A. 1991, ApJ, 351, L51

The EFAR Project: Monte Carlo Testing of the Fundamental Plane Distance Estimator

R.P. Saglia[1], Matthew Colless[2], G. Baggley[3], Edmund Bertschinger[4], David Burstein[5], Roger L. Davies[3], Robert K. McMahan, Jr.[6] and Gary Wegner[7]

[1] Institut für Astronomie und Astrophysik, Scheinerstraße 1, D-81679 Munich, Germany
[2] Mount Stromlo and Siding Spring Observatories, The Australian National University, Weston Creek, ACT 2611, Australia
[3] Dept of Physics, University of Durham, South Road, Durham, DH1 3LE, UK
[4] Dept of Physics and Astronomy, Arizona State University, Tempe, AZ 85287-1504
[5] Dept of Physics, MIT, Cambridge, MA 02139
[6] Dept of Physics and Astronomy, University of North Carolina, CB#3255 Phillips Hall, Chapel Hill, NC 27599-3255
[7] Dept of Physics and Astronomy, Dartmouth College, Wilder Lab., Hanover, NH 03755

Abstract. The EFAR project is a long-term study of 736 candidate early-type galaxies in 84 clusters lying in two directions towards Hercules-Corona Borealis and Perseus-Pisces-Cetus at distances 6000-15000 km/s. In the following we discuss the goals of the project and the properties of the photometric and spectroscopic database. We present a new method, based on maximum likelihood, to determine the parameters of the Fundamental Plane and describe the distribution of galaxies in it. The algorithm takes into account the effects of errors and selection and is superior to least-squares procedures.

1 Introduction

The EFAR (Ellipticals FAR away) project started in 1986 to construct an accurate and homogeneous photometric and spectroscopic database for a large sample of early-type galaxies with redshifts 6000-15000 km/s, distributed in 84 clusters in the Hercules-Corona Borealis and Perseus-Pisces-Cetus regions. Our goals were to study the structural properties of elliptical galaxies and determine their peculiar motions in two independent regions outside the local supercluster using the Fundamental Plane distance estimator. Both scientific goals are important, since they provide powerful tests of theories of galaxy evolution and of the formation of large scale structure in the Universe.

We have completed the observing and data reduction phases of the project and are in the process of publishing the database and analyzing it. The following papers describe the results obtained so far. Colless et al. (1993) present the photoelectric photometry database, Saglia et al. (1993) discuss the effects of seeing on the photometric properties of early-type galaxies, Wegner et al. (1996, Paper I) introduce the EFAR cluster and galaxy sample and its selection properties, Wegner et al. (1997, Paper II) present the spectroscopic database, Saglia et al.

(1997a, Paper III) the photometry database, and Saglia et al. (1997b, Paper IV) the photometric fitting technique. In the following we summarize the properties of the photometric and spectroscopic database (Section 2) and describe the maximum likelihood method we have developed to determine the EFAR Fundamental Plane (Section 3). Section 4 gives our conclusions.

2   The Photometric and Spectroscopic Database

We invested 185 nights of photoelectric and CCD observations with 1m class telescopes and more than 100 nights of spectroscopic observations with 2m to 4m class telescopes to build up the EFAR database.

We collected 2846 R band CCD images for 776 galaxies with 87% of the galaxies having more than one profile. A total of 1278 photometric calibrations were obtained using photoelectric photometry (in both the R and the B bands) and CCD R band photometry. Repeated calibrations show a zero-point precision of 0.03 mag per observation (see Paper III). We computed the circularly averaged surface brightness profiles and the photometric parameters of the EFAR galaxies using a seeing-convolved, $R^{1/4}$ plus exponential fitting algorithm which optimally combines multiple profiles and corrects for sky subtraction errors (see Paper IV).

The precision of the derived D_n diameters, half-luminosity radii R_e, total magnitudes m_T and average effective surface brightness $\langle SB_e \rangle$ is estimated with the help of extensive Monte Carlo simulations to test for systematic effects. We find that the systematic errors on m_T and R_e are less than 0.15 mag and 25% respectively for 90% of our sample, while the systematic errors on the combined quantity $\log R_e - 0.3\langle SB_e \rangle$ which enters the Fundamental Plane equation always remain smaller than 0.03 dex, and remain below 0.015 dex for 90% of the galaxies. The morphological classification based on CCD images shows that 32% of the sample objects, visually selected from photographic images as early-type, are in fact spirals or barred S0 galaxies. The remaining 68% classified as early-type can be subdivided into cD (8%), E (11% with a simple $R^{1/4}$ law best fit) and E/S0 (49% with a disk plus bulge best fit). Therefore most of the profiles deviate significantly (at the 3σ level and in 90% of the cases at more than the 5σ level, see Paper III) from a de Vaucouleurs law, needing a non-negligible exponential component, with a typical "disk-to-bulge" ratio $D/B \approx 0.5$. In these cases, photometric parameters derived using a simple $R^{1/4}$ law (as commonly done in the literature) are affected by large systematic errors, easily larger then 0.3 mag on m_T or 0.2 dex on $\log R_e$.

A total of 1327 spectra (see Paper II) have been collected for 659 EFAR galaxies (1250 spectra) plus 48 galaxy standards (77 spectra). We used both long-slit and fiber-fed spectrographs, spanning MgI b, the Fe lines at 5207 Å and 5269 Å, and (usually) Hβ for cz = 6000 − 15000 km/s, with some spectra going from CaI H+K (3933 Å) to NaI D (5892 Å). The typical resolution was $\Delta\lambda = 1 - 2$ Å/pixel, or $\sigma_{instr} = 80 - 150$ km/s, and we reached a typical signal-to-noise ratio S/N$\approx$ 26 per Å. We derive galaxy redshift and central velocity dispersions σ using the cross-correlation algorithm and we estimate the precision of the results by means of Monte Carlo simulations. The median error on the combined redshift measurements for each galaxy is 18 km/s; 90% of the redshifts have errors less than 36 km/s. Likewise, the median error on the combined σ measurements is 9% per galaxy, and 90% have errors less than 18%. We can determine reliably the Mgb index, while we find that the Mg$_2$ index is subject to significant systematic effects arising from continuum shape variations. We apply aperture corrections to both σ and Mgb values which take into account the slit width, the galaxy distance and the galaxy size. The Mgb–σ relation for the EFAR galaxies does not deviate from previous determinations. The histogram of the zero-points of the relation for the sample of the EFAR clusters has an rms of 0.02 dex. This suggests that environmental effects, if present, are small.

3   A Maximum Likelihood Fundamental Plane Algorithm

Figure 1 shows the distribution of EFAR galaxies in the $\log R_e, \log \sigma, \langle SB_e \rangle$ space which was constructed using redshift distances. Early-type galaxies do not fill this three-dimensional space, rather they occupy a narrow region around a Fundamental Plane defined by the equation $\log R_e = a \log \sigma + b \langle SB_e \rangle + c$ (Dressler et al. 1987, Djorgovski & Davis 1987). The canonical methods used to determine the coefficients a and b are of the least-squares type: one minimizes the orthogonal (absolute or squared) residuals from the plane (Jørgensen et al. 1996, Pahre 1996, Scodeggio et al. 1996) or the (square) residuals from one of the variables (Smith et al. 1996).

We find that this approach can lead to biased coefficients, if selection effects, sizeable random errors and well-defined exclusions (e.g., due to resolution we cannot observe galaxies with $\sigma \leq \sigma_{cut}$ km/s) are present, as in the case of the EFAR sample. Fig. 2, top, shows the histograms of the coefficients recovered using the regressions on one of the variables, for 99 Monte Carlo realizations of the EFAR data sample, which take into account the actual photometric and spectroscopic errors. We assumed that only $\sigma > \sigma_{cut} = 120$ km/s could be measured. The three regressions (and any sort of mean of them) perform very poorly. The mean values of the reconstructed parameters are $a = 0.90, 1.47, 0.93$ and $b = 0.33, 0.34, 0.37$ for the regressions against $\log R_e$, $\log \sigma$, and $\langle SB_e \rangle$ respectively, while the true values are $a = 1.3, b = 0.35$. Fig. 2, bottom, shows the histograms of the coefficients recovered using an improved orthogonal minimization algorithm. The algorithm excludes points falling below an orthogonal cut determined by the largest $\sigma < \sigma_{cut}$, with an optional selection weighting, where every point is inversely weighted by the sample selection function, based on galaxy diameter (see Eq. 6). While the improved orthogonal minimization with selection weighting gives the best result, still the mean of the recovered coefficients is biased.

Better results can be obtained by modeling the distribution of galaxies within the FP. Let us define the coordinates relative to the cosmic averages as $r = \log R_e - \overline{\log R_e} + \log(1 + z_j)$, $s = \log \sigma - \overline{\log \sigma}$, $u = \langle SB_e \rangle - \overline{\langle SB_e \rangle}$, where z_j are

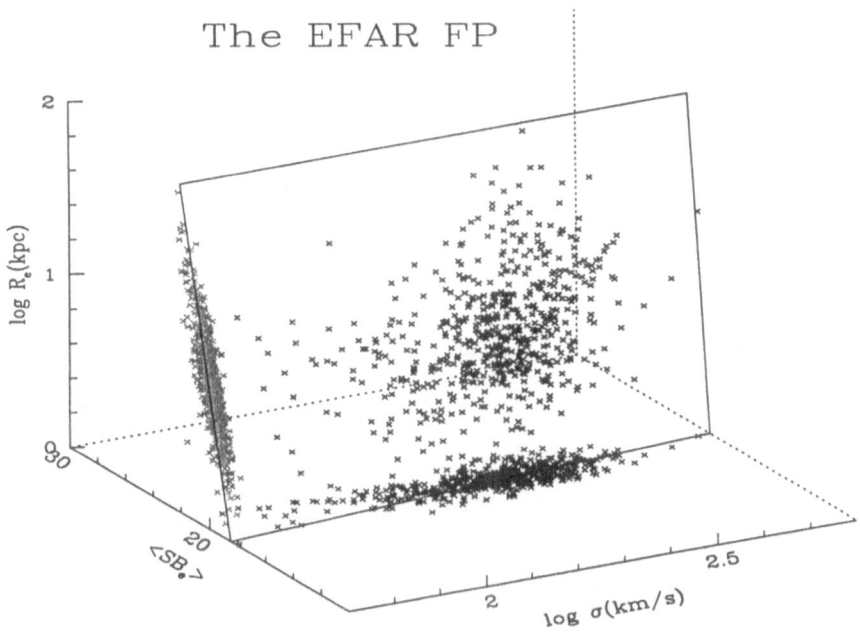

Fig. 1. The distribution of EFAR galaxies in the $\log R_e, \log \sigma, \langle SB_e \rangle$ space, using red-shift distances.

the cluster peculiar redshifts, and the set of rotated coordinates v_1, v_2, v_3:

$$v_1 = \frac{r - as - bu}{\sqrt{1 + a^2 + b^2}},$$

$$v_2 = \frac{br + u}{\sqrt{1 + b^2}}, \qquad (1)$$

$$v_3 = \frac{-ar - \sqrt{1 + b^2}\,s + abu}{\sqrt{1 + b^2}\sqrt{1 + a^2 + b^2}},$$

where v_1 measures the displacement from the FP, and v_2 and v_3 are two or-thogonal vectors in the FP. A look at the distribution of EFAR galaxies (Fig-ure 3) shows that with this choice of coordinates the (normalized) distribu-tion P of galaxies in the FP can be factorized as $P(|v>=(v_1, v_2, v_3)) = f_1(v_1)f_2(v_2)f_3(v_3)$. Moreover, it is not unreasonable (for example, see Willick 1994) to assume that f_1, f_2, f_3 can be well approximated by gaussians:

$$P(|v>) = \frac{1}{(2\pi)^{3/2}\sigma_1\sigma_2\sigma_3} \exp\left[-\left(\frac{v_1^2}{2\sigma_1^2} + \frac{v_2^2}{2\sigma_2^2} + \frac{v_3^2}{2\sigma_3^2}\right)\right], \qquad (2)$$

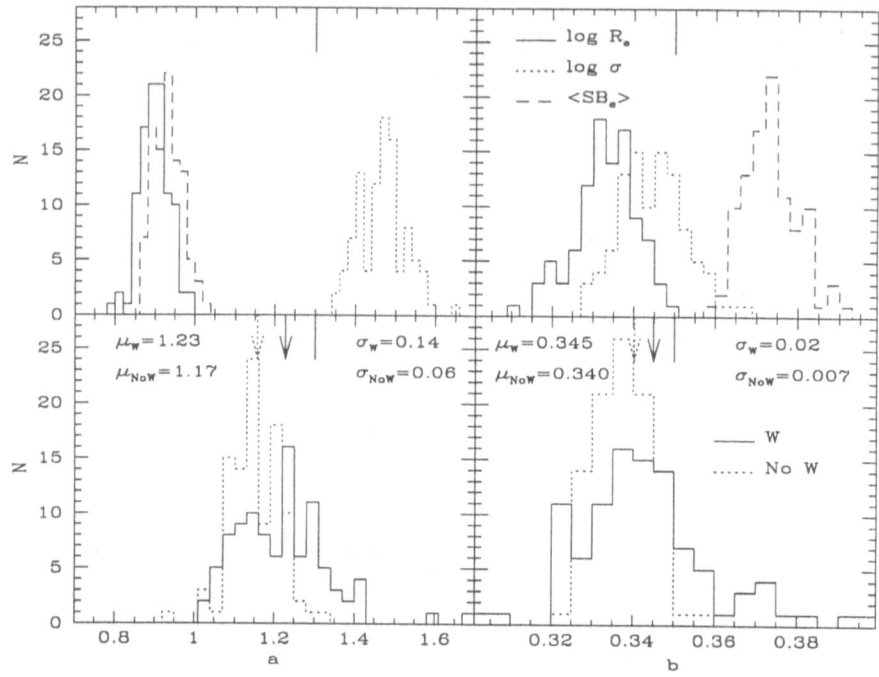

Fig. 2. The histograms of the FP coefficients a and b recovered with different least-square procedures for 99 Monte Carlo realizations of the EFAR data sample, without peculiar velocities. The top panels show the results of the regressions against $\log R_e$ (full line), $\log \sigma$ (dotted line), and $\langle SB_e \rangle$ (dashed line). The bottom panels show the results of the orthogonal residual minimization, with selection function weighting (full line) and without (dotted line). The bars at the top show the true values of a (1.3) and b (0.35), the arrows the mean values recovered. The mean μ and dispersion σ of the distribution of the recovered values are also given for the case with selection weighting (label W) and without (label NoW).

where σ_1, σ_2, σ_3 are the relative sigmas of the distributions, with $\sigma_1 << \sigma_2, \sigma_3$. When written as a function of the original coordinates $|x> = (r, s, u)$ the distribution P becomes:

$$P(|x>) = \frac{|F|^{1/2}}{(2\pi)^{3/2}} \exp(- <x|F|x> /2), \qquad (3)$$

where we have adopted the quantum mechanical formalism for the scalar product, so that $|v> = R|x>$ and R is the rotation matrix defined in Eq. 1. In addition one has $F = R^T DR$, $D_{ij} = 1/\sigma_i^2 \delta_{i,j}$, and $|M|$ denotes the determinant of the matrix M. F is called the covariance matrix and is symmetric and positively defined; its inverse $V = F^{-1}$ is the variance matrix.

The observed galaxy distribution deviates systematically from the true one (which we model by Eq. 2), since measurement errors and selection effects are

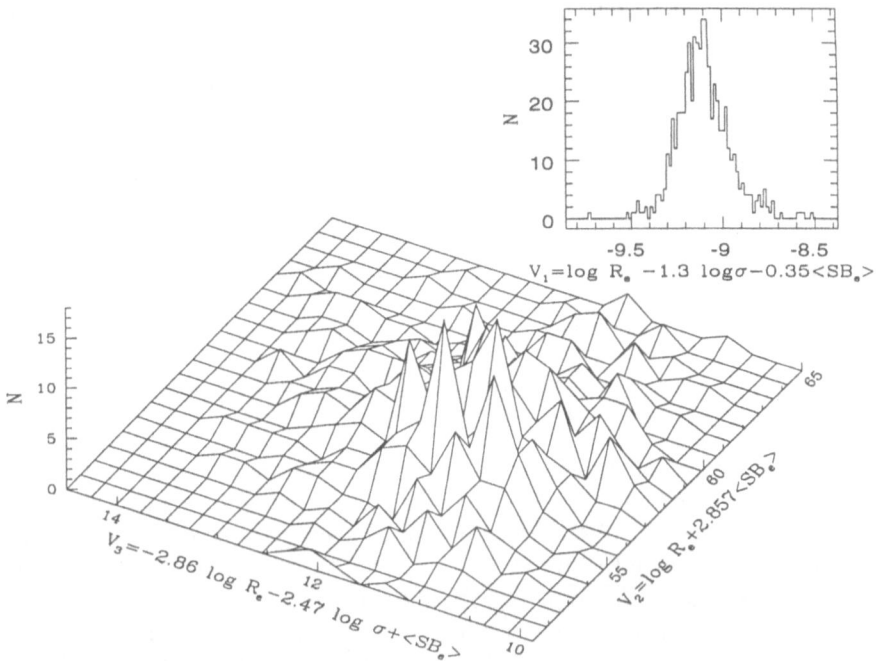

Fig. 3. The distribution of EFAR galaxies within the FP of coefficients $a = 1.3$, $b = 0.35$. The inserted plot to the upper right shows the distribution orthogonally to the plane.

present. Assuming gaussian error distributions, measurement errors act as a convolution of Eq. 3 with a gaussian function of variance matrix E:

$$E = \begin{pmatrix} (1+\alpha^2)\delta r^2 & 0 & \frac{(1+\alpha^2)\delta r^2 - \delta FP^2}{\alpha} \\ 0 & \sigma_s^2 & 0 \\ \frac{(1+\alpha^2)\delta r^2 - \delta FP^2}{\alpha} & 0 & \frac{(\alpha^4-1)\delta FP^2 + (1+\alpha^2)^2\delta r^2)}{\alpha^2(1+\alpha^2)} + \delta ZP^2 \end{pmatrix}, \quad (4)$$

where δr is the error on $\log R_e$, δFP is the error on the combined quantity $FP = \log R_e - \alpha \langle SB_e \rangle$, with $\alpha \approx 0.3$, δZP is the photometric zero-point error (see Paper III and IV), and σ_s the error on $\log \sigma$ (see Paper II). The resulting normalized distribution reads

$$P^c(|x>) = \frac{|C|^{1/2}}{(2\pi)^{3/2}} \exp(- <x|C|x> /2), \quad (5)$$

where $C = (E + V)^{-1}$.

Selection effects operate on all variables (r, s, u). Only velocity dispersions larger than the instrumental resolution σ_{cut} (see Paper II) can be measured, therefore $P^c(|x>) = 0$ if $s \le s_{cut} = \log \sigma_{cut} - \overline{\log \sigma}$. The EFAR galaxy selection

S is described in Paper I and is a function of the D_W diameter measured in arcsec:

$$S_j(\log D_W) = 0.5\left\{1 + \mathrm{erf}\left[\frac{\log D_W - \log D^0_{Wj}}{\delta_{Wj}}\right]\right\},\qquad(6)$$

where $\log D^0_{Wj}$ is the midpoint, δ_{Wj} the width of the cutoff in the selection function, and j is the cluster index. The diameter $\log D_W$ correlates with $FP = \log R_e - 0.3\langle SB_e\rangle$ with 0.09 dex scatter in FP (see Paper III). Taking into account this correlation, one can write the observed probability distribution as:

$$P^o(|x>) = \exp(- < x|C|x > /2)\theta(s - s_{cut})$$
$$\left[1 + \mathrm{erf}\frac{1.28FP + d_0 - \log D^0_W}{\sqrt{\delta^2_W + 2\sigma^2_W}}\right]/N,\qquad(7)$$

where d_0 is a constant depending on the distance D, σ_W the dispersion in the relation between FP and $\log D_W$, $s_{cut} = \log \sigma_{cut} - \overline{\log \sigma}$, $\theta(s - s_{cut}) = 0$ if $s \leq s_{cut}$, $\theta = 1$ otherwise, and N is the normalization constant so that $\int P^o d^3x = 1$.

Given the sample of EFAR datapoints ($|x_i>$), the maximum likelihood choice of parameters $a, b, \overline{\log R_e}, \overline{\log \sigma}, \langle SB_e\rangle, \sigma_1, \sigma_2, \sigma_3$, and the vector of cluster peculiar redshifts z_j can be determined by maximising the likelihood function $L = \prod P_i^o$. We adopt a simplified version of this approach, which yields an unbiased estimate of the parameters (Eadie et al. 1971), by maximizing the following quantity:

$$\ln Prob = \sum_{\sigma > \sigma_{cut}} \frac{1}{S_j(\log D^i_W)} \ln \frac{P^c(|x_i>)}{f_i}\qquad(8)$$

$$= -\sum_{\sigma > \sigma_{cut}} \frac{1}{S_j(\log D^i_W)}\left[0.5 < x_i|C_i|x_i> + \ln f_i + 1.5\ln(2\pi) - 0.5\ln|C_i|\right],$$

where $f_i = \int P_i^c \theta(s - s_{cut})d^3x = 0.5(1 - \mathrm{erf}(y_i))$ with $y_i = s_{cut}/\sqrt{2|C_i|/|C_i^{(2,2)}|}$ is the fraction of galaxies with $\sigma > \sigma_{cut}$ and $C_i^{(2,2)}$ is the matrix obtained by dropping the second column and row of the matrix C_i.

Fig. 4 shows the histograms (full lines) of the FP distribution parameters recovered maximizing Eq. 8 for the same 99 Monte Carlo realizations of the EFAR data sample of Fig. 2. The plots at the top show the recovered coefficients of the Fundamental Plane a and b, the ones in the middle the values of the cosmic averages $\overline{\log R_e}, \overline{\log \sigma}, \langle SB_e\rangle$ and of the normalizing coefficient f_i for one of the simulated galaxies. The plots at the bottom show the reconstructed values of σ_1, σ_2 and σ_3 (see Eq. 2) and the values of Eq. 8. On the mean, the parameters are recovered without bias. As a second step, we add a random "peculiar redshift field" $\log(1 + z_j)$ of rms=0.05 dex, which corresponds to peculiar velocities of the order of 12% of the distance, or to the (rather large for testing purpose) peculiar velocity of 1200 km/s at 10000 km/s (a typical redshifts of the EFAR clusters). The same result as before holds (Fig. 4, dotted lines) if the parameters

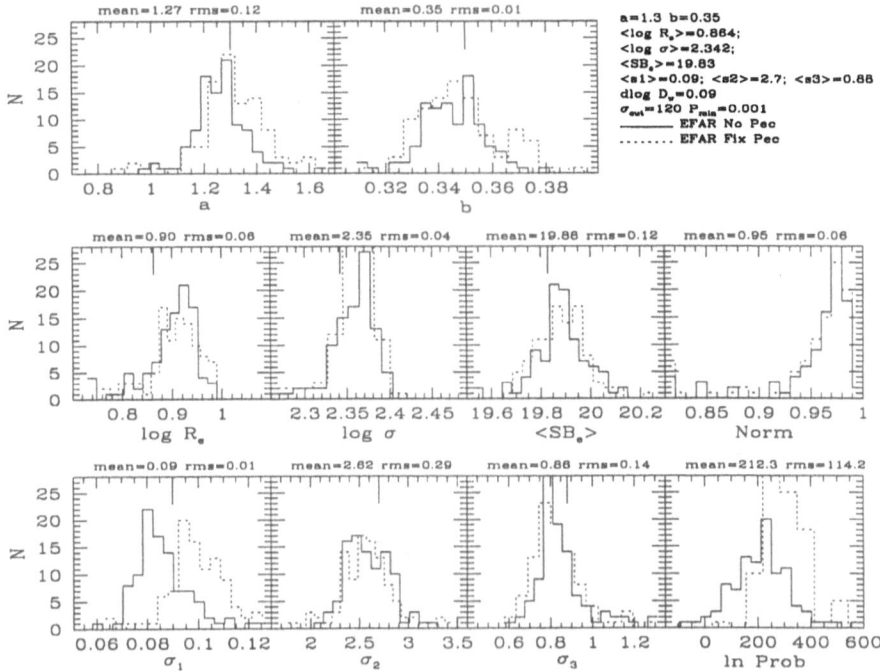

Fig. 4. Full lines: the histograms of the FP distribution parameters recovered with the maximum likelihood algorithm for the 99 Monte Carlo realizations of the EFAR data sample of Fig. 2. Mean and rms values are given at the top. Dotted lines: the same quantities reconstructed from simulations with a random "peculiar redshift field" $\log(1 + z_j)$ of rms=0.05 dex.

are reconstructed ignoring first the peculiar redshifts. The rms of the mean values are however larger. In addition, the reconstructed σ_1 are systematically larger than the input value.

The Monte Carlo simulations can also be used to estimate how well we can determine the peculiar redshifts of the galaxy groups (z_j) if we fix the FP slope to the determined values. Fig. 5(a) shows the values of $\log(1+z_j)$ we obtain from the simulations by taking the average over the 99 Monte Carlo realizations, versus the input values. The relative rms are also given. In order to fix the FP zero-point, we assume that one cluster (our best-observed "reference cluster") has zero peculiar redshift, although in our simulation we have $\log(1 + z_{ref}) = 0.15$ dex to provide a severe test for our distance predictions. Fig. 5(a) shows that the method determines the peculiar redshifts accurately, even when the best-observed cluster has the second largest peculiar redshift. The mean systematic residuals are of the order of 0.01 dex (Fig. 5(b)), or $\approx 2\%$ of the distance (200 km/s at 10000 km/s redshift). The clusters with the largest residuals (≈ 0.03) have ≤ 3 objects only; smaller systematic residuals are achieved for clusters with larger numbers of galaxies (Fig. 5(c)). The random errors are of the order of 0.005 dex for clusters with four or more galaxies (Fig. 5(d)).

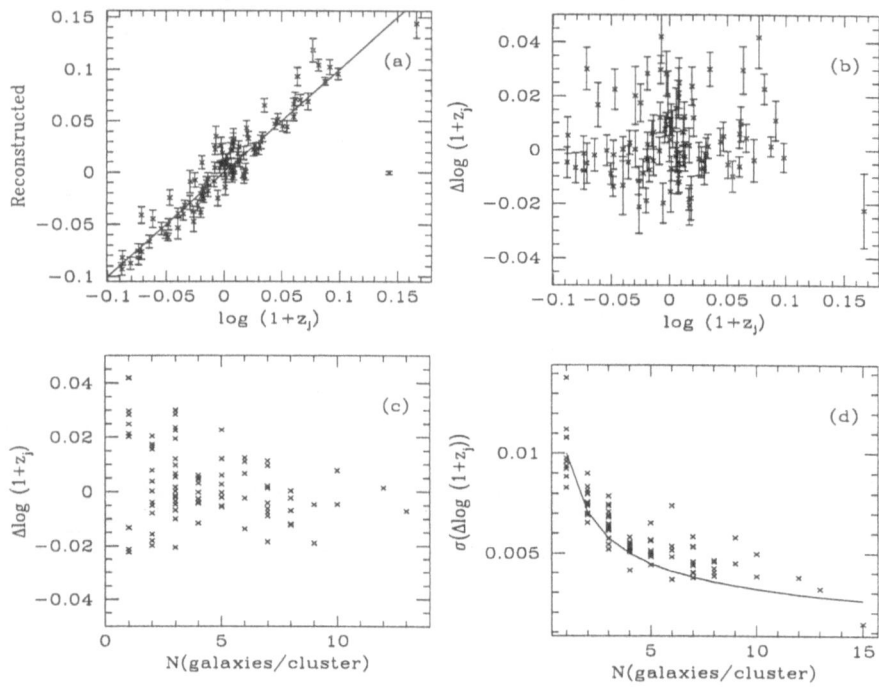

Fig. 5. (a) Peculiar redshifts $\log(1+z_j)$ as reconstructed with the maximum likelihood method, averaged over 99 Monte Carlo realizations, as a function of the input peculiar velocities. The peculiar redshift of the reference cluster has been set to zero. The bars show the random errors.(b) Residual mean $\log(1+z_j)$ with their random errors. (c) Residual mean $\log(1+z_j)$ as a function of the number of galaxies per cluster. (d) Random errors as a function of the number of galaxies per cluster. The line shows the expected poissonian decline.

4   Conclusions

We described the first results of the EFAR project, which measured the structure parameters of 434 early-type galaxies at redshifts 6000-15000 km/s in 84 clusters in the Hercules-Corona Borealis and Pisces-Perseus-Cetus regions. We discussed the properties of the photometric and spectroscopic data, together with new accurate methods used to determine these quantities. We examined the usual least-squares based algorithms to determine the parameters of the Fundamental Plane. Simulations show that these methods can provide biased answers due to the combined effects of selection, random errors and exclusions. These potential problems can be solved by modeling the distribution of galaxies in the

Fundamental Plane in a maximum likelihood sense. We tested our algorithm on simulated EFAR catalogues and found that it provides unbiased estimates of the searched parameters and of the underlying peculiar velocity field.

RPS acknowledges the financial support by the Deutsche Forschungsgemeinschaft under SFB 375. Support was also received from the Durham Visitors grant PPARC GR/K44992 and Durham University.

References

Colless M., Burstein D., Wegner G., Saglia R.P., McMahan R., Davies, R.L., Bertschinger E., & Baggley G., 1993, MNRAS, 262, 475

Djorgovski, S., Davis, M., 1987, ApJ, 313, 59

Dressler, A., Faber, S.M., Burstein, D., Davies, R.L., Lynden-Bell, D., Terlevich, R.J., Wegner, G., 1987, ApJ 313, L37

Eadie, W.T., Dryard, D., James, F.E., Roos, M., Sadoulet, B., 1971, Statistical Methods in Experimental Physics, North-Holland Publishing Company, Amsterdam

Jørgensen, I., Franx, M., Kjærgaard, P., 1996, MNRAS, 280, 167

Pahre, M.A. 1996, this book

Saglia R.P., Bertschinger E., Baggley G., Burstein D., Colless M., Davies R.L., McMahan R., & Wegner G., 1993 MNRAS, 264, 961

Saglia R.P., Bertschinger E., Baggley G., Burstein D., Colless M., Davies R.L., McMahan R., & Wegner G., 1997a, MNRAS, in press (Paper III)

Saglia R.P., Bertschinger E., Baggley G., Burstein D., Colless M., Davies R.L., McMahan R., & Wegner G., 1997b, ApJS, in press (Paper IV, astro-ph/9609089)

Scodeggio M. et al. 1996, this book

Smith R. et al. 1996, this book

Wegner G., Colless M., Baggley G., Davies, R.L., Bertschinger E., Burstein D., McMahan R., & Saglia R.P., 1996, ApJS, 106, 1 (Paper I)

Wegner G., Davies, R.L., Baggley G., Saglia R.P., McMahan R., Colless M., Burstein D., & Bertschinger E., 1997 in preparation (Paper II)

Willick, J.A., 1994, ApJS, 92, 1

Part 4

POSTERS

Quantitative Morphology and Color Gradients of E+A Galaxies in Distant Galaxy Clusters

Paola Belloni

Universitätssternwarte, Scheinerstraße 1, D–81679 München, Germany

Three basic scenarios have been invoked to explain the sudden rise and decline of star formation in distant clusters that lead to the transformation of "active" galaxies into the elliptical– and S0–types which today dominate rich clusters, i.e. the so called Butcher-Oemler effect, (Butcher & Oemler, 1978). These are: galaxy interactions or mergers with nearly equal mass neighbors (Lavery & Henry, 1988), gas-rich field galaxies running for the first time into the hot intercluster gas which ignites a brief but energetic episode of star formation (Bothun & Dressler, 1986), and high speed close encounters of gas-rich galaxies resulting in non-disruptive interactions or "galaxy harassment", (Moore, 1996).

Although neither of the proposed mechanisms for triggering star formation has a definitive answer in its favour, the high resolution of the HST images of distant clusters has recently provided a unique tool to face this question. Indeed, it allows us to study the spatial distribution of the starburst and thus to distinguish among the physical mechanisms at its origin. In the case of an infalling or harassed spiral the enhanced star formation would likely be a galaxy-wide phenomenon, or confined to the disk. If, on the other hand, the original galaxy was an elliptical that accreted gas from a dwarf galaxy, the burst signatures should be detectable as a bluer color of the nuclear region. Finally, in interacting or merging disk galaxies both behaviours have been observed: galaxies with starburst concentrated to the very center (Scoville et al.,1991) and others that show enhanced star formation on a galaxy-wide scale (Standford, 1991).

We focus on one class of "active" galaxies, the post-starburst or E+A galaxies, the latter being *only* a description of the spectra which appear to have an A-star component added to an old elliptical like component. These spectral features are interpreted as evidence of a recent (< 1.5 Gyr) burst of star formation. Ground based narrow-band photometry of 4 intermediate redshift clusters at z=0.4-0.5 has provided us with a sample of 73 E+A galaxies being secure cluster members (Belloni et al., 1995; Belloni & Röser, 1996). This is the largest sample of such galaxies found to date in distant clusters. Cluster membership and spectral type have been obtained by fitting the observed low-resolution spectral energy distributions with template spectra built up with Bruzual & Charlot (1997) population synthesis models. They represent the temporal evolution of a strong star formation episode (involving 20% of the original galactic mass) in an elliptical galaxy or in a spiral galaxy with star formation truncated after the burst. Our approach allows us to detect bursts younger than 2 Gyr while galaxies with an older burst will not be distinguished from a passively evolving elliptical galaxy.

Due to the small HST field of view only for 34 E+A galaxies HST images are available (WFPC2). We have retrieved them and determined: a) **surface brightness profiles** for all E+A galaxies but those showing highly irregular morphology. A fit with an exponential or a $r^{-1/4}$ law has been performed and for disk galaxies a Hubble type classification has been attempted on the basis of the bulge to disk ratio, b) **color gradients** for the 10 E+A galaxies in Cl0016+16 (observed in the F555W and F814W filters).

Table 1. Results of the surface brightness analysis of 33 E+A galaxies in intermediate redshift clusters. The E+A fraction is given with respect to the secure cluster members, about 120 galaxies brighter than $m_R=22.5$ per cluster. For merging/interacting galaxies we indicate if possible whether they appear disk or bulge dominated.

Cluster	E+A	E+A (HST)	Morph. Elliptical	Morph. S0-Sa	Morph. Sb-Sc	Merger Interact.
Cl0016+16 (z=0.54)	20 (21%)	11	1	3	6	1 Disk-dominated
Cl0939+47 (z=0.41)	35 (22%)	11			7	4
Cl0303+17 (z=0.41)	22 (22%)	7		1	5	1 Irregular
Cl1447+26 (z=0.38)	8 (9%)	4		1	1	2 Disk-dominated

We found that **most of the galaxies and all those showing signs of interaction are disk systems** judging from the exponential nature of the profiles (Tab.1). In our sample only one E+A galaxy has a regular elliptical profile. We also found that the E+A galaxies in Cl0016+16 show little spatial variation in V-I colors (restframe U-V) (Fig.1). Only one young E+A galaxy with estimated post-starburst age 0.8 Gyr has a starburst predominantly located in the nucleus whereas a second one has a weaker evidence of the same behaviour. However, the E+A color gradients are flatter compared to those of cluster ellipticals (Belloni et al., 1997) that are consistent with the Δ(U-R)/Δlog r=-0.23 mag observed in the nearby ones (Franx et al.,1993). Thus, our large and homogeneous sample of distant E+A galaxies confirms the claim that their **enhanced star formation is a galaxy-wide phenomenon.** Indeed, similar results have been obtained for the post-starburst galaxies in the Coma cluster by Caldwell et al. (1996). We speculate that strong interactions did not create most of the post-starburst galaxies in our representative sample of galaxy clusters at z=0.4−0.5. Non-disruptive interactions among galaxies or interactions between galaxies in the cluster or subcluster structures and the ICM are likely to be the dominant mechanism.

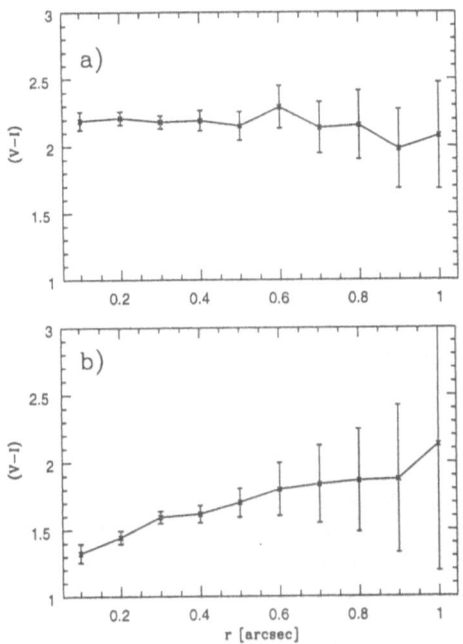

Fig. 1. a) Typical color gradients of E+A galaxies in CL0016+16 (z=0.54). b) One of the two E+A galaxies in Cl0016+16 showing strong color gradients. The very blue central colors (typical of spirals) indicate that the starburst is located in the nucleus. $1'' = 7.3$ Kpc (H_o=50 km sec^{-1} Mpc^{-1} and q_0=0)

References

Belloni, P., Bruzual G., Thimm, G., Röser H.-J., (1995): A&A 297,61
Belloni, P. & Röser (1996): A&AS 118, 65
Belloni, P. et al. (1997): ApJL, submitted
Butcher, H. & Oemler, A. (1978): ApJ 219, 18
Bothun, G. & Dressler A. (1986): AJ 301, 57
Caldwell et al. (1996): AJ.111, 78
Franx, M., Illingworth, G., Heckman,.T. (1993): AJ.98, 2
Lavery, R. & Henry P. (1988): ApJ 330, 596
Moore, et al. (1996): Nature 379, 613.
Scoville, N. et al. (1991): ApJ. 336, L5
Standford, S., (1991): ApJ. 381, 409

Comparing the Mark III and Abell/ACO Density and Velocity Fields

Enzo Branchini[1], Manolis Plionis[2], Idit Zehavi[3] and Avishai Dekel[3]

[1] Department of Physics, University of Durham, South Road, Durham DH1 3LE, UK
[2] National Observatory of Athens, Lofos Nimfon, Thesio, 18110 Athens, Greece
[3] Racah Institute of Physics, The Hebrew University, Jerusalem 91904, Israel

Abstract. We compare the mass density and velocity fields derived by applying the POTENT procedure to the Mark III velocity catalogue to those predicted by linear theory from the distribution of Abell/ACO clusters. The quantitative comparison between the cluster and POTENT density fields allows us to constrain $\beta_c (\equiv \Omega_o^{0.6}/b_c) = 0.26 \pm 0.11$, where b_c is the bias parameter for clusters. This value is also confirmed when the two velocity fields are considered.

1 Introduction

The present knowledge of the cosmic density and velocity fields is largely based on the distribution of galaxies and therefore is restricted to a relatively small region (~ 6000 km/s). A possible way of expanding the present limits is to select objects luminous enough to be observed at much larger distances. Clusters of galaxies are the ideal candidates since they can be identified up to large distances while their physical reality can be verified from their strong X-ray emission and/or compact redshift histograms. Recent efforts have been put in the direction of deriving their real space density and velocity field from their z-space distribution within the linear gravitational instability and linear biasing framework (Branchini & Plionis 1996, Branchini, Plionis & Sciama 1996). The comparisons with the CMB dipole and with the galaxy bulk flow (Dekel 1994 and Giovanelli et al. in this volume) yield a value of $\beta_c \approx 0.2$, with an uncertainty of ~ 50 %. Here we extend these previous attempts by comparing the cluster density and velocity fields with the Mark III POTENT-derived underlying ones, which allows to improve the statistical significance of the recovered β_c value.

2 Qualitative Comparison

Both the smoothed density (δ-) and velocity (**v**-) fields have been reconstructed within a region of 80 h^{-1} Mpc centred on the Local Group location using linear theory for clusters and quasi linear theory for Mark III POTENT. Various tests performed demonstrated that a Gaussian window with a 15 h^{-1} Mpc radius (15G) provides the optimal smoothing for comparing the fields.

The figure displays the 15G smoothed δ- and **v**- fields onto the supergalactic plane for the cluster (top plot) and Mark III POTENT (bottom plot) cases.

The overdensity contour levels are in step of $\Delta\delta = 0.2$ for positive δ (continuous lines) and for negative overdensities (dashed lines). The heavy continuous line shows the volume used for the quantitative comparison. The amplitude of the cluster δ- field and that of the velocity vectors have been scaled by $\beta_c = 0.21$. The similarity between the two δ- fields is evident, showing in both cases two main peaks (Great Attractor on the left and Perseus Pisces region on the right hand side) separated by an extended underdense region.

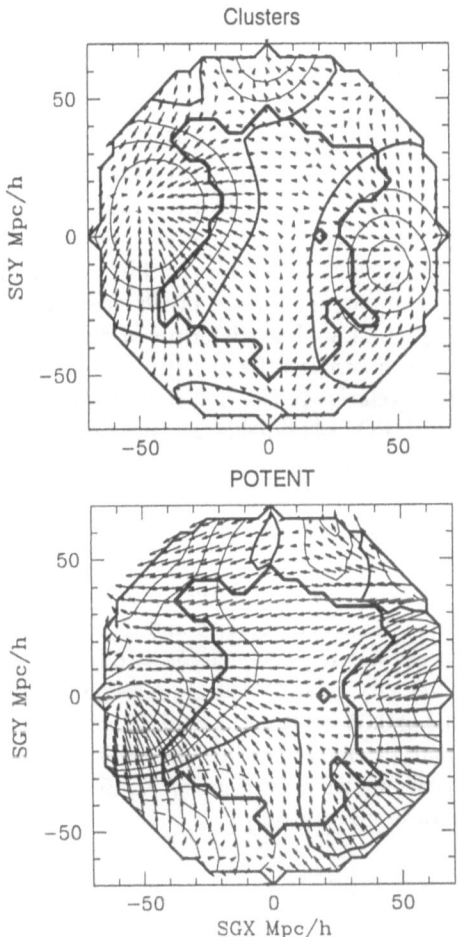

Fig. 1. The density and velocity field onto the supergalactic plane

The resemblance between the projected peculiar velocity fields is less remarkable but still present: the main feature common to the two plots is the infall onto the great attractor, while the infall onto the Perseus Pisces overdensity, predicted from the cluster distribution, is substituted by general motion of this region towards the LG in the POTENT velocity map. It should be noted though that the largest discrepancies lie near the volume borders where the number of Mark III objects is small and the errors large, resulting in a less reliable reconstruction of the fields.

3 Determination of β_c

We have performed two self-consistent determinations of β_c from δ–δ and v–v comparisons of the smoothed cluster and POTENT fields. The assumption underlying these simple analyses is that the clusters indeed trace mass with a linear biasing scheme. These comparisons are affected by different sources of uncertainties: the POTENT and cluster reconstruction errors, the cosmic variance due to the limited volume considered and the intrinsic scatter in the deterministic linear biasing relationship. It is worth stressing that the two analyses are complementary since the v–v comparison, being non-local, allows one to account for the mass distribution up to very large scale and uses all the cluster information; while the δ–δ comparison, being local, is not so much affected by poorly sampled regions, but suffers from the paucity of clusters within the volume considered.

After having defined a comparison volume in which the uncertainties in the various fields are reasonably small, we have determined β_c by minimising a generalized χ^2 merit function (which accounts for the reduced number of independent points due to the smoothing) relative to the expected linear relationship between the cluster and the Mark III/POTENT fields on gridpoints.

The reliability of the comparisons and possible biases affecting them have been carefully estimated independently for the clusters and Mark III fields using mock catalogues extracted from the Kolatt et al. (1996) N-body simulation and mock catalogues. The catalogues are aimed at mimicking the actual data sets that compose the Mark III catalogue, and therefore are suitable to serve as benchmarks for testing Mark III based velocity and density analyses.

The β_c resulting from the δ–δ regression is $\beta_c = 0.26 \pm 0.11$ while the v–v comparison, restricted to velocity vectors that are reasonably well aligned in the two fields, yields a β_c value of 0.31 ± 0.09. Although these results cannot provide us with a firm β_c determination, due to the large uncertainties, they lead toward a value of β_c which is consistent with an Einstein de Sitter universe for a very reasonable cluster bias parameter of $b_c \approx 4$.

References

Branchini, E., Plionis, M. (1996), Reconstructing Positions and Peculiar Velocities of Galaxy Clusters within 25,000 Km/sec: the Cluster Real Space Dipole. ApJ, **460**, 569–583

Branchini, E., Plionis, M., Sciama, D.W. (1996), Reconstructing Positions and Peculiar Velocities of Galaxy Clusters within 25,000 Km/sec: the Bulk Flow. ApJ, **461**, L17–L20

Dekel, A. (1994), Dynamics of Cosmic Flows. ARA&A, **32**, 371–418

Kolatt, T., Dekel, A., Ganon, G., Willick, J.A. (1996), Simulating Our Local Neighborhood: Mock Catalogues for Velocity Analysis. ApJ, **458**, 419–434

Impact of SNIa on SED
of High Redshift Galaxies

E. Brocato[1,3], S. Savaglio[2], G. Raimondo[1,3]

[1] Osservatorio Astronomico di Collurania, Via M. Maggini, I–64100 Teramo, Italy
[2] European Southern Observatory, Schwarzschild Str. 2 Garching, D–85748, Germany
[3] Istituto Nazionale di Fisica Nucleare, LNGS, I–67100 L'Aquila, Italy

Abstract. We present preliminary results on the effects of SNIa explosions on the Spectral Energy Distribution (SED) of distant galaxies and the possible modifications which may occur in integrated spectra, magnitudes and colours of simple galaxy models of different ages and metallicities few days after a SNIa event.

1   Introduction

Recent observations allowed to derive spectra of very high redshift galaxies ($z \gtrsim 3$) providing information on their earlier stage of evolution. Several authors infer galaxy ages by using the population synthesis technique and ages of the order of 1 Gyr or less have been suggested. We are interested in investigating the impact of SNe on the total emitted light of a galaxy in the range of ages running from 0.1 Gyr to few Gyr. The SNIa events should appear for ages comparable to the time scale of the first generation of white dwarf (WD) stars ($\simeq 0.05 - 0.1$ Gyr). We suggest that SNIa may have a non–negligible impact on the SED of a galaxy. We assume few models for the stellar population synthesis which might be representative of the observed SED of high redshift galaxies and do not consider obscuration by dust. We stress that reproducing all the possible models which explain observed SEDs is not the goal of this work, while we focus our attention on the variations of the observable parameters when SNIa events are taken into account.

2   Photometric Impact

The present knowledge of SN rates is still far from being firm. At high redshift, the effect of the expanding Universe will decrease the SN rate of a factor $(1 + z)$ due to time dilution, but the same effect will stretch the light curve of a SN by the same factor. Finally, one expects to observe a *larger* number of SNe Ia in the young Universe where we should observe the primeval galaxy population.

To evaluate the variation in the SED of the parent galaxy in case of a SNIa explosion, we assume a Simple Stellar Population (SSP) originated by a unique burst star formation (Brocato & Romaniello 1997, in preparation) and the stellar atmosphere models by Kurucz 1995 are adopted.

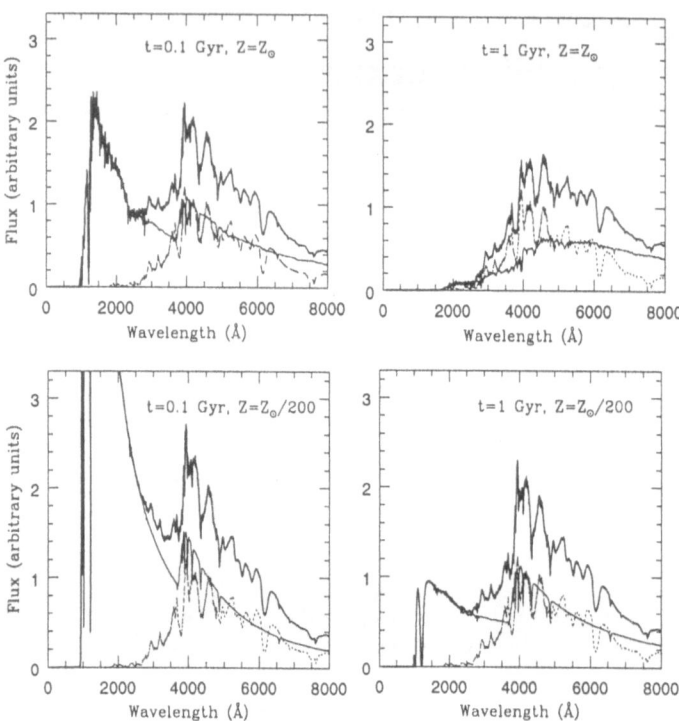

Fig. 1. Spectral Energy Distribution for 4 different galaxy models (thin lines). The dotted line is the SNIa spectrum 5 days after the explosion normalized to the V magnitude of the parent galaxy. Thick lines show the composite spectra (galaxy + SNIa).

The SNIa spectrum refers to SN 1992A (the optical part is kindly provided by B. Leibundgut whereas the UV part has been retrieved from the HST archive) and it was obtained 5 days after the explosion. We simulated a SNIa event in 4 different galaxy models, namely, (i) $t = 0.1$ Gyr, $Z = Z_\odot$; (ii) $t = 1.0$ Gyr, $Z = Z_\odot$; (iii) $t = 0.1$ Gyr, $Z = Z_\odot/200$; (iv) $t = 1.0$ Gyr, $Z = Z_\odot/200$. We assume the V magnitude of the parent galaxy to be equal to that of the SN at the maximum ($V_{SN}^{max} = -19$ for $H_o = 75$ km s^{-1} Mpc^{-1}, Wheeler & Benetti, 1996). The results are shown in Fig. 1 where we plot the SNIa spectrum, the SED of the parent galaxy and the composite spectrum (galaxy + SNIa). The some strong spectral features of the SNIa can be clearly seen in large details making the SNIa detection in distant galaxies a possible task. A rather interesting absorption is the SiII$\lambda\lambda5972, 6355$ doublet.

3   Moving Toward High Redshift Galaxies

We computed some significant colours for 4 galaxy models and plot the differences in these colours when the SNIa is included (Fig. 2). With the exception of the galaxy with $t = 1$ Gyr and solar metallicity, the remaining 3 models have the same behaviour. The J–K colour shows significant differences at high redshift, being around 0.5 at $z > 3$. The difference is even higher in the I–K colour. In the optical, the colour differences are significant only at low redshift ($z < 1$).

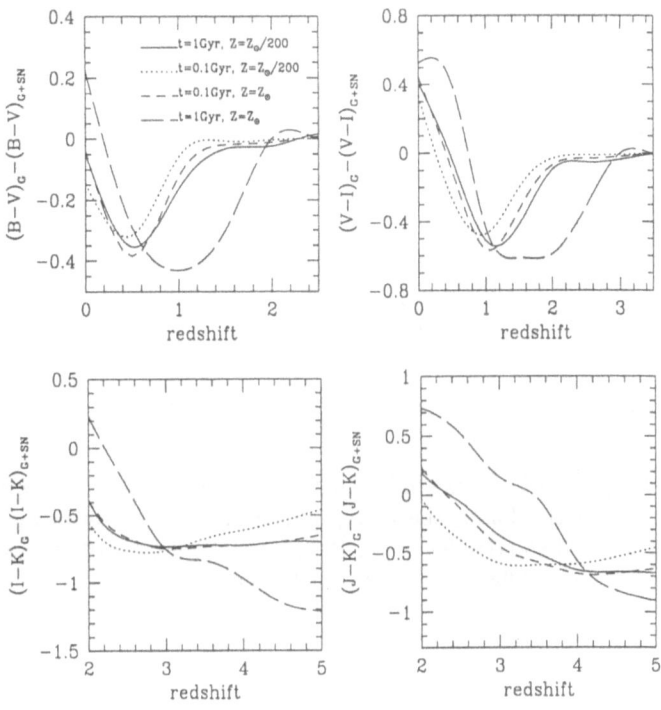

Fig. 2. Colour variations for 4 galaxy models due to the SNIa contribution.

One single SNIa event can dramatically change the SED of young galaxies. In general, for each given redshift there is a preferred colour in which the magnitude variation due to the SNIa event is larger. On the other hand, the SN rates at high redshift have never been tested. Simple considerations on the theoretical expectations and on the galaxy density at high redshift may allow the probability of finding SN events to be determined and to be compared with observations. This will also depend on the morphological galaxy type while the absorption features will be function of the galaxy metallicity.

References

Ciotti L., Pellegrini S., Renzini A., D'Ercole A., 1991, ApJ, 376, 380
Kirshner R. P., Jeffery D. J., Leibundgut B. et al., 1993, ApJ, 415, 589
Kurucz R. L., 1995, ApJ, 452, 102
Wheeler J. C., Benetti S., 1996, Astrophysical Quantities, IVth edition, ed. Arthur N. Cox, in press

Where Does the 'Tilt' Come From?

G. Busarello[1], M. Capaccioli[1], B. Lanzoni[1], G. Longo[1], E. Puddu[1]

Osservatorio Astronomico di Capodimonte, Via Moiariello 16, Napoli, Italy

1 Introduction

The problem of the tilt of the Fundamental Plane (FP), i.e. of the apparent departure of the FP from the prediction of Virial Theorem (VT), is a crucial one for the understanding of the physical meaning of the scaling laws that link the global properties of elliptical galaxies.

Written in terms of central velocity dispersion σ_0, effective radius R_e, and mean surface brightness inside R_e: I_e, the FP and the VT take the form:

$$R_e = c\sigma_0^a I_e^b \ (FP) \ ; \ R_e = C\sigma_0^2 I_e^{-1} \left(\frac{M}{L}\right)^{-1} \ (VT) \ . \tag{1}$$

The parameters a and b assume different values depending on the sample under study, the photometric band, the fitting algorithm, and so on [some examples are: $a = 1.4$, $b = -0.85$ (Bender et al. 1992); $a = 1.04$, $b = -0.88$ (Saglia et al. 1993); $a = 1.56$, $b = -0.94$ (Djorgovski et al. 1995)].

Two different interpretations of the tilt start from two assumptions on the 'structure term' C in the expression of the VT:

i) $C = const$, i.e. ellipticals are assumed to be homologous systems. As a consequence, the mass-to-light ratio scales with the luminosity as $M/L \sim L^\beta I^\gamma$;

ii) the tilt is due to non-homology, so that the non constancy of C reflects some systematic departures of the global quantities entering in the VT (i.e. kinetic and potential energies) from simple scalings of the observed local quantities which are used in the FP.

Here we explore the possible contribution of dynamical non-homology to the tilt of the FP.

2 Kinetic Energy of Random Motions

The (specific) kinetic energy of random motions T_σ is derived in the following way. The observed velocity dispersion profile $\sigma_p(R)$ is deprojected from the integration along the line of sight in order to obtain the intrinsic velocity dispersion $\sigma(r)$. Then σ^2 is integrated on the ellipsoidal region laying inside the effective semiaxes to obtain T_σ. An 'equivalent' velocity dispersion, representing the kinetic energy, is then defined as $\sigma_{eq} = (\frac{2}{3}T_\sigma)^{\frac{1}{2}}$.

The equivalent velocity dispersion scales with σ_0 as $\sigma_{eq} \sim \sigma_0^{0.8}$, so that the kinetic energy of random motions scales with σ_0 as $\underline{T_\sigma \sim \sigma_0^{1.6}}$ instead of σ_0^2.

3   Rotational Kinetic Energy

The rotational kinetic energy T_v is derived from the rotation curve in a way analogous to the derivation of T_σ. Cylindrical symmetry is assumed for the rotational field (as can be inferred from the few cases in which the velocity field has been sufficiently mapped by the data).

The ratio of the total kinetic energy $T = T_\sigma + T_v$ to T_σ turns out to be correlated with the residuals R to the FP: $R \sim 1.1(\pm 0.3)log(3 + v_{eq}^2/\sigma_{eq}^2)$, (in close agreement with the previous result by Prugniel & Simien 1994).

4   FP Solutions

The table lists the FP solutions for a sample of 40 elliptical galaxies, computed with different 'dynamical' terms. In the first column we give the kinematical parameter v which may assume the following forms: central velocity dispersion σ_0; velocity dispersion derived from the kinetic energy σ_{eq}; total kinetic energy including rotation (defined as $T = T_\sigma + T_v$).

The symbols refer to the parameters of the equations: $R_e = cv^a I_e^b$ and $\frac{M}{L} \sim L^\beta I_e^\gamma$. δ_d and δ_{R_e} are the r.m.s. deviations of the distances and of the residuals in R_e with respect to the FP.

ϕ is the angle between the FP and a plane defined by the VT.

v	a	b	δ_d	δ_{R_e}	β	γ	ϕ
σ_0	1.11 ± 0.2	-0.91 ± 0.1	0.08	0.13	0.40	0.23	13°
							dyn. non-h
σ_{eq}	1.53 ± 0.2	-0.92 ± 0.1	0.06	0.13	0.15	0.05	6°
							rotation
T	1.68 ± 0.2	-0.89 ± 0.1	0.05	0.12	0.09	−.03	4°
							other
VT	2	-1			0	0	0°

5   On the Scaling of M/L

Dynamical non-homology and rotational support can account for a large fraction ($\sim 70\%$) of the tilt of the FP.

On the other hand, a considerable fraction of the tilt can also be accounted for by other effects, like the color and/or Mg2 index variations and the non-homology of the brightness profiles (Graham & Colless 1996, Prugniel & Simien 1996, 1997).

In view of these evidences, the interpretation of the tilt in terms of a scaling of M/L with L seems not to be strictly needed (see also Fig. 1).

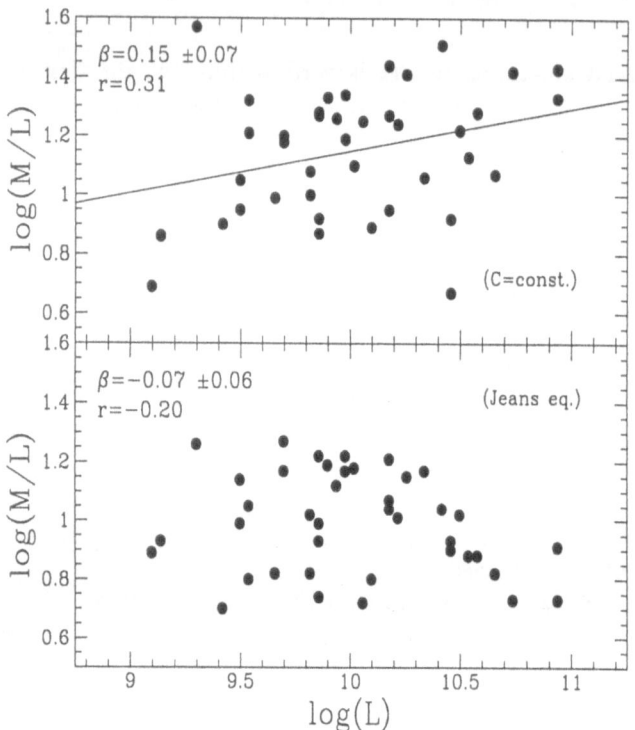

Fig. 1. Mass-to-light ratio vs. luminosity for the 40 galaxies in the present sample, derived in the B waveband and with $H_0 = 75$ km s^{-1} Kpc^{-1}. Masses are computed by means of the Virial Theorem under the assumption C=const (upper panel), and by means of the Jeans equation (lower panel). The values of β are indicated together with the linear correlation coefficient r. Once the non-homology is accounted for, the apparent correlation between M/L and L is cancelled.

References

Bender R., Burstein D., Faber S.M., 1992, ApJ **399**, 462
Busarello G., Longo G., Feoli A., 1992, A&A **262**, 52
Busarello G., Capaccioli M., Capozziello S., Longo G., Puddu E., 1996, A&A in press
Djorgovski S.G., Pahre M.A., de Carvalho R.R., 1995, "Fresh Views of Elliptical Galaxies", Buzzoni A. *et al.* eds., ASPCS, in press
Graham A., Colless M., 1996, MNRAS preprint
Prugniel Ph., Simien F., 1994, A&A **282**, L1
Prugniel Ph., Simien F., 1996, A&A in press
Prugniel Ph., Simien F., 1997, A&A in press
Saglia R.P., Bender R., Dressler A., 1993, A&A **279**, 75

The Fundamental Plane of Ellipticals: The Role of Nonhomology

Hugo V. Capelato[1], Reinaldo R. de Carvalho[2], and Ray G. Carlberg[3]

[1] DAS-INPE, Brazil
[2] CALTECH/DAF-ON, Brazil
[3] University of Toronto, Canada

Abstract. We report the status of our continuing program aimed to study the physical properties of elliptical galaxies. Previously (Capelato et al., 1990) we have shown that the characteristic parameters of the remnants of dissipationless mergers of model galaxies closely reproduce the observed slope of the fundamental plane (FP) and found that this effect could be related to the non-homology of the central mass-velocity distributions of the remnants. Here we discuss the results of a set of "hierarchical" simulations which merged our past remnants among themselves. We show that the hierarchical scheme is able to produce remnants which span the range of the observed FP, although with a possible small bending at higher hierarchical levels. This effect may be related to some recent observations which suggest a small evolution of the FP with redshift. We also discuss a new set of simulations designed to study the origin of the observed thickness of the FP suggesting that the origin of the observed scattering of the FP may be explained if the very first generation of E galaxies spanned a limited range of equilibrium mass distributions. We conclude with a rediscussion of the non-homology effect of dissipationless mergers remnants.

1 The Fundamental Plane: Slope and Thickness

1.1 The Slope

The results presented in Capelato et al. (1990) demonstrated that the characteristic parameters of mergers remnants produced by identical initial progenitors follow very closely the observed Fundamental Plane (FP) of elliptical galaxies. A new set of simulations suggests that subsequent merging of these remnants, as in hierarchical formation scenarios, maintains the FP correlations although with a slightly different slope (1st generation: $1.36 \pm 0.08 \rightarrow$ 2st and 3st generations: 1.50 ± 0.10). These are preliminary results since we do not populate the upper part of the FP with hierarchical mergers as much as we do for the lower part. New simulations are in progress to fill this gap in our experiment.

It is interesting to note that the bending of the FP with the hierarchy level is consistent with the results reported by van Dokkum and Franx (1996) who measured a shallower FP for a cluster at z~0.4, compared to the FP for nearby cluster ellipticals. This is suggestive of a redshift dependence in the FP slope. A direct comparison with our results is however problematic due to the very small intrinsic scatter around the FP, as discussed bellow. In fact, should our results be affected by a scatter as large as the observed one (about 0.07 in $\log r_e$, Pahre et al., 1995), it would be impossible to detect the slope variation reported above.

1.2   The Thickness

One of the most important results presented in Capelato et al. (1990) was the very small scatter around the FP displayed by the merger remnants. However their the objects were produced by merging identical progenitors under different initial orbital conditions. Now we have performed a new set of 16 merger simulations designed to study the influence of different initial galaxy models. A total of 12 initial model galaxies were then constructed as realizations of spherical King models of varying masses and central potentials. The initial orbital conditions were characterized by their dimensionless energy and angular momenta and the remnants were analysed as previously (see Capelato et al., 1995 for details). Most of the mergers tend to stay closer to their primary progenitors. Mergers with orbital energy higher than $\hat{E} = -8$ tend to move away, outlining a sequence depending on their initial orbital energy. Neither the mass-ratio of the progenitors nor the initial model of the secondary seem to constrain the final position of the end-product. Thus, we conclude that the central potential, W_0, of the primary progenitor is driving the observed scatter of the FP. This seems to confirm the suggestion of Capelato et al. (1990) that the observed scatter of the FP could be originated on a first generation of progenitors that spanned a range of mass distributions consistent with a range of merger sequences, all parallel among themselves.

2   Non-Homology of the Remnants

We examine the non-homology effect already discussed by Capelato et al. (1990) by studying the dimensionless peak potential of the remnants, defined as $\psi = r_h \Phi_0 / GM$, where r_h is the half-mass radius and Φ_0 is the central gravitational potential of the merger remnant, as a function of $\hat{E}$.

Figure 1 shows a plot of this quantity (ψ) as a function of the initial orbital energy. As it can be seen there is a small but clear positive correlation between these quantities. Moreover the 2nd generation mergers seem to follow a separate but similar relationship, an effect which was already noticeable in Capelato et al. (1990) (see their fig. 4). This may be related to the bending of the FP discussed before. The bottom panel of this figure shows no correlation between the reduced central potential, $\Psi_0 = \Phi_0 / \sigma_0^2 \simeq cte = 6 - 8$. In fact, the constancy of this quantity may be understood as a manifestation of the virial equilibrium prevailing at the center of the systems: $\sigma_0^2 \propto \Phi_0$. Because mergers are globally homologous then $\sigma_0 / \sigma \doteq 1/C_v \simeq \psi$ where C_v is a geometric factor (Capelato et al., 1990). We thus see that the positive correlation found here is a manifestation of the non-homologous nature of the merger remnants.

Collapse simulations (Londrillo et al, 1991) have shown that ψ is correlated to the initial virial ratio of the collapse, β. This suggests that under appropriate initial conditions collapsed systems may also follow a FP relationship.

Hjorth and Madsen (1995) have recently proposed a physical explanation to the non-homology effect based on the statistical mechanics of violent relaxation

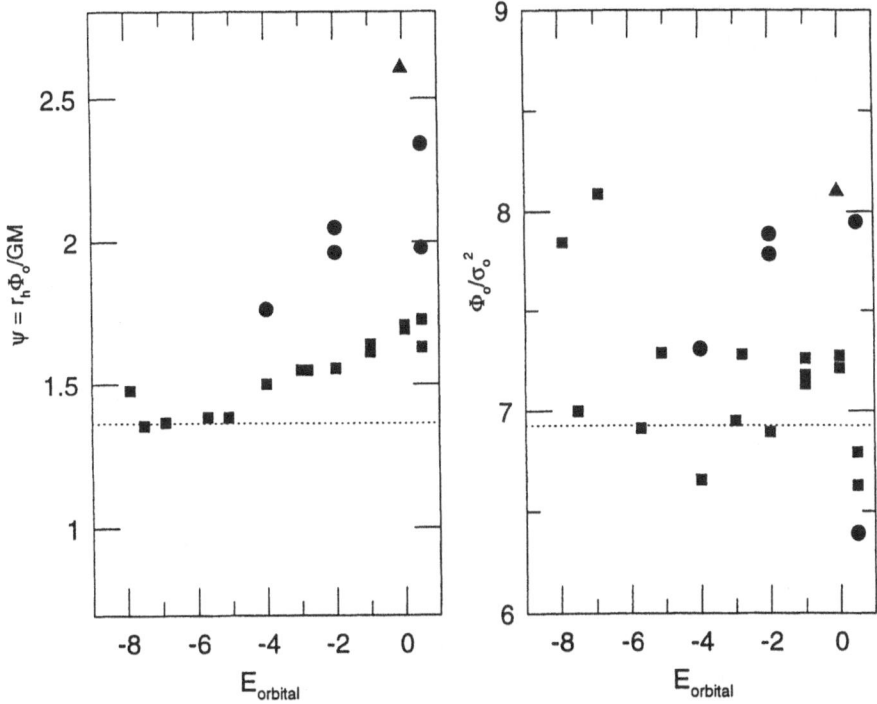

Fig. 1. *left*: the dimensionless peak potential of the remnants, ψ, plotted against the initial orbital energy of the merger. Squares: 1st generation mergers; circles: 2nd generation "hierarchical" mergers; triangle: 3th generation; the dotted line gives the value of ψ of the progenitor of the 1st generation mergers. *right*: the reduced central potential Ψ_0

occurring in a finite region in space. Their theory predicts the resulting equilibrium configuration to be described by a distribution function which has only one free shape parameter given by the dimensionless peak potential defined above and which approximates an isothermal sphere at the center.

3   Discussion

We have presented a progress report of a continuing program which has been insofar successful to demonstrate that dissipationless mergers in hierarchical scenarios are able to reproduce the observational properties of elliptical galaxies embodied in the Fundamental Plane correlation. A counterproof for this hypothesis may be provided in the near future by a study, similar to the one discussed here, dealing with variable initial β's collapses. We are presently undertaking this way as well as performing new higher resolution simulations with much larger number of particles and initial orbits drawn from large scale cosmological simulations. Although we do not expect that these changes will alter our main results, it would help us better understand some of the questions discussed here, like the slope and the thickness of the FP.

References

Capelato, H.V., de Carvalho, R.R., & Carlberg, R.G. 1995, ApJ, 451, 525
Hjorth, J., & Madsen, J. 1991, MNRAS, 253, 703
Londrillo, P., & Messina, A., & Stiavelli, M. 1991, MNRAS, 250, 54
Pahre, M.A., & Djorgovksi, S.G., & de Carvalho, R.R. 1995, ApJ, 453, L17
van Dokkum, P.G., & Franx, M. 1996, astro-ph 9603063

The Fundamental Plane of Galaxy Clusters

Alberto Cappi[1] and Sophie Maurogordato[2]

[1] Osservatorio Astronomico di Bologna, via Zamboni 33, I-40126 Bologna, Italy
[2] CNRS, LAEC, Observatoire de Paris-Meudon, 5 Place J. Janssen,
   F-92125 Meudon Cedex, France

Abstract. In the three–dimensional space defined by the logarithms of central velocity dispersion σ, effective radius R_e and mean effective surface brightness I_e, elliptical galaxies are confined in a narrow plane (Dressler et al. 1987; Djorgovski & Davis 1987). Here we discuss the observational evidence for the existence of an analogous relation for galaxy clusters (Schaeffer et al., 1993).

The relations between global observables in stellar systems are important both from a theoretical and a practical point of view. The Tully–Fisher relation for spirals and the Faber–Jackson relation for ellipticals involve two observables. In the case of ellipticals, the residual scatter suggested the introduction of a third parameter, resulting in the definition of the so–called fundamental plane (Djorgovski & Davis 1987; Dressler et al. 1987). From the virial theorem we expect a relation involving the three observables R, L and σ, which can be expressed in the following way (see Djorgovski & Santiago 1993): $L \propto KR\sigma^2(M/L)^{-1}$ where K is a structural parameter. Therefore the existence of the FP for a given class of objects is not a trivial consequence of the virial theorem; it requires also that the class of objects under study has a similar dynamical structure and a tight mass to light ratio with a small dispersion.

We decided to search for relations between global observables in galaxy clusters (Schaeffer et al. 1993). We used the effective radii and total luminosities for a sample of 29 *regular* galaxy clusters, measured after an accurate and homogeneous reduction of high–quality photometric data by West et al. (1989; WOD), who had found a well–defined radius–luminosity relation, $R \propto L^{0.5}$. For 16 of these clusters we found reliable measures of velocity dispersion (Struble & Rood 1991). Combining these data we showed the existence of a relation between velocity dispersion and luminosity, $L \propto \sigma^{1.9}$. The relation L–σ for clusters is the equivalent of the Faber–Jackson law for ellipticals, but in clusters the luminous matter in stars is a small fraction of the total mass, dominated by dark matter and the hot gas. We realized also that the relations R–L and L–σ had a scatter larger than one could expect from observational errors. Introducing σ as a third parameter, we found that these clusters define a Fundamental Plane (fig.1), as many stellar systems from globular clusters (Nieto et al. 1990) to elliptical galaxies (see fig.4 in Schaeffer et al. 1993). We found $L \propto R^{0.89\pm0.15}\sigma^{1.28\pm0.10}$ or, defining a surface brightness $I_e = L/R_e^2$, $R_e \propto I_e^{-0.81}\sigma^{1.15}$, with the best fit parameters quite similar to the elliptical ones. The cluster FP is obviously shifted relatively to the elliptical one, because of the different M/L ratio. Therefore all gravitationally bound systems appear to define their FP. The common

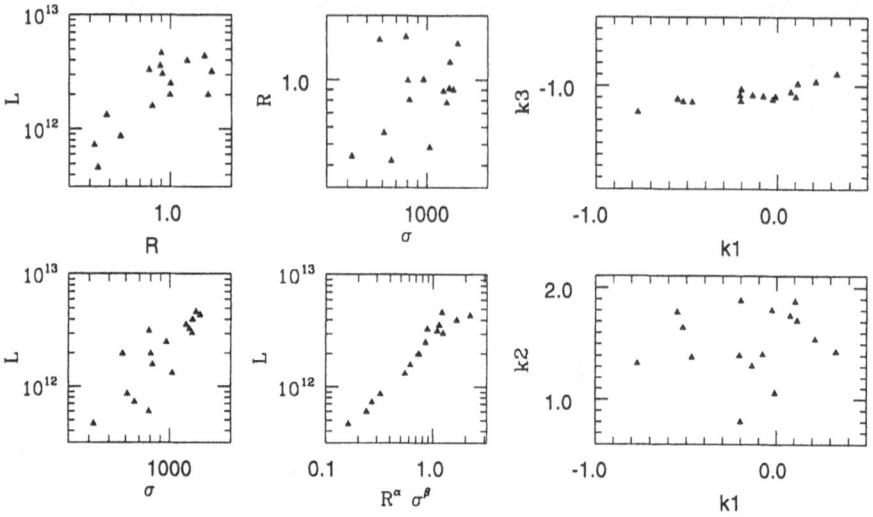

Fig. 1. Relations between L, R and σ, and between k_1, k_2 and k_3

link is the virial theorem, but of course the slope and dispersion of each plane will depend on the class of objects taken into account.

Might the cluster FP be the result of spurious effects? Discussing their $R-L$ relation for galaxy clusters, WOD excluded a bias due to a selection in surface brightness, as one Abell radius includes most of the cluster luminosity. We also note that cluster peculiar velocities cannot produce the observed relation. Furthermore, tests with numerical simulations (Pentericci et al. 1996) show that the cluster FP is conserved after merging and that it cannot be spuriously generated by the procedure used to find R_e and L.

One can also use the coordinates defined by Bender et al. (1992) (fig.1). These coordinate system, which does not correspond exactly to the FP, gives $k_3 \propto \log(M/L)$ vs. $k_1 \propto \log(M)$. It is clear that the M/L of clusters has a small dispersion (see the discussion in Renzini & Ciotti 1993). We find a small trend of M/L with L, with $M/L \propto L^{0.3\pm0.1}$.

We can also have a rough estimate of the cluster peculiar velocity $V_p = V_{obs} - H_0 D$ (assuming that the deviation from the FP is entirely due to V_p, and neglecting the scatter of the M/L ratio). Uncertainties become very large beyond $z = 0.05$; for the 10 clusters at $z \leq 0.05$ we find (Cappi et al. 1994) $V_p \leq 1000$ km/s, with a trend as a function of the cosine angle between clusters and the GA direction consistent with the results of Han & Mould (1992); the upper limit $\Delta H/H \leq 15\%$ is consistent with Lauer & Postman (1992).

We conclude that regular clusters define a FP comparable to that of elliptical galaxies, which can give us information about formation time dispersions, M/L ratios, or cluster v_p. A critical point is the quality of the data, especially the photometry. For example, a number of recent data give measures of virial radii significantly underestimated, because based on a limited region of the cluster (as discussed by Carlberg et al. 1996). One should remind that typical effective radii are around 1 h^{-1} Mpc, while virial radii are more than 2 times R_e. Therefore a large observational effort is still required to address many important issues, and we need CCD photometric surveys covering a large part of each cluster and not only its central regions. Last but not least, the combination with X–ray data will be extremely useful to understand the role of the different matter components in galaxy clusters.

References

Bender R., Burstein D., Faber S.M., 1992, ApJ 399, 462

Cappi A., Maurogordato S., Schaeffer R., Bernardeau F., 1994, in the 9th IAP Meeting on *Cosmic Velocity Fields*, F.R.Bouchet & M.Lachièze-Rey eds., Paris, p.527

Carlberg R.G., Yee H.K.C., Ellingson E., Abraham R., Gravel P., Morris S., Pritchet C.J., 1996, ApJ 462, 32 (CNOC)

Djorgovski S., Davis M., 1987, ApJ 313, 59

Djorgovski S., Santiago B.X., 1993, in ESO workshop on Structure, Dynamics and Chemical Evolution of Elliptical Galaxies, eds. I.J.Danziger, W.W.Zeilinger, K.Kjär, Garching, p.59

Dressler A., Lynden–Bell D., Burstein D., Davies R., Faber S., Wagner M., Terlevich R., 1987, ApJ 313, 42

Han M., Mould J.R., 1992, ApJ **396**, 453

Lauer T.R., Postman M., 1992, ApJ **400**, L47

Nieto J.-L., Bender R., Davoust E., Prugniel Ph., A&A 230, L17

Pentericci L., Ciotti L., Renzini A., 1996, III congresso nazionale di cosmologia, Grado, Astr. Letters & Comm. 33, 213

Renzini A., Ciotti L., 1993, ApJ 416, L49

Schaeffer R., Maurogordato S., Cappi A., Bernardeau F., 1993, MNRAS 263, L21

Struble M.F., Rood H.J., 1991, ApJS 77, 363

West M.J., Oemler A., Dekel A., 1989, ApJ 346, 539 (WOD)

Anomalous Scaling Laws in Lyman–α Clouds

Vincenzo Carbone[1] & Sandra Savaglio[2]

[1] Dipartimento di Fisica, Università della Calabria, Roges di Rende I–87036, Italy
[2] European Southern Observatory, Karl-Schwarzschild-Str. 2, Garching, D–85748, Germany

Abstract. We present some statistical features of the large number of Lyα absorption lines detected in the high redshift QSO 0055–26 obtained by using the correlation integral. The Lyα forest shows the presence of clusters and voids for different scaling exponents.

1 Introduction

The numerous Ly–α absorption lines seen in quasar spectra can be considered a very deep window on the nature of the young Universe. Clustering properties have been studied basically using the two point correlation function. A positive signal was detected in high resolution spectra only for small scales (up to a few hundred km s^{-1}). The redshift distribution of HI column densities can be considered an intermittent process where the important features are the index of the scaling laws and not the amplitudes at every scale. We propose to analyse it by using mathematical tools which are most suitable for their determination.

2 The Correlation Integral Approach

When we are dealing with non–gaussian processes, the two–point correlation function gives only a first order description of structures. A statistical description of any point process requires higher–order N–points correlation functions. We then assume that clusters are organized in the framework of processes where a generic multifractal structure underlies the cloud distribution. In the following we will use the correlation integral as estimator (Grassberger & Procaccia, 1986). In the algorithm applied to the QSO $0055 - 26$ ($z_{em} = 3.67$), we calculated the *conditional probability*

$$p_i(\Delta v|\log N^*) = \frac{1}{M-1}\sum_{j\neq i}\Theta\left(\Delta v - |v_{ij}|\right) \tag{1}$$

where v_{ij} is the velocity difference between two lines in the absorption line list $N(z)$ and $\Theta(x)$ is the usual Heaviside function. The sum is extended to *all the M lines whose HI column density is* $\log N > \log N^*$. The conditional probability has been calculated here because, as evidenced by Chernomordik (1995) and Cristiani et al. (1997), the correlation between lines disappears for small values of $\log N$ (namely for $\log N < 13.6$) and increases for higher $\log N$ values. From this

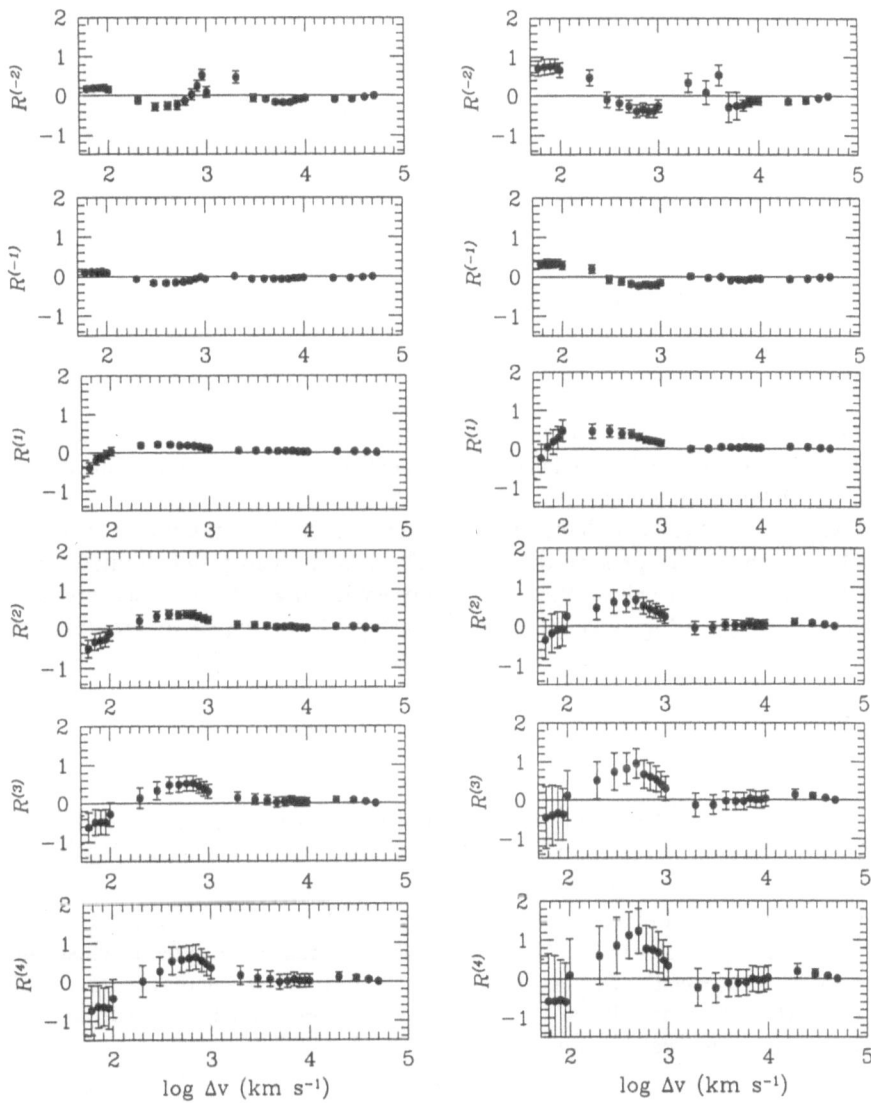

Fig. 1. Ratios $R^{(q)}(\Delta v | \log N^*)$ vs. $\log \Delta v$ for some values of q ($q = -2, -1, 1, 2, 3, 4$). We show the results relevant to two values of $\log N^* = 13.3$ (left panel) and 13.8 (right panel).

probability we calculate the partition functions $Z^{(q)}$ and their scaling exponents ϕ_q through

$$Z^{(q)}(\Delta v|\log N^*) = \frac{1}{M}\sum_i [p_i(\Delta v|\log N^*)]^q \sim [\Delta v]^{\phi_q(\log N^*)} . \qquad (2)$$

The *clustering paradigm* (Martinez et al. 1990) is related to the role played by q which acts like a microscope. For each value of $q > 0$ we probe "regions" (in the velocity space) with different clustering properties, whereas negative values of q probe voids. Our aim is to investigate the departure of clouds from a random distribution. Therefore, by using the same number of lines, we applied our algorithm both to the observed lines of the QSO and simulated distributions. Such a comparison reveals the departure from a random distribution. The quantities of main interest are then the scaling exponents $\phi_q(\log N^*)$ and the ratios

$$R^{(q)}(\Delta v|\log N^*) = \frac{Z^{(q)}_{obs}(\Delta v|\log N^*)}{Z^{(q)}_{exp}(\Delta v|\log N^*)} - 1 \qquad (3)$$

where the subscripts *obs* and *exp* refer to the observed and simulated lines respectively. This last quantity is the obvious generalization of relation of the two point correlation function.

We have applied this algorithm to the lines of the high resolution spectrum of QSO 0055-26. We present the results obtained by using 100 simulated lists $N(z')$ with random values of z', in the observed range Δz, for each observed column density. In Fig. 1 we report $R^{(q)}(\Delta v|\log N^*)$ vs. $\log \Delta v$ for two different values of the cutoff. In a range of two decades (approximately $2 \leq \log \Delta v \leq 4$) two different scaling laws can be recovered with the crossover placed at about $\Delta v \simeq 2 \times 10^3$ km s^{-1}. This is mainly visible for values of $q > 1$.

We can firmly conclude that the multi–scaling analysis of the Lyα forest is a very promising approach for the study of the large scale structure of the Universe at high redshift. In what way the galaxy distribution represents the real mass distribution of the Universe is one of the most controversial problems of extragalactic astronomy. In this respect, the study of one–dimensional high redshift distribution of Lyα clouds can provide a larger picture of the Universe in the early stage of evolution. They probably experienced a different clustering history with respect to galaxies, especially at high redshift where it is reasonable to think that the first collapsed and emitting galaxies formed mainly around the peaks of the mass distribution.

References

Chernomordik V. V. 1995, ApJ, 440, 431

Cristiani S., D'Odorico S., D'Odorico V.,Fontana A., Giallongo E., Savaglio S. 1997, MNRAS, *in press*

Grassberger P., Procaccia I., Phys. Rev. A, 28, 259

Martinez V. J., Jones B. J. T., Dominguez–Tenreiro R. & van de Weygaert R., 1990, ApJ, 357, 50

The I–Band Tully–Fisher Relation at Intermediate Redshifts

Daniel A. Dale[1], Riccardo Giovanelli[1], Martha P. Haynes[1], Marco Scodeggio[1], Luis Campusano[3], and Eduardo Hardy[4]

[1] Astronomy Department, Cornell University, Ithaca, NY 14853, USA
[2] Observatory of Cerro Calán, Universidad de Chile, Casilla 36-D, Santiago, Chile
[3] Department of Physics, Laval University, Ste–Foy, P.Q., Canada G1K 7P4

Abstract. We present first results of all all–sky observing program designed to 1) improve the quality of the I band Tully-Fisher (TF) template and 2) obtain a bulk flow dipole solution for a volume of radius 18,000 km s^{-1}. We are obtaining between 5 and 15 TF measurements per cluster on a sample of $\sim$ 50 clusters at intermediate redshifts ($0.02 \lesssim z \lesssim 0.06$) which should determine the offset in the I–band TF template to within $\sim$ 0.02 mags, or bulk flows of this sample to within $\simeq$ 150 km s^{-1}.

1 Introduction

Claims of large amplitude, coherent flows on scales up to $100h^{-1}$ Mpc (Lauer & Postman 1994: LP; Courteau et al. 1993) present a scenario difficult to accommodate with current cosmological scenarios (Bahcall & Oh 1996). These claims have been challenged by the SN Ia work of Riess et al. (1995) and by the TF work of Giovanelli et al. (1996), both of which use methods independent from the Brightest Cluster Galaxy technique used by LP. Unfortunately, the challenges originate from data samples that are either too sparse or too shallow to adequately verify the amplitude and extent of any bulk flow. The Giovanelli et al. sample derives a TF template relation based on 555 galaxies in 24 clusters to $cz \simeq 9000$ km s^{-1} (Giovanelli et al. 1997a,b: G97a,b). We are extending this work by measuring the redshift–independent distances of some 50 Abell clusters with radial velocities $6000 \lesssim cz \lesssim 18,000$ km s^{-1}. Not only will this enlarge the volume over which bulk flows can be measured, but it will also significantly increase the accuracy to which bulk flows can be measured; we predict we will be able to determine the amplitude of any bulk flow for our sample to within $\simeq$150 km s^{-1}.

2 The Sample

We are obtaining TF measurements for 5 to 15 galaxies per cluster for a total of $\sim$ 50 clusters. Figure 1 shows the distribution of the 36 clusters observed thus far. An attempt to sample the sky uniformly has been made, though regions near the Galactic disk are avoided due to the large foreground extinction present there.

Table 1 lists the main parameters of the first set of completed clusters. Standard names are listed in column 1. Adopted coordinates of the cluster center

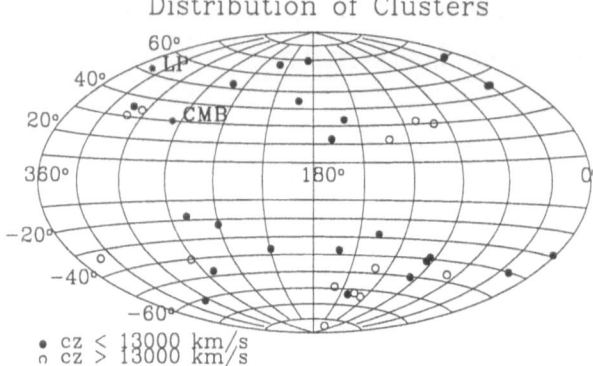

Fig. 1. The distribution of the Abell clusters selected thus far. We have avoided clusters with $|b| \lesssim 20°$ where Galactic extinction is significant. The apices of the CMB and the LP dipole motions have been plotted for reference.

are listed in columns 2 and 3, for the epoch 1950; they are in general obtained from Abell et al. (1989). An exception is a cluster slightly offset from Abell 1983 in both RA–Dec and redshift. We designate this cluster "A1983b." For all the clusters we have compiled our own systemic velocities (column 4 with errors parenthesized) using n_z redshifts (column 5).

Table 1. Cluster Coordinates

Cluster	RA	Dec	V_{cmb}	n_z
	h m s	d m s	km s^{-1}	
A 168	011234	–000144	13067(69)	43
A 397	025412	+154500	9651(76)	38
A 569	070524	+484200	6032(55)	36
A1139	105530	+014600	12218(59)	24
A1228	111848	+343600	10830(39)	32
A1983	145024	+165700	13721(46)	76
A1983b	144724	+170600	11521(102)	14

3   Observations

The I band CCD photometry for the project is being obtained at the KPNO and CTIO 0.9m telescopes and is nearly completed. Only apparent magnitudes with photometric calibrations accurate to 0.02 mags or better are used.

We use the H_α spectral line to obtain rotation curves of the galaxies. This is being done at the Mt. Palomar 5m and the CTIO 4m telescopes using long–slit spectroscopy with high dispersion gratings ($1200 \, l \, mm^{-1}$; $\sim 7 \, km \, s^{-1} \, pixel^{-1}$). Typically, 30–45 minute exposures are sufficient to trace the outer disk regions.

4   Results

Figure 2 gives the TF plots for the first Abell clusters in our project, using data that has not been corrected for sample incompleteness. The dashed line plotted is the relation derived from the G97a,b study of 24 clusters, the relative proximity of which enables an accurate determination of the TF slope due to the broad stretch in galactic parameters observed.

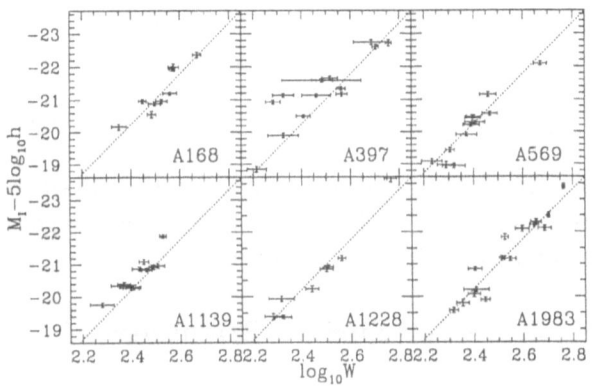

Fig. 2. "Raw" TF plots for the first six clusters in our sample. The dashed line is the TF relation from G97a,b. The A1983 panel includes A1983b. The rotational velocity width, W, has been corrected for rotation curve shape, disk inclination, and cosmological broadening.

This research was supported by NSF grant AST94–20505 to RG.

References

Abell, G., Corwin, H., Olowin, R. (1989): ApJS **70**, 1–138

Bahcall, N., Oh, S. (1996): ApJ **462**, L49–L52

Giovanelli, R., Haynes, M., Wegner, G., da Costa, L., Freudling, W., Salzer, J. (1996): ApJ 464, L99–L102

Giovanelli, R., Haynes, M., Herter, T., Vogt, N., Wegner, G., Salzer, J., da Costa, L., Freudling, W. (1997a): AJ, in press

Giovanelli, R., Haynes, M., Herter, T., Vogt, N., da Costa, L., Freudling, W., Salzer, J., Wegner, G. (1997b): AJ, in press

Lauer, T., Postman, M. (1994): ApJ, **425**, 418–438

Riess, A., Press, W., Kirshner, R. (1995): ApJ, **445**, L91–L94

Tully, R., Fisher, J. (1977): A&A, **54**, 661-673

The Kormendy Relation for 3 Abell Clusters

Giovanni Fasano[1], Daniela Bettoni[1], Per Kjærgaard[2], and Mariano Moles[3]

[1] Osservatorio Astronomico di Padova
[2] Copenhagen University Observatory
[3] Observatorio Astronomico Nacional de Madrid

1 Introduction

As part of our ongoing study of the Fundamental Plane (FP) for cluster ellipticals we present here the Kormendy relation for the Abell clusters A1878 (z=0.254), A2111 (z=0.229) and A2151 (z=0.037). We apply our results to the Surface Brightness Test (SBT) for the expansion of the universe. Our final aim is to carry out this test using the FP for approx. 30 clusters out to z=0.3, see Kjærgaard et al. (1993).

Observations (in Gunn r and Bessel B) were done at the Nordic Optical Telescope in excellent seeing conditions (0".6÷0".8) and we generally reach an isophotal surface brightness of 26.0 mag/$''^2$ in Gunn r.

2 Deriving Radii and Surface Brightness

The luminosity equivalent profiles were deconvolved using the technique by Bendinelli (1991) and were extrapolated with a de Vaucouleurs law. As a test of the procedure we convolved the derived deconvolved profiles and found close fits to the observed profiles.

A number of different radii and the corresponding mean surface brightness were derived from both the observed and the deconvolved profiles.

The different radii, together with the corresponding average surface brightness were plotted against each other for both the observed and the deconvolved profiles. This analysis, together with a similar one carried out on a number of synthetic galaxies, allows us to conclude that the deconvolution technique is very efficient in recovering the true radii and surface brightness from the corresponding *seeing–biased* quantities down to $r_e \simeq 1 \times FWHM$. In particular, the half light radius turned out to be the best suited one to obtain well defined Kormendy relations.

Figure 1 shows that, if we use the observed profiles the Kormendy relations of high redshift clusters is almost completely washed out by the seeing effect. On the contrary, when the deconvolved data are used the scatter is comparable with that found for nearby clusters. The slopes of the Kormendy relations for our clusters are consistent each other within the scatter and are also consistent with that derived from the photometry of 10 nearby clusters by Jørgensen et al. (1995). Therefore, we decided to adopt in the fit this last value of the slope (2.5 in *log r*).

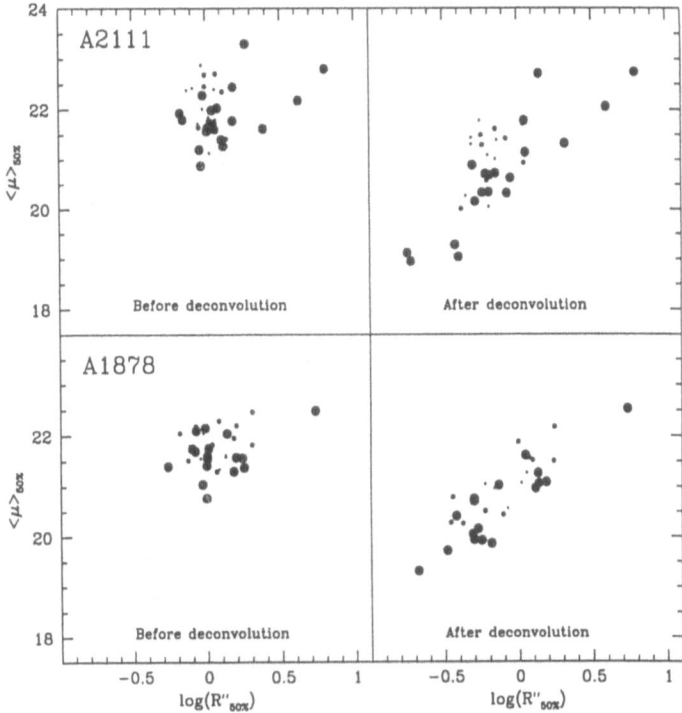

Fig. 1. Kormendy relations, before and after deconvolution, for the clusters A2111 (z=0.229; upper panels) and A1878 (z=0.254; lower panels). Big dots refer to the *bona fide* ellipticals.

3   The Surface Brightness Test

The Kormendy relation can be used to perform the SBT, if the surface brightness is referred to a standard linear radius. This was done by Pahre et al. (1996) who used the metric radius $R_e = 1$ *kpc*. They found in the surface brightness a trend compatible with an expanding universe and a passive stellar evolution. They claim that the static hypothesis can be discarded at the 5 σ level.

However, when using this approach, it is important to use the appropriate metric in each case. I.e. when testing the static model the static metric should be used. For a given metric radius the static metric will yield a smaller radius in arcseconds than for the expanding metric (for the redshifts considered here). Therefore the surface brightness which follows from the Kormendy relation is *brighter* for the the static case at a given redshift. At a redshift of $z = 0.4$ the effect is $0^m.37$. The effect goes always in the direction to favor the given model.

In order to perform the SBT, after correction for galactic extinction, we derived the surface brightness from the three clusters by determining the baricenter of each relation and then using the above given fixed slope of 2.5.

Our values for the surface brightness is compared with the prediction from the expanding universe and from the static universe in Figure 2, where we also plotted the values for A851 using the difference in the R_C value between A851 and Coma, given by Pahre et al. (1996). For each metric we also give two evolutionary models (including the K-effect and the cosmological dimming) computed by Campos (1996) using the Bruzual code. Both models have $H_o=75$, $q_o=0.1$ a Salpeter IMF and an exponentially decreasing star formation rate. Model 1 has an e-folding time of 0.5 Gyr and a time of formation of $z=50$. Model 2 has an e-folding time of 3.0 Gyr and a time of formation of $z=5$. The models were chosen to bracket the evolutionary effects. Actually model 1 gives an evolution similar to more realistic models which are in accordance with faint number counts and colours (Campos 1996). It is seen that passive evolution is important in interpreting the results. The evolution correction applied should eventually be chosen to match the observed evolution of colours in galaxies.

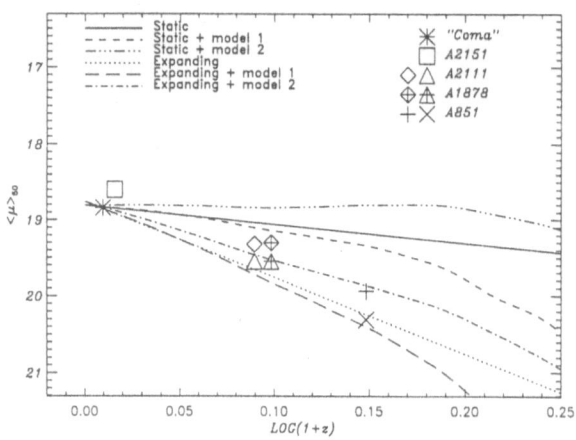

Fig. 2. The surface brightness derived from the Kormendy relation for the metric radius of 1kpc as function of $log(1 + z)$. "Coma" means the mean of 10 nearby clusters. For each galaxy cluster we plot two points, the upper corresponds to the static metric, the lower to the expanding metric.

References

Bendinelli, O. (1991): ApJ **366**, 599

Campos, A. (1996): Private communication

Jørgensen, I., Franx, M., Kjærgaard, P. (1995): MNRAS **273**, 1097

Kjærgaard, P., Jørgensen, I., Moles, M. (1993): ApJ **418**, 617

Pahre, M.A., Djorgovski, S.G., de Carvalho, R.R. (1996): ApJ **456**, L79

On the Upper Limit of Surface Brightness of Starburst Galaxies

Asao Habe[1] and Colin Norman[2,3]

[1] Hokkaido University, Sapporo, JAPAN, e-mail:habe@phys.hokudai.ac.jp
[2] Johns-Hopkins University, Baltimore, USA
[3] STSCI, Baltimore, USA, e-mail:norman@stsci.edu

Abstract. We discuss the possible physical basis for a limiting surface brightness for starburst galaxies. Using basic models for star formation and the interstellar medium, we examine various process that might disrupt the interstellar medium including: radiation driven outflows with opacities given by Thomson scattering, and dust; supernovae driven outflows; cooling processes; and the photodissociation of molecular clouds in the intense radiation field. We find that these processes can roughly explain the observed upper limit although a detailed fit is not yet achievable. We generalize our results to spherical systems and find that luminosity density is a fundamental parameter.

1 Introduction

Lehnert and Heckman (1996) have shown the upper limit of surface brightness of star burst regions as $\sim 10^{11} L_\odot/\text{kpc}^2$ and that this upper limit is independent on size and luminosity of starburst regions. They compiled observational results of starburst regions with a wide range of size (10pc-10kpc) and luminosity (10^{40-46}erg/sec). What and how regulate the upper limit in starburst regions ?

Our idea of regulation mechanism of the starburst phenomena is as follows: Strong radiation from many OB stars and/or huge energy released by SNe in starburst regions can eject gas from the starburst region. If gas is ejected from a starburst region, mass of gas decreases. As a result, star formation rate decreases. In this way, burst star formation is regulated. We examine, as possible regulation process of starburst, a) a radiative wind driven by Thomson scattering, b) a radiative wind driven by dust absorption, c) supernova explosions, and d) supperbubbles.

2 Model

We assume a Schmidt law for star formation (Kennicutt 1989) and estimate the relation between surface brightness and star formation rate as $\Sigma_L = K' \Sigma_{gas}^n$. We also estimate a relation between star formation rate and supernova rate as, $\dot{\Sigma}_{SN} = f_{SN} \dot{\Sigma}_{SF}$. Our normalization is referred to M82 (Young et al. 1986, Heckman, Armus, and Miley 1994). We estimate the condition of gas ejection as $P > P_{gr}$, where $P_{gr} \equiv 2\pi G \Sigma_{mass} \rho_0 \Delta z$.

2.1   Radiation Driven Winds

For the radiation driven outflows, we estimate a condition of outflow is $\Sigma_L >$ $\Sigma_{L,cr}$ where $\Sigma_{L,cr}$ is given in Thomson scattering case as,

$$\Sigma_{L,Th} = \left(\frac{2\pi G c m_p \mu_e \Sigma_{mass}}{\sigma_{Th}}\right) = 1.9 \times 10^{13} \Sigma_{mass,9} L_\odot/kpc^2,$$

and in dust scattering case as

$$\Sigma_{L,d} = 2 \times 10^{10} \frac{(\sigma_{Th}/\mu_a m_p)/(\sigma_d/m_d)}{1.05 \times 10^{-5}} \frac{\xi_{g,d}}{10^2} \Sigma_{mass,9} L_\odot/kpc^2.$$

$\xi_{g,d}$ is the gas dust ratio.

2.2   The Regulation by Supernovae and Superbubbles

Many supernova remnants create hot gas regions with large volume fraction. Gas can be ejected as hot wind, if $P_{h-c} > P_{gr}$. We estimate state of hot gas by assuming (1) the mass balance between SNR sweeping of ISM and evaporation of clouds and (2) overlapping SNRs in the radiative phase (McKee and Ostriker 1979, Habe et al. 1980). From the results and the condition of gas ejection we obtain the gas ejection condition as,

$$\Sigma_L > 3.5 \times 10^{11} \frac{\left(\frac{m_{cl}}{10^3 M_\odot}\right)^{0.4}\left(\frac{\Sigma_{gas}}{10^9 M_\odot/kpc^{-2}}\right)^{0.34}\Sigma_{mass,9}^{3.6}\left(\frac{\Delta z}{2kpc}\right)^{3.6}\left(\frac{\tau_{OB}}{10^7 yr}\right)}{\left(\frac{H_{cl}}{100pc}\right)^{0.34}\left(\frac{(M/L)_{SB}}{1/1000}\right)\left(\frac{f_{SN}}{0.01 SN/M_\odot}\right)\left(\frac{H_{SN}}{200pc}\right)} L_\odot/kpc^2.$$

We also estimate the surface fraction by superbubbles. From the formula of the maximum radius of a superbubble (Tomisaka and Ikeuchi 1987), we get the surface fraction of superbubbles as $Q_{SB} = 2\pi R_{SB}^2 \dot{n}_{OB} t_{SB} H_{SN}$. We relate the rate of OB association $\dot{n}_{OB}$ and the supernova rate as $\dot{n}_{OB} \sim \dot{n}_{SN} f_{OB}/N_{OB}$, where f_{OB} : the fraction of early-type stars in OB associations, and N_{OB} : the typical number of OB stars in an association. If we assume the condition of gas ejection as $Q_{SB} > Q_{CR} = 0.9$, this condition is

$$\Sigma_L > 1.8 \times 10^{11} \frac{\left(\frac{Q_{cr}}{0.9}\right)^{1.35}\left(\frac{N_{OB}}{100}\right)^{1.35}\left(\frac{H_{SN}}{200pc}\right)^{1.35}}{\left(\frac{H_{cl}}{100pc}\right)^{0.7}\left(\frac{f_{OB}}{0.7}\right)^{1.35}\left(\frac{f_{SN}}{0.01 SN/M_\odot}\right)^{1.35}} L_\odot/kpc^2.$$

3   Results and Discussion

We have studied the possible regulation process of the upper limit of surface brightness of the starburst regions. We compare our results with the observational results of Lehnert and Heckman (1996) in fig. 1. The superbubble model is most probable for the regulation process of the constant upper limit of the surface brightness of starburst regions. We generalize our results to spherical systems and find that luminosity density is a fundamental parameter (c.f. Djorgovski and

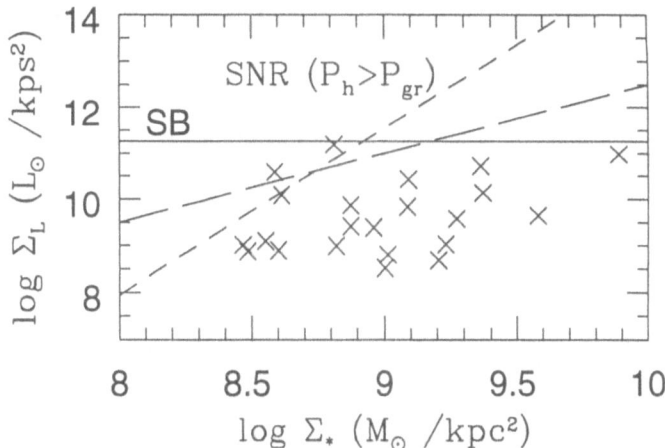

Fig. 1. The limit of the surface brightness of the supernova case (the short dashed line) and the superbubble cases (the solid line). The limit of the surface brightness of the Thomson scattering, the dust absorption and the radiation driven cases. The long dashed line are the Schmidt star formation low of $\Sigma_{gas} = \Sigma_*$ of our model. Crosses are estimated from the observational results of Lehnert and Heckman (1996).

Davis 1987). We will study the relation of the upper limit of surface brightness of high red shift starburst galaxies and current observational properties of galaxies.

Acknowledgements
We are extremely grateful to Drs. T. Heckman, G. Meurer, H. Ferguson and M. Spaans for many important discussions on this problem.

References

Djorgovski, S., and Davis M. 1987, ApJ, 313, 59
Habe, A., Ikeuchi, S. and Tanaka, Y.D. 1980, PASJ, **33**, 23
Heckman, T.M., Armus, L. and Miley, G.K. 1990, ApJ.Suppl, 74, 833
Kennicutt, R.C., Jr.1989, ApJ, 344, 685
Lehnert, M. and Heckman, T.M. 1996, ApJ, in press .
McKee, C. and Ostriker, J.1977, ApJ, **218**, 148
Meurer, G. and Heckman, T.M. 1996, ApJ, in press
Tomisaka, K. and Ikeuchi,S. 1987, PASJ, 38, 697
Young, J.S., Schloerb, P., Kenny, J.D. and Load, S.D. 1986, ApJ, 304, 458

The Fundamental Plane in the Leo-I Group and an Estimate of H_0

Jens Hjorth[1] and Nial R. Tanvir[2]

[1] NORDITA, Blegdamsvej 17, DK–2100 Copenhagen Ø, Denmark
[2] Institute of Astronomy, Madingley Road, Cambridge CB3 0HA, UK

Abstract. New effective radii and intensities for 5 E and S0 galaxies in the nearby Leo-I group are used in conjunction with recent literature velocity data to construct the Fundamental Plane. The rms scatter that we find is only 6 % in distance. The zero point of this relation provides a calibration of the FP as a distance indicator and directly determines the angular diameter distance ratio between the Leo-I group and more distant clusters. Combining this with the Cepheid distance to M96, located close to the core of Leo-I, we find the Hubble constant to be $H_0 = 67 \pm 8$ km s^{-1} Mpc^{-1} in the Coma cluster frame or $H_0 = 70 \pm 7$ km s^{-1} Mpc^{-1} relative to a frame of 9 clusters.

1 The Fundamental Plane in Leo-I

The Fundamental Plane (FP) is an excellent distance indicator relating distances between E and S0 galaxies particularly in clusters and groups. The work of Jørgensen, Franx & Kjærgaard (1996; hereafter JFK) demonstrates that a single global FP yields unbiased distance estimates to early-type galaxies in very different environments with a formal intrinsic scatter of 14 % in distance per galaxy.

New photometric parameters were determined from wide-field observations of the 5 E and S0 galaxies in the Leo-I group (Fig. 1) carried out in the R passband using the INT prime focus camera. To evaluate accurate velocity dispersions we examined recent, high-quality, long-slit spectroscopic data from the literature, as described in detail by Hjorth & Tanvir (1997).

The FP that we fit is

$$\log R_e = \alpha \log \sigma + \beta \log \langle I_r \rangle_e + \gamma, \tag{1}$$

where R_e is measured in arcsec, σ in km s^{-1}, $\langle I_r \rangle_e$ is the mean intensity inside the effective radius in Gunn r measured in L$_\odot$ pc^{-2}, and γ is the distance-dependent zero point of the relation. The coefficients $\alpha = 1.24$ and $\beta = -0.82$ are fixed to the global values adopted by JFK in which case biases above 1% due to magnitude and morphological selection are excluded. To correct from Cousins R_C to Gunn r we adopt $r - R_C = 0.37$.

Figure 2 shows the FP for the Leo-I galaxies. Given the distance from the group center ($\sim 3°$) of NGC 3489 and its comparatively "late-type" properties, we decided *a priori* that it should not be included in the construction of the FP, but include it in the plots for completeness. Indeed, NGC 3489 clearly falls below

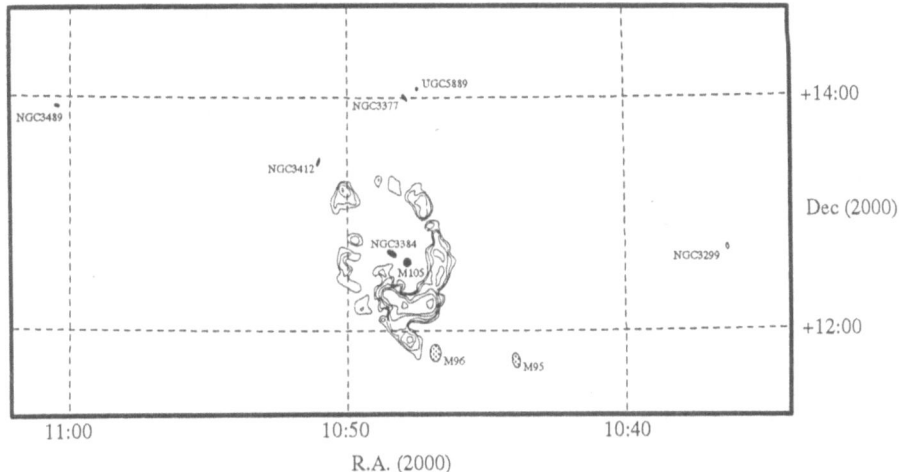

Fig. 1. The Leo-I group is the nearest group of galaxies containing both bright spirals and early type galaxies and is as such an important stepping stone in the determination of the Hubble constant. The centre of the group is defined by the E1 galaxy NGC 3379 (= M105) and the S0 galaxy NGC 3384 which form a close pair. The E5 galaxy NGC 3377, the S0 galaxy NGC 3412 and the S0/a galaxy NGC 3489 are also likely members (Garcia 1993). The distances to the core of Leo-I and the spiral galaxy M96 are tied together in a unique way, due to the giant 200 kpc diameter H I ring (Schneider 1989) which orbits the central early type galaxies and interacts with M96. There is no such direct link to the spiral galaxy M95 which appears to be slightly in the foreground relative to M96 (Graham et al. 1997).

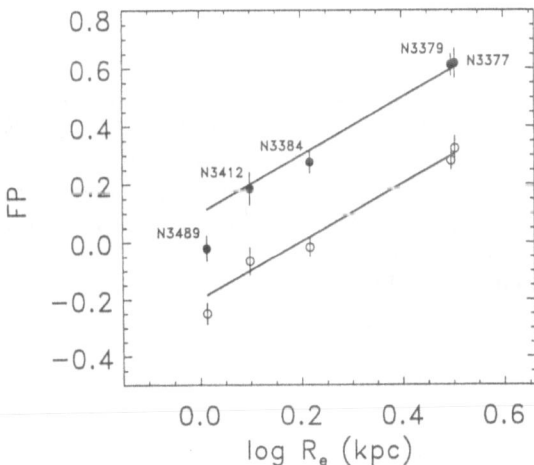

Fig. 2. The Fundamental Plane in Leo-I. The filled circles are 'classic' FP = $1.24 \log \sigma - 0.82 \log \langle I_r \rangle_e$ values for Leo-I galaxies, the solid line through the points is the JFK FP for the four larger galaxies. The open circles are Mg_2-FP = $1.05 \log \sigma - 0.78 \log \langle I_r \rangle_e - 0.40 \Delta Mg_2$ values for Leo-I galaxies and the solid line through these points is the JFK Mg_2-FP for the four larger galaxies.

the line defined by the other galaxies. Thus, the FP zero point is computed as the median offset of the four larger galaxies. We obtain $\gamma_{Leo-I} = 1.150$. The observed formal rms scatter in γ is 0.024 or 6 % in R_e. This is significantly smaller than expected from the 14 % intrinsic scatter plus observational errors found by JFK. From $\gamma_{Coma} = 0.182\pm0.009$ (JFK) we find a zero-point difference between Leo-I and Coma of 0.968 ± 0.015.

Environmental differences may be important given that Leo-I is a relatively poor group. Guzmán & Lucey (1993) and JFK formulated an 'environment-independent' FP which incorporated an Mg_2 term to correct for possible systematic differences. This should provide some robustness to differences in stellar populations between galaxies, which may in turn vary systematically with environment, whether such differences are due to variations in age or metallicity. For this Mg_2-FP $\alpha = 1.05$, $\beta = -0.78$ and $-0.40\Delta Mg_2$ is added to the rhs of the FP equation (1) (JFK). We find $\gamma(Mg_2)_{Leo-I} = 1.449 \pm 0.014$ and an observed formal scatter of 6 %. From $\gamma(Mg_2)_{Coma} = 0.497 \pm 0.009$ we find a zero-point difference between Leo-I and Coma of 0.952 ± 0.017.

2   The Hubble Constant

The differences between the two FP zero-point differences are insignificant and we adopt a value of 0.96 ± 0.03 where we now assume a conservative uncertainty of $0.057/\sqrt{4}$ which is estimated from the expected intrinsic scatter of 14 % in distance per galaxy. It should be noted that this uncertainty is much higher than the estimated individual uncertainties and makes the derived distance ratio a robust value. We thus determine an angular diameter distance ratio of 9.12 ± 0.64 or a luminosity distance ratio of 9.51 ± 0.67 between Coma and Leo-I.

For a Leo-I distance of 11.3 ± 0.9 Mpc based on the Cepheid distance to M96 (Tanvir et al. 1995; Hjorth & Tanvir 1997) the derived distance ratio between Leo-I and Coma implies a Coma distance of 108 ± 12 Mpc. For the recession velocity of Coma we use 7200 ± 400 km s^{-1}. Adding the uncertainties in quadrature we arrive at an uncertainty of 12 % in the Hubble constant: thus we get $H_0 = 67 \pm 8$ km s^{-1} Mpc^{-1} fixing to Coma. Alternatively, tying Leo-I to the larger frame of 9 clusters in the sample of JFK, which has a median redshift of ~ 5000 km s^{-1}, we find a Hubble constant of $H_0 = 70 \pm 7$ km s^{-1} Mpc^{-1}.

References

Garcia, A. M. 1993, A&AS, 100, 47
Graham, J. A. et al. 1997, ApJ, in press
Guzmán, R., & Lucey, J. 1993, MNRAS, 263, L47
Hjorth, J., & Tanvir, N. R. 1997, ApJ, in press
Jørgensen, I., Franx, M., & Kjærgaard, P. 1996, MNRAS, 280, 167 (JFK)
Schneider, S. 1989, ApJ, 343, 94
Tanvir, N. R., Shanks, T., Ferguson, H. C., & Robinson, D. R. T. 1995, Nature, 377, 27

The Fundamental Plane and the Choice of Photometric Parameters

Inger Jørgensen*

McDonald Observatory, The University of Texas, Austin, TX 78712, USA

Abstract. It is investigated which methods for determination of total magnitudes and effective radii for E and S0 galaxies give accurate determinations of the parameters. It is also studied which of the photometric parameters are best used in applications of the Fundamental Plane.

1 Definition of the Photometric Parameters

The photometric parameters usually used in applications of the Fundamental Plane (FP), $\log r_e = \alpha \log \sigma + \beta <\mu>_e + \gamma$, are the effective parameters derived from a fit with a growth curve for an $r^{1/4}$ profile (e.g., Dressler et al. 1987; Djorgovski & Davis 1987; Jørgensen et al. 1996). However, the luminosity profiles of E and S0 galaxies are known to show significant deviations from the $r^{1/4}$ profile (e.g., Caon et al. 1993). It is discussed here which photometric parameters are the best to use in the FP, and which give the best determinations of the "true" parameters. Photometry for E and S0 galaxies in Virgo (52 galaxies, $B_T \leq 14^m$, 80% complete) and Fornax (28 galaxies, $B_T \leq 15^m$, 95% complete) is used (Caon et al. 1990, 1993, 1994). The following parameters are defined. *"True":* Based on the asymptotic magnitudes from Caon et al. These parameters depend the least on model assumptions. $r^{1/n}$: Derived from a fit with an $r^{1/n}$ profile (Caon et al.). $r^{1/4}$: Derived from a fit with the growth curve for an $r^{1/4}$ profile. $r^{1/4}+disk$: Derived from a fit with the growth curve for the sum of an $r^{1/4}$ profile and an exponential profile. *Petrosian* (1976): At the radius r_η, $\mu - <\mu> = 1.39$. The total magnitude is $m_\eta = <\mu>_\eta - 5 \log r_\eta - 1.99$, where $<\mu>_\eta$ is the mean surface brightness within r_η. For an $r^{1/4}$ profile these Petrosian parameters are identical to the $r^{1/4}$ parameters.

2 The Best Choice for the FP as a Distance Estimator

The central velocity dispersions are available for 31 of the 80 galaxies (Faber et al. 1989). The FP was fitted by minimization of the sum of the absolute residuals perpendicular to the relation. The "true" parameters result in slightly smaller scatter around the FP than the other tested parameters. The lower scatter may give more accurate distance determinations. The FP based on the Petrosian parameters has significantly larger scatter than the FP for any of

---

* Hubble Fellow

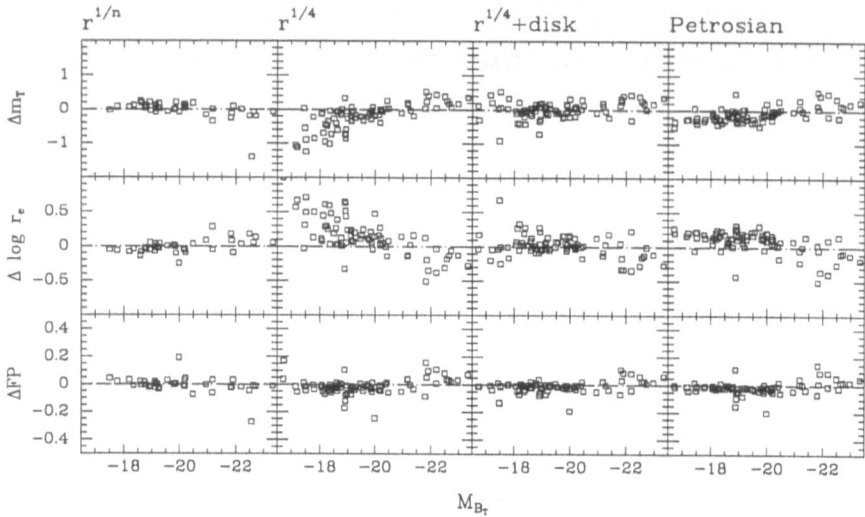

Fig. 1. Comparisons of the various photometric parameters (see labels above the panels) with the "true" parameters. "FP"= $\log r_e - 0.328 <\mu>_e$. The differences are derived as "other"–"true", and plotted versus the absolute asymptotic magnitudes.

Table 1. Comparison of photometric parameters

Method	m_T $\langle\Delta\rangle$	rms	α	$\log r_e$ $\langle\Delta\rangle$	rms	α	"FP" $\langle\Delta\rangle$	rms	α
$r^{1/n}$ (a)	−0.014	0.13	−0.16	−0.008	0.10	0.11	0.001	0.04	0.00
$r^{1/4}$	−0.27	0.53	0.60	−0.16	0.30	−0.44	−0.016	0.06	0.08
$r^{1/4}$+disk	0.08	0.24	0.18	0.03	0.21	−0.17	−0.019	0.07	0.05
Petrosian	−0.10	0.23	0.37	0.08	0.16	−0.28	−0.015	0.05	0.06

Notes – The differences are derived as "other"-"true". "FP"=$\log r_e - 0.328<\mu>_e$. $\langle\Delta\rangle$ – the mean difference. α – the slope of a least squares fit of the difference as function of $\log r_e$. (a) NGC4406, the most deviating galaxy on Fig. 1, is omitted.

the other parameters, presumably because Petrosian parameters are sensitive to small local variations in an otherwise smooth luminosity profile. The tests need to be repeated on a larger sample of galaxies with available velocity dispersions.

3   The Best Estimators of the "True" Parameters

The "true" parameters, based on asymptotic magnitudes, are assumed to be the best available estimators of the real magnitudes and radii. However, the other methods for determination of the total magnitudes are less demanding in terms of how faint surface brightnesses are needed in the mapping of the

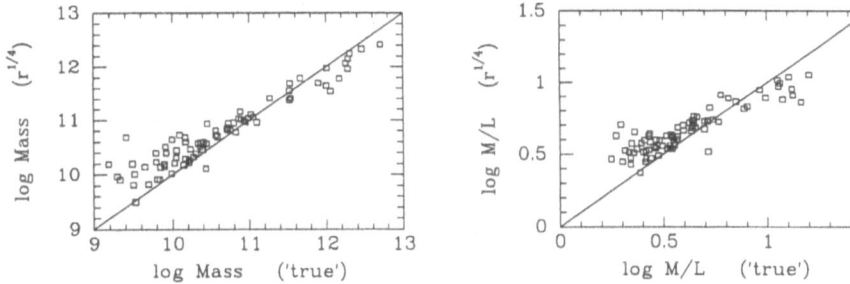

Fig. 2. Comparison of masses and M/L ratios derived from the $r^{1/4}$ parameters and from the "true" parameters. The masses and M/L ratios are based on velocity dispersions predicted from the FP. Compared to the "true" parameters the use of the $r^{1/4}$ parameters leads to larger masses and M/L ratios for the least massive galaxies, and smaller masses and M/L ratios for the massive galaxies.

luminosity profiles. Fig. 1 and Table 1 summarize the comparisons of the "true" parameters with the other photometric parameters. The total magnitudes and effective radii are best estimated either with a fit to a growth curve for an $r^{1/4}$ profile plus an exponential disk, or by fitting an $r^{1/n}$ profile. A fit with a growth curve for an $r^{1/4}$ profile gives for most galaxies rather poor estimates of the total magnitude and the effective radius. The combination "FP"=$\log r_{e} - 0.328 <\mu>_{e}$, which enters the FP, is in general well-determined. There are small systematic offsets in "FP" between determinations from different methods. Thus, parameters based on different methods should not be mixed.

The FP is usually interpreted as a relation between the M/L ratio and the mass, $M/L \propto M^{\kappa}$. In Fig. 2 masses and M/L ratios derived from the $r^{1/4}$ parameters and from the "true" parameters are compared. The determinations of the masses and the M/L ratios are based on predicted velocity dispersions for the galaxies, derived under the assumption that the galaxies follow the FP for the $r^{1/4}$ parameters (Jørgensen et al. 1996). The results shown in Fig. 2 are of importance for the slope and interpretation of the relation $M/L \propto M^{\kappa}$. Note, however, that determinations of true M/L ratios require more detailed kinematic information than the central velocity dispersions provide.

References

Caon, N., Capaccioli, M., D'Onofrio, M., 1993, MNRAS, 265, 1013
Caon, N., Capaccioli, M., D'Onofrio, M., 1994, A&AS, 106, 199
Caon, N., Capaccioli, M., Rampazzo, R., 1990, A&AS, 86, 429
Djorgovski, S., Davis, M., 1987, ApJ, 313, 59
Dressler, A., et al., 1987, ApJ, 313, 42
Faber, S. M., et al., 1989, ApJS, 69, 763
Jørgensen, I., Franx, M., Kjærgaard, P., 1996, MNRAS, 280, 167
Petrosian, V., 1976, ApJ, 209, L1

The Fundamental Plane and Dark Haloes of Ellipticals

Alexei G. Kritsuk

Institute of Astronomy, University of St Petersburg
Stary Peterhof, St Petersburg 198904, Russia

Abstract. A two-component isothermal equilibrium model is applied to reproduce basic structural properties of dynamically hot stellar systems immersed in their massive dark haloes. The origin of the fundamental plane relation for giant ellipticals is naturally explained as a consequence of dynamical equilibrium in the context of the model. The existence of two galactic families displaying different behaviour in the luminosity–surface-brightness diagram is shown to be a result of a smooth transition from dwarfs, dominated by dark matter, to giants dominated by the luminous stellar component near the centre. The model provides some restrictions on the properties of dark haloes implied by the fundamental scaling laws.

1 Introduction

The term 'dynamically hot galaxies' (DHGs) was introduced by Bender, Burstein & Faber (1992) for a class of stellar systems in which random motions provide most of the energy for support of the system. This class includes all varieties of elliptical galaxies from giant to dwarf and compact ellipticals, bulges of S0s and spirals, as well as dwarf spheroidal (dSph) galaxies.

It is generally suggested that the properties of early-type galaxies are determined by dissipative collapse and then modified by mergers. The details of cooling, star formation, and feedback processes are considered to be important to produce the luminosity–surface-brightness relation. However, each particular physical model usually contains a large number of free input parameters and therefore the interpretations do not appear to be unique.

An important source of uncertainties in the models is our poor knowledge of the nature and distribution of dark matter in DHGs (Kormendy 1988, Ashman 1992, Gallagher & Wyse 1994, de Zeeuw 1995). It is clear that the significant content of dark matter will control the brightness profiles of the galaxies and therefore may play a major role in determining the observed global structural properties of DHGs. The purpose of this study is to check whether or not the fundamental properties of DHGs can be explained solely as due to variations in dark matter content and its systematic spatial redistribution in the galaxies along the sequence gEs–dEs–dSphs. The preliminary answer is given here with the aid of an equilibrium galaxy model, which includes the stellar component and the dark component (both are isothermal, but with different temperatures). Such a two-component model implies the existence of a 'conspiracy' between the luminous and dark matter distributions, and may provide some hints about the initial conditions and the nature of star formation processes operating in DHGs.

The paper is organized as follows. Section 2 gives a description of the model. Section 3 discusses the fundamental plane (FP) for giant ellipticals and the deviations of dSph galaxies from it as they originate in the context of the model. In Sect. 4 the distribution of DHGs in the luminosity–surface-brightness diagram is compared with the model predictions. The results are summarized in Sect. 5.

2 The Two-Component Galaxy Model

A spherically symmetric equilibrium configuration of gravitating mass, including the isothermal stellar (luminous) component of a galaxy and its isothermal dark halo, can be described by a combination of Jeans and Poisson equations. After some straightforward transformations these can be reduced to the 'hydrostatic β-law' for the stellar and dark matter densities, $\rho_{\text{DM}} = \mathcal{C}\rho_*^\beta$ (where $\mathcal{C}$ is the integration constant and $\beta \equiv \sigma_*^2/\sigma_{\text{DM}}^2$ is the ratio of their 'temperatures'), and the modified Poisson equation

$$\frac{1}{r^2}\frac{\mathrm{d}}{\mathrm{d}r}\left(r^2\frac{\mathrm{d}\ln\rho_*}{\mathrm{d}r}\right) = -\frac{4\pi G}{\sigma_*^2}(\rho_* + \rho_{\text{DM}}), \tag{1}$$

see (Kritsuk 1997) for more details.

Substituting dimensionless variables $x = r/r_0$, and $y = \rho_*/\rho_{*,0}$, where

$$r_0 = \frac{\sigma_*}{\sqrt{4\pi G\rho_{*,0}}}, \tag{2}$$

and introducing a new parameter as the ratio of the densities at the centre, $\delta = \rho_{\text{DM},0}/\rho_{*,0}$, one gets a generalized equation of Emden-Fowler type:

$$\frac{1}{x^2}\frac{\mathrm{d}}{\mathrm{d}x}\left(x^2\frac{\mathrm{d}\ln y}{\mathrm{d}x}\right) = -(y + \delta y^\beta). \tag{3}$$

It yields a biparametric family of solutions $y(x; \beta, \delta)$ for the initial conditions $y(0) = 1$, $y'(0) = 0$, see Fig. 1.

Each particular model galaxy can be characterized by two dimensional parameters (e.g., σ_* and r_0) that control the scaling, and two dimensionless parameters β and δ that determine the shape of the light distribution in the galaxy. In this framework the structural homology generally assumed for the class of E galaxies implies the constancy of β and δ. Since there are indications that the homology is broken (see, e.g., Djorgovski 1995; Hjorth & Madsen 1995), all four parameters are treated as free in the following.

Parameter β defines the so-called 'conspiracy' of the dark and luminous matter, while δ controls the dominant mass component in the centre of a galaxy, i.e. the galactic central mass-to-light ratio. When $\beta = 1/2$ and $\delta \lesssim 0.1$, the stellar system has a density distribution of an isothermal sphere in the inner region and a cut-off at the core radius of the isothermal dark halo. In this case the

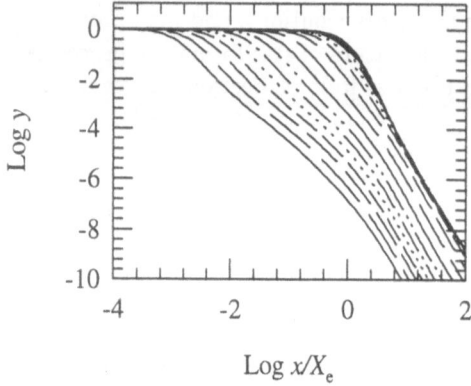

Fig. 1. Luminosity density profiles for $\beta = 0.5$ and δ varying from 0.001 (the leftmost curve) to 10^3; the intervals between the curves $\Delta \log \delta = 0.25$. The radial distance is given in units of the half-light radius, see equation (5) below.

density of luminous material varies as r^{-2} in the intermediate range of r outside the core region and asymptotically as r^{-4} in the outer region where the dark component dominates by mass. It is this behaviour that is required by the de Vaucouleurs surface brightness profile (Bertin & Stiavelli 1993). Figure 2 illustrates the variety of profiles responding to different values of δ and $\beta = 0.5$. While for $\delta \lesssim 0.1$ the profiles closely follow the empirical $R^{1/4}$ law in a range of radii $0.1R_e \leq r \leq 1.5R_e$ (Burkert 1993, 1994), they look more like exponential at $\delta \gtrsim 10$. The farther β deviates from $1/2$, the worse fit the de Vaucouleurs formula yields in the limit of small δ.

The physical meaning of the condition $\beta = 1/2$ can be readily understood in terms of a tentative dissipative galaxy formation scenario. When baryonic gas is initially thermalized in the dark halo potential well, its specific thermal energy equals the energy of dark matter particles and its density distribution follows that of the dark matter. If the efficiency of subsequent star formation follows the Schmidt law $\dot{\rho}_* \propto \rho_{gas}^m$ with $m = 2$, the stellar density distribution will automatically satisfy the $\beta = 1/2$ condition. The observed light profiles of gEs show a preference for such a choice of β.

3   An Edge-On View of the Fundamental Plane

The fundamental plane for ellipticals is argued to contain clues to the initial conditions and processes of galaxy formation, being not just a consequence of the virial theorem. A simple reason behind this statement is that scale-free pure dynamics cannot explain the origin of the dimensional scaling laws (Bertin & Stiavelli 1993). If the existence of dark haloes is allowed, the homology hypothesis appears to be violated. Then a 'conspiracy' between the luminous and dark matter is able to introduce fractional power-law indices into the scale-free dynamical relationships and thereby into some of the observed correlations.

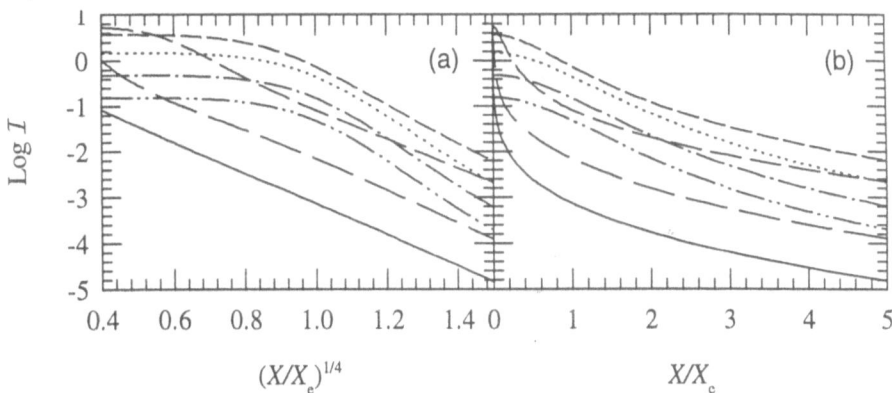

Fig. 2. Surface brightness profiles for $\beta = 0.5$. The sequence of line styles: solid, dashed (long, medium, short), dotted, dashed dotted, dashed double dotted corresponds to $\delta = 0.001, 0.01, 0.1, 1, 10, 100, 1000$, respectively. Panels (a) and (b), having different horizontal scales, display $R^{1/4}$ and exponential profiles as straight lines. The surface brightness $\mathcal{I}$ is given in dimensionless units defined by equation (7).

Let $\mathcal{L}$, X_e, and $\mathcal{I}_e$ be *dimensionless* luminosity, half-light radius, and mean effective surface brightness, respectively, so that

$$\mathcal{L} \equiv 4\pi \int_0^\infty x^2 y \mathrm{dx} = \frac{4\pi G (M/L)_*}{r_0 \sigma_*^2} L, \qquad (4)$$

$$\mathcal{L}/2 = 2\pi \int_0^{X_e} \mathcal{I}(X) X \mathrm{dX}, \qquad (5)$$

$$\mathcal{I}_e = \frac{\mathcal{L}}{2\pi X_e^2}, \qquad (6)$$

where

$$\mathcal{I}(X) \equiv 2 \int_X^\infty \frac{y x \mathrm{dx}}{\sqrt{x^2 - X^2}} = \frac{4\pi G r_0 (M/L)_*}{\sigma_*^2} I(R) \qquad (7)$$

is the dimensionless surface brightness, $X = R/r_0$ is the radial distance in projection, whereas L, I, and $(M/L)_*$ are galaxy luminosity, surface brightness, and stellar mass-to-light ratio in physical units.

In order to study the structural properties of dynamically hot galaxies, Bender et al. (1992) have defined an orthogonal coordinate system, which is termed as the κ-space. While κ_1 and κ_2 coordinates are essentially dimensional, κ_3 is proportional to the central mass-to-light ratio and can be expressed as a combination of the dimensionless variables defined above:

$$\kappa_3 = (\log X_e - \log \mathcal{L} + const)/\sqrt{3}. \qquad (8)$$

This allows for identification of a scale-free counterpart to the edge-on projection of the FP in terms of dimensionless variables $(\mathcal{I}_e, \mathcal{L}, \delta)$, which is shown in Fig. 3.

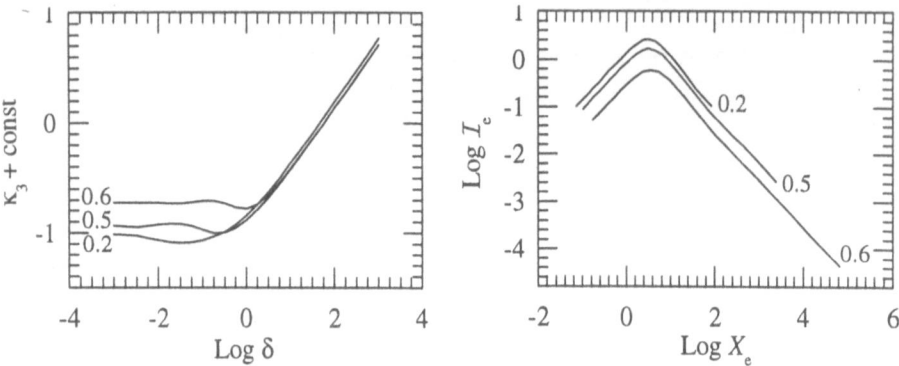

Fig. 3. Plot of κ_3 versus δ for the set of models with $\beta = 0.2, 0.5$, and 0.6 (left panel).

Fig. 4. Mean effective surface brightness versus effective radius for the set of models with $10^{-3} < \delta < 10^3$ and $\beta = 0.2, 0.5, 0.6$ (right panel); axes are scaled in dimensionless units.

As the realistic range for δ is undefined at this stage, the graphs are given for a wide interval from 10^{-3} to 10^3; also for completeness three cases of β are shown. The asymptotics $\kappa_3 \sim const$ for $\delta \ll 1$ (the virial limit) and $\kappa_3 \sim (\log \delta + const)/\sqrt{3}$ for $\delta \gg 1$ are common to all represented values of β. In the intermediate range of δ ($-1.5 < \log \delta < 0.5$ for $\beta = 0.5$), where the solution undergoes a transition from one asymptotic behaviour to another, one can see a slight negative tilt ($\Delta \kappa_3 \simeq -0.1$) for $-1.5 < \log \delta < -0.4$, a minimum at $\delta \simeq 0.4$ and a rise at higher δ. These changes are implied by the non-trivial restructuring of the model galaxies in the course of the transition from luminous to dark matter dominated systems, see Figs 1 and 2.

If there is a one-to-one mapping between the luminosity (or 'the mass of the galaxy within the luminous confines' κ_1) and the dark matter content δ of DHGs, a comparison of the galaxy distribution in (κ_1, κ_3) projection [see fig. 1 from Burstein et al. (1993)] and the $\log \delta - \kappa_3$ relation makes sense. The FP of gEs can be identified with a slightly tilted part of the $\beta = 0.5$ curve in Fig. 3 (cf. Renzini & Ciotti 1993; Ciotti, Lanzoni & Renzini 1996). Dwarf ellipticals, having progressively higher dark matter content, show small positive deviations from the FP. Finally, the dark matter dominated dSphs are characterized by extremely high mass-to-light ratios. Small observed scatter of gEs about the FP can be interpreted in this framework as an argument for a universal value of β, indicating possible similarity of star formation processes and conservation of the effective dark-to-luminous specific kinetic energy ratio in mergers.

Thus the scale-free dependences implied by the two-component model are able to explain the gross features of edge-on projections for the distribution of DHGs about the FP for giant ellipticals. Similar analysis for the face-on view of the FP, which cannot be reduced to scale-free considerations, will be a subject of the following section.

4   The Luminosity–Surface-Brightness Diagram

The scale-free counterpart for the distribution of galaxies in the effective-radius–surface-brightness plane is shown in Fig. 4. The curves $\beta = const$ display a characteristic shape of the empirical $R_e- <\mu>_e$ diagram [Capaccioli et al. 1993, fig. 2], changing the sign of their slopes at $\delta \simeq 1$. However, a direct comparison of this scale-free diagram with the one based on observables is not possible since $I_e \propto r_0^{-1} \sigma_*^2 \mathcal{I}_e$, and $R_e \propto r_0 X_e$, while r_0 and σ_* are unknown functions of δ.

These hidden parametric relationships can be recovered, assuming a unique value of β for all model galaxies ($\beta = 0.5$ will be used hereafter), and using the observational scaling relations; details can be found in (Kritsuk 1997). The adopted fitting procedure returns two approximate power-law dependencies: $\delta \propto L^F$ and $\rho_{*,0} \propto \delta^H$, with indices $F \simeq -0.5$ and $H \simeq -1$, which can be used as a zero-order approximation for both galaxy branches. These allow one to delimit the range of realistic values of δ for the giants and dwarfs.

If $\delta \propto L^{-0.5}$ the range of absolute magnitudes of gEs, $-25.5 < M_B < -19.5$ (Capaccioli et al. 1993), corresponds to a difference of ~ 1.3 dex in δ. The intersection point of two galaxy branches in the $(R_e, \mathcal{I}_e)$ plane is located at $\log \delta \simeq -0.2$. Thus the brightest galaxies may have $\log \delta \simeq -1.5$, which means that they contain ~ 30 times more mass in stars than in dark matter in their core regions. [Note that when $\delta \approx 10^{-2}$ the masses of stars and dark matter enclosed inside R_e are comparable, cf. (Saglia, Bertin & Stiavelli 1992).]

For dwarf galaxies the observed range of luminosities of 4.6 dex from $M_B = -8$ to $M_B = -19.5$ implies the value of δ at the faint end of this branch to be about 60, however, with very large uncertainties due to poorly known slope of the luminosity–velocity-dispersion relation for dwarfs. Yet this estimate is in rough agreement with the mass-to-light ratios determined for the extreme dSph galaxies of $M_V \simeq -9$.

The value of $H \approx -1$ implies a weak dependence of $\rho_{DM,0}$ on δ along the gE sequence, cf. Kormendy (1988). In this case the central stellar density and velocity dispersion would be related via $\rho_{*,0} \propto \sigma_*^2$ and the characteristic radius r_0 would be nearly constant (see equation 2). This makes it easy to imagine the mapping of the scale-free diagram shown in Fig. 4 onto $R_e- <\mu>_e$ plane: one simply has to deform the curves, correcting the $\mathcal{I}_e$ values for $\log \sigma_*$ variations along the galaxy sequence dSph–dE–gE.

5   Summary

A two-component spherically symmetric galaxy model with a stellar system immersed in an isothermal massive dark halo reproduces the essence of changes in galaxy structure along the morphological sequence dSph–dE–gE (including bulges of S0 and spiral galaxies).

The conclusions can be formulated as follows.

1. Dark haloes of dynamically hot galaxies can play an important role in controlling the shape of their surface brightness profiles.

2. The existence of the fundamental plane for giant ellipticals and observed deviations of dwarf spheroidal galaxies from it follow naturally from the dynamical equilibrium condition in the framework of the two-component model.

3. The major difference in the empirical luminosity–surface-brightness relation for dwarf and giant families of galaxies could be explained in the context of a smooth transition from dark matter dominated dwarfs to luminous matter dominated (in the centre) giants.

Acknowledgments

This contribution has been partly supported by the Russian Foundation for Basic Research, project 96-02-19670.

References

Ashman K. (1992): PASP **104**, 1109

Bender R., Burstein D., Faber S.M. (1992): ApJ **399**, 462

Bertin G., Stiavelli M. (1993): Rep. Prog. Phys. **56**, 493

Burkert A. (1993): A&A **278**, 23

Burkert A.M. (1994): Reviews in Modern Astronomy, Vol. 7. Astronomische Gesellschaft, Hamburg, p. 191

Burstein D., Bender R., Faber S.M. (1993): in Danziger I.J., Zeilinger W.W., Kjär K., eds, Proc. ESO/EIPC Workshop, *Structure, Dynamics and Chemical Evolution of Elliptical Galaxies*. ESO, Garching, p. 31

Capaccioli M., Caon N., D'Onofrio M. (1993): *ibid.*, p. 43

Ciotti L., Lanzoni B., Renzini A. (1996): MNRAS **282**, 1

de Zeeuw P.T. (1995): in van der Kruit P.C., Gilmore G., eds, Proc. IAU Symp. 164, *Stellar Populations*. Kluwer, Dordrecht, p. 215

Djorgovski S. (1995): ApJ **438**, L29

Gallagher J.S., III, Wyse R.F.G. (1994): PASP **106**, 1225

Hjorth J., Madsen J. (1995): ApJ **445**, 55

Kormendy J. (1988): in Fang L.Z., ed., *Origin, Structure and Evolution of Galaxies*. World Scientific, Singapore, p. 252

Kritsuk A.G. (1997): MNRAS, in press (astro-ph/9611037)

Renzini A., Ciotti L. (1993): ApJ **416**, L49

Saglia R.P., Bertin G., Stiavelli M. (1992): ApJ **384**, 433

Line-Strength Indices and Kinematics in Fornax

Harald Kuntschner[1], Roger Davies[1] and Ralf Bender[2]

[1] University of Durham, South Road, Durham DH1 3LE, England
[2] Universitäts-Sternwarte, Scheinerstr. 1, 81679 München, Germany

Abstract. The gross morphological similarity of elliptical galaxies masks a large diversity in their detailed structure, kinematics and star formation history. Luminous ellipticals usually show high velocity dispersion and above solar abundances which makes the measurement & interpretation of line strength indices difficult. On the other hand low luminosity ellipticals (LLEs) are close to solar abundance and their lines are less blended, hence the interpretation of their spectra is more straightforward. To investigate the differences between luminous galaxies and LLEs we have measured absorption line strength in spectra of five LLEs and in four bright elliptical galaxies in the Fornax cluster. A preliminary assessment of our data suggests that the elliptical galaxies in Fornax represent a metallicity sequence rather than an age sequence.

1 Observations and Data Reduction

The spectra were taken with the ESO NTT at a resolution of a 4.5 Å ($\sigma \simeq 110km/s$). The definitions introduced by Burstein et al. (1984) were used to compute, from the deredshifted spectra, line-strength indices. Furthermore line-strengths were standardized to zero velocity dispersion by correcting for velocity dispersion broadening. The correction factors were derived from stellar template spectra convolved with a Gaussian to mimic the sigma range from 0 - 500 km/s in steps of 20km/s.

The velocity dispersion data were derived from our own measurements with the Fourier correlation quotient (FCQ) method (Bender et al. 1990) and for the outer parts a simple Fourier-Correlation method (FXCOR in IRAF). Finally we adopted the results from D'Onofrio et al. (1995) when they had better S/N.

The line-strengths data were **not** corrected to match the Lick system (Burstein et al., 1984) (i) because only two comparison galaxies are available for each observation; (ii) because we decided to keep our resolution of 4.5 Å instead of Lick's ~ 9 Å. Preserving our relatively high resolution matches better the line-broadening due to *low* velocity dispersions and allows us to resolve weak Fe-features.

2 Central Mg - σ Relation in Fornax Ellipticals

In Figure 1 we present the Mg-σ Relation measured from the central $1/10$ r_e regions of each galaxy. Plotted are Mg_2 against $log\sigma$ and a linear χ^2 fit to it. The cloud of dots represents the sample from Bender et al. (1993). As expected

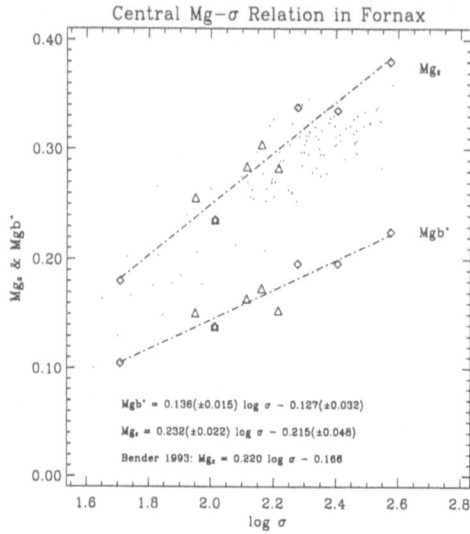

Fig. 1. The central $1/10\ r_e\ Mg - \sigma$ relation for Fornax ellipticals.

there is a tight relation between Mg_2 & σ and no significant differences can be seen between LLEs and bright ellipticals.

Due to the fact that the sidebands of the Mg_2 index are ~ 200 Å to the red and blue, this index is sensitive to continuum shape changes. In particular, changes in the response function of the detector and/or under-sampling the steep radial light profile can change this index significantly. Having the empirically very good $Mgb - Mg_2$ relation (Burstein et al. 1984) in mind, we used Mgb instead of Mg_2. The Mgb index is a compact index with a ~ 22 Å difference between index-band and sidebands. Hence it is less sensitive to continuum variations.

In order to compare with Mg_2 we constructed a 'new' molecular Mgb index (Baggley 1996) defined in the same way as Mg_2:

$$Mgb^+ = -2.5\ log\ \left(1 - \frac{Mgb_{old}}{32.5}\right) \tag{1}$$

As can be seen from Figure 1 the new Mgb^+ index shows also a **tight linear relation** with no worse spread around the fit.

3   Central $Fe - Mgb$ Relation

With higher resolution and better S/N spectra it becomes possible to measure not only the strong Fe5270 & Fe5335 features but also weaker indices such as Fe5406, Fe5015, Fe4531 and Fe4668. It is interesting to investigate whether all Fe-indices follow the same relations with Mgb ie. whether they are measuring the same properties. Figure 2 shows the central $1/10 \ r_e$ relation for several Fe-indices in our wavelength range. There is a good general agreement in slope and linearity between all measured Fe-indices except Fe4668.

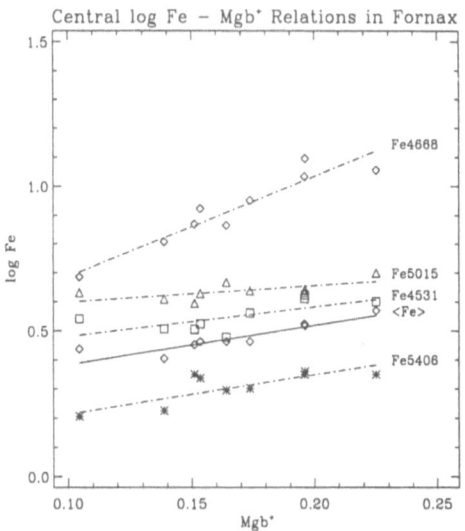

Fig. 2. Central $1/10 \ r_e$ $log \ Fe - Mgb^+$ in Fornax.

A preliminary assessment of this diagram suggests that the steep rise in Fe4668 is evidence for a rather strong mean metallicity gradient. Using Worthey (1996) an interpretation of this feature in terms of a change in age between bright and faint galaxies calls for unrealistically small ages for the low luminosity systems. Therefore assuming a single population model, the Fornax sequence of Es is more likely to be a sequence of metallicity rather than of age.

References

Baggley, G. (1996), PhD thesis, Univ. Oxford, England
Bender, R. (1990), *AA*, **229**, 441
Bender, R., Burstein, D. & Faber, S.M. (1993), *ApJ*, **411**, 153.
Burstein, D., Faber, S.M., Gaskell, C.M. and Krumm, N. (1984), *ApJ*, **287**, 586
D'Onofrio, M. , Zaggia, R., Longo, G., Caon, N. & Capaccioli, M. (1995), *AA*, **296**, 319
Worthey, G. (1996), 'Fresh Views of Elliptical Galaxies', ASP Conf. Series, Vol 86

Anisotropic $R^{1/m}$ Models. Velocity Profiles, and the FP of Elliptical Galaxies

Barbara Lanzoni[1]* and Luca Ciotti[2]

[1] Dipartimento di Astronomia di Bologna, Via Zamboni 33, Bologna, Italy
[2] Osservatorio Astronomico di Bologna, Via Zamboni 33, Bologna, Italy

Abstract. We study the dynamical properties of spherical galaxies with surface luminosity profile described by the $R^{1/m}$-law, in which a variable degree of orbital anisotropy is allowed. The stability of the models against radial–orbit instability is studied, and the limits on the maximum anisotropy allowed for each model are determined. The consequent constraints on the the projected velocity dispersion imply that no fine-tuning for anisotropy is required along the fundamental plane (FP) in order to maintain its small thickness. Finally, the velocity profiles are constructed, and their deviations from a gaussian discussed.

1 The Models

The models surface brightness profile is described by the $R^{1/m}$ law (Sersic 1968):

$$I(R) = I_{\circ} \exp\left[-b(m)\,(R/R_{\rm e})^{1/m}\right], \tag{1}$$

which seems to give a better fit to the spheroidal galaxies surface brightness profiles (e.g., Caon et al. 1993; Graham et al. 1996, and references therein) than the "standard" $R^{1/4}$ law [de Vaucouleurs 1948; Eq.(1) for $m = 4$].

The velocity dispersion tensor of our models is described by the Osipkov–Merritt parameterization (Osipkov 1979; Merritt 1985), and characterized by the *anisotropy radius* $r_{\rm a}$: in the limit $r_{\rm a} \to \infty$ the velocity dispersion tensor is globally isotropic, while in general the radial anisotropy increases with radius.

2 Stability

The stability of the anisotropic models is investigated in a semi-quantitative way using the radial-orbit instability indicator ξ (e.g., Fridman and Polyachenko 1984). A model is likely to be unstable if

$$\xi \equiv \frac{2K_{\rm r}}{K_{\rm t}} \gtrsim 1.5 \div 2, \tag{2}$$

where $K_{\rm r}$ and $K_{\rm t}$ are the radial and the tangential kinetic energies, respectively. This parameter is quite independent of the assumed density distribution profile,

---

* Present address: Osservatorio Astronomico di Capodimonte, Via Moiariello 16, Napoli, Italy.

and for any globally isotropic system $\xi = 1$, while in presence of radial anisotropy $2K_r > K_t$, and so $\xi > 1$.

For the investigated models, $\xi = \xi(r_a, m)$, and for a fixed m it decreases towards unity for increasing r_a (see Fig.1), according to the previous discussion. So, assuming a fiducial critical value of ξ for stability (e.g., $\xi = 1.7$), a minimum value for the anisotropy radius $(r_a)_\xi$ is obtained, i.e., all models with $r_a < (r_a)_\xi$ are unstable.

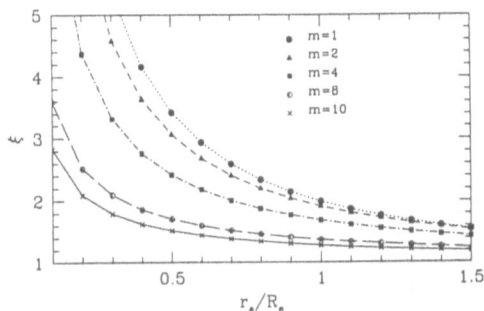

Fig. 1.

For stable models we compute the spatial velocity dispersion profile by integration of the Jeans equation. After the operations of projection and mean, we obtain the *aperture* velocity dispersion σ_a, which mimics the observed central one (see Ciotti and Lanzoni 1997).

3   Implications for the FP

In principle, one of the possible origins of the FP tilt could be a systematic increase of radial anisotropy from faint to bright galaxies (Ciotti et al. 1996). If this is the case, a variation of factor 3 from isotropic to anisotropic squared central velocity dispersion σ_a^2 is required, in the assumption of structural homology (i.e., the same m for all galaxies). On the contrary, for our models we find that, if the maximum degree of anisotropy allowed by the stability requirement is considered, the *radial anisotropy cannot produce the tilt* (a conclusion already reached for different galaxy models in Ciotti et al. 1996).

As concerns the problem of the very small thickness of the FP, the limits imposed by the stability requirement imply that the variations between the isotropic and the maximum anisotropic velocity dispersion are so small that the *anisotropy is not required to be fine-tuned with the galaxy luminosity in order to maintain the small observed FP scatter* (see Ciotti and Lanzoni 1997).

4   Velocity Profiles

The velocity profile (VP) at a certain projected distance from the galaxy center is the distribution of the stars line–of–sight velocities at that point. The analysis

of the deviations of VPs from gaussianity may give important insights on the
dynamical structure of a galaxy (e.g., van der Marel 1994).

We numerically recover the VPs of our models (as in Carollo et al. 1995),
and expand them on the Gauss-Hermite basis (Gerhard 1993; van der Marel and
Franx 1993), thus obtaining the values of the coefficient h_4 at various distances
from the galaxy center. Generally, a negative h_4 indicates a flat–topped VP,
while a positive one indicates a VP more centrally peaked than a Gaussian.

In Fig.2 the radial behaviour of h_4 for various m and for isotropic (left panels)
and anisotropic ($\xi = 1.7$, right panels) models is shown.

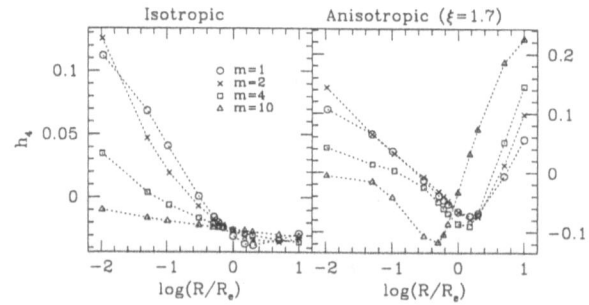

Fig. 2.

Because of the differences in the trend of h_4 between the isotropic and the
anisotropic cases, and because of the growing evidence that $R^{1/m}$ law appropriately describe the surface brightness profiles of elliptical galaxies, we suggest
that a detailed study of the VPs along the FP could be in principle a tool to
study the effect of orbital anisotropy on its tilt and thickness.

References

Caon N., Capaccioli M., D'Onofrio M., (1993): MNRAS, **265**, 1013
Carollo C.M., de Zeeuw P.T., van der Marel R.P., (1995): MNRAS, **276**, 1131
Ciotti L., Lanzoni B., (1997): A& A, in press (astro-ph/9610251)
Ciotti L., Lanzoni B., Renzini, A., (1996): MNRAS, **282**, 1
de Vaucouleurs G., (1948): Ann. d'Astroph., **11**, 247
Fridman A.M., Polyachenko V.L., (1984): Physics of Gravitating Systems, 2 vols., New
    York: Springer
Gerhard O.E., (1993): MNRAS, **265**, 213
Graham A., Lauer T.R., Colless M., Postman M., (1996): ApJ, **465**, 534
Merritt D., (1985): AJ, **90**, 1027
Osipkov L.P., (1979): Pis'ma Astron.Zh., **5**, 77
Sersic J.L., 1968, Atlas de Galaxias Australes, Cordoba: Observatorio Astronomico
van der Marel R.P., (1994): Ph.D. Thesis
van der Marel R.P., Franx M., (1993): ApJ, **407**, 525

Surface Brightness Parameters and Entropy of Elliptical Galaxies

Gastão B. Lima Neto[1,2], Daniel Gerbal[2,3], Isabel Márquez[2,4], Huub Verhagen[2,5]

[1] Observatoire de Lyon, Av. Charles André, 69561 St Genis Laval Cedex, France
[2] Institut d'Astrophysique de Paris, CNRS, 98bis Bd Arago, 75014 Paris, France
[3] DAEC, Observatoire de Paris, Univ. Paris 7, CNRS UA 173, 92195 Meudon, France
[4] Instituto de Astrofísica de Andalucía, CSIC, Apdo 3004, 18080 Granada, Spain
[5] Sterrenkundig Instituut, 'Anton Pannekoek', Universiteit van Amsterdam, Nederland

Abstract. We present a new parametrization of elliptical galaxies based on the specific entropy. An application is presented using galaxies of the Coma cluster.

1 Introduction – The Fundamental Plane

Elliptical galaxies are known to populate a surface (the so-called fundamental plane, FP) in the [effective radius R_e, effective surface brightness μ_e, central velocity dispersion σ_0]-space (Dressler et al. 1987, Djorgovski & Davis 1987, Bender et al 1992). In principle, the existence of a FP is a direct consequence of the virial equilibrium of hot stellar systems. In the [gravitational radius r_g, total mass M, mean velocity dispersion $\overline{v^2}$]-space the virial theorem defines the surface $\overline{v^2} = G\,M/r_g$. These quantities are not accessible to observation but one can use the following relations to obtain observable quantities: $L = c_1(M/L)\,\mu_e\,R_r^2$, $\overline{v^2} = c_2\sigma_0^2$, $r_g = c_3 R_e$. Comparison with the observed FP implies that:

- if c_1, c_2, c_3 are constants (the galaxies are homologous) then $M/L \propto L^{0.2}$;
- if $M/L \approx$ cte, then there is a non-homology in velocity or/and mass distribution.

There are a number of effects that may account for the deviation of the observed FP to the virial plane: the increase of rotational support in smaller galaxies, the variation of stellar population, and the non-homology in mass distribution.

The understanding of the origin of the FP would provide a valuable clue to theories of galaxy formation and evolution. Moreover, the FP, as a scaling relation, can be used to determine distances independently of the redshift and the Hubble law.

2 Entropy Relation

The FP is not the only parametrization of elliptical galaxies. Given the uniformity of elliptical galaxies, they are supposed to be in a quasi-equilibrium state. Like any dynamical system, an equilibrium state implies a condition on maximum entropy. However, elliptical galaxies are self-gravitating, and for such systems it is not clear what is the maximum entropy. In principle, self-gravitating systems have no defined maximum entropy.

Following White and Narayan (1987), we assume that the entropy of a hot self-gravitating system may be written as an ideal gas entropy. Keeping only the term that depends on the mass distribution of the galaxy one has the specific entropy s:

$$s = \frac{S}{M_{\text{tot}}} = \frac{1}{M_{\text{tot}}} \int_V \ln(\rho^{-5/2} P^{3/2}) \rho \, \mathrm{d}V \,, \tag{1}$$

where ρ and P are respectively the density and pressure, and the integral is done on the whole volume V of the galaxy.

2.1 3D Sérsic Law – the General Mellier-Mathez Law

As a working model of an elliptical galaxy we have adopted the Sérsic (1968) profile,

$$\mu(R) = \mu_0 \exp(-(R/a)^\nu). \tag{2}$$

This model provides a non-homologous generalization of the de Vaucouleurs $R^{1/4}$ law. The above empirical law describes satisfactorily the projected light distribution of a broad range of elliptical galaxies (e.g. Caon et al. 1993, Graham et al. 1996). In order to compute the entropy s, however, we need a 3-dimension mass distribution. We use an analytical approximation of the Sérsic law deprojection (Mellier & Mathez 1987, Gerbal et al. 1997):

$$\rho(r) = \rho_0 \left(\frac{r}{a}\right)^{-p} \exp\left(-\left(\frac{r}{a}\right)^\nu\right) \text{ with } p = 0.9976 - 0.5772\nu + 0.03242\nu^2. \tag{3}$$

The normalization is given as $\rho_0 = \mu_0 \, (M/L) \, \Gamma(2/\nu)/(2a \, \Gamma(3-p)/\nu)$. From now on, we assume constant (M/L). The pressure is obtained from the hydrostatic equilibrium condition and can be written as a single quadrature:

$$P(r) = 4\pi G \left(\frac{\rho_0 \, a}{\nu}\right)^2 \int_{(r/a)^\nu}^\infty \gamma\left(\frac{3-p}{\nu}, x\right) e^{-x} x^{\frac{-p-1}{\nu}-1} \mathrm{d}x \tag{4}$$

Substitution of equations (3) and (4) in eq.(1) yields a relation between μ_0, a, and ν. In this parameter space, a constant specific entropy defines a surface. The question is how real elliptical galaxies populate this space.

3 Galaxies in the $[\mu_0, a, \nu]$-Space

It turns out that the surface defined by $s(\mu_0, a, \nu) = \text{cte}$ is almost edge on when seen perpendicular to the plane a, ν. This implies a well defined correlation between these two parameters.

In order to verify this relation with real galaxies, we have fitted a sample of 79 elliptical galaxies pertaining to the Coma cluster using the Sérsic law

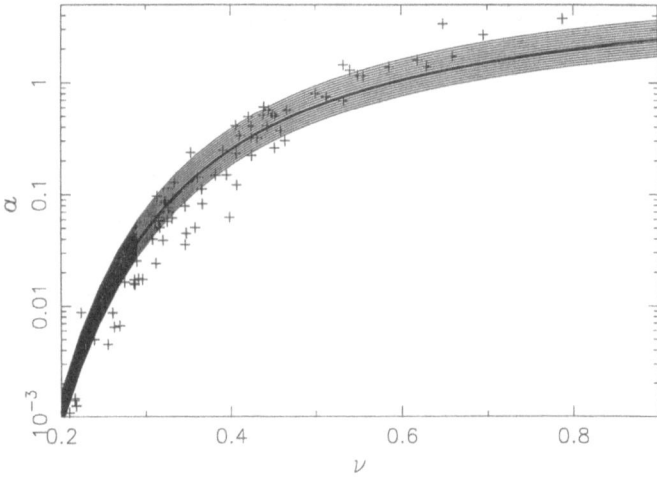

Fig. 1. Distribution of a versus ν for galaxies of the Coma cluster. The lines are the projection in the $a - \nu$ plane of constant specific entropy

(catalogue of Lobo et al. 1997, cf. Gerbal et al. 1997 for details). Figure 1 shows the a and ν parameters obtained from the fits superimposed to a surface of constant specific entropy. An important consequence of the observed correlation between the scale parameter a (which is distance dependent) and the shape parameter ν (distance independent) is possibility of using it as an independent distance indicator. Notice that in contrast with the FP, the distance independent parameter here is the shape of the light profile (ν), while in the former case it is the velocity dispersion.

References

Bender R., Burstein D., Faber S.M., 1992, Ap.J. 399, 462
Caon N., Capaccioli M., D'Onofrio M., 1993, MNRAS 265, 1013
Dressler A., Lynden-Bell D., Burstein D., Davies R.L., Faber S.M., Terlevich R.J., Wegner G., 1987, Ap.J. 313, 42
Djorgovski S., Davis M., 1987, Ap.J. 313, 59
Gerbal D., Lima Neto G.B., Márquez I., Verhagen H., 1997, MNRAS *in press*
Graham A., Lauer T.R., Colless M., Postman, M., 1996, ApJ 465, 534
C. Lobo, A. Biviano, F. Durret, D. Gerbal, O. Le Fèvre, A. Mazure, E. Slezak, 1997, A&A Suppl. *in press*
Sérsic J.L., 1968, "Atlas de galaxias australes", Observatorio Astronómico de Córdoba
White S.D.M., Narayan, R., 1987, MNRAS 229, 103

Stellar Populations in Galaxies

Claudia Maraston

Dipartimento di Astronomia, via Zamboni 33, 40126 Bologna, Italy

Abstract. An innovative tool for the construction of Evolutionary Synthesis models of Stellar Populations is presented. It is based on three independent matrices giving respectively 1) the fuel consumption during each evolutionary phase as a function of stellar mass, 2) the typical temperatures and gravities during such phases, and 3) colors and bolometric corrections as a function of gravity and temperature. The modular structure of the code allows to easily assess the impact, on the synthetic spectral energy distribution, of the various assumptions and model ingredients. As an illustrative example, a solar model (Y=0.27,Z=0.02), with an age ranging between 30 Myr and 15 Gyr is presented and synthetic broad band colours are compared with Magellanic Clouds Globular Clusters data. Then, the evolution of the stellar mass-to-light ratios has been computed and the implications with respect to the properties of the Fundamental Plane (FP) of elliptical galaxies are briefly discussed.

1 Introduction

In the last years, many authors devoted themselves to the Evolutionary Synthesis of Stellar Populations and two main approaches have been followed. The first, known as "Isochrone Synthesis" technique (e.g.Bruzual & Charlot, 1996), simply consists in summing up the monochromatic luminosities contributions of all the mass-points along an isochrone, having assumed an Initial Mass Function. The integration terminates to the latest point on the isochrone itself, that usually coincides with the end of the so-called Early Asymptotic Giant Branch phase: later stellar phases are then added, in an analytical or empirical way, with some sort of individual receipts. An alternative approach is that introduced by Renzini & Buzzoni (1986) and based on the so-called *Fuel Consumption Theorem*. In this case, the main ingredient for the synthesis is the nuclear fuel available for a star in a peculiar evolutionary stage. This quantity, obtained from the stellar tracks, is then converted in the various luminosities, using a temperatures/gravities-colours set of transformations. At present, this method has been followed by Buzzoni (1989) and Worthey (1992), but they both offered models only in a narrow age range, missing young and intermediate age populations. In this work we have computed simple population models in the age range 30 Myr-15 Gyr and compared them with the globular cluster family of the Magellanic Clouds; then, the evolution of the global stellar mass-to-light ratios and its consequences with regards to the tilt/thickness of the Fundamental Plane of cluster elliptical galaxies are discussed. For a complete description of the technique see Maraston & Renzini (1997).

2   A Test for Simple Stellar Populations Models

The synthetic broad band colors have been compared with the data relative to the Magellanic Clouds Globular Clusters and with the results of some other authors (Bruzual & Charlot, 1996 and Tantalo et al. 1996). We have chosen two-colours diagrams instead of age-colours ones just because these don't make use of any age-relations, that are extremely stellar tracks dependent and so a good agreement can be achieved simply varying the set in using. While, in the case of UBV indexes, the three sets of models evolve in much the same way in time and show a quite good agreement with the observations, (Fig. 1a), the infrared colours behave very differently. In particular, the other models fail in reproducing the observed locus for the Magellanic Clouds clusters in the (V-K) vs. (U-B) diagram (Fig. 1b). This is due to our correct inclusion of the *TP-AGB* phase, because this stellar stage has a strong impact on the integrated $(V - K)_0$ colour at intermediate age. The disagreement between models and observational data at $(B - V) > 0.5$ in panel (a) or at the later SWB types in panel (b) is due to the increasing difference in metallicity between the older clusters and the solar model.

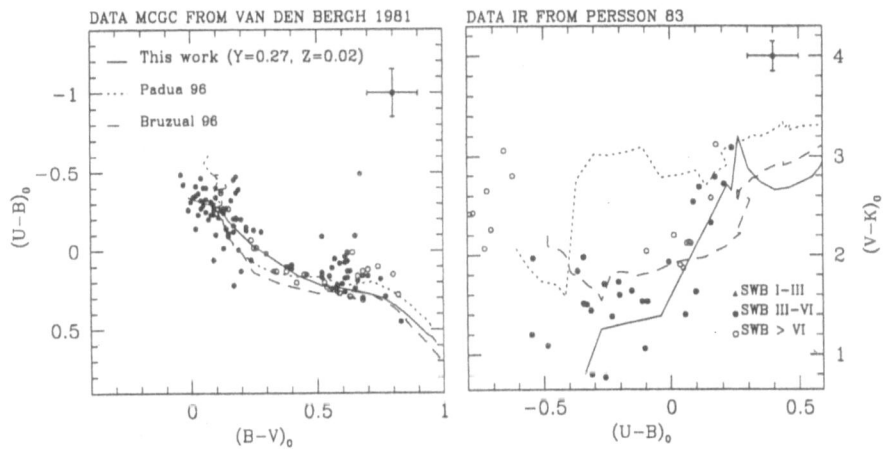

Fig. 1.

3   The Role of the IMF on the FP Properties

To couple the topic of this conference, we have re-analysed the discussion of Renzini & Ciotti (1993, RC93), concerning the role of the IMF slope (x), of the lower Main Sequence cut-off (M_{inf}) and of the luminosity time evolution, on the apparent trend of M_*/L_B for cluster ellipticals (CE). In our computations, following RC93, mass of "living" stars plus mass in remnants has been considered. Fig. 2 shows M_*/L_B, for a 15 Gyr population, as a function of M_{inf} for several choices of a Scalo multislope IMF, with x being the exponent for the low MS ($M < 0.3M_\odot$) component. The *observable narrowness* of the FP at low luminosities (which results in M_*/L_B spread less than $\simeq 12\%$), is easily reproducible as low M_*/L_B values are quite insensitive to variations both in M_{inf} and x. For what it concerns the FP *tilt*, which implies an increase of a factor of $\simeq 3$ in M_*/L_B when passing from faintest to brightest CE's, a simple variation of M_{inf}, with fixed x, doesn't suffice, but a major change in slope is required. So, to simultaneously obtain a very small galaxy-galaxy dispersion (at the same luminosity/mass) and the tilt, seems to require a low-MS IMF slope that changes very much with L_B, but again fine-tuned with M_{inf}. Finally M_*/L_B evolve in time, because so do luminosities: Fig. 2 (b) shows that the FP thickness impose a very narrow range in galaxy ages, with a dispersion of $\simeq 1.3$ Gyr (for $t > 10Gyr$). So we confirm, in a more quantitative way, the preliminary results of RC93.

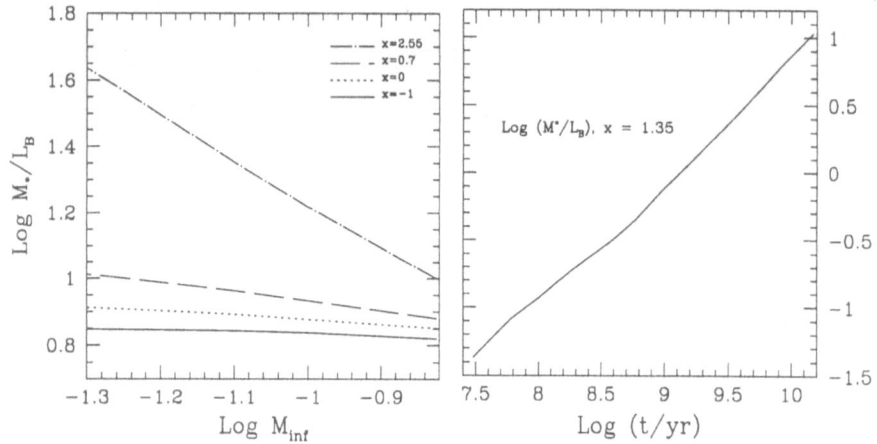

Fig. 2.

References

Bruzual, G., Charlot, S. (1996): ApJ, *in preparation*
Buzzoni, A. (1989): ApJS, **71**, p. 817 (Paper I)
Maraston, C., Renzini, A. (1997): *in preparation*
Renzini, A., Buzzoni, A. (1986) in *Spectral Evolution of Galaxies* ed. C. Chiosi, A. Renzini (Dordrecht: Reidel), p.135
Renzini, A., Ciotti, L. (1993): ApJ, **416**, L49, RC93
Tantalo, R., Chiosi, C., Bressan, A., Fagotto, F. (1996): A&A, **311**, 361
Worthey, G. (1992): Ph.D. Thesis, Univ. California, Santa Cruz

The Fundamental Plane for Hydra I and Coma

Bo Milvang-Jensen[1] and Inger Jørgensen[2,3]

[1] Copenhagen University Observatory, DK-2100 Copenhagen, Denmark
[2] McDonald Observatory, University of Texas, Austin, Texas 78712, USA
[3] Hubble Fellow

Abstract. We have analyzed the the Fundamental Plane (FP) (in Gunn r) and the Mg_2–σ relation for E and S0 galaxies in the central parts of the clusters HydraI and Coma. For the HydraI sample, we have also studied the FP in Johnson B and Johnson U.

1 Introduction

The Fundamental Plane (FP) for E and S0 galaxies, $\log r_e = \alpha \log \sigma + \beta \log <I>_e + \gamma$, (Djorgovski & Davis 1987; Dressler et al. 1987) may provide constraints on galaxy formation and evolution. Further, the FP can be used to determine relative distances, if it is universal at least to some accuracy. In both contexts comparison of the FP for large samples of galaxies in different clusters provide essential knowledge on how similar (or different) the stellar populations of E and S0 galaxies are in different cluster environments. The Mg_2–σ relation (Burstein et al. 1988) gives additional information about the stellar populations.

Observational data for the clusters HydraI and Coma are used in this paper. **HydraI:** 45 E and S0 galaxies within the central $68' \times 83'$. CCD surface photometry in Gunn r and Johnson B, and for 19 galaxies also in Johnson U (Milvang-Jensen & Jørgensen, in prep.). Velocity dispersions and Mg_2 indices from Jørgensen et al. (1995b) and Milvang-Jensen & Jørgensen (in prep.). The sample is 80% complete to m_T (Gunn r) = 14^m5 ($M_T = -20^m0$ with $H_0 = 50\,\mathrm{km\,s^{-1}\,Mpc^{-1}}$).

Coma: 116 E and S0 galaxies within the central $64' \times 70'$. CCD surface photometry in Gunn r from Jørgensen et al. (1995a). Velocity dispersions and Mg_2 indices from the literature and Jørgensen (in prep.). The literature data have been calibrated to a consistent system and aperture corrected (cf. Jørgensen et al. 1995b). The sample is 93% complete to m_T (Gunn r) = 15^m05 ($M_T = -20^m75$).

2 The Fundamental Plane

We fit the FP by minimization of the sum of the absolute residuals perpendicular to the plane. For photometry in Gunn r we find

$$
\begin{aligned}
\text{HydraI}: \quad \log r_e &= \underset{\pm\,0.17}{1.51 \log \sigma} - \underset{\pm\,0.05}{0.80 \log <I>_e} + \gamma_{cl} \qquad \sigma_{fit} = 0.108 \\
\text{Coma}: \quad \log r_e &= \underset{\pm\,0.04}{1.28 \log \sigma} - \underset{\pm\,0.03}{0.83 \log <I>_e} + \gamma_{cl} \qquad \sigma_{fit} = 0.095
\end{aligned}
\tag{1}
$$

The uncertainties were derived by a bootstrap procedure. σ_{fit} is the rms scatter in the $\log r_e$ direction. Figure 1 shows the FP edge-on. The coefficients of the two FPs are not significantly different. If we fix β at the value -0.82 and only fit α, we find $\alpha_{\text{HydraI}} = 1.64 \pm 0.16$ and $\alpha_{\text{Coma}} = 1.32 \pm 0.05$. The difference, $\Delta\alpha = 0.32 \pm 0.17$, is significant at the 2 sigma level.

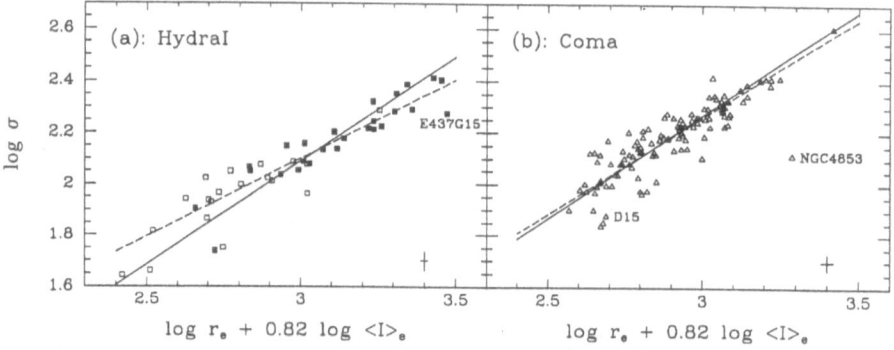

Fig. 1. The FP in Gunn r seen edge-on. r_e is in arcsec. Solid line: $\alpha = 1.24$ (Jørgensen et al. 1996). Dashed lines: (a) $\alpha = 1.64$, (b) $\alpha = 1.32$. Filled boxes – HydraI galaxies $M_T \leq -20.^{\text{m}}75$. Open boxes – HydraI galaxies $M_T \geq -20.^{\text{m}}75$. Triangles – Coma galaxies. E437G15, D15 and NGC4853 deviate strongly from the Mg_2–σ relation.

The apparent difference is mostly due to the difference in limiting magnitude for the two samples. If the galaxies below the Coma completeness limit are omitted from the two samples, the difference is non-significant, $\Delta\alpha = 0.06 \pm 0.25$. Thus, no significant differences in the coefficients for the FP can be detected. Differences in α on the 10% level cannot be ruled out. A fit to the two clusters together gives $\alpha = 1.34 \pm 0.05$, $\beta = -0.83 \pm 0.03$. This is not significantly different from the coefficients found by Jørgensen et al. (1996), $\alpha = 1.24 \pm 0.07$, $\beta = -0.82 \pm 0.02$, based on 226 E and S0 galaxies in 10 clusters.

From the photometry in Johnson B and Johnson U for galaxies in HydraI we find that the scatter around the FP is 0.110 in Johnson B, and 0.068 in Johnson U (for 19 galaxies). The scatter in Gunn r for the same sub-sample of 19 galaxies is 0.089. These results mean that the intrinsic scatter in the three passbands is the same. This sets limits on the internal extinction in the galaxies. It may also be used to set limits on age and metallicity variations as sources of the scatter.

3   The Mg_2–σ Relation

The Mg_2 line index has been measured for 42 of the HydraI galaxies, and 113 of the Coma galaxies. The slopes of the Mg_2–σ relations for the two clusters are very similar. We find 0.171 ± 0.037 for HydraI and 0.186 ± 0.013 for Coma; cf. Fig. 2. This is also in agreement with the results from Jørgensen et al. (1996),

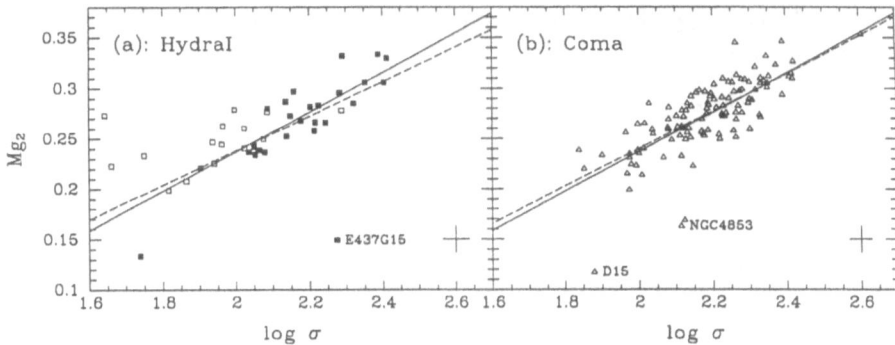

Fig. 2. The $Mg_2-\sigma$ relation. Three values of the slope are shown: Solid line: 0.196 (Jørgensen et al. 1996). Dashed lines: (a) 0.171, (b) 0.186. Filled boxes – HydraI galaxies $M_T \leq -20^m75$. Open boxes – HydraI galaxies $M_T \geq -20^m75$. Triangles – Coma galaxies. E437G15 has a strong disk and $v_{rot} > 150\,km\,s^{-1}$. D15 and NGC4863 have post starburst spectra.

0.196 ± 0.016. Using the latter, we find that also the zero points for the two clusters are the same: -0.155 ± 0.007 for HydraI and -0.153 ± 0.003 for Coma. We conclude that the galaxies in the central parts of the HydraI cluster and the Coma cluster follow the same $Mg_2-\sigma$ relation.

4 Conclusions

The FP for E and S0 galaxies in the HydraI cluster and the Coma cluster have been compared. Also, the $Mg_2-\sigma$ relation has been studied. No significant differences in the coefficients for the FP can be detected, though differences in α on the 10% level cannot be ruled out. The distributions of galaxies within the FP are similar for the two clusters. The galaxies in the two clusters follow the same $Mg_2-\sigma$ relation. From surface photometry of the galaxies in the HydraI sample in Gunn r, Johnson B and Johnson U we find, that the scatter of the FP does not depend significantly on which passband is used. This shows that internal extinction in the galaxies cannot be a major source of the scatter. It may also be used to set limits on variations in mean age and metal content.

References

Burstein, D., et al. 1988, in Towards Understanding Galaxies at High Redshifts, eds. Kron, R. G., & Renzini, A., Kluwer Academic Publishers, Dordrecht, p. 17

Djorgovski, S., & Davis, M. 1987, ApJ, 313, 59

Dressler, A., et al. 1987, ApJ, 313, 42

Jørgensen, I., Franx, M., & Kjærgaard, P. 1995a, MNRAS, 273, 1097

Jørgensen, I., Franx, M., & Kjærgaard, P. 1995b, MNRAS, 276, 1341

Jørgensen, I., Franx, M., & Kjærgaard, P. 1996, MNRAS, 280, 167

The Relation Between X-ray Emission, Galaxy Structure and Internal Kinematics in Early-Type Galaxies

Silvia Pellegrini

Dipartimento di Astronomia, Università di Bologna,
via Zamboni 33, I-40126 Bologna, Italy

Abstract. Using data in the literature plus new spectroscopic observations, we have built the first large sample of X-ray emitting early-type galaxies with known kinematics (central velocity dispersion σ_c and maximum rotational velocity v_{rot}). With this sample we investigate the effect of rotation on the X-ray emission, particularly with regard to the X-ray underluminosity of flat systems. The role of structural parameters as the ellipticity ϵ, the isophotal shape parameter a_4/a, and the inner slope of the surface brightness profile γ recently measured by HST are also investigated.

1 The Problem

The large scatter in the $L_X - L_B$ diagram is the most striking feature of the X-ray properties of early-type galaxies: it well reaches two orders of magnitude in L_X at any fixed $L_B > 3 \ 10^{10} L_\odot$ (Fabbiano et al. 1992). It was originally explained by environmental differences (White & Sarazin 1991), or by different dynamical phases for the hot gas flows, ranging from winds to subsonic outflows to inflows (Ciotti et al. 1991). Recent observational results have produced a new debate on the explanation of the scatter. Eskridge et al. (1995a,b) showed that on average S0s have lower L_X and L_X/L_B at any fixed L_B than do Es. Moreover, galaxies with axial ratio close to unity span the full range of L_X, while flat systems all have $L_X \lesssim 10^{41}$ erg s^{-1}; this holds both for Es and S0s separately. Has the shape of the mass distribution any role in determining L_X? Since flat galaxies exhibit a higher rotation level, what is the effect of galactic rotation?

Finally, recent HST results show that hot galaxies can be divided into two types: *core* galaxies, described by a surface brightness profile $I \propto r^{-\gamma}$ ($\gamma \lesssim 0.3$) at the center, and *power-law* galaxies, with steep, featureless profiles (Faber et al. 1996). These central properties correlate with global ones as rotation and isophotal shape. Is there any correlation also with global L_X?

2 Our Observations and Sample

We obtained measurements of velocity dispersion σ and radial velocity v along the major axis for 7 Es and S0s belonging to the *Einstein* sample (Fabbiano et al. 1992), the largest homogeneous sample of early-type galaxies with measured X-ray emission. The main characteristics of these galaxies, the details of the spectroscopic observations, the final radial velocity and velocity dispersion profiles, and the kinematic properties of the final sample obtained by collecting data also from the literature are given in Pellegrini et al. (1996, hereafter PHC).

3   Results

3.1   X-Ray Emission, Rotation and Flattening

The trend between the X-ray to optical ratio L_X/L_B, a measure of the hot gas content of the galaxies, and v_{rot}/σ_c is L-shaped, with the X-ray brightest objects confined at $v_{rot}/\sigma_c \lesssim 0.4$, both for Es and S0s (Fig. 1, with L_X and L_B derived as in PHC). The trend between L_X/L_B and the ellipticity ϵ is also L-shaped, even though less sharp: there are no high L_X/L_B objects with high ϵ. Instead, there is no clear trend between L_X/L_B and the anisotropy parameter $(v/\sigma)^*$. So, the observations suggest that the major effect on L_X/L_B is given by the relative importance of rotation and random motions (see also PHC).

3.2   X-Ray Emission, Isophotal Shape, and Central Brightness Profile

Faber et al. (1996) show that core galaxies tend to be boxy and slowly rotating, whereas power-law galaxies tend to be disky and rapidly rotating. Bender et al. (1989) showed that disky objects are low X-ray emitters, while boxy and irregular objects show a large range of L_X.

The results of our investigation (Fig. 1) show a striking relationship between L_X and γ: core galaxies span the whole range of L_X, while power-law galaxies are confined within $\log L_X \lesssim 41$. The origin of this new threshold effect should be looked for in the connection between central structural properties (a central black hole? different formation histories?) and global L_X.

4   Conclusions

Rotation and flattening are predicted to have different effects in the pure inflow and in the wind/outflow/inflow scenarios. Rotation plays the major role in the first one (Brighenti & Mathews 1996), but the expected correlation of L_X/L_B, or of the residuals of the $L_X - L_B$ correlation, with v_{rot}/σ_c is not present (PHC). Flattening is determinant in the second scenario (Ciotti & Pellegrini 1996; D'Ercole & Ciotti 1996), but it remains to be explained why the L-shape is more pronounced with respect to v_{rot}/σ_c rather than to ϵ.

The most significant effect found is the L-shape in the $L_X - \gamma$ plot: the division of early-type galaxies into two types corresponds to a difference also in their X-ray properties. This represents another link between the central structures of hot galaxies and their global properties, that might also be useful to better understand their formation history and evolution.

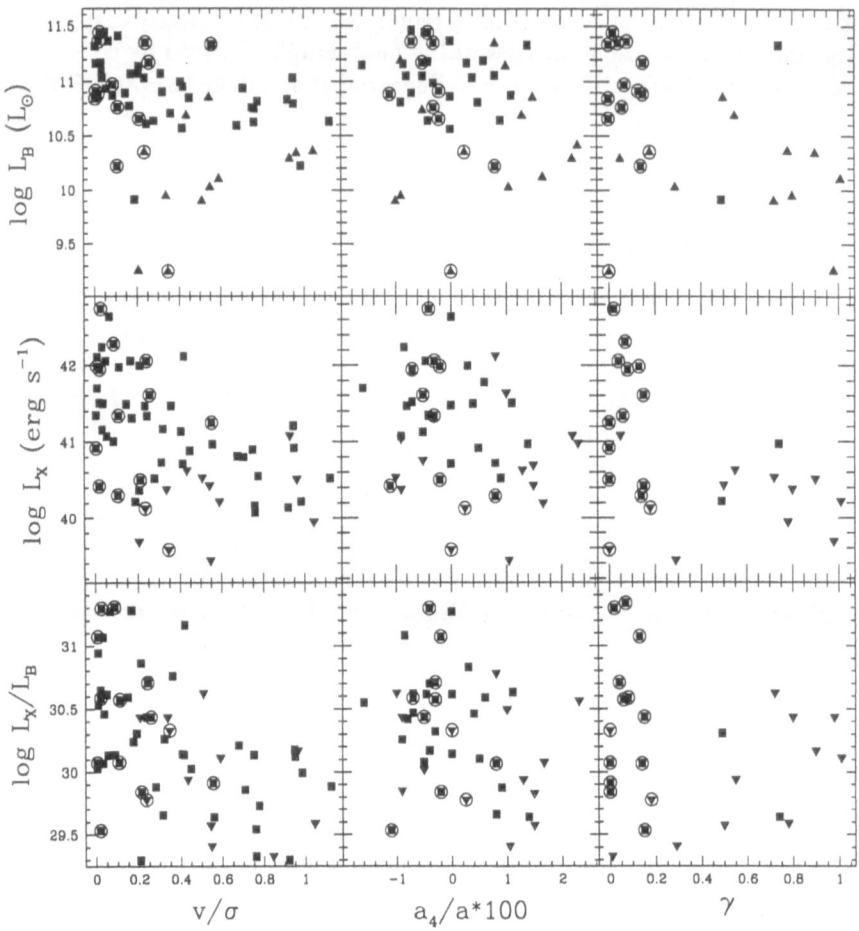

Fig. 1. a_4 is from Eskridge et al. (1995b), γ from Faber et al. (1996). Boxes are X-ray detections, triangles upper limits; *core* galaxies have been circled.

References

Bender, R., et al. (1989): A&A **217**, 35
Brighenti, F., Mathews, W.G. (1996): ApJ **470**, 747
Ciotti, L., Pellegrini, S. (1996): MNRAS **279**, 240
Ciotti, L., D'Ercole, A., Pellegrini, S., Renzini, A. (1991): ApJ **376**, 380
D'Ercole, A., Ciotti, L. (1996): submitted to ApJ
Eskridge, P., Fabbiano, G., Kim, D.W. (1995a): ApJS **97**, 141
Eskridge, P., Fabbiano, G., Kim, D.W. (1995b): ApJ **442**, 523
Fabbiano, G., Kim, D.W., Trinchieri, G. (1992): ApJS **80**, 531
Faber, S.M., et al. (1996): submitted to AJ
Pellegrini, S., Held, E., Ciotti, L. (1996): in press on MNRAS (PHC)
White, R.E., III, Sarazin, C.L. (1991): ApJ **367**, 476

The Exponential Disk Parameters in the Near-Infrared

Daniele Pierini[1,2] and Giuseppe Gavazzi[1,2]

[1] Universitá di Milano, Dipartimento di Astrofisica, I-20133 Milano, Italy
[2] Osservatorio Astronomico di Brera, I-20121 Milano, Italy

Abstract. We present exponential disk parameters (central surface brightness μ_{0H} and scalelength h_H) for 224 normal giant galaxies later than S0a, as derived from H-band (1.6μm) imaging. We show that: 1) Freeman's law does not apply in the Near-IR (NIR); 2) μ_{0H} and h_H do not depend on the Hubble classification; 3) μ_{0H} and h_H correlate with the NIR luminosity; 4) the disk central surface brightness becomes dimmer in bulge+disk systems. We argue that these observational evidences are determined by the different interplay of gravitational collapse and viscous dissipation in galaxy formation, with respect to the galaxy total mass.

1 The Working Hypothesis

NICMOS3 H-band (1.6μm) images of 400 CGCG (Zwicky et al. 1961-68) spiral galaxies in the A262, Virgo, Cancer, Coma+A1367 clusters and in the isolated regions of the Coma Supercluster (selected by Gavazzi & Boselli 1996) were obtained with the TIRGO 1.5 m and the Calar Alto 2.2 m telescopes (Gavazzi et al. 1996a,b). Thanks to these NIR observations, Gavazzi et al. (1996c) have shown that the NIR light is a reliable tracer of the dynamic mass in rotationally supported systems and that the photometric and metric properties of late-type galaxies depends on the total mass, at the first order (see Gavazzi, this Conference). Gavazzi & Scodeggio (1996) claimed that the star formation exponential timescale depends on dynamic mass. Different star formation histories must affect the interplaying between gravitational collapse (violent relaxation) and viscous dissipation during the disk formation in late-type galaxies spanning a wide range of masses (or NIR luminosities). Accordingly, we conclude that the exponential disk parameters depend on the H-band absolute magnitude, in contrast with Freeman (1970), who claimed that the B-band central surface brightness is constant (21.65 ± 0.3 mag arcsec^{-2}).

2 Results

Monodimensional H-band surface brightness profiles are extracted by azimuthally integrating counts in elliptical annuli along the major axis and decomposed into disk and bulge (if present), by fitting respectively: (1) an exponential law: $\mu(r) = \mu_0 + 1.086 \times (r/h)$, where μ_0 is the central surface brightness and h the scalelength; (2) a de Vaucouleurs' law: $\mu(r) = \mu_e + 8.325 \times ((r/r_e)^{1/4} - 1)$ with μ_e and r_e the effective surface brightness and radius.

Only those fits with a reduced $\chi^2 < 4$ were used (165 disk + 59 bulge+disk galaxies). Typical uncertainties are 0.1-0.2 mag arcsec^{-2} for μ_{0H} and 10% for h_H. Brightnesses and scales are not corrected for inclination.

Figure 1 shows the distribution of μ_{0H} vs. $R_{21.5}/h_H$ ($R_{21.5}$ is the isophotal radius at 21.5 H-mag arcsec^{-2}). The straight line represents eq.(1) for the H-band limit in surface brightness. If Freeman's law were true in the NIR, μ_{0H} and $R_{21.5}/h_H$ would be constant (see eq. (1)). This is not our case!

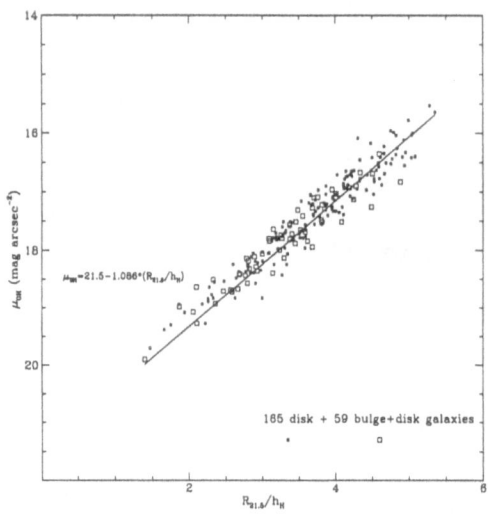

Fig. 1. The disk central surface brightness (μ_{0H}) vs. the ratio between the 21.5 H-mag arcsec^{-2} isophotal radius and the disk scalelength ($R_{21.5}/h_H$). Filled squares represents pure disks and open squares bulge+disk systems.

Disk parameters change by a factor of forty (for μ_{0H}) and ten (for h_H), with a large spread within each Hubble type (Figs.2a,b). The optical morphological classification does not account for the NIR disk properties (and the old stellar population distribution - see Rix 1993). Instead, the NIR scalelength and central surface brightness scale with the absolute magnitude (dynamic mass), as shown in Figs.2c,d. Furthermore, objects with relevant bulges are luminous (Gavazzi et al. 1996c) and have central surface brightnesses fainter than pure disks at any given H-band absolute magnitude.

The mechanisms of spiral galaxy formation (gravitational collapse and viscous dissipation) might determine different distributions of matter into the two (bulge and disk) components, according to the total mass, angular momentum

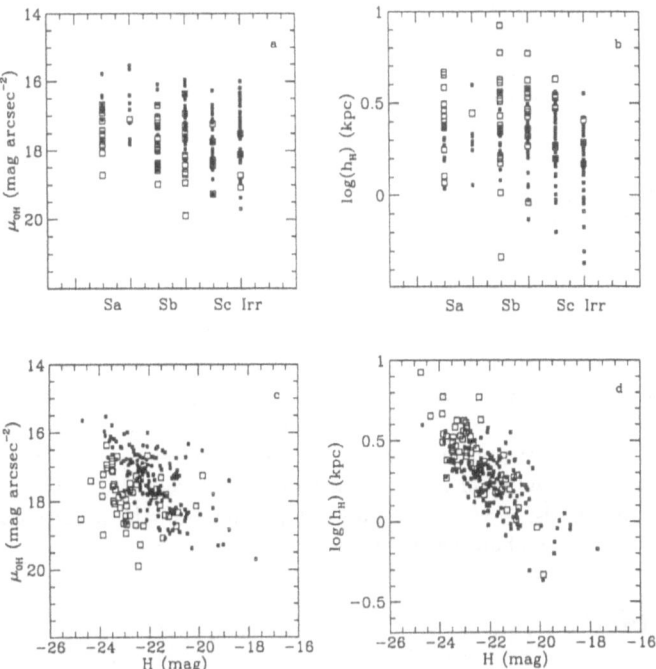

Fig. 2. The disk parameters as a function of the Hubble classification (a,b) and the total H-band luminosity (c,d). (Same symbols as in Fig.1.)

and initial star formation. Spheroidal structures are associated with a dissipationless collapse and a rapid star formation (Gott & Thuan 1976), opposite to disks. This could explain the central faintness of disks in bulge | disk galaxies with respect to pure disk galaxies of similar luminosity.

References

Freeman, K. (1970): ApJ, 160, 811
Gavazzi, G., Boselli, A. (1996): Astroph. Lett. & Comm., 35, 1
Gavazzi, G., Scodeggio, M. (1996): A&A, 312, L29
Gavazzi, G., Pierini, D., Boselli, A., Tuffs, R. (1996a): A&A, 120, 489
Gavazzi, G., Pierini, D., Baffa, C., Lisi, F., Hunt, L., Randone, I., Boselli, A. (1996b): A&A, 120, 521
Gavazzi, G., Pierini, D., Boselli, A. (1996c): A&A, 312, 397
Gott, J., Thuan, T. (1976): ApJ, 204, 649
Rix, H. (1993): PASP, 105, 1005
Zwicky, F. et al. (1961-68): *Catalogue of Galaxies and Clusters of Galaxies* (California Institute of Technology Press, Pasadena)

Near-IR Mass-to-Light Ratio Profiles of Disk Galaxies

D. Pierini[1,2], G. Gavazzi[1,2], A. Boselli[3] and F. Casoli[4]

[1] Universitá di Milano, Dipartimento di Astrofisica, I-20133 Milano, Italy
[2] Osservatorio Astronomico di Brera, I-20121 Milano, Italy
[3] Laboratoire d'Astronomie Spatiale, Traverse du Siphon, F-13376 Marseille Cedex 12, France
[4] DEMIRM, Observatoire de Paris, 61 Av. de l'Observatoire, F-75014 Paris, France

Abstract. Optical rotation curves (RC's) and mass-to-light ratio (M/L) profiles are given for 30 CGCG late-type galaxies with H-band (1.6μm) surface photometry. Although the present sample is not complete, some preliminary results are given: 1) the central steepness of the normalized velocity profile correlates with the luminosity, "concentration index" and inner surface brightness; 2) the Near-IR (NIR) mass-to-light ratio increases outwards; 3) the M/L gradients or shapes of the profiles do not show any clear trend with the photometric properties.

1 Introduction

Gavazzi et al. (1996c) have shown that the NIR light is a reliable tracer of the dynamic mass in rotationally supported systems and that the photometric properties of normal late-type galaxies mainly correlate with the total mass, disregarding the morphological classification and environment (see Gavazzi, this Conference). Moreover, the most luminous galaxies have centrally condensed surface brightness distributions, as proved by the relation between the H-band luminosity and "concentration index" (c_{31}). The latter is defined as the ratio between the radii containing, respectively, 75% and 25% of the luminosity within the 21.5 H-mag arcsec^{-2} isophote. Spiral galaxies with $c_{31} > 3$ have either bulges or centrally condensed disks. As a consequence, the dynamic properties should be linked to the NIR light (mass) distribution.

2 The Inner Rising of the Optical Rotation Curves

High S/N optical spectra of 13 disk galaxies were taken at the 1.93 m telescope of the OHP. High dispersion spectroscopy is available in the literature only for 39 galaxies of our H-band data-set (5 spectra are in common). We disregard 17 objects because their RC's do not reach far enough radial distance, at one or both sides from the nucleus; some galaxies are clearly disturbed by tidal interaction, or show strong velocity dissimmetries or have inclinations not sufficient to derive reliable corrected velocities.

The unfolded RC's are normalized to the optical major axis (r_{opt} at 25 B-mag arcsec^{-2}) and the maximum rotational velocity (V_{Max}) (the HI measure, if available, otherwise the average of the maximum velocities of the two arms of the optical RC). The normalized rotation curves are fitted with a χ^2 algorithm, using a simple analytical curve (called "sigmoid"): $V_n = a[1 + \exp(b\, r_n)]^{-1} - a/2$, where V_n and r_n are the normalized velocity and radius, $a/2$ (the maximum velocity) and $1/b$ (the exponential scale of the RC) are free parameters. The arctangent of the first derivative of the sigmoid, computed at $r_n = 0$, (α) provides us with a measure of the central slope of the velocity profile.

In spite of a huge scatter, a positive correlation exists between α and the H-

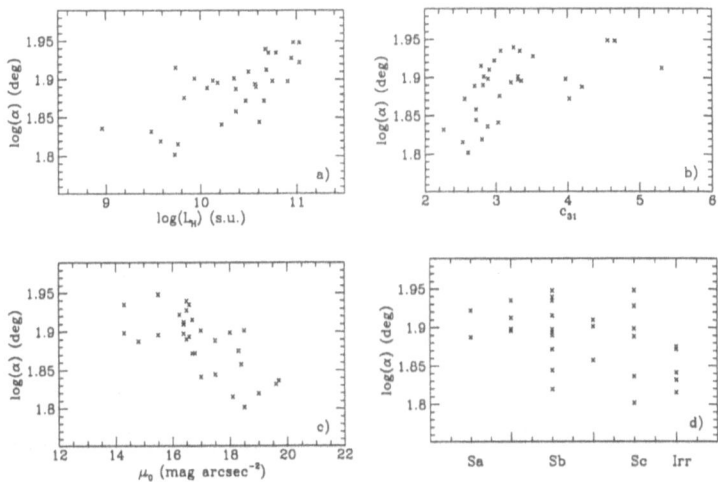

Fig. 1. The parameter α vs. the H-band luminosity (a), concentration index (b), central surface brightness (c) and Hubble type (d).

band luminosity (Fig.1a). α correlates with the concentration index c_{31} (Fig.1b) and central surface brightness (Fig.1c). These results suggest that in rotationally supported systems, the greater the central luminous mass density, the steepest the inner RC shape (cf. Rubin et al. 1985; Persic et al. 1996). A mild correlation is found between α and the morphological type (Fig.1d).

3   The H-Band Mass-to-Light Ratio Profiles

M/L radial profiles are determined for the present sample. $L_H(r)$ is derived integrating up to r the surface brightness profiles and is corrected to the face-on value. The dynamical mass is derived from the centrifugal equilibrium law. We estimate an uncertainty of about 20% for M/L. Figure 2 shows an example of three radial profiles of L_H, in solar unit (s.u.) (a); the fitted rotation velocity V_c, in km s^{-1} (b); M_{dyn}, in s.u. (c); M/L_H, in s.u. (d).

We find that: 1) the M/L increases outwards (see Forbes 1992); 2) the NIR M/L values at (or extrapolated to) r_{opt} are consistent with 4.6 $L_{\odot H}/M_\odot$ (Gavazzi

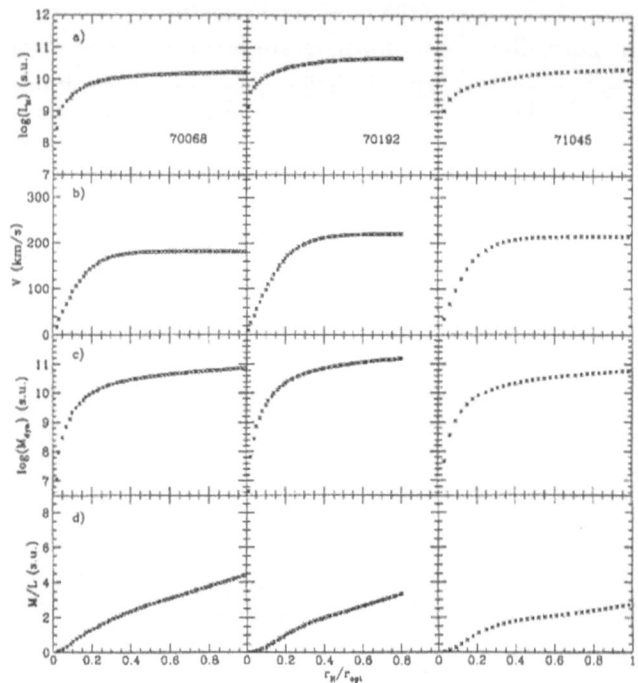

Fig. 2. Integrated radial profiles of L_H (a), V_c (b), M_{dyn} (c) and M/L_H (d) for three CGCG galaxies of the sample.

et al. 1996c); 3) no obvious relation between the shape of the profiles and the NIR photometric properties is devised. The behavior of the M/L could be due to either changes in the age and metallicity of the stellar populations and/or in the internal extinction and/or in the luminous-to-dark mass contents.

The incompleteness of the present samples prevents finding any relation between M/L gradients and luminosity (see Persic et al. 1996).

References

Forbes, D. (1992): A&AS, 92, 583
Gavazzi, G., Pierini, D., Boselli, A. (1996c): A&A, 312, 397
Persic, M., Salucci, P., Stel, F. (1996): MNRAS, 281, 27
Rubin, V., Burstein, D., Ford, W. Thonnard, N. (1985): ApJ, 289, 81

Tidal Thickening of Galaxy Disks

Vladimir Reshetnikov[1,2], Françoise Combes[2]

[1] Astronomical Institute of St. Petersburg State University, St. Petersburg, Russia
[2] DEMIRM, Observatoire de Paris, 61 Av. de l'Observatoire, F-75014 Paris, France

Abstract. We have studied a sample of 24 edge-on interacting galaxies and compared them to edge-on isolated galaxies, to investigate the effects of tidal interaction on disk thickening. We found that the ratio h/z_0 of the radial exponential scalelength h to the scale height z_0 is about twice smaller for interacting galaxies. This is found to be due both to a thickening of the plane, and to a radial stripping or shrinking of the stellar disk. If we believe that any galaxy experienced a tidal interaction in the past, we must conclude that continuous gas accretion and subsequent star formation can bring back the ratio h/z_0 to higher values, in a time scale of 1 Gyr.

1 Motivation

Galaxy disks are very sensitive to tidal interactions, from the formation of tidal tails and bridges up to the complete disruption of initial disks in mergers. Even non-merging interactions or minor mergers can thicken and destroy a stellar disk, and this has been advanced as an argument against frequent interactions in a galaxy life, or formation of the bulge through minor mergers in spiral galaxies.

Toth & Ostriker (1992) have used the argument of the fragility of disks to constrain the frequency of merging and the amount of accretion, and draw implications on cosmological parameters. They claim that the thickness of the Milky Way disk implies that no more than 4% of its mass can have accreted within the last $5\,10^9$ yrs; moreover they question the currently fashionable theory of structure growth by hierarchical merging, which would not be supported by the presence of thin galactic disks, cold enough for spiral waves to develop.

2 Sample and Observations

Our sample consists of 24 interacting systems containing at least one edge-on galaxy. We also observed a control sample of 7 edge-on isolated galaxies for comparison.

Observations were carried out at the OHP 1.2 m telescope in the B, V and I passbands. General photometric results of the observations (including isophotal maps of all objects) are presented in Paper I.

3 Results

For most galaxies, we find a constant scale height with radius, within 20%.

From the direct comparison of scale height z_0 and h/z_0 distributions in both samples of interacting and isolated spirals we find evidence for **thickening** of galactic disks in interacting systems, **by a factor 1.5 to 2** (see Fig.1). This thickening refer to the region of exponential disk between 1 and 2.4 of exponential scalelength (or between 0.6 and 1.4 of effective radius).

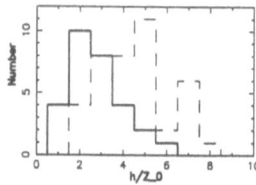

Fig. 1. Distribution of scalelength to scale height ratios for interacting (thick solid line) and normal (dashed line) galaxies. The data for normal spirals are from our observations and literature.

The mean characteristics of edge-on interacting galaxies in our sample are: absolute blue luminosity $M_B = --19.6 \pm 1.0$ (so "face-on" magnitude must be about $--21$), exponential scalelength $h = 4 \pm 2$ kpc ($H_0 = 75$ km/s/Mpc). Therefore, the typical galaxy in our sample is comparable with the Galaxy and M 31. Most edge-on galaxies in the interacting sample have comparable luminosity companions within one optical diameter.

One can conclude that tidal interaction between large spiral galaxies, like the Milky Way and the Andromeda galaxy, at a relative distance of about one optical diameter, leads to thickening by a factor 1.5-2 of their stellar disk.

4   Discussion

The h/z_0 ratio is 1.5-2 times smaller in interacting galaxies: this is found to be due not only to a higher average scale height z_0 in the interacting sample, but also to a somewhat smaller scalelength h.

This corresponds quite well to the predictions of N-body simulations (Quinn et al 1993, Walker et al 1996): the gravity torques induced by the tidal interaction produce a central mass concentration, while the outer disk spreads out radially, leading to a decrease of h. Most of the heating is expected to be vertical, since the planar heating is taken away by the stripped stars either in the primary or in the satellite. The quantitative agreement between observations and simulations is rather good, given the large dispersion expected due to the initial morphology of the interacting galaxies: a dense satellite will produce much more heating than a diffuse one, where stripped stars take the orbital energy away; a mass-condensed primary will inhibit tidally-induced spiral and bar perturbations, that are the source of heating both radially and vertically.

The fact that tidal interactions and minor mergers must have concerned every galaxy in a Hubble time, and therefore also the presently isolated and undisturbed galaxies, tells us that the lower values of h/z_0 observed for the interacting sample must be transient. Radial gas inflow induced by the interaction may have contributed to **reform a thin young stellar disk**, while the vertical thickening has formed the thick disk components now observed in the Milky Way and many nearby galaxies. This process might be **occuring all along the interaction**, so that the galaxy is never observed without a thin disk. One cannot therefore date back the period of the last interaction by the age of the thin disk, as has been proposed by Toth & Ostriker (1992) and Quinn et al (1993). The Milky Way, experiencing now interactions with the Magellanic Clouds and a few dwarf spheroidal companions, has still a substantial gaseous and stellar thin disk. Further self-consistent simulations, including gas and star-formation, must be performed to derive more significant predictions.

The fundamental role of the re-formation of a thin stellar disk is obvious in Fig.2: there are correlations between the HI content of a galaxy and its stellar scale height and relative thickness h/z_0. Dashed lines in Fig.2 show double regression fits for normal spirals: $z_0 \, (\text{kpc}) = 0.84 \times [\text{M(HI)}/\text{L}_\text{B}]^{-1/2}$ and $h/z_0 = 5.0 \times [\text{M(HI)}/\text{L}_\text{B}]^{1/2}$, where M(HI) is the total HI mass (in $\text{M}_\odot$) and L_B is the total luminosity of the edge-on galaxy (in $\text{L}_\text{B}^\odot$) uncorrected for internal absorption. Interacting galaxies in general follow the same relations as normal spirals but with larger scatter. It is interesting that the Milky Way is also satisfying the above relations. Adopting for the absolute luminosity of "edge-on" Milky Way $M_B \approx -20.5 + 1.5 = -19.0$ and $\text{M(HI)} = 8\,10^9\,\text{M}_\odot$, we obtain from the above correlations $z_0 = 0.74 \pm 0.27$ kpc and $h/z_0 = 5.7 \pm 1.7$. These values are in agreement with current estimates of the Milky Way parameters (e.g. Sackett 1997).

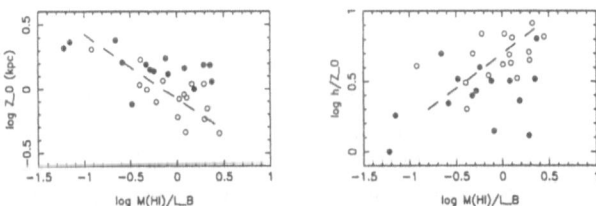

Fig. 2. Distribution of normal (open circles) and interacting (solid circles) galaxies in the plane M(HI)/L_B - z_0 (left) and M(HI)/L_B - h/z_0 (right).

References

Quinn, P.J., Hernquist, L., Fullagar, D.P. (1993): ApJ **403**, 74–93
Reshetnikov, V., Combes, F. (1996): A&AS **116**, 417–428 (Paper I)
Sackett, P.D. (1997): ApJ, submitted (astro-ph/9608164)
Toth, G., Ostriker, J.P. (1992): ApJ **389**, 5–26
Walker, I.R., Mihos, J.C., Hernquist, L. (1996): ApJ **460**, 121–135

A Purely Photometric Fundamental Plane ?

Marco Scodeggio[1,2], Riccardo Giovanelli[1], and Martha P. Haynes[1]

[1] Astronomy Department, Cornell University, Ithaca, NY 14853, USA
[2] ESO, Karl-Schwarzschild-Str.2, D-85748 Garching b. München, Germany

1 Economical Distance Indicators

The Tully–Fisher (TF) and Fundamental Plane (FP) relation are the most commonly used distance-indicator relations. Observationally, their ingredients are analogous: magnitudes, radii and surface brightnesses require relatively easy to obtain optical photometric data. Rotational widths and velocity dispersions are instead more demanding, both in terms of telescope aperture and of longer exposures. An economical distance-indicator, based entirely on photometric parameters, would be an extremely valuable tool for observational cosmology, if it were to achieve an accuracy comparable to those of the TF and FP relations.

The Kormendy (1977) relation between r_e and μ_e is one such candidate. The scatter in the relation is fairly large, but this disadvantage can be partly offset, when the relation is applied to clusters of galaxies, by the availability of a large number of objects, that can statistically compensate for the large scatter in the derivation of the cluster distance. In Fig.1a we present the Kormendy relation for 405 E and S0 galaxies in the Coma, A1367, and A2634 clusters. The rms scatter in $\log R_e$ is 0.19, equivalent to an uncertainty of 0.95 mag. in the distance modulus of a single galaxy. However, the residuals from the Kormendy relation are not random: they correlate very well with the galaxy magnitude, or, equivalently, with the galaxy velocity dispersion, as shown in Fig.1b. Because of the combination of a large scatter and of residuals correlated with the galaxy luminosity, the Kormendy relation is severely affected by the cluster population incompleteness bias (Sandage 1994, and references therein): both the zero-point and the observed dispersion depend on the limiting magnitude of the sample. It is therefore very important to derive accurate bias corrections before using the Kormendy relation for cosmological applications.

Other economical photometric distance indicators are based on the determination of a characteristic magnitude for a cluster sample. Sandage & Hardy (1973) have used the luminosity of the first ranked galaxy in a cluster, that they measured to be approximately constant, with a dispersion of only 0.28–0.32 magnitudes. A modified version of this method was proposed by Hoessel (1980), and has been used most recently by Lauer & Postman (1994). The uncertainty associated with this method is approximately 16%, or 0.35 magnitudes (Lauer & Postman 1994). Also luminosity functions (LF) have been used as distance indicators. Globular clusters and planetary nebulae LF (Jacoby et al. 1992, and references therein) have been used to derive redshift-independent distances for

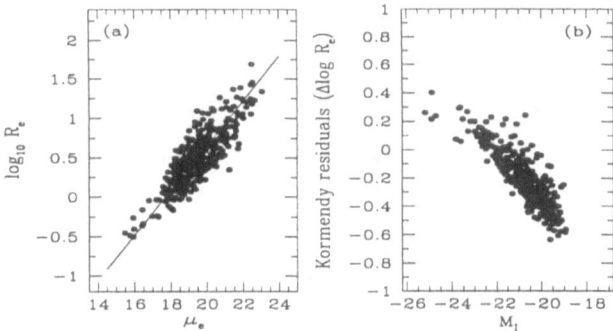

Fig. 1. (a) The Kormendy relation between effective radius and effective surface brightness. (b) The residuals from the Kormendy relation, as a function of the galaxy magnitude. The photometric parameters in this and the following figure have been derived from I band CCD observations obtained with the 0.9m telescope at KPNO.

individual galaxies. The same technique can be applied to the galaxy LF in a cluster, provided that such LF is to a large extent a universal one. Since giant E and S0 galaxies and spiral galaxies have been shown to have a Gaussian LF (Sandage et al. 1985), these galaxies would provide the most accurate determination of a characteristic magnitude, through the measurement of the peak of the Gaussian.

2   The Photometric Fundamental Plane

Here we propose to use jointly the Kormendy relation and one of the methods that can provide a characteristic magnitude for a cluster sample, to derive a modified version of the FP relation, based entirely on photometric parameters. We define for each galaxy ΔM as the difference between the galaxy's magnitude and the sample characteristic magnitude, and use ΔM, a distance–independent parameter, in substitution of the velocity dispersion in the FP relation. We illustrate the characteristics of such a Photometric Fundamental Plane (PFP), using the peak of the Gaussian LF of E and S0 galaxies to derive the characteristic magnitude. We combined the sample of 405 galaxies already used in Fig.1, assuming the clusters have negligible peculiar motions, and derived a Gaussian LF using a maximum likelihood method. Figure 2a shows the edge-on projection of such a PFP. The rms scatter in $\log R_e$ about the PFP plane is 0.096, which is approximately half the scatter in the Kormendy relation, and very similar to the scatter shown by the FP relation. This scatter, however, is not the only source of uncertainty in the determination of a cluster distance with the PFP, because the uncertainty in the determination of the characteristic magnitude introduces a systematic uncertainty in the determination of the PFP zero-point. For large samples this systematic uncertainty becomes the dominant source of error in a cluster distance determination. However, because the ΔM coefficient α_M in the PFP is significantly smaller than 0.2 (0.13 for the sample used here), the effect of this systematic uncertainty on the distance determination is reduced by the ratio $\alpha_M/0.2$, which is $\simeq 0.65$ in our case. A comparison of the accuracy in a distance determination that can be obtained using the Fundamental Plane, the PFP, and the Kormendy relation, as a function of sample size, is shown in Fig. 2b. If the characteristic magnitude can be measured with an uncertainty of 0.3 magnitudes or less, the PFP can offer an accuracy in distance measurements comparable to that of the FP, for a typical FP cluster sample, requiring only one fifth to one tenth of the telescope time.

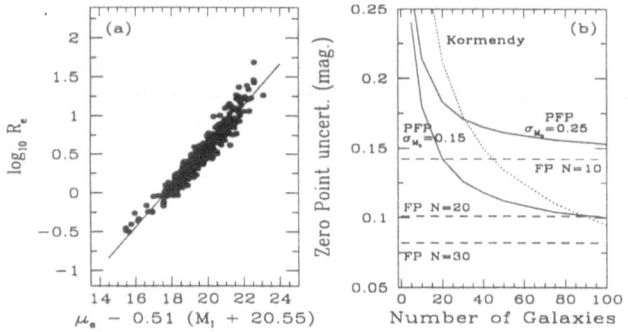

Fig. 2. (a) Edge-on projection of the PFP. (b) Comparison of the accuracy that can be obtained using the PFP, Kormendy, and FP relation, vs sample size. The horizontal dashed lines give the accuracy achieved using the FP (fixed sample size, assuming a scatter in the FP of 0.085, or 0.43 mag.). The solid line curves give the accuracy achieved using the PFP, for two different values of the uncertainty in the determination of the characteristic magnitude. The dotted line gives the accuracy obtained with the Kormendy relation.

References

Hoessel, J.G. 1980, ApJ, 241, 493
Kormendy, J. 1977, ApJ, 218, 333
Lauer, T.R., & Postman, M. 1994, ApJ, 425, 418
Sandage, A. 1994, ApJ, 430, 1
Sandage, A., & Hardy, E. 1973, ApJ, 183, 743
Sandage, A., Binggeli, B., & Tammann, G.A. 1985, AJ, 90, 1759

Streaming Motions of Abell Clusters Within $12000\,\mathrm{km\,s^{-1}}$

R.J. Smith[1], M.J. Hudson[1,2], J.R. Lucey[1], D.J. Schlegel[1],
R.L. Davies[1] and G. Baggley[1]

[1] Department of Physics, University of Durham, South Road, Durham DH1 3LE, U.K.
[2] Department of Physics and Astronomy, University of Victoria, P.O. Box 3055, Victoria BC V8W 3PN, Canada

Abstract. We describe the objectives and current status of an ongoing all-sky survey of peculiar motions for 54 Abell clusters within $cz = 12000\,\mathrm{km\,s^{-1}}$. Preliminary Fundamental Plane plots are presented for three newly-studied clusters.

1 Introduction

The $D_n-\sigma$ and Fundamental Plane (FP) relations (Dressler et al. 1987, Djorgovski & Davis 1987) can be employed as efficient distance indicators for early-type galaxies. The application of these techniques to clusters of galaxies makes possible distance measurements with random errors of a few per cent, given data for $\gtrsim 5$ galaxies per cluster. Such small random errors help also to beat down the effects of homogeneous and inhomogeneous Malmquist biases (see, for example, Strauss & Willick 1995).

We describe here a survey of cluster peculiar motions which employs the FP method as a distance indicator, for 54 Abell clusters within $cz = 12000\,\mathrm{km\,s^{-1}}$.

2 Sample and Current Status

We have obtained new data for 30 clusters in the sample. Together with these observations, we will employ published data for 6 clusters from the work of Jørgensen, Franx & Kjærgaard (1996), 3 clusters from Lucey et al. (1997) and 3 clusters from Hudson et al. (1997). Also to be included are the forthcoming data for 9 clusters in the EFAR survey (Wegner et al. 1996), and unpublished observations, by Lucey and collaborators, for 3 clusters. The resulting merged catalogue will consist of complementary spectroscopic and photometric parameters for approximately 700 galaxies in 54 clusters. The cluster sample will have excellent sky coverage to a median depth of $90\,h^{-1}$ Mpc (see Figure 1).

A major bugbear of merged catalogues of peculiar velocity data is the need to ensure homogeneity between datasets from various sources. This is particularly difficult for velocity dispersion measurements, which are subject to random errors of 5–10%. For the work described here, we will apply the homogenisation scheme described by Smith et al. (1997), based on a set of more than 1500 observations of galaxies common to more than one dataset. This method allows the estimation

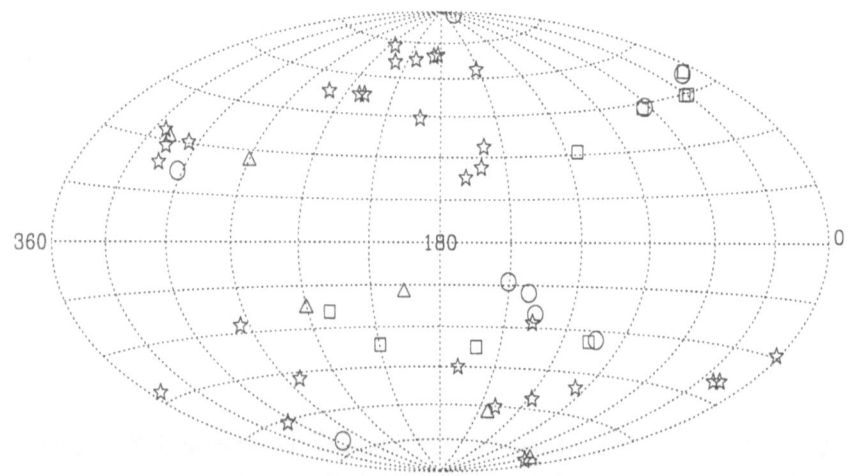

Fig. 1. Distribution of the 54 clusters studied, in galactic coordinates. Symbol types indicate the principal data source for each cluster, viz. *Stars*: This study ; *Circles*: previous work by Lucey and collaborators ; *Squares*: EFAR project ; *Triangles*: Jørgensen et al. 1996.

of systematic uncertainties in the matching of datasets, reflecting the degree of 'rigidity' of the merged catalogue.

At the time of writing, data collection for the survey has been completed. For the 30 'new' clusters, spectroscopic and photometric reductions are in progress. Figure 2 presents FP plots for 3 southern clusters for which new data have been obtained. Peculiar motion results for ~ 45 clusters are expected by Autumn 1997.

The data issuing from this project will be well suited to the following analyses:

1. To determine the local bulk motion (to ~ 150 km s^{-1} error) on scales intermediate between those of Lauer & Postman (1994) and more local surveys (eg Giovanelli et al. 1996).
2. To constrain $\beta = \Omega^{0.6}/b$, through comparison with the velocity field predicted from redshift surveys (eg IRAS PSCz).
3. To constrain the range of viable cosmological models by means, for instance, of the distribution function of cluster velocities (eg Bahcall & Oh 1996).

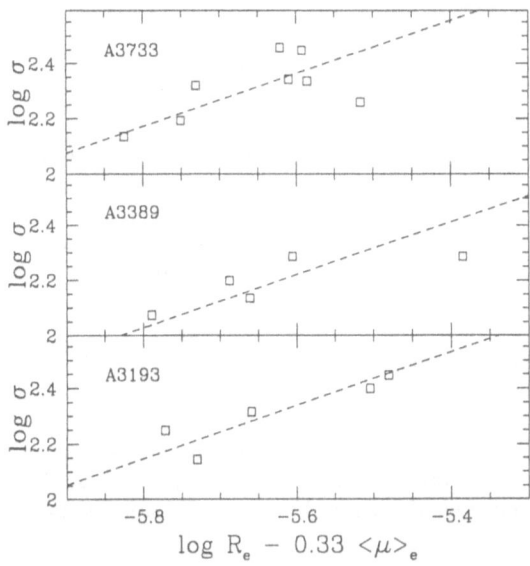

Fig. 2. Preliminary Fundamental Plane plots for three new clusters A3733 ($cz \sim 12000$ km s^{-1}), A3389 ($cz \sim 8000$ km s^{-1}), A3193 ($cz \sim 10000$ km s^{-1}). The scatters are equivalent to distance errors of 25%, 22% and 14% per galaxy, respectively.

4. To provide an extensive comparison of FP distances with those derived from the Tully–Fisher and Brightest Cluster Galaxy methods.

References

Bahcall, N.A., Oh, S.P. 1996, ApJ, 462, L49

Djorgovski, S., Davis, M. 1987, ApJ, 313, 59

Dressler, A., Faber, S.M., Burstien, D., Davies, R.L., Lynden-Bell, D., Terlevich, R.J., Wegner, G. 1987, ApJ, 313, L37

Giovanelli, R., Haynes, M.P., Wegner, G., da Costa, L.N., Freudling, W., Salzer, J.J. 1996, ApJ, 464, L99

Hudson, M.J., Lucey, J.R., Smith, R.J., Steel, J. 1997, MNRAS, submitted

Jørgensen, I, Franx, M., Kjærgaard, P. 1996 MNRAS, 280, 167

Lauer, T.R., Postman, M. 1994, ApJ, 425, 418

Lucey, J.R., Guzmán, R, Steel, J., Carter, D. 1997a, MNRAS, in press

Smith, R.J., Lucey, J.R., Hudson, M.J., Steel, J. 1997, MNRAS, submitted

Strauss, M.A., Willick, J.A. 1995, Physics Reports, 261, 271

Wegner, G., Colless, M., Baggley, G., Davies, R.L., Bertschinger, E., Burstein, D., McMahon, R.K., Saglia, R.P. 1996, ApJS, 106, 1

Models of Elliptical Galaxies: the Gradients

Rosaria Tantalo[1], Cesare Chiosi[1], Alessandro Bressan[2]

[1] Department of Astronomy, Vicolo dell'Osservatorio 5, 35122 Padua, Italy
[2] Astronomical Observatory, Vicolo dell'Osservatorio 5, 35122 Padua, Italy

Abstract. Elliptical galaxies possess gradients in broad-band colors and line strength indices (cf. Davies et al. 1993; Carollo et al. 1993; Carollo & Danziger 1994), whose interpretation is basic to cast light on the process of galaxy formation and evolution. In this paper, we present new chemo-spectro-photometric models of elliptical galaxies by Tantalo et al. (1997) which are specifically designed to explain the gradients in question. The models stand on the previous studies by Bressan et al. (1994) and Tantalo et al. (1996) but include the radial dependence of gas density and star formation, and adopt a simple scheme to simulate the collapse of luminous material into the potential well of dark matter. The theoretical gradients in broad-band color and line strength indices are presented and compared with their observational counterparts.

1 Outline of the Models

1.1 Geometrical Description

Elliptical galaxies are described as spherically symmetric systems whose mass density decreases outward. The density of luminous material in each shell and the corresponding gravitational potential are derived from Young (1976). We assume that each shell contains about 5% of the total mass in the luminous material.

The mass distribution and gravitational potential of the dark-matter as a function of the radial distance are derived from the density profile of Bertin et al. (1992)

$$\rho(r) = \frac{\rho_0 r_0^4}{(\rho_0^2 + r^2)^2} \qquad (1)$$

however adapted to the Young formalism. The total mass of the dark-matter component is assumed to be 5 times the total mass in luminous material.

Finally, we adopt a mass (luminous) - effective radius relationship $M_L(R_e)$ derived from the data of Carollo et al. (1993) for a selected sample of elliptical galaxies.

1.2 Infall and Star Formation Rates

In order to simulate the collapse of luminous material into the potential well of dark-matter (whose mass is assumed to be constant in time) the infall scheme is adopted and the density of luminous material (gas) is let grow with time at the rate:

$$\frac{d\overline{\rho_L}(r,t)}{dt} = \rho_{L0}(r)e^{-\frac{t}{\tau(r)}} \tag{2}$$

where $\tau(r)$ is the time scale of gas accretion (in principle it can be a function of the radial distance), and $\rho_{L0}(r)$ is fixed by imposing that at the present-day age of the galaxy T_G the density of luminous material in each shell has grown to the value given by the Young profile.

A successful description of the gas accretion phase is possible adapting to galaxies the radial velocity law describing the final collapse of the core in a massive star, i.e. free-fall in all regions external to a certain value of the radius ($v \propto r^{-\frac{1}{2}}$) and homology inside ($v \propto r$). This picture is also confirmed by numerical calculations of dynamical models with the Tree-SPH technique (cf. Carraro et al. 1997). This simple scheme allows us to derive the radial dependence of $\tau(r)$ as a function of some arbitrary time scale. For this latter we adopt the mean free-fall time scale of Arimoto & Yoshii (1987).

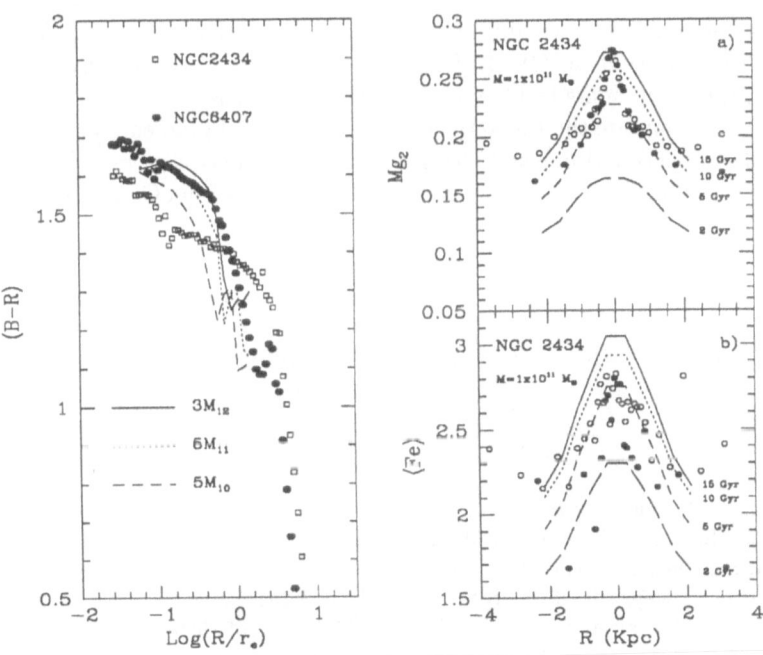

Fig. 1. Gradients in colors and line strength indices for the galaxies NGC 2434 and NGC 6407 (Carollo & Danziger 1994), filled and open circles, respectively. *Left panel*: comparison with the theoretical gradients in (B–R) for models of different mass and same age (15 Gyr). *Right panel*: comparison with the theoretical gradients in line strength indices for the $1 \times 10^{11} M_\odot$ model which has nearly the same M/L_B ratio as NGC 2434.

The rate of star formation (SFR) follows the standard Schmidt law

$$\Psi(r,t) = \nu(r)\overline{\rho_{Lg}}(r,t) \tag{3}$$

where $\rho_{Lg}(r,t)$ is the local gas density and $\nu(r)$ is the specific efficiency. For this latter, we have adopted the formulation by Arimoto & Yoshii (1987).

The chemical evolution of elemental species is governed by the same set of equations as in Tantalo et al. (1996) however adapted to the density formalism and improved as far as the ejecta and the contribution from Type Ia and Type II supernovae are concerned.

The models allow for galactic winds triggered by the energy deposit from Type I and II supernova explosions and stellar winds from massive stars. To evaluate the amount of energy stored into the interstellar medium by supernova explosions we adopt the same cooling law as in Gibson (1994, 1996).

For all other details see Tantalo et al. (1997).

2   The Gradients

These galactic models are able to reproduce (i) the slope of colour-magnitude relation by Bower et al. (1992); (ii) the UV excess as measured by the colour (1550-V), (iii) the mass to blue luminosity ratio M/L_B.

In Fig. 1 we compare the theoretical and observational gradients in broad-band colors (left panel) and line strength indices (right panel). The radial distance is expressed in units of the effective radius. The data are from Carollo & Danziger (1994).

This study has been financed by the Italian Ministry of University, Scientific Research and Technology (MURST), the Italian Space Agency (ASI), and the TMR grant ERBFMRX-CT96-0086 from the European Community.

References

Arimoto, N., Yoshii, Y. 1987: A&A 173, 23
Bertin, G., Saglia, R.P., Stiavelli, M. 1992: ApJ 384, 423
Bressan, A., Chiosi, C., Fagotto, F. 1994: ApJS 94, 63
Bower, R.G., Lucey, J.R., Ellis, R.S. 1992: MNRAS 254, 589
Carollo, C.M., Danziger, I.J., Buson, L. 1993: MNRAS 265, 553
Carollo, C.M., Danziger, I.J. 1994: MNRAS 270, 523
Carraro, G., Lia, C., Chiosi, C. 1997: MNRAS, submitted
Davies, R.L., Sadler, E.M., Peletier, R.F. 1993: MNRAS, 262, 650
Gibson, B.K. 1996: MNRAS, submitted
Gibson, B.K. 1996: MNRAS 271, L35
Tantalo, R., Chiosi, C., Bressan, A. 1997: A&A, to be submitted
Tantalo, R., Chiosi, C., Bressan, A., Fagotto, F. 1996: A&A 311, 361
Young, P.J. 1976: AJ 81, 807

The Fundamental Plane at $z = 0.3 - 0.6$

Pieter G. van Dokkum

Kapteyn Astronomical Institute, P.O. Box 800, 9700 AV Groningen, The Netherlands

Abstract. We present results on the Fundamental Plane (FP) of early-type galaxies at intermediate redshifts. Structural parameters and velocity dispersions were measured of galaxies in three clusters, at $z = 0.33$, $z = 0.39$, and $z = 0.58$. The data are compared to published data for the Coma cluster at $z = 0.02$. The form of the FP is very similar at intermediate redshifts. The evolution of the mass-to-light (M/L) ratio with redshift is derived from the FP. The mass-to-light ratio evolves as $\Delta \log(M/L_V) \sim -0.3z$. This evolution is low when compared with models for stellar populations, and consistent with a high formation redshift of the stars in the galaxies. However, this analysis may suffer from the 'progenitor bias': we may be preferentially selecting the oldest galaxies at each redshift.

1 Introduction

It has become clear that the evolution of early-type galaxies in clusters is a complicated process. This makes the interpretation of observations of distant clusters difficult. Starbursts and encounters with other galaxies can substantially change both the luminosities and morphologies of the galaxies, thus rendering the comparison of local galaxies with their high redshift counterparts ambiguous. Luminosities and colors alone are not sufficient to differentiate between the mechanisms that drive galaxy evolution.

An important additional constraint is the measurement of galaxy masses, and mass-to-light ratios. These measurements are essential for a correct interpretation of the luminosity function, and help constrain the importance of merging, infall, and starbursts.

As first shown by Faber et al. (1987), the existence of the Fundamental Plane (Dressler et al. 1987, Djorgovski & Davis 1987) implies that the M/L ratios of early-type galaxies are well behaved: $M/L \propto r_e^{0.22}\sigma^{0.49}$, with low scatter. Therefore, observations of the FP at higher redshifts directly constrain the evolution of the M/L ratio. Here, we use the Fundamental Plane relation in three clusters at intermediate redshift (CL1358+62 at $z = 0.33$, CL0024+16 at $z = 0.39$, and MS2053−04 at $z = 0.58$) to derive the evolution of the M/L ratio with redshift.

2 Data

High resolution, high S/N spectroscopic data were obtained at the Multiple Mirror Telescope (for CL0024+16) and the W.M. Keck Telescope, equipped with LRIS (for CL1358+62 and MS2053−04). We used Hubble Space Telescope data for the determination of the structural parameters. Details of the reduction and analysis may be found in van Dokkum & Franx (1996), and Kelson et al. (1997).

3   The Fundamental Plane

The Fundamental Planes for the three clusters are shown in Fig. 1, along with the FP for Coma at $z = 0.02$ from Jørgensen et al. (1996) [JFK] ($q_0 = 0.05, H_0 = 50$). The FP relation is well defined in the intermediate redshift clusters. The scatter is very similar in all four clusters, ~ 0.065 in $\log r_{\mathrm{e}}$. The current sample of galaxies is too small to determine whether there is a small change in slope of the FP in the intermediate z clusters (cf. Kelson et al. 1997). The offsets of the FPs with respect to the Coma cluster are primarily due to the $(1 + z)^4$ cosmological surface brightness dimming.

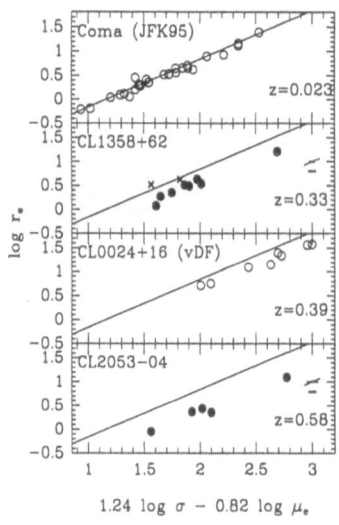

Fig. 1. The FP for the three clusters, and the FP of Coma, from JFK.

After correction for this effect, we determined the evolution of the M/L_V ratio directly from the FP zero-point. The resulting evolution is shown in Fig. 2. The M/L ratios are lower at higher redshift, which is expected from the evolution of stellar populations. Single-burst model predictions for the evolution of the M/L_V are plotted in the figure. The data are consistent with a high formation redshift of the stars.

4   Discussion

The observed evolution of the M/L_V ratio is consistent with a high formation redshift of the bulk of the stars in early-type galaxies. This result seems hard to reconcile with the starbursts and enhanced star formation seen in distant clusters. However, the type of study presented here may suffer from the 'progenitor

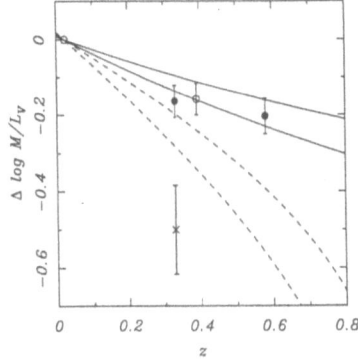

Fig. 2. The evolution of the M/L_V ratio with redshift. The cross indicates the offset of the two 'E+A' galaxies in Cl1358+62.

bias': the selection criteria may ensure that the galaxies we study are at least a few Gyr old, at every redshift (cf. Franx & van Dokkum 1996, van Dokkum & Franx 1996). If a present-day elliptical was, e.g, actively forming stars at $z = 0.5$, it would not be included in the FP sample, thus biasing the intermediate red-shift samples towards the oldest galaxies. As a case in point, two 'E+A' galaxies were included in the CL1358+62 sample. They are indicated by crosses in Fig. 1 and Fig. 2. The M/L ratios of these galaxies are consistent with a luminosity weighted formation redshift of $z \sim 0.5$, in agreement with the hypothesis that they have undergone a burst of star formation $1 - 2$ Gyr prior to the epoch of ob-servation. However, it is still uncertain whether these 'E+A' galaxies will evolve into galaxies that *are* included in the low z FP samples. Detailed studies of the morphological evolution of cluster galaxies are needed to address that issue.

References

Djorgovski S., Davis M., 1987, ApJ, 313, 59

Dressler A., Lynden-Bell D., Burstein D., Davier R. L., Faber S. M., Terlevich R. J., Wegner G., 1987, ApJ, 313, 42

Faber S. M., Dressler A., Davies R. L., Burstein D., Lynden-Bell D., Terlevich R.J., Wegner G., 1987, Faber S. M., ed., Nearly Normal Galaxies. Springer, New York, p. 175

Franx M., van Dokkum P. G., 1996, Davies R., and Bender R., ed., New Light on Galaxy Evolution. Kluwer, Dordrecht, p. 233

Jørgensen I., Franx M., Kjærgaard P., 1996, MNRAS, 280, 167 [JFK]

Kelson D. D., van Dokkum P. G., Franx M., Illingworth G. D., Fabricant D., 1997, ApJL, in press

van Dokkum P. G., Franx M., 1996, MNRAS, 281, 985

Author Index